METHODS IN MOLECULAR BIOLOGY™

Series Editor
John M. Walker
School of Life Sciences
University of Hertfordshire
Hatfield, Hertfordshire, AL10 9AB, UK

For further volumes:
http://www.springer.com/series/7651

Cell Migration

Developmental Methods and Protocols

Second Edition

Edited by

Claire M. Wells

Division of Cancer Studies, King's College London, London, UK

Maddy Parsons

Randall Division of Cell and Molecular Biophysics, King's College London, London, UK

Humana Press

Editors
Claire M. Wells, Ph.D
Division of Cancer Studies
King's College London
London, UK
claire.wells@kcl.ac.uk

Maddy Parsons
Randall Division of Cell
and Molecular Biophysics
King's College London
London, UK
maddy.parsons@kcl.ac.uk

Please note that additional material for this book can be downloaded from
http://extras.springer.com

ISBN 978-1-4939-6193-1 ISBN 978-1-61779-207-6 (eBook)
DOI 10.1007/978-1-61779-207-6
Springer New York Dordrecht Heidelberg London

Printed on acid-free paper

Humana Press is part of Springer Science+Business Media (www.springer.com)

Preface

Cell migration is a key component of many biological processes, including embryonic development, immune responses, wound healing, organ regeneration, and cancer cell metastasis. Such cell movements can involve migration of individual cells or groups of cells, or the movement of large sheets of conjoined cells. Researchers have spent many years unravelling the highly complex molecular mechanisms involved in the initiation and maintenance of migration using a range of biochemical and functional assay systems. Furthermore, as the study of cell migration is an intrinsically visual science, recent efforts have also focused on the development of microscopy techniques to precisely define the contribution of specific signalling pathways to different aspects of motility.

Considerable advances have been made in our understanding of the cellular machinery that drives cell movement, the signalling pathways involved, and how these may be regulated to achieve specific modes of migration in different contexts. Recently, much of this work has been driven by the use of RNAi technology and the expression of fluorescently tagged proteins. Moreover, large-scale screening programmes have helped to identify novel proteins not previously thought to be involved in regulating migration.

Significant advances have also been made in the study of cell migration within three dimensional (3D) environments, whether that is imaging migration of cells in contact with a cell-derived matrix, monitoring the movement of cells through an artificial extracellular matrix, or using organotypic models of cancer progression. Such studies have identified key differences in the behaviour of cells migrating in 2D and 3D environments. This underscores the need for the translation of findings from in vitro studies to cells within more complex, physiologically relevant model environments.

In addition to the study of migration within 3D matrices, there has been significant development in the use of model organisms, many of which are optically transparent in the embryonic phase, to study cell migration in vivo in real time. These model organisms are also genetically tractable within relatively short generation times, thus providing a powerful genetic tool for the study of cell migration in vivo.

The combined efforts of many researchers in the field of cell migration have led to the development of a wealth of cell migration assays which can be used to elucidate the proteins, pathways, and systems that may be involved in cell movement.

The aim of *Cell Migration: Developmental Methods and Protocols, Second Edition* is to bring together a wide range of these techniques from the more basic migration assays, which are still the foundation of many cell migration studies, to state-of-the-art techniques and recent technical advances. Each chapter is presented by a leading expert in the field who has considerable experience with the protocol described and has applied that method to his or her own research. The protocols are presented in a user-friendly format previously used in the highly successful *Method in Molecular Biology*™ series with a valuable troubleshooting notes section.

Cell Migration: Developmental Methods and Protocols begins with an overview of cell migration that discusses our current understanding of the biological process involved. Part I (Chapters 2–16) describes a number of basic in vitro migration assays, including measurements of wound healing, cell scattering, invasion, and chemotaxis. Part I also includes more complex measurements of transendothelial migration, the use of microfluidic chambers, and imaging cell migration in 3D. Part II (Chapters 16–24) focuses on the imaging and measurement of cell migration in vivo, including protocols for the use of chick, drosophila, and zebrafish embryos, as well as methods to measure metastatic spread and angiogenesis in mice. Part III (Chapters 25–30) introduces protocols for some of emerging techniques in the cell migration field, including the use of TIRF, FRAP, and FRET microscopy.

The overall aim of this publication is to provide a comprehensive catalogue of techniques for the study of cell migration that can be used as a useful reference source for any researcher who wishes to embark upon such experiments. It is designed to be used by both new and experienced researchers, particularly cell and developmental biologists, but also researchers from genetic, biochemical, clinical, and molecular backgrounds who are interested in this field of investigation.

This book would not have been possible without the outstanding contributions from all of the authors. We are extremely grateful for their generosity in taking the time to document these protocols in a way that is easily accessible to the wider scientific community. We would also like to thank colleagues who took part in the review process, in particular the series editor, Dr. John Walker, who gave us valuable input during the editing process. Rigorous review and editing has greatly strengthened this publication. We hope that this book will inspire current and future scientists to explore the world of cell migration.

London, UK *Claire M. Wells*
London, UK *Maddy Parsons*

Contents

Contributors

Kurt I. Anderson • *The Beatson Institute for Cancer Research, Glasgow, UK*
Yulia Artemenko • *Department of Cell Biology, Johns Hopkins University School of Medicine, Baltimore, MD, USA*
Christoph Ballestrem • *Faculty of Life Sciences, The University of Manchester, Manchester, UK*
Silvia Batista • *Centre for Tumour Biology, Queen Mary, Barts and The London School of Medicine and Dentistry, Barts Institute of Cancer, London, UK*
Hans de Bont • *Leiden/Amsterdam Center for Drug Research, University of Leiden, Leiden, The Netherlands*
Gary M. Box • *Cancer Research UK Cancer Therapeutics Unit, The Institute of Cancer Research, Sutton, Surrey, UK*
Bárbara Borda d'Água • *Randall Division of Cell and Molecular Biophysics, King's College London, London, UK*
Robert J. Cain • *Randall Division of Cell and Molecular Biophysics, King's College London, London, UK*
Alex Carisey • *Faculty of Life Sciences, The University of Manchester, Manchester, UK*
Sanjay Chaubey • *Division of Cardiovascular Medicine, King's College London, London, UK*
Olivier Collin • *Functional Imaging of Transcription, Institut de Biologie de l'Ecole Normale Supérieure (IBENS), Paris cedex, France*
Christa L. Cortesio • *Department of Biomolecular Chemistry, University of Wisconsin-Madison, Madison, WI, USA*
Giles Cory • *Peninsula College of Medicine and Dentistry, University of Exeter, Exeter, UK*
Peter N. Devreotes • *Department of Cell Biology, Johns Hopkins University School of Medicine, Baltimore, MD, USA*
Suzanne A. Eccles • *Cancer Research UK Cancer Therapeutics Unit, The Institute of Cancer Research, Sutton, Surrey, UK*
Philip M. Elks • *Department of Infection and Immunity, The Medical School, Sheffield University, Sheffield, UK*
Marco Falasca • *Blizard Institute of Cell and Molecular Science, Barts and The London School of Medicine and Dentistry, London, UK*
Sally T. Fram • *Division of Cancer Studies, King's College London, London, UK*
Shawn A. Galdeen • *Bio-Imaging Resource Center, The Rockefeller University, New York, NY, USA; Olympus America, Corporate Parkway Center Valley, PA, USA*
Li He • *Johns Hopkins School of Medicine, Baltimore, MD, USA*
Mélina L. Heuzé • *Immunity and Cancer, Institut Curie, Paris cedex, France*

KAIRBAAN HODIVALA-DILKE • *Barts Cancer Institute, Barts and The London, Queen Mary's School of Medicine and Dentistry, Queen Mary, University of London, London, UK*
ELIAS HORN • *ibidi GmbH, Martinsried, München, Germany*
ALAN RICK HORWITZ • *Department of Cell Biology, University of Virginia School of Medicine, Charlottesville, VA, USA*
ANNA HUTTENLOCHER • *Department of Medical Microbiology and Immunology, University of Wisconsin-Madison, Madison, WI, USA*
ANTONIO JACINTO • *Instituto de Medicina Molecular, Faculdade de Medicina da Universidade de Lisboa, Unidade de Morfogenese, Lisboa, Portugal*
VERONIKA JENEI • *Division of Cancer Sciences, School of Medicine, University of Southampton, Southampton, UK*
GARETH E. JONES • *Randall Division of Cell and Molecular Biophysics, King's College London, London, UK*
SAMANTHA J. KING • *Randall Division of Cell and Molecular Biophysics, King's College London, London, UK*
IREEN KÖNIG • *The Beatson Institute for Cancer Research, Glasgow, UK*
TIM LÄMMERMANN • *Laboratory of Systems Biology, Lymphocyte Biology Section, National Institute of Allergy and Infectious Diseases, National Institutes of Health, Bethesda, MD*
RESHMA LALAI • *Leiden/Amsterdam Center for Drug Research, University of Leiden, Leiden, The Netherlands*
SYLVIA E. LE DÉVÉDEC • *Leiden/Amsterdam Center for Drug Research, University of Leiden, Leiden, The Netherlands*
ANA-MARIA LENNON-DUMÉNIL • *Immunity and Cancer, Institut Curie, Paris cedex, France*
CATHERINE A. LOYNES • *Department of Infection and Immunity, The Medical School, Sheffield University, Sheffield, UK*
TANIA MAFFUCCI • *Blizard Institute of Cell and Molecular Science, Barts and The London School of Medicine and Dentistry, London, UK*
JOHN MARSHALL • *Barts Cancer Institute, Barts and The London, Queen Mary's School of Medicine and Dentistry, Queen Mary, University of London, London, UK*
MARIA MARTYNOVSKY • *Department of Molecular Biology, Princeton University, Princeton, NJ, USA*
ROBERTO MAYOR • *Department of Cell and Developmental Biology, University College London, London, UK*
DENISE J. MONTELL • *Johns Hopkins School of Medicine, Baltimore, MD, USA*
CAROLINA G.A. MOREIRA • *Instituto de Medicina Molecular, Faculdade de Medicina da Universidade de Lisboa, Unidade de Morfogenese, Lisboa, Portugal*
PENNY E. MORTON • *Randall Division of Cell and Molecular Biophysics, King's College London, London, UK*
ANDREA MÜNSTERBERG • *School of Biological Sciences, University of East Anglia, Norwich, Norfolk, UK*
ALISON J. NORTH • *Bio-Imaging Resource Center, The Rockefeller University, New York, NY, USA*

MARIA L. NYSTROM • *Division of Cancer Sciences, School of Medicine, University of Southampton, Southampton, UK*
MADDY PARSONS • *Randall Division of Cell and Molecular Biophysics, King's College London, London, UK*
MATTHIEU PIEL • *Subcellular Structure and Cellular Dynamics, Institut Curie, Paris, Cedex, France*
SOREN PRAG • *Instituto de Medicina Molecular, Faculdade de Medicina da Universidade de Lisboa, Unidade de Morfogenese, Lisboa, Portugal*
MOHIT PRASAD • *Johns Hopkins School of Medicine, Baltimore, MD, USA*
CHANTAL PONT • *Leiden/Amsterdam Center for Drug Research, University of Leiden, Leiden, The Netherlands*
CLAUDIO RAIMONDI • *Blizard Institute of Cell and Molecular Science, Barts and The London School of Medicine and Dentistry, London, UK*
JENNIFER C. REGAN • *Instituto de Medicina Molecular, Faculdade de Medicina da Universidade de Lisboa, Unidade de Morfogenese, Lisboa, Portugal*
STEPHEN A. RENSHAW • *MRC Centre for Developmental and Biomedical Genetics, The University of Sheffield, Sheffield, UK*
ANNE J. RIDLEY • *Cell Motility and Cytoskeleton Group, Randall Division of Cell and Molecular Biophysics, King's College London, London, UK*
WIES VAN ROOSMALEN • *Leiden/Amsterdam Center for Drug Research, University of Leiden, Leiden, The Netherlands*
JULIANE P. SCHWARZ • *The Beatson Institute for Cancer Research, Glasgow, UK*
JEAN E. SCHWARZBAUER • *Department of Molecular Biology, Princeton University, Princeton, NJ, USA*
MICHAEL SIXT • *Institute of Science and Technology Austria, Am Campus 1, Klosterneuburg, Austria*
JUNFANG SONG • *School of Biological Sciences, University of East Anglia, Norwich, Norfolk, UK*
TAYLOR W. STARNES • *Department of Medical Microbiology and Immunology, University of Wisconsin-Madison, Madison, WI, USA*
MATTHEW STROUD • *Faculty of Life Sciences, The University of Manchester, Manchester, UK*
KRISTEN F. SWANEY • *Department of Cell Biology, Johns Hopkins University School of Medicine, Baltimore, MD, USA*
DANIEL M. SUTER • *Department of Biological Sciences, Bindley Bioscience Center, Purdue University, West Lafayette, IN, USA*
BERNARDO TAVORA • *Centre for Tumour Biology, Queen Mary, Barts and The London School of Medicine and Dentistry, Barts Institute of Cancer, London, UK*
EMMANUEL TERRIAC • *Subcellular Structure and Cellular Dynamics, Institut Curie, Paris cedex, France*
ERIC THEVENEAU • *Department of Cell and Developmental Biology, University College London, London, UK*
GARETH J. THOMAS • *Division of Cancer Sciences, School of Medicine, University of Southampton, Southampton, UK*
RICKY TSANG • *Faculty of Life Sciences, The University of Manchester, Manchester, UK*

MIGUEL VICENTE-MANZANARES • *Department of Cell Biology, University of Virginia School of Medicine, Charlottesville, VA, USA*
XIAOBO WANG • *Johns Hopkins School of Medicine, Baltimore, MD, USA*
BOB VAN DE WATER • *Leiden/Amsterdam Center for Drug Research, Division of Toxicology, University of Leiden, Leiden, The Netherlands*
CLAIRE M. WELLS • *Division of Cancer Studies, King's College London, London, UK*
BEATA WOJCIAK-STOTHARD • *Department of Experimental Medicine and Toxicology, Imperial College London, London, UK*
MING-CHING WONG • *Department of Molecular Biology, Princeton University, Princeton, NJ, USA*
QIAOYUN YUE • *School of Biological Sciences, University of East Anglia, Norwich, Norfolk, UK*
ANNA ZAIDMAN-RÉMY • *Instituto de Medicina Molecular, Faculdade de Medicina da Universidade de Lisboa, Unidade de Morfogenese, Lisboa, Portugal*
ROMAN ZANTL • *ibidi GmbH, Martinsried, München, Germany*
SANDRA ZOVKO • *Leiden/Amsterdam Center for Drug Research, University of Leiden, Leiden, The Netherlands*

Chapter 1

Cell Migration: An Overview

Miguel Vicente-Manzanares and Alan Rick Horwitz

Abstract

Cell migration is a fundamental process that controls morphogenesis and inflammation. Its deregulation causes or is part of many diseases, including autoimmune syndromes, chronic inflammation, mental retardation, and cancer. Cell migration is an integral part of the cell biology, embryology, immunology, and neuroscience fields; as such, it has benefited from quantum leaps in molecular biology, biochemistry, and imaging techniques, and the emergence of the genomic and proteomic era. Combinations of these techniques have revealed new and exciting insights that explain how cells adhere and move, how the migration of multiple cells are coordinated and regulated, and how the cells interact with neighboring cells and/or react to changes in their microenvironment. This introduction provides a primer of the molecular and cellular insights, particularly the signaling networks, which control the migration of individual cells as well as collective migrations. The rest of the chapters are devoted to describe in detail some of the most salient technical advances that have illuminated the field of cell migration in recent years.

Key words: Cell migration, Adhesion, Molecular biology, Biochemistry, Imaging, Microscopy, Signaling

1. Introduction

Cell migration is a cornerstone of development, homeostasis, and disease. Embryogenesis involves the coordinated migration, proliferation, and differentiation of precursor cells to generate the different layers of the embryo and the tissues and organs derived from them. Immune cells patrol the body, exiting the bloodstream and migrating into the tissue in response to insult and infection. Migrating fibroblasts and epithelial cells collaborate in wound closure and tissue regeneration. Finally, deregulated cell migration can lead to autoimmune diseases, and is central to tumor cell dissemination and metastasis (1–3).

Claire M. Wells and Maddy Parsons (eds.), *Cell Migration: Developmental Methods and Protocols*, Methods in Molecular Biology, vol. 769, DOI 10.1007/978-1-61779-207-6_1, © Springer Science+Business Media, LLC 2011

This book provides an authoritative update on the techniques that have enabled the recent advances in cell migration. It is divided into three major sections: the first section describes "classic" methods to analyze cell migration in toto, the second section describes microscopy-based approaches that have brought enormous success in delineating the component processes of cell migration, and the third is devoted to novel approaches that will surrender crucial information to understand cell migration.

Cellular migrations across metazoan organisms adopt different modes of migration: (a) *single* migrating cells adopt morphologies that depend on the adhesive receptors they express, the microenvironment in which they migrate, and the intrinsic contractility of the cell: they can be small and fast moving, such as immune cells moving from the bloodstream to the target tissue that summons them by secreting pro-inflammatory molecules; or they can be larger, slow moving and display elaborate actin structures and large adhesions, as in fibroblasts migrating on a dish (4); (b) *chain linked*, in which a leading cell directs the migration of a "train" of cells physically linked to one another; this has been seen in the dissemination of different tumor cells during invasion and metastasis (5); (c) *guided by heterotypic cell–cell contacts*, such as in neural progenitors that migrate to specific layers of the brain using radial glia as physical "tracks" (6); (d) *coordinated and multicellular*, forming sheet- or wave-like layers of cells that form a front row of moving cells and multiple, tightly bound layers of cells that trail after the front row, as in embryogenesis and wound healing (7).

Cell migration requires the fine spatial-temporal integration of the hundreds of proteins that comprise or regulate the fundamental processes that drive cell migration; the proteins include membrane receptors, signaling kinases, phosphatases, and adapters, and cytoskeletal and adhesion components. It is useful to parse cell migration into its highly interrelated, component processes. Since many regulatory processes are common to migration, in general, we will focus our discussion on the best studied model: a single, mesenchymal cell moving across a planar substratum, but will point out the differences observed in other modalities where appropriate.

2. Protrusion

Protrusions are extensions of the cellular membrane. In most cells, there is a leading protrusion that points in the direction of movement and is part of a polarity axis (see Subheading 3). Protrusions occur in response to chemoattractant signals from the extracellular medium (e.g., migrating amoeba that respond to the feeding cue cAMP, or leukocytes that follow the "scent" of inflammatory cytokines in injured tissue), although some cells extend them in a probing, exploratory manner in the absence of directional stimulation.

Protrusions require contributions from several individual, interrelated cellular systems. Force for membrane deformation is provided by the actin cytoskeleton, which polymerizes dynamically pushing the membrane forward. The membrane itself expands under tension, which requires membrane traffic and the fusion of membrane-containing vesicles to support the increase in membrane surface. Finally, protrusions must adhere to the substratum. If they do not attach, protrusions are unproductive and tend to move rearward in waves in response to the tension generated in the cell ("membrane ruffling").

Actin filaments are the physical backbone of protrusion. Their polymerization and organization also determine the overall shape of the cell and contribute to its internal organization. Actin filaments are comprised of arrays of monomeric globular, G-actin subunits that bind in a process catalyzed by ATP hydrolysis. During the polymerization process, ATP bound to G-actin is hydrolyzed to ADP+P_i as the bond between the two monomers form. In filamentous, F-actin, ADP+P_i-actin forms stable filaments, whereas release of the phosphate destabilizes the structure, prompting depolymerization (8–10).

Actin filaments are polarized, reflecting the oriented, head-to-tail structure of the binding monomers, and display a fast-growing "barbed" end, where new monomers are incorporated, and a "pointed" end, which constitutes the origin of the growing filament and exhibits filament aging (bound to ADP) (9, 10).

Actin filaments associate laterally to other actin filaments, forming bundles. This relies on the activity of actin-binding proteins with cross-linking activity, such as myosins, α-actinins, filamins, and others, and determines a higher order polarity of the actin bundles: if two filaments display their barbed ends in the same direction, the resulting bundle is parallel; if the barbed end and the pointed end point to opposite directions, the bundle is antiparallel.

Actin filaments adopt different morphologies depending on the number of filaments and the type and number of actin-binding proteins that associate with the filaments; the usual results are filament bundles or a branched, dendritic structure. The details of these associations lead to different cellular structures:

1. *Filopodia* are long, thin protrusions that emerge from the cellular membrane. They are made of long, unbranched, parallel actin bundles, often decorated with tropomyosin and fascin; their elongation is mediated by formins (11); and they serve an exploratory function, enabling the cell to probe its local environment (12, 13).

2. *Lamellipodia* are broad, sheet-like protrusions that contain a branched, "dendritic" network of thin, short actin filaments (14, 15). They are often followed by another structure, termed the lamellum, which is different from the lamellipodium in that the actin is bundled rather than branched (14–16). Most motile protrusions display a thin (~1 μm) lamellipodium closer to the membrane, followed by a wider lamellum (5–10 μm) closer to the nucleus. Some uncertainty remains about the dendritic structure of lamellipodia and its relationship to the lamellum (16–19).
3. *Stress fibers* are thick actin filament bundles containing many antiparallel actin filaments heavily decorated with myosin II, which endows them with contractile properties (20). They are often found along the basal portion of the cell and terminate in large adhesive structures (21).
4. *Arcs* (dorsal/ventral) are large actin filament bundles that provide transverse structural support for the cell. They are tethered to the dorsal and side portions of the cell and sometimes terminate in adhesions (21).

In this section, we will only discuss the mechanisms that control formation of filopodia and the sheet-like protrusions comprising both the lamellipodium and the lamellum.

Elongation of parallel bundles in filopodia and dendritic networks in lamellipodia is actively driven by actin polymerization, which is regulated by multiple regulatory proteins. To simplify this discussion, we will break the regulatory molecules based on their function.

2.1. Monomeric (G-) Actin-Binding Proteins

Two molecules regulate the availability of actin monomers for polymerization, profilin and thymosin β4. Both bind specifically to G-actin, but profilin can bind to different actin polymerization nucleators and stimulators (e.g., formins, see below), thus being an actin monomer "shuttle" (22). On the contrary, thymosin β4 does not bind to any nucleator, being believed to sequester G-actin and maintain an appropriate reservoir that can exchange G-actin with profilin to promote actin filament growth (9, 10, 22, 23).

2.2. Maintenance of the Pool of G-Actin

Monomeric actin mainly comes from two sources. One is from de novo synthesis. It is worth noting that actin synthesis is concentrated at sites of protrusion, as revealed by the presence of β-actin RNA in protrusions (24). The other is monomer recycling driven by depolymerization of preformed structures. This is a spontaneous process that occurs at the pointed ends due to filament aging (bound to ADP alone; see above) and also due to filament severing, which is catalyzed by ADF (ADP depolymerization factor)/cofilin. Due to its severing function, cofilin serves a dual purpose, creating new free ends from which polymerization can be reinitiated, and also breaking down and depolymerizing old networks for monomer recycling (25, 26). Cofilin is regulated by phosphorylation by LIM kinase (LIMK), which is in turn regulated by PAK- or Rho-kinase/ROCK-mediated phosphorylation (26, 27). On the contrary, the actin-binding protein coronin binds to the sides of the filaments and protects the actin filament from cofilin-mediated severing and depolymerization (28). There is also evidence that coronin regulates cofilin activity through recruitment of a specific phosphatase (29).

2.3. Actin Polymerization at the Barbed End

Polymerization occurs at the barbed end much more efficiently than at the pointed end. Several proteins regulate this, which can be divided into two antagonistic groups: barbed-end polymerization promoters and capping proteins (30). Promoters include direct nucleators, for example, formins (31). Formins are multimodular proteins that bind to the barbed end, to profilin, and, in some cases, to microtubules, regulating their stability. They function as processive polymerization devices that initiate de novo actin filament assembly by recruiting actin monomers and using them to nucleate new filaments. They are then able to maintain filament elongation by remaining bound to the fast-growing end and promoting addition of further monomers, simultaneously preventing binding of capping proteins (see below) (31, 32). There are many formins, the best characterized are mDia1, mDia2, and mDia3, whose activity is regulated by binding of small Rho GTPases (e.g., RhoA → mDia1, Cdc42 → mDia2), which relieves head-to-tail autoinhibitory interactions in the formins (32). Other proteins such as vasodilator-stimulated phosphoprotein (VASP) and its relatives Mena and EVL also prevent binding of capping proteins and promote actin polymerization through profilin, but cannot nucleate filaments de novo (33, 34). On the contrary, capping proteins terminate elongation, thereby limiting polymerization of new filaments (35). CapZ is an α/β heterodimeric capping protein found in most eukaryotic cells, but there are others, such as adducin, that cap the barbed end of short actin filaments. Perhaps the best known is gelsolin/brevin, which is one of a series of family members that includes capG, severin and villin, adseverin, advillin, and supervillin (36). These bind to the barbed ends with high affinity in the presence of calcium and are also regulated by phosphoinositides (36).

2.4. Actin Nucleation at the Pointed End

In the lamellipodium, actin polymerization is also promoted by actin nucleators that bind to the pointed end, providing a scaffold to support polymerization at the barbed end. This is mainly catalyzed by the Arp2/3 complex, which is a hetero-heptamer that binds to the side of a preformed actin filament and promotes actin polymerization, forming a 70° angle with the pre-existing filament (37).

This imposes a dendritic, branched geometry in the actin network. The Arp2/3 complex is activated locally at the cell membrane by proteins of the WASP/WAVE family. The WASP family contains two main members: WASP, which is expressed exclusively in hematopoietic cells (38), and N-WASP, which is more widely expressed. Recent studies suggest that while both Arp2 and Arp3 are required for nucleation in the presence of WASP, the loss of Arp2 does not interfere with binding to the N-WASP tail (39). WASP/N-WASP/WAVE regulation of the Arp2/3 complex is tightly regulated by phosphorylation (WASP), small Rho GTPases (Cdc42 → WASP/N-WASP; Rac → WAVE/Scar and N-WASP), and phosphoinositides (30, 38, 40). The WAVE/Scar complex (which includes WAVE, Abi, Nap125, Sra-1, and HSPC-300) is activated by binding of activated Rac, which induces the dissociation of the trimer Abi-Nap125-Sra-1 from WAVE, resulting in its activation (41). Additionally, Abi has been shown to connect the WAVE/SCAR complex with formins (42), suggesting that the formation of the two distinct actin protrusions – Arp2/3-complex-generated branched networks filopodia – may be linked (43).

2.5. Actin Reorganization and Bundling in the Lamellum

In the lamellum, actin reorganization, rather than polymerization, is the leading mechanism thought to create the architecture of this region. Actin filaments are organized and bundled together in thicker, linear, and mostly antiparallel arrays by the action of two families of actin-binding proteins: the α-actinins and the class II myosins. These molecules act as actin cross-linkers, forming supramolecular aggregates that bind simultaneously and bring together multiple individual actin filaments into stable bundles (23). In addition to its function as an actin cross-linker, myosin II provides contractile force through an ATPase-driven conformational change that induces filament sliding (44). In motile cells, myosin II activity is regulated mainly by phosphorylation on its regulatory light chain by Rho GTPase, particularly RhoA, regulated kinases. Other actin-binding proteins with cross-linking function are also implicated in the creation of actin bundles in this region, particularly tropomyosins and filamins.

2.6. Other Modes of Protrusion

In some specific cases, protrusion seems to occur in the absence of actin assemblies. For example, cells invading a 3D matrix generate protrusive "blebs" driven initially by hydrostatic pressure from inside the cell (45); actin assembly is implicated at later stages to provide stability to the blebs, filling the bleb and evolving into a psuedopod/lobopod, or retracting back into the cell membrane. In this model, myosin II-based contraction is thought to provide the contractile power that generates the hydrostatic extension of the blebs (2, 46).

3. Polarity

Cells polarize in a number of ways to serve different purposes. In developmental biology, it refers to the asymmetric divisions that result in two daughter cells with different fate and purpose (47). This is similar to yeast, in which the daughter cell buds asymmetrically from one pole of the mother cell (48). On the contrary, many cells (e.g., gut epithelia or vascular endothelial cells) display striking spatial differences in function and composition. These cells develop a marked, stable apical and basal polarity (49). In cell migration, polarity refers to the front–rear polarity, e.g., the molecular and functional differences between the cell front and rear. Interestingly, all these types of polarization share many common effectors and adaptors (50, 51). Our focus is on front–back polarity.

3.1. Where Is the Front and Where Is the Rear? Breaking Symmetry

Cell polarization can arise from spontaneous, intrinsic mechanisms, or it can be dictated by the extracellular environment. Adhesion to the substratum or lateral contacts with other cells provides polarization cues. Often, a polarized influx of soluble, extracellular signals dictates the asymmetric recruitment or activation of specific signaling proteins within the cell, creating a functional asymmetry that translates into a morphological one. In the developing embryo, gradients of soluble morphogens induce an asymmetric distribution of partitioning-defective (PAR) proteins and other scaffolds (52, 53). Such asymmetric translocation translates into the formation of an anterior and a posterior pole that is crucial for proper development (51). Similarly, gradients of chemotactic or haptotactic stimuli promote the asymmetric activation of otherwise evenly distributed receptors, which induce polarized signals that generate a protrusion in the front (closer to the highest concentration of the gradient) and a well-defined rear (closer to the lowest concentration of the gradient) (54, 55). It is worth noting that chemoattractants often induce cellular polarization and the development of migratory, polarized morphologies even when presented in a non-polarized, homogeneous manner. This phenomenon, called chemokinesis, reflects the intrinsic ability of migratory cells to self-polarize in response to intracellular signals or homogeneous extracellular stimuli (56).

These polarization examples have in common the cellular response to an extracellular stimulation that initially generates the front of the cell. However, there is strong evidence that cells can also polarize by creating the rear first (56–58). Some cells, such as keratocytes, fibroblasts, and leukocytes, polarize in the absence of extracellular stimulation by generating subcellular "zones of no protrusion," in which protrusive signaling is inhibited and the actin cytoskeleton is reorganized into non-protrusive actin bundles

(58, 59). This mode of polarization is driven by myosin II, which has been postulated as the symmetry breaker in this model.

3.2. Chemoattraction and Actin Polymerization: Forming a Front

Gradients of morphogens (in the embryo), pheromones (in yeast), or chemoattractants provide spatial cues that generate cellular asymmetry (55). They do so by activating specific receptors. These receptors often belong to the superfamily of seven-transmembrane, heterotrimeric G-protein-coupled receptors (GPCR) (60), which are distributed homogeneously in the cellular membrane (61, 62) (although the initial wave of activation may induce signaling-dependent recruitment and clustering of receptors and signaling components, see below). Asymmetric activation of these receptors is amplified through the asymmetric recruitment and activation of signaling adaptors, in an exquisite process that amplifies very shallow differences in the gradient as perceived by the front and the rear of the cell (55). Among the effectors, asymmetrically recruited and activated are heterotrimeric G-proteins, which activate, among other enzymes, PLC and PKCs, inducing the local formation of second messengers such as DAG and IP3, and protein phosphorylation (63). G-proteins also activate the phospholipid enzyme PI3K, which generates PIP3 (64, 65), an important second messenger in the amplification of the response to the gradient and the asymmetric activation of Rho GTPases via recruitment of GEFs containing PH domains (66). This has been best described for PKB/Akt, an important kinase implicated in cell survival that contains a PIP3-binding PH domain, but whose role in chemotaxis remains unclear (55, 64). Interestingly, PIP3 seems confined to the cellular front, which is oriented toward the maximum concentration of the gradient, whereas it is rapidly hydrolyzed elsewhere by the action of the lipid phosphatase/tumor suppressor PTEN to maintain a plasma membrane gradient of PIP3 (67, 68).

Cdc42 is among the initial GTPases implicated in response to polarizing signals; it controls the recruitment of the Par3/6 proteins, aPKCs, and actin polymerization machinery to the leading edge (69). Cdc42 participates in additional polarity-related events, such as the positioning of the nucleus and the orientation of the microtubules (see below). Genetic evidence shows that many cells can actually migrate in the absence of PI3K activity (70), whereas Cdc42 is mandatory in most systems (71, 72), suggesting that PI3K constitutes a polarity amplification mechanism (55). It is unclear how Cdc42 is initially activated. The small Ras-related GTPase Rap1 is required for activation of Cdc42 in some cells (73), as is the β-PIX (which is a Cdc42 GEF)-Scribble complex (74). Interestingly, Scribble is downstream of the Par3/Par6/aPKC complex activated by Cdc42, suggesting the existence of a positive feedback loop for Cdc42 activation that probably enhances and amplifies polarization (51).

The Rho family protein Cdc42 also regulates actin polymerization. It can directly promote nucleation of dendritic networks via its effect on WASP and N-WASP, and it also promotes the activation of Rac through an ill-defined mechanism (75). The conjoined activation of Cdc42 and Rac promotes the extension of protrusions containing lamellipodia (see Subheading 2) in the direction of migration, and also regulates the formation of additional recruitment and clustering of adhesive and chemotactic receptors and signaling molecules at the leading edge. Finally, it is worth noting that in some cases, the composition of the plasma membrane (cholesterol-rich lipid rafts) reflects the functional polarization and acquisition of a leading and a trailing edge (76).

3.3. Myosin and Actin Bundling: Bringing (Up) the Rear

The activation of Cdc42 and PI3K at the front of the cell facing the maximum concentration of the chemotactic gradient drives the activation of Rac, which generates lamellipodia-containing protrusions. Rac activity correlates inversely with the activation of the small GTPase RhoA through a poorly understood feedback loop (77). RhoA activation organizes and contracts actin filaments through the activation of myosin II to form larger, thicker actomyosin bundles (44). These bundles are stable (i.e., they disassemble slowly) and prevent protrusion once formed, likely because they produce stable adhesions that do not signal to Rac (78). The combination of large, stable actomyosin bundles and large adhesions results in the formation of "no protrusion" areas that define the cell rear and sides that comprise it.

Cells do not always form morphologically identical rears. This likely reflects different amounts of myosin II, its isoforms, and activation state, and the adhesions they produce (79). For example, in wounded epithelial monolayers, the "no protrusion" areas are defined by the contacts, on the sides and the rear, with the cells that surround the leading cell. Isolated fibroblasts and migrating endothelial cells display an extended rear that lends a triangular appearance to the migrating cell. In some cases, transformed variants of these cells may display less organized rears, which often correlate with more than one leading edge caused by over-activated protrusive signaling. Keratocytes and melanocytes often display a fan-shaped morphology with a flat rear characterized by thick actomyosin bundles parallel to the leading edge (79). Leukocytes and amoeba have short, stubby rears that accumulate small actomyosin bundles and microtubules; are often elevated over the plane of the substratum ("uropod"); and accumulate adhesion receptors (80).

3.4. Other Intracellular Polarity Systems

Additional cellular organelles are also polarized in migrating cells. The nucleus, Golgi apparatus, and microtubules seem to act in concert to guide cell migration. Nuclear translocation and microtubule polarization define the direction of polarization in most mesenchymal cells but likely not in small, amoeboid-like cells (such as leukocytes) (81, 82). Microtubules are polarized in part through regulation of their dynamic plus end. Microtubule capture at the cell cortex involves stabilization and capture of the plus end, and may lead to pulling forces on the microtubules (83). There are specific proteins associated with the plus end of microtubules, collectively termed the plus-end-tracking proteins (+TIPs), which are involved in cortical capture. End-binding protein-1 (EB-1) is an example of a +TIPS. In addition to its role in controlling the chemotactic response, Cdc42, acting through its kinase effector MRCK, positions the nucleus through myosin-dependent rearward flow of cortical actin (84). MTOC positioning is regulated by Cdc42 via Par proteins, PKCζ, and its downstream effectors GSK-3 and adenomatous polyposis coli (APC) (85). These pathways regulate, among others, the dynein–dynactin complex, and are thought to keep the MTOC immobilized with respect to the moving nucleus (84, 86). Thus, Cdc42 is a master regulator of polarity through its role in controlling protrusion at the leading edge, nuclear positioning, and microtubule organization in response to motile cues.

4. Adhesion

Adhesion (i.e., the physical interaction of a cell with another cell or the extracellular matrix) is essential for cell migration and tissue integrity. Cell–cell adhesion maintains epithelial tissues, supports functional contacts between specialized cells (e.g., neuronal synapse or T cell:APC contacts), and provides the scaffold for directed migration (e.g., radial glia in the nervous system). Cell–matrix adhesions are the best studied adhesions that mediate cell migration and are the focus of this outline.

4.1. Cell–Matrix Adhesions

Cell–matrix adhesions were first observed in the fibroblasts and other cells in culture (87–91). They are sites of convergence between the actin cytoskeleton and the fibers of the extracellular matrix. Due to their highly localized nature, they were initially called focal adhesions. Over the years, multiple types of cell–matrix adhesions have been described, including the following:

1. *Nascent adhesions* are the first observable adhesive structures, emerging within the lamellipodium. Their formation and stability are linked to the dendritic actin that forms lamellipodium (92, 93). They are small and highly transient – either going on to mature or disassemble (turnover) – and therefore are not easily observed in every cell type (93, 94).
2. *Focal complexes* are adhesions in their early stages of maturation. They were first observed in cells expressing a constitutively

active Rac. They are larger than nascent adhesions, myosin dependent, and resided at the boundary of the lamellum and lamellipodium (95, 96). Like nascent adhesions, they also tend to either disassemble ("turn over") or grow and elongate into focal adhesions. The elongation occurs along a template of bundled actin (93). Nascent adhesions and focal complexes are related to motile cells and subcellular structures; their fast appearance and turnover correlates directly with high velocities of protrusion and movement.

3. *Focal adhesions.* This term is currently reserved for mature adhesions that evolve slowly over time (97). They are usually linked to large, contractile actomyosin stress fibers (21), and their appearance correlates inversely with motility. They are conspicuously absent in cells migrating in 3D. They are similar compositionally to nascent adhesions and focal complexes. However, there are several adhesion proteins that appear specifically in focal adhesions, such as tensin. The term "fibrillar" or "matrix" adhesions can also be found in the literature, often referring to terminally grown adhesions that cooperate in the reorganization of the underlying extracellular matrix into elongated fibrils.
4. *Podosomes* are ring-shaped adhesions often found in fast-moving cells, such as macrophages and dendritic cells (98). Although their molecular composition is very similar to that of adhesions, their spatial distribution is radial, forming dot-like structures similar to "suction cups" (99). Podosomes are structurally divided into a core, which mainly contains proteins involved in actin polymerization, such as WASP, the Arp2/3 complex, and cortactin, and a surrounding ring populated with integrin receptors and adhesion proteins, e.g., paxillin and FAK/Pyk2 (100). Whereas focal adhesion formation requires actomyosin force downstream of the small GTPase RhoA, podosome formation is stimulated by local loss of contractility and recruitment of RhoA inhibitors, such as p190RhoGAP (101).
5. *Invadopodia* are specialized actin-based structures very similar morphologically to podosomes (the nomenclature is ambiguous, but a consensus seems to be to use "podosome" for normal cells, and "invadopodium" for cancer cells). Like podosomes, they are mainly radial; however, adhesion proteins are not restricted to the peripheral ring, but appear intermixed with the structural core, where most actin-related proteins reside (102). In addition to their adhesive nature, invadopodia often concentrate proteolytic components that degrade the matrix around it to permit eventual crossing of the cell across the matrix. These structures are often found in invading cells, such as metastatic tumor cells (99, 103).

4.2. Some Major Adhesion Components

4.2.1. Integrins

The different types of matrix adhesions share many common components. They include adhesion receptors, which constitute the physical link between the cell and the extracellular matrix. The best studied receptors are the integrin family, which is comprised of an α and a β subunit. Both subunits are type I transmembrane proteins and associate to form a heterodimer. Although not every combination of α and β chain is possible, there are many combinations and they give rise to multiple heterodimers that interact with specific ligands (104, 105). The integrins also organize signaling molecules through direct or indirect associations with the C-terminal, cytoplasmic tails.

Integrins are finely regulated receptors; some are designed to engage in only appropriate circumstances (e.g., platelet or leukocyte adhesion). They are activated by inside-out signaling. Talin and kindlin bind to the intracellular portion of the integrin receptor, inducing a conformational change that causes the molecule to extend and expose the sites of ligand binding ("affinity modulation") (106–109). Talin-induced conformational extension, in turn, allows other receptors, e.g., chemokines or growth factor receptors, to control integrin activation through a "receptor cross talk" (110).

Integrins also modulate the adhesiveness of a cell in several ways. One is by forming large clusters that increase the avidity. The clusters are usually driven by molecules that bind to the cytoplasmic tails of the integrins and serve to cross-link them (111). Another is by conformation change in the integrin that creates a high affinity state (108). The affinity changes arise through conformation changes induced by the binding of talin or kindlin to the cytoplasmic tails, as described above. In some cases, membrane microdomains or other transmembrane proteins (e.g., tetraspanins) may modulate avidity (112–114).

4.2.2. Actin–ECM Linkage

Some of the proteins that bind to the cytoplasmic tail of the integrins also bind directly or indirectly to actin. These include talin, vinculin, and α-actinin. Talin binds directly to integrin and actin simultaneously (115, 116); vinculin does not bind to integrins directly, but does so through talin (117, 118). Notably, vinculin not only binds to actin, but also to the actin nucleator Arp2/3 (119). α-Actinin, which binds to actin, is also reported to bind directly to β1 integrins (120) and also binds through vinculin (121). Another actin cross-linker, filamin, also binds to β1 and β7 integrin tails (122). These linkages define a molecular clutch that shunts retrograde forces due to actomyosin contraction or resistance from the leading edge membrane to the extracellular matrix. The efficiency of this linkage varies among cells and thus regulates signaling through adhesions and protrusion. In addition to its function in controlling integrin affinity through conformational change (see above), talin also regulates integrin avidity by controlling integrin clustering through its association with actin; similarly, vinculin and α-actinin also control integrin clustering. The binding strength of vinculin, talin, and α-actinin to integrin and/or actin is tightly regulated. Mechanisms of regulation include conformational change induced by binding to their partners, e.g., vinculin to talin or actin (123), phosphorylation, e.g., α-actinin (124), or mechanical force, e.g., talin (125, 126). There is evidence that integrins themselves modulate their activation in response to mechanical forces (127).

4.2.3. Signaling Molecules

A plethora of signaling molecules also associate with integrin-containing adhesions. Some are kinases, whereas others function as scaffolds that recruit additional molecules to adhesions. A key signature of several adhesion proteins is their prominent phosphorylation in Tyr (128).

- *FAK* is a Tyr kinase. It is recruited to adhesions early and is activated by autophosphorylation and Src family kinases (129). The FAK/Src module is implicated in the disassembly of adhesion complexes (130). FAK is involved in the recruitment and phosphorylation of other adhesion proteins, such as paxillin, and potentially controls RhoA activation through binding of GEFs and GAPs. For example, FAK can bind p190RhoGEF (a RhoA activator) and/or p190RhoGAP (a RhoA inhibitor) via p120RasGAP (131), and the spatial-temporal regulation of these interactions likely determine the local activation of the GTPase.
- *Paxillin* is a phosphoprotein localized to adhesions. Through multiple phosphorylated Tyr and Ser/Thr residues, it creates binding sites for recruitment of other adhesion proteins and signaling regulators. Paxillin is emerging as a Rac signaling hub due to its ability to recruit effectors implicated in activating Rac (132).
- *p130CAS* is a phosphoprotein localized to adhesions, through binding to paxillin and other molecules, such as FAK or Nck (133). It is implicated in the recruitment and activation of CrkII/DOCK180 to adhesions to activate Rac (134). A novel property is that it reacts to mechanical force, becoming phosphorylated in response to stretching (135).
- *Zyxin* is a scaffold protein, distantly related to paxillin. It interacts with α-actinin and may play a role in regulating actin polymerization close to adhesions via association to VASP (136). Zyxin is mechanoreactive, shuttling between actin/α-actinin and adhesions upon stretching (137).
- *Tensin* is a phosphoprotein that appears only in mature adhesions (94) and plays a prominent role in signaling through its interaction with protein phosphatases (138).

The above only presents a sample of the best characterized adhesion molecules. However, the actual list is much larger, with more than 180 proteins participating in a complicated network of interactions called the "adhesome" (139, 140). These components and their interactions depend on the type of integrin, adhesion, presence of other receptors, and cell type. These complicated signaling networks generate a plethora of signals that control not only cellular motility (by controlling actin polymerization and disassembly; adhesion formation and assembly; etc.), but also many other cellular functions, such as proliferation, gene expression, survival, etc.

5. Retraction

Cell migration requires protrusion of the front, translocation of the cell body, and retraction of the rear. Rear retraction requires the coordinated contraction of the actin cytoskeleton and disassembly of the adhesions at the trailing edge. Several mechanisms converge to promote adhesion disassembly: actomyosin contraction, microtubule-induced adhesion relaxation, endocytic traffic of adhesion receptors, and proteolytic cleavage.

- *Actomyosin contraction.* Early evidence pointed to a role for the actomyosin cytoskeleton (myosin II) in promoting retraction of the rear of migrating cells (20, 141). Different lines of evidence have demonstrated that myosin IIA downstream of the RhoA/ROCK axis is required for retraction in a variety of cell lines (142–147).
- *Microtubule-induced adhesion relaxation.* Some adhesions are repeatedly targeted by microtubules, especially adhesions at the trailing edge; in many cases, microtubule targeting promotes adhesion disassembly (148). This requires dynamin and FAK (149). However, the precise relaxation signal(s) are still unknown.
- *Endocytosis of adhesion receptors.* Emerging evidence points to integrin endocytosis and recycling as a key mechanism in regulating adhesion turnover (150). In neutrophils, endocytosis drives the internalization and recycling of the α5 β1 integrin (151). More specifically, clathrin-mediated endocytosis has been shown to mediate disassembly of a specific subset of adhesions by promoting internalization and recycling of β1 integrins (152).
- *Calpain-dependent cleavage.* Calpain is involved in the retraction of the leading edge of migrating cells (153). Several focal adhesion proteins have been identified as targets of the calcium-dependent calpain. Cleavage of two of these, talin and FAK, is critical for retraction of the rear of migrating cells (154, 155).

6. Signaling, Regulation, and Integration

In most cases, cells migrate, proliferate, or differentiate in response to the conditions they find around them (the "microenvironment"). The ME informs the cell of the current needs of the tissue or the availability of resources for the cell to prosper. In both cases, this information triggers the cellular response, e.g., migration, proliferation, and so on. Such information may take different forms, e.g., nutrient availability, growth factor gradients, and the pliability of the surrounding tissue. A very good example is the social amoeba *Dictyostelium discoideum*. When nutrients are aplenty, the cells divide actively, lead unicellular "lifestyles," and are actively chemotactic, "sniffing" for cues to nutrient-rich environments (156); on the contrary, when nutrients become scarce, cells aggregate, form multicellular bodies that culminate with the death of the cells, and spore production and dissemination (reviewed in, for example, (157)). This simple example portrays the power of the cellular surroundings in dictating the behavior of the cell.

The ME directs cell behavior by triggering the activation of different signaling networks that dictate the appropriate cellular response. The cell must then interpret these signals, which are both mechanical and chemical. Cell migration is a particular response that is regulated by multiple intertwined signaling networks that are initiated and organized at the cell surface. Signaling can be initiated through GPCR, integrins, growth factors, and other receptors (e.g., cadherins, and neuronal and immune receptors). Many of these receptors aggregate in response to the extracellular stimulus and ultimately organize signaling scaffolds. Responses at the level of the receptor include an intrinsic kinase activity in some growth factor receptors, which leads to either autophosphorylation or phosphorylation of proximal effectors. In other cases, activation occurs via conformational change upon binding to the ligand, which triggers the binding (or dissociation) of the downstream effectors. These initial signals are then propagated along one or more branches of a complex, interconnected network of intermediates that eventually lead to the Rho family of small GTPases, which in turn regulate the migration machinery, e.g., polarity, protrusion, and adhesion. The complexity of this network is seen in the synergy between growth factor and adhesive signaling, for example.

Integrin-mediated migration is particularly well studied and reveals the complexity of these networks. They also recruit multiple proteins to adhesions. Well-studied examples are FAK, which in turn recruits GEFs and GAPs that regulate the Rho GTPases (e.g., p190RhoGEF and p190RhoGAP (131)), and paxillin, which recruits Rac effectors such as CAS/CRK-II/DOCK180, PI3K, and β-PIX (132). Many other proteins involved in the adhesome (139, 140) can anecdotally recruit and activate different Rho GEFs and GAPs that regulate Rho family GTPases.

In addition to the Rho GTPases, other small GTPases are implicated in migration. *Ras* activates PI3K, which is involved in chemotaxis. A close analog of Ras, *Rap*, controls integrin affinity (and thus cell adhesion) via its effector RapL (158). The *Rab/ARF* family controls vesicular traffic and is implicated in the recycling of adhesion receptors from and to the plasma membrane during cell migration (150). Internalization and recycling of migratory receptors are important steps in controlling cell migration, and Rab GTPases play a critical role in this. Internalization of integrins and other receptors implicated in motility, such as GPCR, involves the formation of clathrin-coated pits with FAK, and dynamin (149, 152), which delivers them to endosomes that are transported toward the center of the cell; Rab5 is required for this. Some of these receptors are recycled back to the plasma membrane, a process that requires Rab11 and/or Arf6.

Small *Rho* GTPases are central to cell migration and act through a myriad of effectors on a relatively small number of endpoints that include actin polymerization, organization, and contraction; microtubule polymerization and/or stability; and transcriptional regulation of motogenic gene products.

Rho GTPases are small Ras-like (*R*as-*HO*mology) proteins. There are ~20 Rho GTPases, which can be roughly distributed into RhoA-like, Rac-like, TC10/Cdc42-like, and atypical Rho GTPases (71). What makes these GTPases different from each other are their activation/inactivation and effector protein complements. There are multiple GTP exchange factors (GEFs), each one capable of activating one or more Rho GTPases specifically (159). Likewise, there are many GTPase-activating proteins (GAPs), each one capable of catalyzing GTP hydrolysis by one or more Rho proteins. Finally, GDP-dissociation inhibitors (GDIs) can sequester-specific Rho GTPases in the cytoplasm. In a similar fashion, specific Rho GTPases control specific effectors (kinases, actin polymerization nucleators, etc.), which results in each GTPase controlling different aspects of the cellular organization required for efficient cell migration. In this section, we will provide a succinct overview of the most relevant Rho GTPases implicated in cell migration, their major regulators, and effectors.

– *RhoA* was originally described as a serum-responsive element that mediated formation of stress fibers in adherent cells (160). RhoA is a key mediator of actin polymerization and reorganization in response to extracellular signals. Multiple GEFs activate RhoA; likewise, multiple GAPs target RhoA. Rho is critically regulated through sequestering at the cellular cytoplasm by RhoGDIs. RhoA can activate different effectors, including p160ROCK/Rho-kinase, which activates myosin II through direct phosphorylation of RLC (161) and phosphorylation and inactivation of the MLC phosphatase (162), resulting in actin organization into large filament bundles (20); mDia1, a formin that promotes actin polymerization (30, 163) and also controls microtubule dynamics (164, 165); citronK, which is a kinase that controls RLC phosphorylation and cytokinesis (166); and some others.

- *Rac* was identified as the key mediator in the membrane ruffling response to growth factors; soon, it was revealed that Rac controls protrusion formation (167). Many GEFs activate Rac: some notable ones are α/β-PIX, Tiam1, DOCK180, and some members of the Vav subfamily. Similarly, multiple GAPs downregulate Rac activity. Rac also activates several specific downstream effectors, for example, PAKs (168), which are S/T kinases that phosphorylate a plethora of substrata implicated in cell migration, e.g., MLCK (inhibitory) (169), RLC (activating) (170), LIMK (activating, which inhibits cofilin, regulating its actin-severing activity) (171), and stathmin (which controls microtubule stability) (172). Perhaps the best studied effect of Rac is its dramatic increase in actin polymerization and the formation of dendritic (branched) actin (173). Rac activates the scaffold protein WAVE/Scar, which in turn activates the Arp2/3, a potent actin nucleator ((174), *see* Subheading 2).
- *Cdc42* is a critical mediator of cell polarization and also drives actin polymerization. The polarity function is conserved from very simple eukaryotes, such as fission or budding yeast, to specialized mammalian cells, such as neurons or leukocytes. An early insight into its function revealed that Cdc42 activation promotes filopodia formation (75). Known GEFs and GAPs are included in Table I. At this point, it remains unclear whether Cdc42 is sequestered in the cytoplasm by GDIs. Key effectors include PAK (see above) (175); the WASP/N-WASP proteins (176), which activate the Arp2/3 similar to WAVE/Scar (177); IQGAP, through which Cdc42 controls actin and microtubules (178); MRCK, which activates myosin II and controls nuclear repositioning during cell migration (84, 179); and Par proteins/aPKC, which also control centrosome reorientation and microtubule polarization through their effect on GSK-3/APC (85).

7. Conclusion

Hopefully, this overview has set the stage for appreciating the methods presented in the book, since they have driven the science outlined here. A great deal is now known about migration, and this knowledge is now ready for translation to understand fundamental biological processes, migration-related diseases, and the development of diagnostics and therapeutics. As these methods are introduced into complex biological systems, they no doubt will reveal important, unexpected observations and fundamental mechanisms.

Acknowledgments

The authors regret the rather large number of studies that were not cited due to space constraints. Rick Horwitz's work is supported by NIH grants GM23244 and the Cell Migration Consortium (U54 GM064346).

References

1. Ridley AJ, Schwartz MA, Burridge K *et al.* Cell migration: integrating signals from front to back. *Science*, **302**(5651), 1704–1709 (2003).
2. Friedl P, Wolf K. Plasticity of cell migration: a multiscale tuning model. *J Cell Biol*, **188**(1), 11–19 (2010).
3. Olson MF, Sahai E. The actin cytoskeleton in cancer cell motility. *Clin Exp Metastasis*, **26**(4), 273–287 (2009).
4. Stephens L, Milne L, Hawkins P. Moving towards a better understanding of chemotaxis. *Curr Biol*, **18**(11), R485–494 (2008).
5. Gaggioli C, Hooper S, Hidalgo-Carcedo C *et al.* Fibroblast-led collective invasion of carcinoma cells with differing roles for RhoGTPases in leading and following cells. *Nat Cell Biol*, **9**(12), 1392–1400 (2007).
6. Hatten ME. New directions in neuronal migration. *Science*, **297**(5587), 1660–1663 (2002).
7. Weijer CJ. Collective cell migration in development. *J Cell Sci*, **122**(Pt 18), 3215–3223 (2009).
8. Korn ED, Carlier MF, Pantaloni D. Actin polymerization and ATP hydrolysis. *Science*, **238**(4827), 638–644 (1987).
9. Pollard TD, Borisy GG. Cellular motility driven by assembly and disassembly of actin filaments. *Cell*, **112**(4), 453–465 (2003).
10. Pollard TD, Cooper JA. Actin, a central player in cell shape and movement. *Science*, **326**(5957), 1208–1212 (2009).
11. Block J, Stradal TE, Hanisch J *et al.* Filopodia formation induced by active mDia2/Drf3. *J Microsc*, **231**(3), 506–517 (2008).
12. Mattila PK, Lappalainen P. Filopodia: molecular architecture and cellular functions. *Nat Rev Mol Cell Biol*, **9**(6), 446–454 (2008).
13. Mejillano MR, Kojima S, Applewhite DA *et al.* Lamellipodial versus filopodial mode of the actin nanomachinery: pivotal role of the filament barbed end. *Cell*, **118**(3), 363–373 (2004).
14. Cramer LP, Siebert M, Mitchison TJ. Identification of novel graded polarity actin filament bundles in locomoting heart fibroblasts: implications for the generation of motile force. *J Cell Biol*, **136**(6), 1287–1305 (1997).
15. Svitkina TM, Borisy GG. Arp2/3 complex and actin depolymerizing factor/cofilin in dendritic organization and treadmilling of actin filament array in lamellipodia. *J Cell Biol*, **145**(5), 1009–1026 (1999).
16. Ponti A, Machacek M, Gupton SL, Waterman-Storer CM, Danuser G. Two distinct actin networks drive the protrusion of migrating cells. *Science*, **305**(5691), 1782–1786 (2004).
17. Koestler SA, Auinger S, Vinzenz M, Rottner K, Small JV. Differentially oriented populations of actin filaments generated in lamellipodia collaborate in pushing and pausing at the cell front. *Nat Cell Biol*, **10**(3), 306–313 (2008).
18. Vallotton P, Small JV. Shifting views on the leading role of the lamellipodium in cell migration: speckle tracking revisited. *J Cell Sci*, **122**(Pt 12), 1955–1958 (2009).
19. Danuser G. Testing the lamella hypothesis: the next steps on the agenda. *J Cell Sci*, **122**(Pt 12), 1959–1962 (2009).
20. Chrzanowska-Wodnicka M, Burridge K. Rho-stimulated contractility drives the formation of stress fibers and focal adhesions. *J Cell Biol*, **133**(6), 1403–1415 (1996).
21. Hotulainen P, Lappalainen P. Stress fibers are generated by two distinct actin assembly

mechanisms in motile cells. *J Cell Biol*, **173**(3), 383–394 (2006).

22. Yarmola EG, Bubb MR. Profilin: emerging concepts and lingering misconceptions. *Trends Biochem Sci*, **31**(4), 197–205 (2006).
23. Le Clainche C, Carlier M-F. Regulation of actin assembly associated with pprotrusion and adhesion in cell migration. *Physiol. Rev.*, **88**(2), 489–513 (2008).
24. Kislauskis EH, Zhu X, Singer RH. beta-Actin messenger RNA localization and protein synthesis augment cell motility. *J Cell Biol*, **136**(6), 1263–1270 (1997).
25. Condeelis J. How is actin polymerization nucleated in vivo? *Trends Cell Biol*, **11**(7), 288–293 (2001).
26. Bamburg JR, Bernstein BW. ADF/cofilin. *Curr Biol*, **18**(7), R273–275 (2008).
27. Maciver SK, Hussey PJ. The ADF/cofilin family: actin-remodeling proteins. *Genome Biol*, **3**(5), reviews3007 (2002).
28. Uetrecht AC, Bear JE. Coronins: the return of the crown. *Trends Cell Biol*, **16**(8), 421–426 (2006).
29. Cai L, Marshall TW, Uetrecht AC, Schafer DA, Bear JE. Coronin 1B coordinates Arp2/3 complex and cofilin activities at the leading edge. *Cell*, **128**(5), 915–929 (2007).
30. Campellone KG, Welch MD. A nucleator arms race: cellular control of actin assembly. *Nat Rev Mol Cell Biol*, **11**(4), 237–251 (2010).
31. Paul AS, Pollard TD. Review of the mechanism of processive actin filament elongation by formins. *Cell Motil Cytoskeleton*, **66**(8), 606–617 (2009).
32. Chesarone MA, DuPage AG, Goode BL. Unleashing formins to remodel the actin and microtubule cytoskeletons. *Nat Rev Mol Cell Biol*, **11**(1), 62–74 (2010).
33. Bear JE, Gertler FB. Ena/VASP: towards resolving a pointed controversy at the barbed end. *J Cell Sci*, **122**(Pt 12), 1947–1953 (2009).
34. Bear JE, Svitkina TM, Krause M *et al.* Antagonism between Ena/VASP proteins and actin filament capping regulates fibroblast motility. *Cell*, **109**(4), 509–521 (2002).
35. Cooper JA, Schafer DA. Control of actin assembly and disassembly at filament ends. *Curr Opin Cell Biol*, **12**(1), 97–103 (2000).
36. Zigmond SH. Beginning and ending an actin filament: control at the barbed end. *Curr Top Dev Biol*, **63**, 145–188 (2004).
37. Mullins RD, Heuser JA, Pollard TD. The interaction of Arp2/3 complex with actin: nucleation, high affinity pointed end capping, and formation of branching networks of filaments. *Proc Natl Acad Sci USA*, **95**(11), 6181–6186 (1998).
38. Thrasher AJ, Burns SO. WASP: a key immunological multitasker. *Nat Rev Immunol*, **10**(3), 182–192 (2010).
39. Nolen BJ, Pollard TD. Structure and biochemical properties of fission yeast Arp2/3 complex lacking the Arp2 subunit. *J Biol Chem*, **283**(39), 26490–26498 (2008).
40. Pollitt AY, Insall RH. WASP and SCAR/WAVE proteins: the drivers of actin assembly. *J Cell Sci*, **122**(Pt 15), 2575–2578 (2009).
41. Vartiainen MK, Machesky LM. The WASP-Arp2/3 pathway: genetic insights. *Curr Opin Cell Biol*, **16**(2), 174–181 (2004).
42. Yang C, Czech L, Gerboth S *et al.* Novel roles of formin mDia2 in lamellipodia and filopodia formation in motile cells. *PLoS Biol*, **5**(11), e317 (2007).
43. Insall RH, Machesky LM. Actin dynamics at the leading edge: from simple machinery to complex networks. *Dev Cell*, **17**(3), 310–322 (2009).
44. Vicente-Manzanares M, Ma X, Adelstein RS, Horwitz AR. Non-muscle myosin II takes centre stage in cell adhesion and migration. *Nat Rev Mol Cell Biol*, **10**(11), 778–790 (2009).
45. Charras G, Paluch E. Blebs lead the way: how to migrate without lamellipodia. *Nat Rev Mol Cell Biol*, **9**(9), 730–736 (2008).
46. Fackler OT, Grosse R. Cell motility through plasma membrane blebbing. *J Cell Biol*, **181**(6), 879–884 (2008).
47. Gonczy P. Mechanisms of asymmetric cell division: flies and worms pave the way. *Nat Rev Mol Cell Biol*, **9**(5), 355–366 (2008).
48. Slaughter BD, Smith SE, Li R. Symmetry breaking in the life cycle of the budding yeast. *Cold Spring Harb Perspect Biol*, **1**(3), a003384 (2009).
49. Mellman I, Nelson WJ. Coordinated protein sorting, targeting and distribution in polarized cells. *Nat Rev Mol Cell Biol*, **9**(11), 833–845 (2008).
50. Li R, Gundersen GG. Beyond polymer polarity: how the cytoskeleton builds a polarized cell. *Nat Rev Mol Cell Biol*, **9**(11), 860–873 (2008).
51. Iden S, Collard JG. Crosstalk between small GTPases and polarity proteins in cell polarization. *Nat Rev Mol Cell Biol*, **9**(11), 846–859 (2008).
52. Goldstein B, Macara IG. The PAR proteins: fundamental players in animal cell polarization. *Dev Cell*, **13**(5), 609–622 (2007).

53. Johnson MH. From mouse egg to mouse embryo: polarities, axes, and tissues. *Annu Rev Cell Dev Biol*, **25**, 483–512 (2009).
54. Iglesias PA, Devreotes PN. Navigating through models of chemotaxis. *Curr Opin Cell Biol*, **20**(1), 35–40 (2008).
55. Swaney KF, Huang CH, Devreotes PN. Eukaryotic Chemotaxis: A Network of Signaling Pathways Controls Motility, Directional Sensing, and Polarity. *Annu Rev Biophys*, (2010).
56. Verkhovsky AB, Svitkina TM, Borisy GG. Self-polarization and directional motility of cytoplasm. *Curr Biol*, **9**(1), 11–20 (1999).
57. Yam PT, Wilson CA, Ji L *et al.* Actin-myosin network reorganization breaks symmetry at the cell rear to spontaneously initiate polarized cell motility. *J Cell Biol*, **178**(7), 1207–1221 (2007).
58. Vicente-Manzanares M, Koach MA, Whitmore L, Lamers ML, Horwitz AF. Segregation and activation of myosin IIB creates a rear in migrating cells. *J Cell Biol*, **183**(3), 543–554 (2008).
59. Mseka T, Coughlin M, Cramer LP. Graded actin filament polarity is the organization of oriented actomyosin II filament bundles required for fibroblast polarization. *Cell Motil Cytoskeleton*, **66**(9), 743–753 (2009).
60. Cotton M, Claing A. G protein-coupled receptors stimulation and the control of cell migration. *Cell Signal*, **21**(7), 1045–1053 (2009).
61. Xiao Z, Zhang N, Murphy DB, Devreotes PN. Dynamic distribution of chemoattractant receptors in living cells during chemotaxis and persistent stimulation. *J Cell Biol*, **139**(2), 365–374 (1997).
62. Servant G, Weiner OD, Neptune ER, Sedat JW, Bourne HR. Dynamics of a chemoattractant receptor in living neutrophils during chemotaxis. *Mol Biol Cell*, **10**(4), 1163–1178 (1999).
63. Van Haastert PJ, Devreotes PN. Chemotaxis: signalling the way forward. *Nat Rev Mol Cell Biol*, **5**(8), 626–634 (2004).
64. Kolsch V, Charest PG, Firtel RA. The regulation of cell motility and chemotaxis by phospholipid signaling. *J Cell Sci*, **121**(Pt 5), 551–559 (2008).
65. Cain RJ, Ridley AJ. Phosphoinositide 3-kinases in cell migration. *Biol Cell*, **101**(1), 13–29 (2009).
66. Garcia-Mata R, Burridge K. Catching a GEF by its tail. *Trends Cell Biol*, **17**(1), 36–43 (2007).
67. Iijima M, Devreotes P. Tumor suppressor PTEN mediates sensing of chemoattractant gradients. *Cell*, **109**(5), 599–610 (2002).
68. Funamoto S, Meili R, Lee S, Parry L, Firtel RA. Spatial and temporal regulation of 3-phosphoinositides by PI 3-kinase and PTEN mediates chemotaxis. *Cell*, **109**(5), 611–623 (2002).
69. Etienne-Manneville S, Hall A. Cell polarity: Par6, aPKC and cytoskeletal crosstalk. *Curr Opin Cell Biol*, **15**(1), 67–72 (2003).
70. Hoeller O, Kay RR. Chemotaxis in the absence of PIP3 gradients. *Curr Biol*, **17**(9), 813–817 (2007).
71. Heasman SJ, Ridley AJ. Mammalian Rho GTPases: new insights into their functions from in vivo studies. *Nat Rev Mol Cell Biol*, **9**(9), 690–701 (2008).
72. Etienne-Manneville S. Cdc42--the centre of polarity. *J Cell Sci*, **117**(Pt 8), 1291–1300 (2004).
73. Gerard A, Mertens AE, van der Kammen RA, Collard JG. The Par polarity complex regulates Rap1- and chemokine-induced T cell polarization. *J Cell Biol*, **176**(6), 863–875 (2007).
74. Osmani N, Vitale N, Borg JP, Etienne-Manneville S. Scrib controls Cdc42 localization and activity to promote cell polarization during astrocyte migration. *Curr Biol*, **16**(24), 2395–2405 (2006).
75. Nobes CD, Hall A. Rho, Rac, and Cdc42 GTPases regulate the assembly of multimolecular focal complexes associated with actin stress fibers, lamellipodia, and filopodia. *Cell*, **81**(1), 53–62 (1995).
76. Gomez-Mouton C, Lacalle RA, Mira E *et al.* Dynamic redistribution of raft domains as an organizing platform for signaling during cell chemotaxis. *J Cell Biol*, **164**(5), 759–768 (2004).
77. Sander EE, ten Klooster JP, van Delft S, van der Kammen RA, Collard JG. Rac downregulates Rho activity: reciprocal balance between both GTPases determines cellular morphology and migratory behavior. *J Cell Biol*, **147**(5), 1009–1022 (1999).
78. Vicente-Manzanares M, Choi CK, Horwitz AR. Integrins in cell migration--the actin connection. *J Cell Sci*, **122**(Pt 2), 199–206 (2009).
79. Keren K, Pincus Z, Allen GM *et al.* Mechanism of shape determination in motile cells. *Nature*, **453**(7194), 475–480 (2008).
80. Sanchez-Madrid F, Serrador JM. Bringing up the rear: defining the roles of the uropod.

Nat Rev Mol Cell Biol, **10**(5), 353–359 (2009).

81. Omelchenko T, Vasiliev JM, Gelfand IM, Feder HH, Bonder EM. Mechanisms of polarization of the shape of fibroblasts and epitheliocytes: Separation of the roles of microtubules and Rho-dependent actin-myosin contractility. *Proc Natl Acad Sci USA*, **99**(16), 10452–10457 (2002).
82. Ratner S, Sherrod WS, Lichlyter D. Microtubule retraction into the uropod and its role in T cell polarization and motility. *J Immunol*, **159**(3), 1063–1067 (1997).
83. Lansbergen G, Akhmanova A. Microtubule plus end: a hub of cellular activities. *Traffic*, 7(5), 499–507 (2006).
84. Gomes ER, Jani S, Gundersen GG. Nuclear movement regulated by Cdc42, MRCK, myosin, and actin flow establishes MTOC polarization in migrating cells. *Cell*, **121**(3), 451–463 (2005).
85. Etienne-Manneville S, Hall A. Cdc42 regulates GSK-3beta and adenomatous polyposis coli to control cell polarity. *Nature*, **421**(6924), 753–756 (2003).
86. Schmoranzer J, Fawcett JP, Segura M *et al.* Par3 and dynein associate to regulate local microtubule dynamics and centrosome orientation during migration. *Curr Biol*, **19**(13), 1065–1074 (2009).
87. Curtis AS. The Mechanism of Adhesion of Cells to Glass. A Study by Interference Reflection Microscopy. *J Cell Biol*, **20**, 199–215 (1964).
88. Lazarides E, Burridge K. Alpha-actinin: immunofluorescent localization of a muscle structural protein in nonmuscle cells. *Cell*, **6**(3), 289–298 (1975).
89. Izzard CS, Lochner LR. Cell-to-substrate contacts in living fibroblasts: an interference reflexion study with an evaluation of the technique. *J Cell Sci*, **21**(1), 129–159 (1976).
90. Heath JP, Dunn GA. Cell to substratum contacts of chick fibroblasts and their relation to the microfilament system. A correlated interference-reflexion and high-voltage electron-microscope study. *J Cell Sci*, **29**, 197–212 (1978).
91. Geiger B. A 130K protein from chicken gizzard: its localization at the termini of microfilament bundles in cultured chicken cells. *Cell*, **18**(1), 193–205 (1979).
92. Alexandrova AY, Arnold K, Schaub S *et al.* Comparative dynamics of retrograde actin flow and focal adhesions: formation of nascent adhesions triggers transition from fast to slow flow. *PLoS One*, **3**(9), e3234 (2008).
93. Choi CK, Vicente-Manzanares M, Zareno J *et al.* Actin and alpha-actinin orchestrate the assembly and maturation of nascent adhesions in a myosin II motor-independent manner. *Nat Cell Biol*, **10**(9), 1039–1050 (2008).
94. Zaidel-Bar R, Ballestrem C, Kam Z, Geiger B. Early molecular events in the assembly of matrix adhesions at the leading edge of migrating cells. *J Cell Sci*, **116**(Pt 22), 4605–4613 (2003).
95. Giannone G, Dubin-Thaler BJ, Rossier O *et al.* Lamellipodial actin mechanically links myosin activity with adhesion-site formation. *Cell*, **128**(3), 561–575 (2007).
96. Rottner K, Hall A, Small JV. Interplay between Rac and Rho in the control of substrate contact dynamics. *Current Biology*, **9**(12), 640–649 (1999).
97. Zaidel-Bar R, Cohen M, Addadi L, Geiger B. Hierarchical assembly of cell-matrix adhesion complexes. *Biochem Soc Trans*, **32**(Pt 3), 416–420 (2004).
98. Calle Y, Anton IM, Thrasher AJ, Jones GE. WASP and WIP regulate podosomes in migrating leukocytes. *J Microsc*, **231**(3), 494–505 (2008).
99. Gimona M, Buccione R, Courtneidge SA, Linder S. Assembly and biological role of podosomes and invadopodia. *Curr Opin Cell Biol*, **20**(2), 235–241 (2008).
100. Buccione R, Orth JD, McNiven MA. Foot and mouth: podosomes, invadopodia and circular dorsal ruffles. *Nat Rev Mol Cell Biol*, **5**(8), 647–657 (2004).
101. Lener T, Burgstaller G, Crimaldi L, Lach S, Gimona M. Matrix-degrading podosomes in smooth muscle cells. *Eur J Cell Biol*, **85**(3–4), 183–189 (2006).
102. Block MR, Badowski C, Millon-Fremillon A *et al.* Podosome-type adhesions and focal adhesions, so alike yet so different. *Eur J Cell Biol*, **87**(8–9), 491–506 (2008).
103. Linder S. Invadosomes at a glance. *J Cell Sci*, **122**(Pt 17), 3009–3013 (2009).
104. Hynes RO. Integrins: bidirectional, allosteric signaling machines. *Cell*, **110**(6), 673–687 (2002).
105. Humphries JD, Byron A, Humphries MJ. Integrin ligands at a glance. *J Cell Sci*, **119**(Pt 19), 3901–3903 (2006).
106. Kim M, Carman CV, Springer TA. Bidirectional transmembrane signaling by cytoplasmic domain separation in integrins. *Science*, **301**(5640), 1720–1725 (2003).

107. Ye F, Hu G, Taylor D *et al.* Recreation of the terminal events in physiological integrin activation. *J Cell Biol*, **188**(1), 157–173 (2010).
108. Shattil SJ, Kim C, Ginsberg MH. The final steps of integrin activation: the end game. *Nat Rev Mol Cell Biol*, **11**(4), 288–300 (2010).
109. Askari JA, Tynan CJ, Webb SE *et al.* Focal adhesions are sites of integrin extension. *J Cell Biol*, **188**(6), 891–903 (2010).
110. Schwartz MA, Ginsberg MH. Networks and crosstalk: integrin signalling spreads. *Nat Cell Biol*, **4**(4), E65–68 (2002).
111. Carman CV, Springer TA. Integrin avidity regulation: are changes in affinity and conformation underemphasized? *Curr Opin Cell Biol*, **15**(5), 547–556 (2003).
112. Krauss K, Altevogt P. Integrin leukocyte function-associated antigen-1-mediated cell binding can be activated by clustering of membrane rafts. *J Biol Chem*, **274**(52), 36921–36927 (1999).
113. Hemler ME. Tetraspanin functions and associated microdomains. *Nat Rev Mol Cell Biol*, **6**(10), 801–811 (2005).
114. Yanez-Mo M, Barreiro O, Gordon-Alonso M, Sala-Valdes M, Sanchez-Madrid F. Tetraspanin-enriched microdomains: a functional unit in cell plasma membranes. *Trends Cell Biol*, **19**(9), 434–446 (2009).
115. Horwitz A, Duggan K, Buck C, Beckerle MC, Burridge K. Interaction of plasma membrane fibronectin receptor with talin--a transmembrane linkage. *Nature*, **320**(6062), 531–533 (1986).
116. Critchley DR. Biochemical and structural properties of the integrin-associated cytoskeletal protein talin. *Annu Rev Biophys*, **38**, 235–254 (2009).
117. Burridge K, Mangeat P. An interaction between vinculin and talin. *Nature*, **308**(5961), 744–746 (1984).
118. Hemmings L, Rees DJ, Ohanian V *et al.* Talin contains three actin-binding sites each of which is adjacent to a vinculin-binding site. *J Cell Sci*, **109** (Pt 11), 2715–2726 (1996).
119. DeMali KA, Barlow CA, Burridge K. Recruitment of the Arp2/3 complex to vinculin: coupling membrane protrusion to matrix adhesion. *J Cell Biol*, **159**(5), 881–891 (2002).
120. Otey CA, Pavalko FM, Burridge K. An interaction between alpha-actinin and the beta 1 integrin subunit in vitro. *J Cell Biol*, **111**(2), 721–729 (1990).
121. Gilmore AP, Burridge K. Regulation of vinculin binding to talin and actin by phosphatidyl-inositol-4-5-bisphosphate. *Nature*, **381** (6582), 531–535 (1996).
122. Pfaff M, Liu S, Erle DJ, Ginsberg MH. Integrin beta cytoplasmic domains differentially bind to cytoskeletal proteins. *J Biol Chem*, **273**(11), 6104–6109 (1998).
123. Bakolitsa C, Cohen DM, Bankston LA *et al.* Structural basis for vinculin activation at sites of cell adhesion. *Nature*, **430**(6999), 583–586 (2004).
124. Izaguirre G, Aguirre L, Hu YP *et al.* The cytoskeletal/non-muscle isoform of alpha-actinin is phosphorylated on its actin-binding domain by the focal adhesion kinase. *J Biol Chem*, **276**(31), 28676–28685 (2001).
125. Jiang G, Giannone G, Critchley DR, Fukumoto E, Sheetz MP. Two-piconewton slip bond between fibronectin and the cytoskeleton depends on talin. *Nature*, **424**(6946), 334–337 (2003).
126. del Rio A, Perez-Jimenez R, Liu R *et al.* Stretching single talin rod molecules activates vinculin binding. *Science*, **323**(5914), 638–641 (2009).
127. Friedland JC, Lee MH, Boettiger D. Mechanically activated integrin switch controls alpha5beta1 function. *Science*, **323**(5914), 642–644 (2009).
128. Ballestrem C, Erez N, Kirchner J *et al.* Molecular mapping of tyrosine-phosphorylated proteins in focal adhesions using fluorescence resonance energy transfer. *J Cell Sci*, **119**(Pt 5), 866–875 (2006).
129. Parsons JT. Focal adhesion kinase: the first ten years. *J Cell Sci*, **116**(Pt 8), 1409–1416 (2003).
130. Webb DJ, Donais K, Whitmore LA *et al.* FAK-Src signalling through paxillin, ERK and MLCK regulates adhesion disassembly. *Nat Cell Biol*, **6**(2), 154–161 (2004).
131. Tomar A, Schlaepfer DD. Focal adhesion kinase: switching between GAPs and GEFs in the regulation of cell motility. *Curr Opin Cell Biol*, **21**(5), 676–683 (2009).
132. Deakin NO, Turner CE. Paxillin comes of age. *J Cell Sci*, **121**(Pt 15), 2435–2444 (2008).
133. Defilippi P, Di Stefano P, Cabodi S. p130Cas: a versatile scaffold in signaling networks. *Trends Cell Biol*, **16**(5), 257–263 (2006).
134. Matsuda M, Ota S, Tanimura R *et al.* Interaction between the amino-terminal SH3 domain of CRK and its natural target proteins. *J Biol Chem*, **271**(24), 14468–14472 (1996).
135. Sawada Y, Tamada M, Dubin-Thaler BJ *et al.* Force sensing by mechanical extension of the Src family kinase substrate p130Cas. *Cell*, **127**(5), 1015–1026 (2006).
136. Drees BE, Andrews KM, Beckerle MC. Molecular dissection of zyxin function reveals

its involvement in cell motility. *J Cell Biol*, **147**(7), 1549–1560 (1999).

137. Yoshigi M, Hoffman LM, Jensen CC, Yost HJ, Beckerle MC. Mechanical force mobilizes zyxin from focal adhesions to actin filaments and regulates cytoskeletal reinforcement. *J Cell Biol*, **171**(2), 209–215 (2005).
138. Hall EH, Daugherty AE, Choi CK, Horwitz AF, Brautigan DL. Tensin1 requires protein phosphatase-1alpha in addition to RhoGAP DLC-1 to control cell polarization, migration, and invasion. *J Biol Chem*, **284**(50), 34713–34722 (2009).
139. Zaidel-Bar R, Itzkovitz S, Ma'ayan A, Iyengar R, Geiger B. Functional atlas of the integrin adhesome. *Nat Cell Biol*, **9**(8), 858–867 (2007).
140. Zaidel-Bar R, Geiger B. The switchable integrin adhesome. *J Cell Sci*, **123**(Pt 9), 1385–1388 (2010).
141. Crowley E, Horwitz AF. Tyrosine phosphorylation and cytoskeletal tension regulate the release of fibroblast adhesions. *J Cell Biol*, **131**(2), 525–537 (1995).
142. Vicente-Manzanares M, Zareno J, Whitmore L, Choi CK, Horwitz AF. Regulation of protrusion, adhesion dynamics, and polarity by myosins IIA and IIB in migrating cells. *J Cell Biol*, **176**(5), 573–580 (2007).
143. Even-Ram S, Doyle AD, Conti MA *et al.* Myosin IIA regulates cell motility and actomyosin-microtubule crosstalk. *Nat Cell Biol*, **9**(3), 299–309 (2007).
144. Worthylake RA, Lemoine S, Watson JM, Burridge K. RhoA is required for monocyte tail retraction during transendothelial migration. *J Cell Biol*, **154**(1), 147–160 (2001).
145. Vicente-Manzanares M, Cabrero JR, Rey M *et al.* A role for the Rho-p160 Rho coiled-coil kinase axis in the chemokine stromal cell-derived factor-1alpha-induced lymphocyte actomyosin and microtubular organization and chemotaxis. *J Immunol*, **168**(1), 400–410 (2002).
146. Smith A, Bracke M, Leitinger B, Porter JC, Hogg N. LFA-1-induced T cell migration on ICAM-1 involves regulation of MLCK-mediated attachment and ROCK-dependent detachment. *J Cell Sci*, **116**(Pt 15), 3123–3133 (2003).
147. Eddy RJ, Pierini LM, Matsumura F, Maxfield FR. Ca2+-dependent myosin II activation is required for uropod retraction during neutrophil migration. *J Cell Sci*, **113** (Pt 7), 1287–1298 (2000).
148. Kaverina I, Krylyshkina O, Small JV. Microtubule targeting of substrate contacts promotes their relaxation and dissociation. *J Cell Biol*, **146**(5), 1033–1044 (1999).
149. Ezratty EJ, Partridge MA, Gundersen GG. Microtubule-induced focal adhesion disassembly is mediated by dynamin and focal adhesion kinase. *Nat Cell Biol*, 7(6), 581–590 (2005).
150. Caswell PT, Vadrevu S, Norman JC. Integrins: masters and slaves of endocytic transport. *Nat Rev Mol Cell Biol*, **10**(12), 843–853 (2009).
151. Pierini LM, Lawson MA, Eddy RJ, Hendey B, Maxfield FR. Oriented endocytic recycling of alpha5beta1 in motile neutrophils. *Blood*, **95**(8), 2471–2480 (2000).
152. Ezratty EJ, Bertaux C, Marcantonio EE, Gundersen GG. Clathrin mediates integrin endocytosis for focal adhesion disassembly in migrating cells. *J Cell Biol*, **187**(5), 733–747 (2009).
153. Huttenlocher A, Palecek SP, Lu Q *et al.* Regulation of cell migration by the calcium-dependent protease calpain. *J Biol Chem*, **272**(52), 32719–32722 (1997).
154. Chan KT, Bennin DA, Huttenlocher A. Regulation of adhesion dynamics by calpain-mediated proteolysis of focal adhesion kinase (FAK). *J Biol Chem*, **285**(15), 11418–11426 (2010).
155. Franco SJ, Rodgers MA, Perrin BJ *et al.* Calpain-mediated proteolysis of talin regulates adhesion dynamics. *Nat Cell Biol*, **6**(10), 977–983 (2004).
156. Willard SS, Devreotes PN. Signaling pathways mediating chemotaxis in the social amoeba, Dictyostelium discoideum. *Eur J Cell Biol*, **85**(9–10), 897–904 (2006).
157. Shaulsky G, Kessin RH. The cold war of the social amoebae. *Curr Biol*, **17**(16), R684–692 (2007).
158. Katagiri K, Maeda A, Shimonaka M, Kinashi T. RAPL, a Rap1-binding molecule that mediates Rap1-induced adhesion through spatial regulation of LFA-1. *Nat Immunol*, **4**(8), 741–748 (2003).
159. Rossman KL, Der CJ, Sondek J. GEF means go: turning on RHO GTPases with guanine nucleotide-exchange factors. *Nat Rev Mol Cell Biol*, **6**(2), 167–180 (2005).
160. Ridley AJ, Hall A. The small GTP-binding protein rho regulates the assembly of focal adhesions and actin stress fibers in response to growth factors. *Cell*, **70**(3), 389–399 (1992).
161. Amano M, Ito M, Kimura K *et al.* Phosphorylation and activation of myosin by Rho-associated kinase (Rho-kinase). *J Biol Chem*, **271**(34), 20246–20249 (1996).
162. Kimura K, Ito M, Amano M *et al.* Regulation of myosin phosphatase by Rho and Rho-associated kinase (Rho-kinase). *Science*, **273**(5272), 245–248 (1996).

163. Watanabe N, Madaule P, Reid T *et al.* p140mDia, a mammalian homolog of Drosophila diaphanous, is a target protein for Rho small GTPase and is a ligand for profilin. *Embo J*, **16**(11), 3044–3056 (1997).

164. Watanabe N, Kato T, Fujita A, Ishizaki T, Narumiya S. Cooperation between mDia1 and ROCK in Rho-induced actin reorganization. *Nat Cell Biol*, **1**(3), 136–143 (1999).

165. Wen Y, Eng CH, Schmoranzer J *et al.* EB1 and APC bind to mDia to stabilize microtubules downstream of Rho and promote cell migration. *Nat Cell Biol*, **6**(9), 820–830 (2004).

166. Yamashiro S, Totsukawa G, Yamakita Y *et al.* Citron kinase, a Rho-dependent kinase, induces di-phosphorylation of regulatory light chain of myosin II. *Mol Biol Cell*, **14**(5), 1745–1756 (2003).

167. Ridley AJ, Paterson HF, Johnston CL, Diekmann D, Hall A. The small GTP-binding protein rac regulates growth factor-induced membrane ruffling. *Cell*, **70**(3), 401–410 (1992).

168. Bokoch GM. Biology of the p21-activated kinases. *Annu Rev Biochem*, **72**, 743–781 (2003).

169. Sanders LC, Matsumura F, Bokoch GM, de Lanerolle P. Inhibition of myosin light chain kinase by p21-activated kinase. *Science*, **283**(5410), 2083–2085 (1999).

170. Chew TL, Masaracchia RA, Goeckeler ZM, Wysolmerski RB. Phosphorylation of non-muscle myosin II regulatory light chain by p21-activated kinase (gamma-PAK). *J Muscle Res Cell Motil*, **19**(8), 839–854 (1998).

171. Yang N, Higuchi O, Ohashi K *et al.* Cofilin phosphorylation by LIM-kinase 1 and its role in Rac-mediated actin reorganization. *Nature*, **393**(6687), 809–812 (1998).

172. Wittmann T, Bokoch GM, Waterman-Storer CM. Regulation of microtubule destabilizing activity of Op18/stathmin downstream of Rac1. *J Biol Chem*, **279**(7), 6196–6203 (2004).

173. Machesky LM, Hall A. Role of actin polymerization and adhesion to extracellular matrix in Rac- and Rho-induced cytoskeletal reorganization. *J Cell Biol*, **138**(4), 913–926 (1997).

174. Machesky LM, Mullins RD, Higgs HN *et al.* Scar, a WASp-related protein, activates nucleation of actin filaments by the Arp2/3 complex. *Proc Natl Acad Sci USA*, **96**(7), 3739–3744 (1999).

175. Manser E, Leung T, Salihuddin H, Zhao ZS, Lim L. A brain serine/threonine protein kinase activated by Cdc42 and Rac1. *Nature*, **367**(6458), 40–46 (1994).

176. Symons M, Derry JM, Karlak B *et al.* Wiskott-Aldrich syndrome protein, a novel effector for the GTPase CDC42Hs, is implicated in actin polymerization. *Cell*, **84**(5), 723–734 (1996).

177. Machesky LM, Insall RH. Scar1 and the related Wiskott-Aldrich syndrome protein, WASP, regulate the actin cytoskeleton through the Arp2/3 complex. *Curr Biol*, **8**(25), 1347–1356 (1998).

178. Osman MA, Cerione RA. Iqg1p, a yeast homologue of the mammalian IQGAPs, mediates cdc42p effects on the actin cytoskeleton. *J Cell Biol*, **142**(2), 443–455 (1998).

179. Leung T, Chen XQ, Tan I, Manser E, Lim L. Myotonic dystrophy kinase-related Cdc42-binding kinase acts as a Cdc42 effector in promoting cytoskeletal reorganization. *Mol Cell Biol*, **18**(1), 130–140 (1998).

Chapter 2

Scratch-Wound Assay

Giles Cory

Abstract

The scratch-wound assay is a simple, reproducible assay commonly used to measure basic cell migration parameters such as speed, persistence, and polarity. Cells are grown to confluence and a thin "wound" introduced by scratching with a pipette tip. Cells at the wound edge polarise and migrate into the wound space.

Advantages of this assay are that it does not require the use of specific chemoattractants or gradient chambers and it generates a strong directional migratory response, even in cell types that do not show robust responses in "single cell" migration assays. It is most reliably analysed when performed using time-lapse imaging, which can also yield valuable cell morphology/protein localisation information (Table 1).

Key words: Polarity, Migration, Scratch-wound, Time-lapse, Actin, Golgi apparatus

1. Introduction

The scratch-wound assay studies the migration of a sheet of cells in two dimensions in response to formation of a crude wound. Cells may break free from the sheet or may maintain an unbroken front. The introduction of a scratch initiates migration from a standing start in a well-defined direction – perpendicular to the wound edge (Fig. 1). Deviations from this path are easily measured, giving values for directional persistence and speed in wild-type and manipulated cells (for example knock-down/over-expression/addition of inhibitors), as described in Subheading 3.1. The assay requires time-lapse imaging of the cells and subsequent analysis with imaging software such as ImageJ. Visual data can also be analysed to assess lamellipodium formation and integrity and other migratory phenotypes.

Large numbers of cells can be tracked in a single field of view, making it easy to obtain high *n* values for analysis.

Claire M. Wells and Maddy Parsons (eds.), *Cell Migration: Developmental Methods and Protocols*,
Methods in Molecular Biology, vol. 769, DOI 10.1007/978-1-61779-207-6_2, © Springer Science+Business Media, LLC 2011

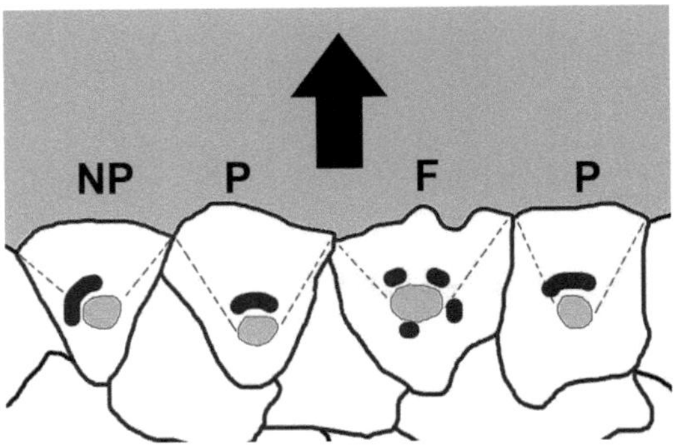

Fig. 1. Scoring of Golgi apparatus polarity. Cells adjacent to the wound edge are assessed for Golgi polarisation. The vast majority of the immunostained marker of the Golgi apparatus (*dark grey*) must be compact and lie between the nucleus and imaginary lines between the nucleus (*light grey circles*) and the leading edge cell contacts (*dotted lines*). *NP* = not polarised, *P* = polarised, *F* = not polarised (fragmented). A similar scoring system can be used for analysing MTOC markers.

Table 1
Pros and cons of the scratch-wound assay

Advantages	Disadvantages
Reproducible, directional, and large-scale migration studied	Confounding parameters: Cell "crowding" Cell/cell adhesion effects Matrix effects
Speed and persistence and lamellipodium formation and morphology easily analysed	Little direct physiological relevance
Multiple cell types amenable	Not good for chemotaxis studies
Simple to run with a time-lapse microscope, no other special equipment needed	
MTOC/Golgi apparatus polarisation easily studied	
Tagged protein localisation possible with fluorescence microscope	

Following scratching, several cell types polarise their microtubule organising centres (MTOCs) and Golgi apparatus to face into the wound. This process exhibits cell-type dependence on actin and microtubule dynamics (1, 2) and can also be easily studied in cells fixed after scratch wounding (Subheading 3.2).

The simplicity of the scratch-wound assay does not come without its problems. As the cells are studied as a confluent monolayer, migratory parameters may well be affected by confounding factors such as changes in the strength of cell–cell adhesions under different conditions and cell crowding. It is important to start with "equally confluent" monolayers when comparing conditions and also to factor in the contribution of cell adhesions (Table 1).

2. Materials

2.1. Migration Assay

Heated time-lapse microscope, cell culture dish, 10–200 μl pipette tip, growth medium, cells of choice (see Note 1), and image analysis software (e.g. ImageJ).

2.2. Polarity Assay

Cells grown on coverslips, 200 μl pipette tip, antibodies to Golgi apparatus marker [e.g. anti-GM130 (Transduction Labs) or anti-giantin (Covance)] or MTOC (e.g. anti-gamma-tubulin or anti-pericentrin), fluorescently labelled phalloidin, DAPI, fluorescent microscope, phosphate-buffered saline (PBS), 0.25% Triton X-100 in PBS, 4% paraformaldehyde in PBS, and 1% bovine serum albumin (BSA) in PBS.

3. Methods

3.1. Migration Assay

1. The day (or two) before the experiment, plate your cells of choice in normal growth medium into a live cell imaging dish or chamber slide (see Note 2). Perform each condition in triplicate.
2. When cells are at a suitable confluence, use a 10–200-μl pipette tip (see Note 3) to scratch a wound through the centre of the well. Try to do this with one flowing movement to give a clean straight edge (see Note 4 for inclusion of stimuli).
3. Place the cells on the microscope and select field or fields of view if using an automated stage. Try to avoid imaging fields with layers of scratched cells that have accumulated at the wound edge. Use of a 20× objective provides sufficient resolution to track cells.
4. Initiate time-lapse microscopy sequence. We film NIH-3T3 fibroblasts for 16 h (roughly the time taken for the wound edges to meet) at 15-min intervals, although these parameters will be dependent on the cell type/speed and the resolution required (see Note 5).
5. Terminate filming upon wound closure.
6. Analysis: For phase images, this is best done by eye. Using an image analysis package such as ImageJ, open the image sequence. One cell at a time, mark the position of the nucleus through successive frames (see Note 6 for ImageJ method). Stop tracking after a predefined length of time or until the cell makes contact with another cell from the other wound edge, which ever occurs first. Repeat this for each cell along the wound edge. Disregard cells that divide during the time course or leave the field of view.

When a cell has been tracked, a set of coordinates should be obtained (see Note 5). Export cell coordinates into a spreadsheet and normalise the starting coordinate of each cell track to (0,0) by subtracting the starting coordinate from all the time point coordinates. Once this has been performed, a scatter graph of time vs. normalised cell coordinate can be plotted for all cells within a condition. These give a ready visual view of the directionality and speed of the cells (3).

In order to obtain numerical values for speed and persistence, it is necessary to do some trigonometry. For each cell, calculate the change in X coordinate and the change in Y coordinate for each time point compared to the previous time point. The distance travelled by the cell between images is estimated as the square root of the sum of $X^2 + Y^2$. In this way, the total distance covered by a cell during the experiment can be estimated and the speed derived (distance = speed × time) (3). This will give a value in pixels. To convert to actual distance travelled, it is necessary to know the pixel size of the image, available in most microscope-driving software. Failing this, use of a graticulated coverslip may be necessary.

The parameter of persistence gives an indication of the directness of a cell's path. It is calculated by dividing the direct "crow flies" distance between a cell's start and finish point, with the total distance travelled by a cell. Cells that change direction often will have a low persistence and those that move in a perfectly straight line will have a persistence of 1 (see Note 5).

An alternative analysis sometimes used is to measure the proportion of the wound area that has been closed after a defined period of time. This is simple and does not require time-lapse microscopy, but it does not provide any information on individual cell speed or persistence.

3.2. Cell Polarity Assay

1. Follow steps 1 and 2 above (Subheading 3.1). Perform each condition in triplicate.
2. After 3 h (see Note 7), remove growth medium and fix cells with 4% formaldehyde or ice-cold methanol, depending on the antibody you will be using (step 3).
3. Permeabilise cells (if formaldehyde fixed) with 0.2% Triton X-100 in PBS, 5 min.
4. Block cells with 1% BSA and stain with desired antibody according to the manufacturer's instructions. Co-stain with fluorescent phalloidin and DAPI.
5. Analysis: Using a fluorescent or confocal microscope, take images along the wound edge. Golgi apparatus or MTOC staining is then assessed by eye: if it lies in the sector between the nucleus and the cell's junctions with neighbouring cells at the leading edge (Fig. 1), it is scored as polarised (3).

4. Notes

1. Microscopes with motorised stages are particularly good for these studies, allowing several fields to be captured per condition. It also allows several conditions (e.g. various knock-downs, inhibitors, etc.) to be studied in parallel, minimising the effects of differences in environmental and cell culture conditions (e.g. passage number, growth rate of cells, and confluence).

   Cell culture dish: When using chamber slides on a microscope stage, beware of "edge" effects when using phase microscopy. This can make only the very central portion of the well amenable to imaging. We find this particularly problematic when using eight-well chamber slides, and now limit ourselves to four-well slides.

   We use CO_2-independent growth medium (i.e. containing serum) for our assays, to avoid the need for gassing. This works fine for all cell lines we have tried including NIH-3T3 and HeLa. One may also add HEPES (pH 7.4–7.6) to a final concentration of 25 mM to bypass the requirement for CO_2 gassing. However, we find that this only maintains pH for a couple of hours or so.

   Cell lines: we have had success with NIH-3T3 (3), HeLa, and C2C12. Others have used astrocytes (1) among others.

2. It is important to ensure that you have a single-cell suspension as clumps will give an uneven cell density. It will be necessary to titrate your cells to find the optimal plating density. At the time of scratching, you want the cells to be uniformly "just" confluent (no gaps) but not overcrowded as this can make tracking single cells difficult and constrain cell movement. Practice is also required to ensure an even dispersion of cells across the well-knocked plates, and vibrating incubators can lead to cells adhering in wave patterns.

3. Ultimately, the choice of weapon is yours. Ideally, you wish to make a clean wound with both sides visible in the field of view, as this allows you to track both edges and also ensure that the wound widths are kept constant between conditions. Wounds should be around 0.5 mm in width.

4. The rupture of cells by scratch wounding causes the release of motogenic stimuli (we see activation of the Erk kinase pathway when we treat unscratched monolayers with medium from wounded cells (data not shown)). In order to study the effect of an exogenous stimulus on migration speed or persistence (e.g. growth factorand serum), cells may be starved prior to scratching, and the medium changed immediately after. However, there will always be some stimulation caused by the

wounding process which should be considered. It is possible to grow cells around a physical barrier (e.g. small metal cylinder) placed in a tissue-culture dish prior to plating. The "barrier" is then removed and the cells are free to migrate into the space, in the absence of "wounded" cell contents. Our experience of this assay using NIH-3T3 cells is that migration is very slow, even in the presence of serum or PDGF.

5. Because the "total distance" travelled by a cell is approximated (as the sum of individual straight-line distances between time points), the value is dependent on the time interval used – closer time points will give a longer "total distance" value. This is important to bear in mind when choosing the time intervals for image capture – for example, fast moving cells will need to be imaged with shorter time intervals. Most importantly, time intervals must be kept the same when comparing different conditions.
6. In ImageJ, this can be achieved by "Files-Import-Image Sequence" followed by "Image-Stacks-Image to Stack". Successive frames can be viewed and the point selections tool used with Shift depressed to mark a position for each time point. "Analyse-Measure" will then give a list of frame coordinates for the cell at each time point.
7. Optimal time for fixation will have to be determined for each cell type. At least 80% of cells should exhibit polarised Golgi/MTOC (see Fig. 1). In the case of the Golgi apparatus, the staining should reveal a compact structure.

References

1. Etienne-Manneville, S., and Hall, A. (2001) Integrin-mediated activation of Cdc42 controls cell polarity in migrating astrocytes through PKCzeta, *Cell* **106**, 489–498.
2. Magdalena, J., Millard, T. H., Etienne-Manneville, S., Launay, S., Warwick, H. K., and Machesky, L. M. (2003) Involvement of the Arp2/3 complex and Scar2 in Golgi polarity in scratch wound models, *Mol Biol Cell* **14**, 670–684.
3. Danson, C. M., Pocha, S. M., Bloomberg, G. B., and Cory, G. O. (2007) Phosphorylation of WAVE2 by MAP kinases regulates persistent cell migration and polarity, *J Cell Sci* **120**, 4144–4154.

Chapter 3

HGF-Induced DU145 Cell Scatter Assay

Sally T. Fram, Claire M. Wells, and Gareth E. Jones

Abstract

Epithelial mesenchymal transition (EMT) is a multi-stage process whereby epithelial cells lose their cell:cell adhesions and acquire the capacity to migrate independently. It is a process that is important in normal development and is thought to be adopted by some invasive cancer cells. EMT requires modifications in cell shape and substratum adhesions and these events are dependent on the reorganisation of the actin cytoskeleton. Hepatocyte growth factor (HGF) is a mitogenic growth factor that is well known to induce such a conversion, termed "cell scattering", in Madin Darby canine kidney (MDCK) cells. Recently, we have developed an alternative model of cell scattering using the human prostate cancer cell line, DU145. Like MDCK cells, DU145 cells normally grow as tight colonies with firm cell:cell junctions, but they can be induced to 'scatter' upon HGF stimulation. Here, we describe the optimised protocol for conducting and analysing an HGF-induced DU145 scatter assay. This model is particularly useful for monitoring changes in actin cytoskeletal organisation and dynamics, cell:cell adhesions, and cell migration in human cells that respond to HGF stimulation.

Key words: Hepatocyte growth factor, DU145 cells, Cell scattering, Cell migration, Actin cytoskeleton

1. Introduction

Adult epithelial cells can be characterised by their non-migratory phenotype combined with strong cell:cell contacts and cell:substratum adhesions. Epithelial mesenchymal transition (EMT) is a multistage process where epithelial cells lose these adhesions and acquire the capacity to migrate independently (1). As well as being seen during normal embryonic development, EMT is also thought to be re-capitulated by some cancer cells during the process of metastatic spread (2).

It has been known for some 20 years that hepatocyte growth factor (HGF) induces EMT or "cell scattering" in colony-forming

Claire M. Wells and Maddy Parsons (eds.), *Cell Migration: Developmental Methods and Protocols*,
Methods in Molecular Biology, vol. 769, DOI 10.1007/978-1-61779-207-6_3, © Springer Science+Business Media, LLC 2011

Madin–Darby canine kidney (MDCK) cells (3, 4). More recently however, a number of groups have reported that DU145 prostate cancer cells are also scattered in response to HGF stimulation (5–7).

The HGF receptor c-Met is known to be mutated or overexpressed in many types of cancer (8). Indeed the increased expression of HGF/c-Met is frequently linked with poor prognosis for cancer patients (9). The binding of HGF to c-Met triggers receptor kinase activation and in turn initiates various signal transduction pathways. In addition to stimulating cell growth, proliferation, and survival, HGF and c-Met are able to contribute to cell migration by stimulating pathways that are responsible for cell adhesion turnover and the reorganisation of the actin cytoskeleton (10), processes regulated by Rho family proteins, RhoA, Rac1, and Cdc42 (11). Indeed, HGF specifically increases the level of active Rho family GTPases in DU145 cells (12).

To date, the HGF-induced DU145 scatter model has been used in live cell imaging to show the migration of individual DU145 cells in response to HGF and to observe the changes that occur in the organisation of the actin cytoskeleton and cell adhesions downstream of growth factor stimulation (12). The DU145 scatter assay has been used to demonstrate that endogenous Cdc42 is activated following HGF stimulation (12) and that knockdown of p-21-activated kinases effects cell migratory behaviour (13, 14) and to monitor the role of p210ctn in adherens junctional regulation downstream of HGF (15) .

2. Materials

2.1. Scatter Assay

1. DU145 grow medium: RPM1-1640 with L-glutamine, 1% penicillin–streptomycin, and 10% heat-inactivated foetal bovine serum.
2. DU145 starve medium: RPM1-1640 with L-glutamine, 1% penicillin–streptomycin, and 0.5% heat-inactivated foetal bovine serum.
3. 13-mm diameter glass coverslips (see Note 1).
4. Forceps.
5. The 4-welled sterile tissue-culture dishes.
6. PBS (minus $CaCl_2$ and $MgCl_2$).
7. Hepatocyte growth factor (HGF, recombinant human) reconstituted according to the manufacturer's instructions and stored at –80°C in small aliquots.

2.2. Cell Staining and Scatter Analysis

1. 4% Paraformaldehyde in sterile PBS (minus $CaCl_2$ and $MgCl_2$).
2. 0.2% Triton X-100 in sterile PBS.
3. TRITC-Phalloidin prepared according to the manufacturer's instructions (20 mg/μl), aliquoted, and then stored at −20°C, protected from light exposure at all times.
4. 3% Bovine serum albumin (BSA) in sterile PBS.
5. FluorSave™ reagent.
6. Glass slides.
7. 1 M hydrochloric acid.
8. 96% Ethanol.
9. Parafilm.
10. Fluorescence microscope with 10× objective; associated imaging accessories optional.

3. Methods

3.1. Preparing the DU145 Cell Coverslips (Day 1)

1. DU145 cells (see Note 2) are cultured at 37°C in a tissue-culture incubator with humidified air, supplemented with CO_2 to 5% over atmospheric levels and maintained with regular growth medium change, every 2–3 days and passaging. We routinely passage and maintain the cells at sub-confluency (see Note 3).
2. Using sterile forceps, place 1× acid-washed coverslip per well in the desired number of wells of the 4-well culture dish (see Note 4). Wash each coverslip twice with sterile PBS. For each DU145 scatter assay, a serum-starved control and HGF-stimulated cell condition are required as negative and positive controls, respectively.
3. Seed the cells onto the coverslips in 1 ml growth medium at a density of 0.5×10^4/ml (see Note 3). Using a 200-μl sterile pipette tip, gently but immediately apply pressure to the coverslip. This ensures that the cells do not adhere beneath the coverslip. Incubate the cells for 48 h to allow them to form colonies. The cells are seeded at a density to favour growth of small cell colonies.

3.2. The DU145 Scatter Assay (Days 3–5)

1. On day 3, remove growth medium and wash twice with sterile PBS. 1 ml of serum-starved medium is added to each well and the cells are incubated for a further 24 h (see Note 5).
2. On day 4, make up a HGF solution containing serum-starved medium and 10 ng/ml HGF. Remove the medium from cells

and add an equal volume of HGF solution to each well (except negative control). The plate is then swirled gently to ensure even distribution of the HGF and returned to the incubator.

3. On day 5, remove the medium, wash with PBS, and then fix in 4% PFA at room temperature for 20 min. Wash the fixed cells three times with PBS. The cells can be stained immediately after this or stored in PBS at 4°C overnight.

3.3. F-actin Staining of DU145 Cells

1. Permeabilise with 0.2% Triton X-100/PBS for 5 min at room temperature.
2. Wash the coverslips again three times with PBS and carefully transfer them onto Parafilm strips laid out on a flat surface (cell side up).
3. Dilute the TRITC-phalloidin 1:1,000 in 3% BSA: PBS for F-actin staining. Add 100 μl of this solution to each coverslip, ensuring that the entire surface of the coverslip is covered. This step should be done as quickly as possible to avoid the coverslips drying out (see Note 6).
4. Incubate the coverslips in a dark and humidified environment at room temperature for 1 h.
5. Wash the coverslips twice with PBS and once with dH_2O. Drain off any residual dH_2O using tissue paper.
6. Pipette 10 μl of FluorSave™ reagent per coverslip to be mounted onto clean glass slides.
7. Carefully mount the coverslip so that the cell side is facing the FluorSave™ reagent and blot any residual FluorSave™ reagent using tissue paper. This should be done with care to avoid moving the coverslip and without allowing the formation of air bubbles between the coverslip and the glass slide.
8. Leave the coverslips to set in a dark environment at room temperature for 24 h. Once the coverslips have set, they can be visualised using a fluorescent microscope. Coverslips can be stored at 4°C and are useable for approximately 2 weeks, although immediate collection of images is advised, especially if using antibody-mediated detection (see Note 6).

3.4. Collecting Images of the DU145 Cells and Analysing Results

1. Use the 10× objective on a fluorescent microscope (we use an Olympus IX71 inverted microscope) to visualise the F-actin-stained coverslips.
2. Manually count the number of cells in five random fields of view per coverslip per treatment and categorise as outlined below and in Fig. 1. The categories are as follows:
   (a) Unscattered – cells still in a colony (Fig. 1, box 1).
   (b) Partial scatter – cells exhibit an elongated phenotype but still maintain some contacts with adjacent cells (Fig. 1, box 2).

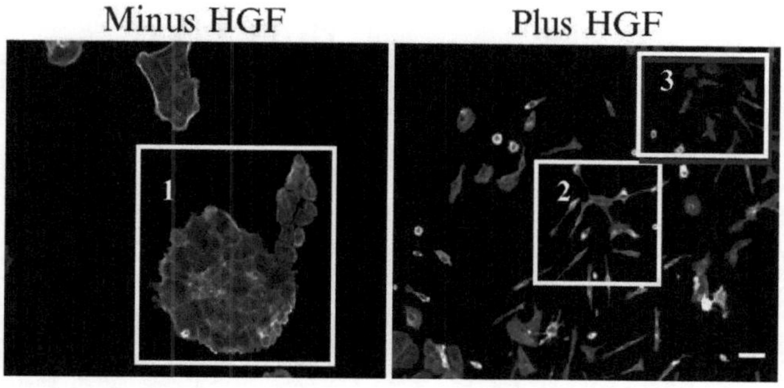

Fig. 1. DU145 cells, in the presence and absence of HGF, stained for F-actin using TRITC-phalloidin. Images illustrate the categorisation of the DU145 cells. Unscattered = cells in a colony as shown in *box 1* (minus HGF). Partial scatter = a colony where cells exhibit an elongated phenotype and there is some loss of cell:cell junctions as shown in *box 2* (plus HGF). Scatter = loss of cell:cell junctions and single cells with an elongated migratory phenotype as shown in *box 3* (plus HGF). *Bar* = 10 μm.

(c) Scatter – loss of cell:cell junctions, single cells with an elongated migratory phenotype as shown in (Fig. 1. box 3).

Disregard the cells at the extreme edge of the image to keep the cell counting consistent across experiments. Calculate the percentage of scattered cells (see Note 7) per experiment using the following formulae:

$$\text{Unscattered} + \text{partial scatter} + \text{scattered} = n,$$

$$\text{Scattered} / n \times 100 = \%\text{ of scattered cells.}$$

3.5. Real-Time Imaging of Cell Scattering

1. To observe live cells scattering in response to HGF requires a microscope adapted for time-lapse recording. We use an Olympus IX71 inverted microscope equipped with a heated stage and environmental chamber, a Prior electronic shutter, and a Q Imaging Retiga monochrome cooled CCD camera. The system is controlled and coordinated by Image Pro AMS software. The system was supplied and installed by Media Cybernetics, UK. Ideally, the microscope should have a heated stage (37°C), but a fan heating curtain system can be used. Phase-contrast microscopy using an inverted or upright microscope will provide good results (Fig. 2).
2. For time-lapse analysis, we seed cells in a 6-well plate and treat as described above for coverslips (Subheadings 3.1 and 3.2). After the HGF solution is added, 25 mM HEPES is added to each well and the plate sealed with Parafilm. The 6-well plate is then placed on the heated stage. The live images from the microscope are then viewed on the computer monitor using Image pro AMS software and cell colonies selected for recording.

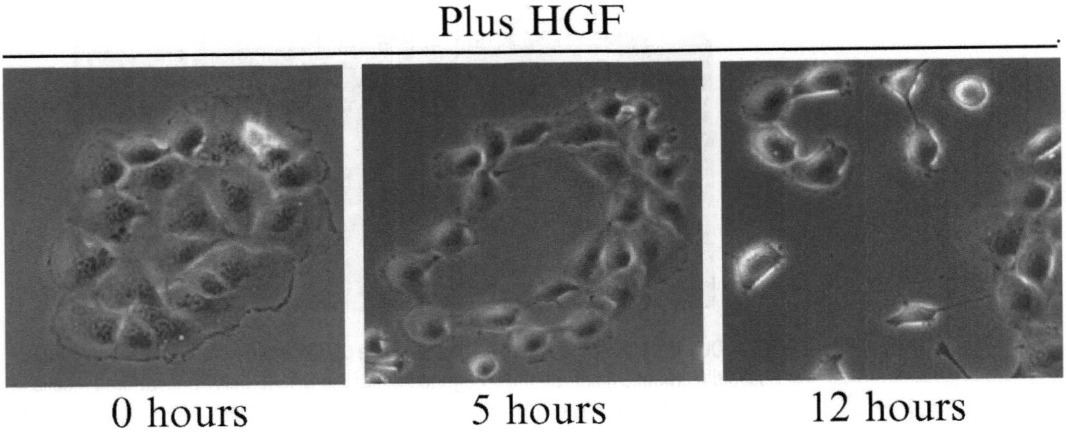

Fig. 2. Still images of HGF-induced DU145 cell scattering filmed using time-lapse microscopy. Here, DU145 cells were seeded onto fibronectin-coated coverslips and filmed following HGF stimulation (10 ng/ml).

3. For analysing the response of DU145 cells to HGF, we use a time-lapse interval of 5 or 10 min and film the cells for 24 h.
4. Once the assay is finished, the recorded sequence is exported from Image Pro as an 8-bit Tiff file and analysed using AQM tracking software (Andor Technology, UK). Each cell must be individually tracked using the AQM software throughout the sequence. In our laboratory, we adopt the following tracking criteria: only cells present in a colony in the first frame are tracked. If any cell present in the first frame divides within the first 60 frames, it is excluded. If a cell present in the first frame divides in the later stage in the film, it is tracked until it ceases to be migratory prior to cell division; for DU145 cells, this is normally 20 min before mitosis occurs. The daughter cells of this division are not tracked.
5. Once all the appropriate cells in the field of view have been tracked, the marked positions of each cell in each frame are saved into a file with the extension ".cel". This file records the cell number, frame number, *x*-coordinate, and *y*-coordinate of every tracked cell in every frame.
6. Finally, we carry out cell speed analysis using a range of software originally written by Prof. Graham Dunn and Dr. Daniel Zicha at the Randall Division of Cell and Molecular Biophysics, Kings College London. This software is not commercially available; however, Kinetic Imaging also has software packages that can be used to calculate cell velocity.
7. Alternatively, cell tracking can be done using ImageJ software (free to download from http://www.rsbweb.nih.gov/ij/).

4. Notes

1. No. 1.5 glass coverslips are acid-cleaned prior to use. Coverslips are immersed in a solution containing 1 M hydrochloric acid and 96% ethanol (2:3) overnight, rinsed three times in dH_2O, and then boiled with six changes of dH_2O. Excess water is drained and the coverslips are incubated in an oven at 150°C overnight for sterilisation before use. The coverslips are stored at 4°C in sterile conditions until required for use.
2. Different colony-forming cell types can be used in the HGF-induced cell scatter assay. These include MDCK cells (12), MCF-7 breast carcinoma cells (16), and HT29 colon carcinoma cells (17). The protocol can be adapted and optimised by, for example, changing the cell density or HGF concentration as required.
3. The efficiency of the DU145 scatter response to HGF varies over time in culture, whereby the optimum time for scattering is defined as the point at which the unscattered cell count is at its lowest and the scatter response is at its highest following HGF addition. An optimum time window has been determined and should be used in all subsequent scatter assay experiments. This optimum time window is between weeks 2 and 5 following recovery of the DU145 cells from cryogenic storage (Fig. 3). It is also necessary for the cells in culture to grow in tight colonies in the absence of HGF at all times in order to investigate HGF-induced scattering in DU145 cells effectively. Therefore, cells should be routinely cultured at sub-confluent densities (never more than 70% confluent).

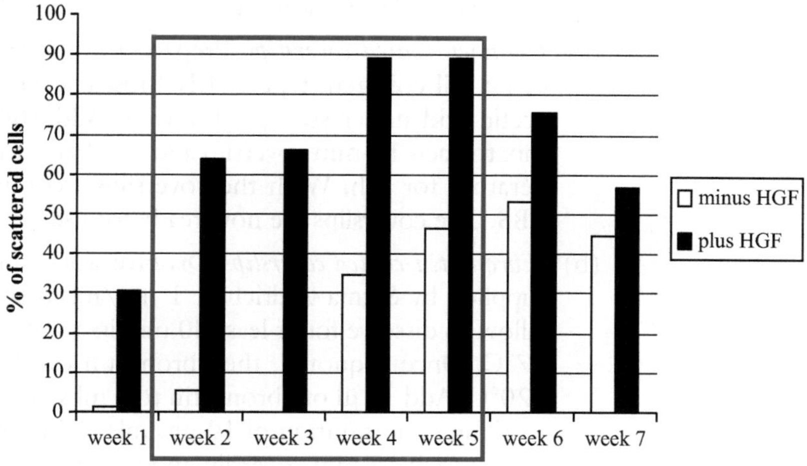

Fig. 3. The HGF-induced DU145 scatter assay was repeatedly conducted over a number of weeks to determine the optimal culture time for investigating HGF-induced cell scattering in DU145 cells as indicated.

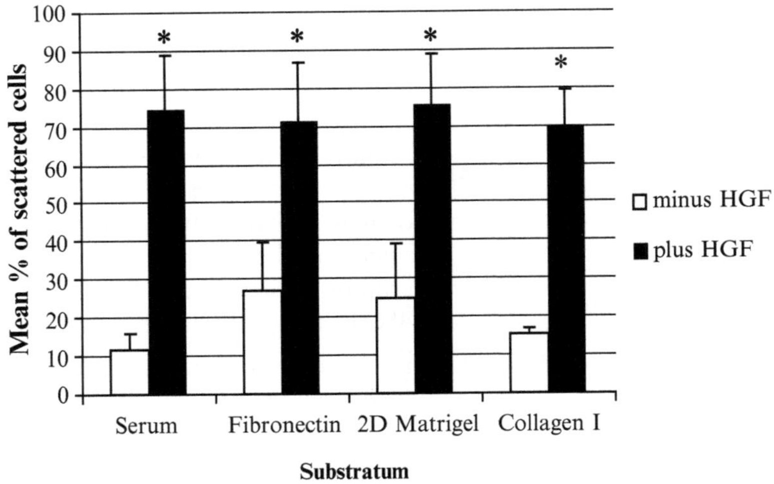

Fig. 4. Cell counts of DU145 cells seeded on different substratum in the presence and absence of HGF. The mean % of scattered cells and standard error of the mean were calculated over three independent experiments for each substratum. Statistical significance compared with minus HGF cells was calculated using Student's *t*-test; $*P<0.05$.

If the cells are allowed to become confluent, the scatter assay will be compromised. Indeed, ensure that the cells are well spaced but not too sparse before using for the scatter assay; otherwise, they will not form tight colonies.

4. There is considerable evidence to suggest that substratum rigidity can influence cell migration behaviour (18). The acid-washed coverslips can be pre-coated to investigate cell migration on different substratum. Once coated, cells can be seeded directly onto the coverslips as normal. All subsequent steps remain the same. The following substrata have been tested and support the HGF-induced scatter assay (Fig. 4):
   (a) *Collagen-coated coverslips.* Prepare a 50 μg/ml solution of rat tail collagen, type I (BD Biosciences), in 20 mM acetic acid under sterile conditions. Add 100 μl of the mix to each 13-mm coverslip and incubate at room temperature for 1 h. Wash the coverslips twice with sterile PBS. The coverslips are now ready for use.
   (b) *Fibronectin-coated coverslips.* Dissolve stock of fibronectin supplied by Sigma–Aldrich at 1 mg/ml in sterile water. Allow to dissolve for at least 30 min in a waterbath set to 37°C. Once aliquoted, the fibronectin can be stored at −20°C. Add 10 μl of fibronectin to 1 ml sterile PBS (final working concentration of 10 μg/ml). Add 100 μl of the mix to each 13-mm coverslip and incubate at room temperature for 45 min. Wash the coverslips twice with sterile PBS. The coverslips are now ready for use.

(c) *2D Matrigel-coated coverslips.* Allow the BD Matrigel™ Matrix (BD Biosciences) to thaw on ice overnight at 4°C. Aliquot the matrigel and store at −20°C. Aliquots should be thawed overnight at 4°C (preferably on ice) prior to use. Set up a 1 in 100 dilution of the Matrigel in serum-free medium. Add 100 μl of the mix to each 13-mm coverslip and incubate at 37°C for 1 h. Wash the coverslips twice with sterile PBS. The coverslips are now ready for use.

5. It is particularly important that the coverslips are washed thoroughly prior to adding starve medium. This is to ensure that there is no residual growth medium and that the cells are sufficiently starved. Otherwise, the scatter response may be affected.

6. Cells can also be stained for cell:substratum adhesion (for example paxillin) or cell:cell adhesion (for example E-cadherin) markers in addition to F-actin. In this instance, permeabilised cells would be pre-blocked with 3%BSA:PBS, washed with PBS, and then incubated with the primary antibody for 2 h. Cells would then be washed with PBS and incubated with the appropriate secondary antibody in addition to TRITC-phalloidin for 1 h. Cells would then be washed and mounted as described in Subheading 3.3 above.

7. In addition to assessing the % of scattered cells upon HGF stimulation, other forms of cell analysis can be conducted. ImageJ software (NIH) can be used to measure cell shape and area parameters of F-actin-labelled cells to give elongation ratios and relative cell areas (12).

References

1. Thiery, J. P., Acloque, H., Huang, R. Y., and Nieto, M. A. (2009) Epithelial-mesenchymal transitions in development and disease, *Cell* **139**, 871–890.
2. Thiery, J. P. (2002) Epithelial-mesenchymal transitions in tumour progression, *Nature Reviews Cancer* **2**, 442–454.
3. Potempa, S., and Ridley, A. J. (1998) Activation of both MAP kinase and phosphatidylinositide 3-kinase by Ras is required for hepatocyte growth factor/scatter factor-induced adherens junction disassembly, *Mol Biol Cell* **9**, 2185–2200.
4. Ridley, A. J., Comoglio, P. M., and Hall, A. (1995) Regulation of scatter factor/hepatocyte growth factor responses by Ras, Rac, and Rho in MDCK cells, *Mol Cell Biol* **15**, 1110–1122.
5. Humphrey, P. A., Zhu, X. P., Zarnegar, R., Swanson, P. E., Ratliff, T. L., Vollmer, R. T., and Day, M. L. (1995) HEPATOCYTE GROWTH-FACTOR AND ITS RECEPTOR (C-MET) IN PROSTATIC-CARCINOMA, *American Journal of Pathology* **147**, 386–396.
6. Nishimura, K., Kitamura, M., Miura, H., Nonomura, N., Takada, S., Takahara, S., Matsumoto, K., Nakamura, T., and Matsumiya, K. (1999) Prostate stromal cell-derived hepatocyte growth factor induces invasion of prostate cancer cell line DU145 through tumor-stromal interaction, *Prostate* **41**, 145–153.
7. Miura, H., Nishimura, K., Tsujimura, A., Matsumiya, K., Matsumoto, K., Nakamura, T., and Okuyama, A. (2001) Effects of hepatocyte growth factor on E-cadherin-mediated cell-cell adhesion in DU145 prostate cancer cells, *Urology* **58**, 1064–1069.
8. Maulik, G., Shrikhande, A., Kijima, T., Ma, P. C., Morrison, P. T., and Salgia, R. (2002) Role

of the hepatocyte growth factor receptor, c-Met, in oncogenesis and potential for therapeutic inhibition, *Cytokine Growth Factor Rev* **13**, 41–59.

9. Gallego, M. I., Bierie, B., and Hennighausen, L. (2003) Targeted expression of HGF/SF in mouse mammary epithelium leads to metastatic adenosquamous carcinomas through the activation of multiple signal transduction pathways, *Oncogene* **22**, 8498–8508.
10. Birchmeier, C., Birchmeier, W., Gherardi, E., and Vande Woude, G. F. (2003) Met, metastasis, motility and more, *Nat Rev Mol Cell Biol* **4**, 915–925.
11. Jaffe, A. B., and Hall, A. (2005) Rho GTPases: Biochemistry and biology, *Annual Review of Cell and Developmental Biology* **21**, 247–269.
12. Wells, C. M., Ahmed, T., Masters, J. R. W., and Jones, G. E. (2005) Rho family GTPases are activated during HGF-stimulated prostate cancer-cell scattering, *Cell Motility and the Cytoskeleton* **62**, 180–194.
13. Bright, M. D., Garner, A. P., and Ridley, A. J. (2009) PAK1 and PAK2 have different roles in HGF-induced morphological responses, *Cell Signal* **21**, 1738–1747.
14. Wells, C. M., Whale, A. D., Parsons, M., Masters, J. R., and Jones, G. E. (2010) PAK4: a pluripotent kinase that regulates prostate cancer cell adhesion, *J Cell Sci* **123**, 1663–1673.
15. Kumper, S., and Ridley, A. J. p120ctn and P-cadherin but not E-cadherin regulate cell motility and invasion of DU145 prostate cancer cells, *PLoS One* **5**, e11801.
16. Marchesini, R., Dasdia, T., Melloni, E., and Rocca, E. (1989) EFFECT OF LOW-ENERGY LASER IRRADIATION ON COLONY FORMATION CAPABILITY IN DIFFERENT HUMAN-TUMOR CELLS-INVITRO, *Lasers in Surgery and Medicine* **9**, 59–62.
17. Herrera, R. (1998) Modulation of hepatocyte growth factor-induced scattering of HT29 colon carcinoma cells - Involvement of the MAPK pathway, *Journal of Cell Science* **111**, 1039–1049.
18. Chun-Min, L., Hong-Bei, W., Dembo, M., and Yu-Li, W. (2000) Cell movement is guided by the rigidity of the substrate, *Biophysical Journal|Biophysical Journal* **79**, 144–152.

Chapter 4

Using the Dunn Chemotaxis Chamber to Analyze Primary Cell Migration in Real Time

Sanjay Chaubey, Anne J. Ridley, and Claire M. Wells

Abstract

The directed migration of cells (chemotaxis) occurs not only during wound healing and inflammatory responses but also during embryonic development. However, the intracellular signaling pathways that enable a cell to detect a chemoattractant and subsequently migrate toward the source are not clearly defined. The Dunn chemotaxis chamber in conjunction with time-lapse microscopy is a powerful tool that enables the user to observe directly the morphological response of cells to a chemoattractant in real time. Here, using the Dunn chemotaxis chamber, we describe the response of murine bone marrow-derived macrophages to colony stimulating factor-1. This is a particularly useful protocol as it can be adapted to study bone marrow-derived macrophages isolated from genetically modified mice and thus study the requirement of a specific protein in cell migration and chemotaxis.

Key words: Chemotaxis, Dunn chemotaxis chamber, Cell migration, Bone marrow-derived macrophages, Time-lapse microscopy

1. Introduction

The directional migratory response of a cell to a gradient of a diffusible signaling factor (chemotaxis) is important during embryogenesis (1), wound healing (2), and inflammatory responses (3). In addition, cell migration is believed to contribute to the metastatic process (4). The technique described here has been developed to study the intracellular signaling pathways and subsequent changes in cell adhesion and morphology that are required for a cell to initiate a migratory response toward a chemoattractant. Chemotaxis can be studied using a Boyden chamber or Transwell assay; however, both these methods are based on scoring cells that have migrated into or through a filter

Claire M. Wells and Maddy Parsons (eds.), *Cell Migration: Developmental Methods and Protocols*,
Methods in Molecular Biology, vol. 769, DOI 10.1007/978-1-61779-207-6_4, © Springer Science+Business Media, LLC 2011

membrane toward a source of putative chemotactic factor. In these assays, the local concentration gradients of chemotactic factor in and around the pores of the filter membrane are variable and unknown. Furthermore, the migratory behavior of the cells is unobservable and can only be deduced from the final distribution of the cell population. To overcome these experimental problems, the Dunn chemotaxis chamber (DCC) was developed. In this chamber, the migratory behavior of cells can be directly observed in a gradient of known direction and magnitude. The DCC is a modification of the Zigmond chamber (5), which allows the direct observation of slowly moving cells in a concentration gradient that is stable over longer periods of time (6, 7). The DCC consists of two concentric circles ground into one face of a glass slide (referred to as the inner and outer wells). An annular ridge (referred to as the bridge) separates the two wells. The bridge is 20 μm lower than the surrounding glass slide. The outer well contains the chemoattractant, whereas the inner well does not (Fig. 1). A linear gradient of chemoattractant forms by diffusion across the bridge between these two wells. Cells seeded onto a glass coverslip are inverted over the chamber and it is the cells that lie directly above the bridge that are viewed during the assay. The DCC chamber has been successfully used to characterize the migratory response of human macrophages to colony-stimulating factor-1 (CSF-1 (8)), fibroblasts to platelet-derived growth factor (7) and TSP1 (9), neutrophils to interleukin-8 (10), and microglia to adenosine triphosphate/adenosine diphosphate (11). In addition, the DCC has been used to study neurite outgrowth (12). We have previously used the DCC to elucidate the role of

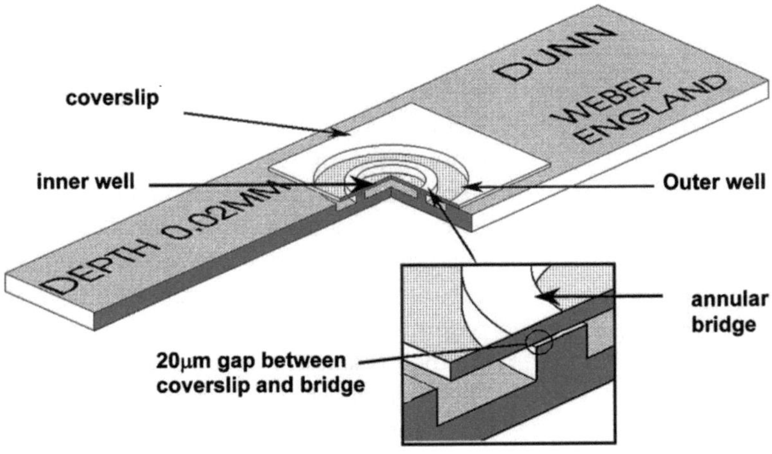

Fig. 1. The Dunn chemotaxis chamber with coverslip in place. The chamber consists of two concentric circles ground into one face of a glass slide (*inner and outer wells*). An annular ridge (*bridge*) separates the two wells. *Inset*, the bridge is 20 μm lower than the surrounding glass slide, resulting in a gap between the coverslip and the bridge. Adapted from a figure kindly provided by Professor Jones.

Rho family proteins in the chemotactic response of BAC1.2F5 macrophages to CSF-1 (13) and to study the response of primary murine bone marrow-derived macrophages (BMMs) to CSF-1 (14). Here, we will describe our BMM protocol, a protocol that can be extended to study the migratory behavior of macrophages and other cell types derived from genetically modified mice.

2. Materials

2.1. Isolation of Macrophages

1. Macrophage starve medium: RPMI1640 with l-glutamine, 1% essential amino acid, 1% sodium pyruvate, 1% penicilin–streptomyosin, 10% heat-inactivated fetal calf serum, and 0.5% β-mercaptoethanol solution (see item 2 below).
2. β-Mercaptoethanol solution (50 mM): 34 μL of 14.3 M β-mercaptoethanol diluted in 10 mL of RPMI1640 containing antibiotics; this can be stored at 4°C.
3. 75-cm^2 cell culture flasks.
4. 15-mL centrifuge tube.
5. Petri dish.
6. 5-mL syringe.
7. 21-Gauge needle.
8. Forceps.
9. Sterile 6-cm bacteriological plates.
10. Dimethyl sulfoxide (DMSO).
11. Cryogenic vials.

2.2. Dunn Chamber Assembly and Washing

1. 18×18-mm glass coverslip (no. 3, BDH laboratory supplies).
2. DCCs supplied by Weber Scientific International.
3. Dental wax, which can be purchased from Agar Scientific.
4. Sable artists' paintbrush.
5. Versene (Gibco) or cell dissociation buffer (Sigma).
6. Recombinant CSF-1 (R&D Systems).
7. 2-cm cell culture dishes.
8. Whatman paper.
9. 30% Hydrogen peroxide.
10. Acetone.
11. Glass Coplin jar (or beaker).
12. 100% Ethanol.

3. Methods

3.1. Isolation and Culture of Mouse Primary BMMs

1. Isolate and clean the murine femoral bone using forceps. Maintain the bones in a small Petri dish containing macrophage starve medium. It is important that all of the tissue surrounding the femur is scraped away to prevent contamination of the BMM preparation.
2. Once the bone has been cleaned, use a 21-gauge needle to piece both ends of the bone. Keep the needle in one end of the bone (fatter end). Using a 5-mL syringe, flush the bone marrow out of the bone with 5 mL of macrophage starve medium into a 15-mL Falcon tube. Add an additional 5 mL of macrophage starve medium and resuspend vigorously using a p1000.
3. Centrifuge the cell suspension for 5 min at $1,000 \times g$ and resuspend the pellet in 5 mL of macrophage starve medium. Count the hematopoietic progenitor cells (do not count the smaller crescent-shaped red blood cells).
4. Seed the cells at 2×10^5 cells/cm^2 on tissue-culture plastic. We routinely seed 15 mL of cells suspended in growth medium in a 75-cm^2 flask at a density of 1×10^6 cells/mL. Incubate the cells for 3 days. Macrophage growth medium consists of macrophage starve medium with the addition of 15% conditioned medium from L-929 fibroblasts (L cells) as a source of CSF-1.
5. Carefully collect the non-adherent population of cells. The cells attached to the bottom of the flask are a mixed population of differentiated hematopoietic cells and fibroblasts and care should be taken not to dislodge these cells when collecting the non-adherent population. Centrifuge the cell suspension and count all the cells present.
6. The cells are then cryogenically frozen in growth medium containing 10% DMSO at a cell density depending on the subsequent application. We routinely place multiples of 5×10^5 cells in each vial.
7. To recover frozen cells, rapidly thaw a cryovial, add 4 mL of warm growth medium dropwise, and centrifuge for 5 min at $1,000 \times g$.
8. Seed the cells on 6-cm bacterial culture plates (Falcon; see Note 1) in 5 mL of growth medium at a density of 10^5 cells/mL (i.e., a vial containing 5×10^5 cells will seed one 6-cm dish). One 6-cm dish will produce enough cells for over 20 Dunn chamber coverslips.
9. Incubate the cells for a further 5 days without refeeding. The cells remain in suspension until approximately day 4.

The differentiated BMMs will then become adherent and can be harvested on day 5 (see Note 1).

10. To harvest the cells, carefully remove the medium as the cells are only loosely attached and add 2.5 mL of Versene. Incubate for approximately 20 min and then collect the cell suspension by vigorous pipetting across the dish. Add 2.5 mL of macrophage starve medium and centrifuge for 5 min at 1,000 × *g*. Resuspend the cells in an appropriate amount of macrophage growth medium.
11. Place a flamed acid-washed Dunn chamber coverslip (see Note 2) in a 2-cm tissue-culture dish and add 2 mL of growth medium containing 2×10^4 cells/mL. Incubate the cells for a further 18 h.
12. To prepare the cells for the DCC, wash twice with starve medium and then incubate in starve medium for 8 h. At the same time, place 8 mL of starve medium in the incubator, so that it becomes warm and gassed, to use in the chamber assembly.

3.2. Assembly of the DCC

1. Place some dental wax in a glass beaker and melt slowly using a low setting on a stage heater. Avoid overheating the wax, as a noxious fume will be produced. Ensure that the time-lapse stage is heated to 37°C and that filming can begin immediately after the chamber is assembled.
2. Aliquot 5 mL of the pre-gassed (see Subheading 3.1, step 12) starve medium into a 7-mL Bijoux tube: this will be used to wash the chambers (wash). Aliquot 1 mL of the starve medium into a 7-mL Bijoux tube (control) and 1 mL of starve medium into 7-mL Bijoux tube (test). Add 30 ng/mL recombinant CSF-1 to the 1 mL of starve medium in the test Bijoux tube.
3. Place the coverslip with cells attached, wash, control, and test Bijoux tubes in an incubator close to the DCC assembly site. Alternatively, 5 or 10% CO_2 can be pumped from a cylinder across the DCC assembly site. It is important to keep the medium used during assembly of the DCC fully gassed, as once the chamber is sealed by wax, there is no opportunity for gas exchange.
4. Place a DCC on tissue paper until the storage ethanol has evaporated. The chamber should be completely dry before proceeding. Do not attempt to wipe the chamber as contact with the bridge can result in damage that may affect the gradient.
5. Initially, the chamber is washed to remove any residual traces of ethanol. Using a Gilson, carefully pipet 100–150 μL of wash medium over the center of the chamber without letting the tip touch the bridge. Both wells should fill but not overflow. Carefully remove the medium and repeat six times.

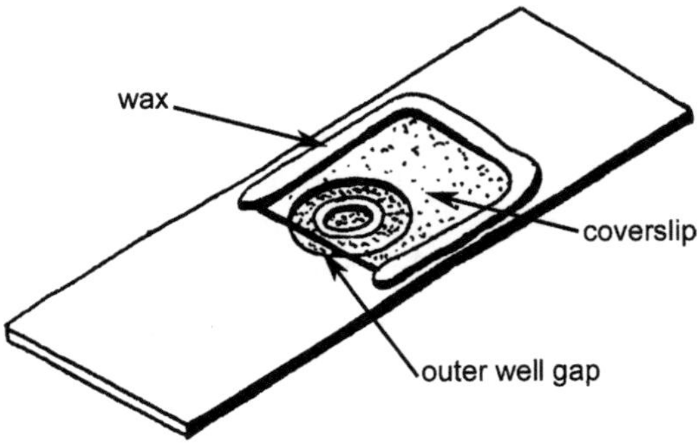

Fig. 2. Part-assembled chamber. Once the coverslip is in place, the chamber is sealed on three sides with wax. The fourth side remains unsealed and the medium is removed from the outer well and replaced with chemoattractant using this outer well gap. Adapted from a figure kindly provided by Dr. Graham Dunn.

6. Place 150 μL of control medium over the chamber until it fills both wells. Quickly but carefully invert the coverslip over the two wells in the center of the chamber. It is important that inverting the coverslip over the chamber does not incorporate any bubbles into the chamber as these can affect the gradient (see Note 3).

   The coverslip should be placed slightly off center so that a small gap remains in the outer well (Fig. 2). However, the coverslip should completely cover the inner well so that the outer well can be drained without any loss of medium from the inner well.
7. Lightly press down the coverslip around the edges and mop up excess medium with Whatman paper. Using a sable paintbrush, wax three sides of the coverslip, leaving the side with the outer well gap unwaxed (Fig. 2).
8. Tear a piece of Whatman paper and place a small corner of the torn edge just inside the outer well gap until it starts to absorb the medium. Leave the filter paper to absorb all the medium in the outer well. It is important not to move or lift the paper as this can introduce air into the DCC.
9. Using a micropipet, add 100 μL of test solution into the outer well through the gap, making sure that it is bubble free, until the well is full (see Note 4 for testing the gradient formation). Quickly wax the remaining side ensuring that the gap is completely sealed. Once the wax has set, wash the remaining coverslip surface with distilled water and dry with Whatman paper or compressed air.
10. Immediately place the DCC on the microscope stage and start filming (see Note 5 for fixing cells at the end of filming).

3.3. Time-Lapse Microscopy and Migration Analysis

1. To observe live cells in the DCC requires a microscope adapted for time-lapse recording. We use an Olympus IX71 inverted microscope equipped with a heated stage and environmental chamber, a prior electronic shutter, and a Q Imaging Retiga monochrome cooled CCD camera. The system is controlled and coordinated by Image Pro AMS software. The system was supplied and installed by Media Cybernetics, UK. Ideally, the microscope should have a heated stage, but a fan heating system can be used (15). Phase-contrast microscopy using an inverted or upright microscope will provide good results. To determine the chemotactic potential of a cell population (using the software described in Subheading 3.3, steps 7–9) requires the analysis of a large number of cells. Therefore, a low-magnification objective lens (routinely 10×) is used whereby the full width of the DCC bridge can be visualized.
2. Place the assembled chamber on the heated stage. The live image from the microscope is then viewed on the computer monitor using the Image Pro AMS software, and a region of the bridge is selected for recording. Ideally, there should be 15–25 cells in the field of view and the cells should be evenly spaced.
3. The high concentration (outer well) end of the chemotactic gradient is then aligned to the top of the monitor screen (Fig. 3). We achieve this by rotating the camera. This is extremely useful for subsequent tracking and mathematical analysis of chemotactic behavior.

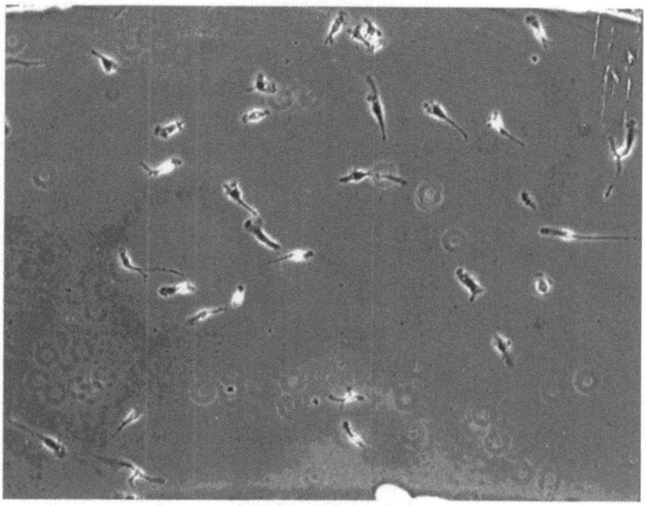

Fig. 3. BMMs in the DCC as imaged by AQM Software. A region of the bridge is selected for recording and the image orientated such that the source of chemoattractant (*outer well*) is at the top of the screen.

4. For analyzing the chemotactic behavior of BMMs in a gradient of recombinant CSF-1, we use a time-lapse interval of 10 min and film the cells for 24 h. BMMs respond slowly to a CSF-1 gradient compared with macrophage cell lines such as Bac 1.2F5 (13).
5. Once the assay is finished, the recorded sequence is exported from Image Pro as an 8-bit tiff file and analyzed using AQM tracking software (Andor Technology, UK). Each cell must be individually tracked using the AQM software throughout the sequence. In our laboratory, we adopt the following tracking criteria: only cells present in the first frame are tracked. If any cell present in the first frame divides within the first 60 frames, it is excluded. If a cell present in the first frame divides in the later stage in the film, it is tracked until it ceases to be migratory; for BMMs, this is normally one to two frames or 20 min before mitosis occurs. The daughter cells of this division are not tracked.
6. Once all the appropriate cells in the field of view have been tracked, the marked positions of each cell in each frame are saved into a file with the extension ".cel." This file records the cell number, frame number, *x*-coordinate, and *y*-coordinate of every tracked cell in every frame.
7. Finally, we conduct trajectory analysis using a range of software originally designed by Dr. Graham Dunn and Dr. Daniel Zicha at the Randall Division of Cell and Molecular Biophysics, Kings College London. This program returns a Rayleigh test for unimodal clustering of direction where the null hypothesis is uniform distribution of cell trajectories. If the null hypothesis is rejected by the Rayleigh test, the cells show a significant directional response to CSF-1. This software is not commercially available; however, Kinetic Imaging has some software analysis to use with the DCC incorporated into the AQM software.
8. The cell trajectories can also be plotted so that their starting point is shifted to the origin in Microsoft Excel to give a preliminary in the indication of chemotaxis. A significant chemotactic response should result in the majority of cell trajectories radiating from the origin in the direction of the highest chemoattractant concentration ((10) and Fig. 4). To do this, the tracking data must also be saved as an .xld file, which can be recognized by Excel. This option is available in the Kinetic Imaging tracking software.
9. Alternatively, cell tracking can be done using ImageJ software (free to download from http://rsbweb.nih.gov/ij/). ImageJ also has a chemotaxis plug-in which will allow the user to import cell tracks generated using ImageJ tracking and measure the degree of directional migration.

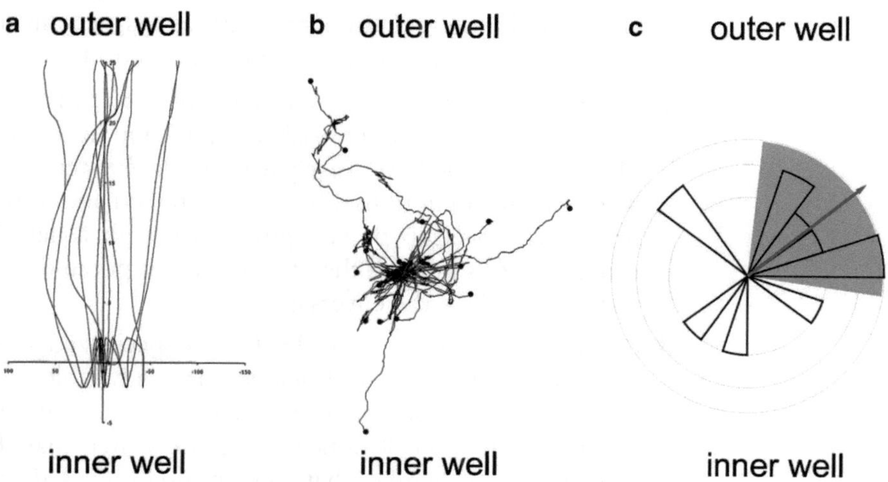

Fig. 4. Examples of migration analysis; in each case, the source of chemoattractant is at the top of the plot. (**a**) A simple trajectory plot of 10 BMM cell trajectories (pixel coordinates) where their starting point is shifted to the origin in Microsoft Excel. (**b**) Cell trajectories' starting points are shifted to 0,0 and plotted using in-house Mathematica software designed by Prof. Graham Dunn and Dr. Daniel Zicha (13). (**c**) Trajectories from (**b**) chemotaxis circular histograms were generated to show the proportion of cells with a direction of migration lying within each 20° segment with the wound site at the top. *Arrow* means direction of cell migration, *highlighted segment* 95% confidence interval calculated from a Rayleigh test using in-house Mathematica software designed by Prof. Graham Dunn and Dr. Daniel Zicha (13).

10. It is prudent to clean the chambers immediately after use. The most important principle to observe when cleaning the DCC is to avoid touching the bridge. All the dental wax is removed carefully with a razor blade, scraping in the opposite direction to the bridge at all times. Using gloved fingers, rub off any residual wax with distilled water and rinse a further six times with distilled water. The chamber is then placed in a glass Coplin jar (or beaker) filled with 100% acetone and sonicated for 10 min using a water bath sonicator. After sonication, the chamber is rinsed six times with distilled water. Once rinsed, place the chamber in a glass Petri dish with the grooved side upward and soak in 30% hydrogen peroxide for 10 min. After acid cleaning, the chamber is rinsed a further six times with distilled water. Store the chamber in 100% ethanol until use.

4. Notes

1. The BMMs must be differentiated and cultured on bacterial plates because they adhere too strongly to tissue-culture plastic and are impossible to passage. BMMs change in morphology and migratory behavior during passaging. They gradually become more elongated and less motile, thus reducing their

chemotactic potential. In our motility assays, we only use cells for 7 days after the 5-day differentiation period.

2. Coverslips are individually washed with detergent, rinsed six times with distilled water, soaked for 10 min in concentrated HCL, then washed extensively with distilled water, and stored in 100% ethanol until use. Before use, the ethanol is removed from each coverslip by sweeping it through a naked flame using tweezers until all the ethanol has evaporated. This also serves to sterilize the coverslip.
3. Successful assembly of the DCC is a specialized technique that requires patience and practice. It is advisable for a new user to practice assembling the chamber initially using a wet coverslip with no cells attached. It is essential that the user can assemble a chamber that contains no bubbles after a 24-h incubation. The presence of bubbles, particularly in the inner well, will seriously interfere with the formation of the gradient and produce unreliable results. Bubbles can result from incorrect sealing of the chamber, but are most commonly the result of the coverslip being lifted away from the chamber during the draining of the outer well. Once the chamber has been assembled correctly, the user should be able to proceed to assembling the chamber with coverslips seeded with cells. If the cells are still viable 24 h later, then the user can confidently begin to conduct chemotaxis assays.
4. The time taken for a gradient to form between the two wells of the DCC depends on the molecular weight of the chemotactic factor. Factors with a molecular weight between 350 and 370 Da will form a gradient in 10 min, whereas factors with a molecular weight between 10 and 20,000 Da will take 30 min (8). Similarly, the molecular weight of the chemoattractant will determine the stability of the gradient; generally, the greater the molecular weight, the longer the gradient will be maintained. It is advisable for a new user to test the formation and stability of the gradient in the DCC using a fluorescein isothiocyanate-conjugated dextran of comparable molecular weight to the intended chemoattractant. Assemble the chamber as normal using the fluorescein isothiocyanate–dextran in the outer well. The formation of the gradient can then be monitored using the standard epifluorescence microscope. A similar method was used to establish that a gradient of CSF-1 forms within 25 min and is stable for at least 17 h in the DCC (16).
5. In addition to studying the response of BMMs isolated from knockout mice, the DCC can be used to characterize the chemotactic response of cells expressing tagged proteins introduced either by microinjection of recombinant protein or deoxyribonucleic acid (17) or by transfection. To identify expressing cells, the coverslip is removed from the chamber

and the cells fixed immediately after the filming period ends. To fix the cells, remove the wax using a razor blade and carefully lift the coverslip away from chamber using Watchmaker forceps. Immediately place the coverslip in fixative (e.g., 4% paraformaldehyde). The coverslip can then be stained for expressed protein following normal protocols. If a GFP tag is used in conjunction with time-lapse fluorescence microscopy, the cells can be directly identified and tracked from the resulting film.

References

1. Dormann D, Weijer CJ. (2003) Chemotactic cell movement during development. *Curr Opin Genet Dev* **13**(4):358–64.
2. Seppa H, Grotendorst G, Seppa S, et al. (1982) Platelet-derived growth factor in chemotactic for fibroblasts. *J Cell Biol* **92**(2):584–8.
3. Jones GE. (2000) Cellular signaling in macrophage migration and chemotaxis. *J Leukoc Biol* **68**(5):593–602.
4. Kassis J, Lauffenburger DA, Turner T, et al. (2001) Tumor invasion as dysregulated cell motility. *Semin Cancer Biol* **11**(2):105–17.
5. Zigmond SH, Hirsch JG. (1973) Leukocyte locomotion and chemotaxis. New methods for evaluation, and demonstration of a cell-derived chemotactic factor. *J Exp Med* **137**(2): 387–410.
6. Dunn GA, Zicha D.(1993) Long-term chemotaxis of neutrophils in stable gradients: preliminary evidence of periodic behavior. *Blood Cells* **19**(1):25–39.
7. Zicha D, Dunn GA, Brown AF. (1991) A new direct-viewing chemotaxis chamber. *J Cell Sci* **99**(Pt 4):769–75.
8. Jones GE, Zicha D, Dunn GA, et al. (2002) Restoration of podosomes and chemotaxis in Wiskott-Aldrich syndrome macrophages following induced expression of WASp. *Int J Biochem Cell Biol* **34**(7):806–15.
9. Orr AW, Elzie CA, Kucik DF, et al. (2003) Thrombospondin signaling through the calreticulin/LDL receptor-related protein co-complex stimulates random and directed cell migration. *J Cell Sci* **116**(Pt 14):2917–27.
10. Zicha D, Allen WE, Brickell PM, et al. (1998) Chemotaxis of macrophages is abolished in the Wiskott-Aldrich syndrome. *Br J Haematol* **101**(4):659–65.
11. Honda S, Sasaki Y, Ohsawa K, et al.(2001) Extracellular ATP or ADP induce chemotaxis of cultured microglia through Gi/o-coupled P2Y receptors. *J Neurosci* **21**(6):1975–82.
12. Maden M, Keen G, Jones GE.(1998) Retinoic acid as a chemotactic molecule in neuronal development. *Int J Dev Neurosci* (5):317–22.
13. Allen WE, Zicha D, Ridley AJ, et al. (1998) A role for Cdc42 in macrophage chemotaxis. *J Cell Biol* **141**(5):1147–57.
14. Wells CM, Walmsley M, Ooi S, et al. (2004) Rac1-deficient macrophages exhibit defects in cell spreading and membrane ruffling but not migration. *J Cell Sci* **117**(Pt 7):1259–68.
15. Jones GE, Ridley AJ, Zicha D. (2000) Rho GTPases and cell migration: measurement of macrophage chemotaxis. *Methods Enzymol* **325**:449–62.
16. Webb SE, Pollard JW, Jones GE. (1996) Direct observation and quantification of macrophage chemoattraction to the growth factor CSF-1. *J Cell Sci* **109**(Pt 4):793–803.
17. Ridley AJ. (1998) Mammalian cell microinjection assay to study the function of Rac and Rho. *Methods Mol Biol* **84**:153–60.

Chapter 5

Imaging Cells Within 3D Cell-Derived Matrix

Samantha J. King and Maddy Parsons

Abstract

Much of what is known about the mechanisms regulating cell adhesion and migration come from in vitro studies of cells plated on 2-dimensional (2D) extracellular matrix (ECM) proteins. The importance of studying these processes in cells within 3-dimensional (3D) environments is becoming increasingly recognised, as a number of studies have now demonstrated that adhesion formation and migration are markedly different in cells within 3D environments (Cukierman et al. Science 1708–1712, 2001). It is also known that the composition and pliability or density of the ECM are important in regulating cell adhesion and migration (Cukierman et al. Science 1708–1712, 2001). Cell-derived matrices (CDM) are naturally deposited fibrillar ECM from fibroblasts that can be used to study the adhesions and migration of cells in a more physiologically relevant environment.

Key words: Cell-derived matrix, Fibroblast, 3-Dimensional, Microscopy, Migration, Adhesion, Extracellular matrix

1. Introduction

Many studies of cell adhesion and migration have been performed in vitro in cells plated on cell culture surfaces coated with a purified extracellular matrix (ECM) protein or ligand of interest. This environment can induce an artificial polarity between the lower and upper surfaces of normally non-polar cells such as fibroblasts (1). The importance of studying cells within a 3-dimensional (3D) environment has been shown in epithelial cells, whereby plating in 3D ECM gels (for example collagen or Matrigel) can induce normal epithelial polarity and differentiation (1). Further-more, it is now becoming clear that cells behave and migrate differently within 3D environments and form altered adhesions compared to those formed by cells in a 2-dimensional (2D) environment (1). Therefore, although it is still important to study signalling events

Claire M. Wells and Maddy Parsons (eds.), *Cell Migration: Developmental Methods and Protocols*,
Methods in Molecular Biology, vol. 769, DOI 10.1007/978-1-61779-207-6_5, © Springer Science+Business Media, LLC 2011

contributing towards cell migration in a "classical" 2D cell culture system, it is also very important to transfer these studies to 3D environments, which may give more relevant answers to the mechanism and regulation of cell migration within an organism.

Cell-ECM contacts of cells in culture have been broadly divided into four groups according to their molecular composition (2). Three of the four groups of adhesions, characterised in fibroblasts but also found in other cell types (3), are formed in cells plated on 2D surfaces and these are focal complexes, focal adhesions, and fibrillar adhesions. The fourth group is 3D-matrix adhesions; these have been identified in fibroblasts adhering to 3D cell-derived fibronectin and collagen containing matrices (1). Focal complexes are small, short-lived adhesions formed at the leading edge of spreading or migrating cells (4). Focal complexes are rich in paxillin and tyrosine phosphorylated proteins; they also contain β3 integrin, talin, and low levels of vinculin and focal adhesion kinase (FAK) but are devoid of zyxin and tensin (3). Focal adhesions are larger, more stable structures that are located at, and associated with, the ends of actin stress fibres (4). They contain focal complex components as well as predominantly α5β1 and zyxin and are regulated by the activity of the small GTPase Rho (3–5). The translocation of α5β1 along bundles of actin filaments generates fibrillar adhesions, which are also tensin dependent (6). Fibrillar adhesions are centrally located elongated adhesions that are enriched in α5β1 and tensin, but lack αVβ3, paxillin, phosphotyrosine markers and other focal adhesion components, e.g. vinculin (5–7). Three-dimensional matrix adhesions exhibit a distinctive molecular composition that is primarily dependent on interactions between α5β1 and fibronectin (1, 8). The pliability of the surrounding ECM is important for the formation of 3D matrix adhesions, which are replaced by smaller adhesions upon rigidifying the matrix (1). The properties of the 3D matrix adhesions seen in cell-derived matrices (CDM) are due both to the 3D structure and composition of the matrix (1). It has been indicated that cells respond differently to complex 3D matrices, e.g. CDM, compared to more simple 3D matrices, e.g. collagen gels, or 3D pure fibronectin environment (1, 4). Humphries and colleagues have shown via liquid chromatography tandem mass spectrometry that the major components of CDM are fibronectin and collagen (9). Yamada and colleagues originally developed the use of cell-derived matrices to facilitate the study of the composition of adhesions in 2D and 3D environments (1). In the original study, the CDM was formed from naturally deposited ECM from NIH3T3 cells, and these cells were then removed and replaced by cells of interest (1) to permit the study of adhesion composition in 3D matrices compared to those formed in a 2D environment (1). Cells in CDM have also been shown to behave similarly to those in vivo (10, 11), and therefore CDM has been used to study proliferation and drug responses of tumour cells (11). Therefore, CDM provides a more physiologically relevant model for studying cell migration and cell adhesions in either live or fixed samples than 2D or 3D matrices formed of only one ECM component.

2. Materials

2.1. Preparation of Coverslips or Dishes for Imaging Live Cells in CDM

1. Glass coverslips (thickness number 1.5 (0.17mm)).
2. Cell culture dishes.
3. Phosphate Buffered Saline without Mg^{2+} or Ca^{2+} (PBS−).
4. 0.2% Gelatin/PBS−: Add 5 ml warm 2% gelatin stock (incubate gelatin in a 37°C water bath until it has liquefied) to 50 ml PBS−.
5. 1% Glutaraldehyde/PBS−: Add 1 vial of glutaraldehyde (25% stock solution) to 25 ml PBS− before use, treat with care as glutaraldehyde solutions are poisonous. This solution should be made up fresh each time.
6. High glucose Dulbecco's modified Eagle's media (DMEM).

2.2. Plating Fibroblasts to Generate CDM

1. Cell culture flasks.
2. Primary cultured human dermal fibroblasts (TCS cellworks Ltd) (see Note 1).
3. Human dermal fibroblast basal media plus supplements (TCS cellworks Ltd).
4. PBS−.
5. Trypsin/EDTA solution.
6. Haemocytometer.
7. Human dermal fibroblast basal media plus supplements with 35 μg/ml ascorbic acid and 5 μg/ml fibronectin: Add 35 μl ascorbic acid (50 mg/ml stock) and 250 μl fibronectin (1 mg/ml stock) to 50 ml Human dermal fibroblast basal media plus supplements. This solution can be stored at 4°C before use.

2.3. Denudation of Fibroblasts from CDM

1. Extraction buffer: Add 20 mM NH_4OH (100 ml of 100 mM stock solution), 0.5% Triton-X100 (2.5 ml) to 500 ml PBS− and then pre-warm in a 37°C water bath (store at 4°C).
2. PBS with Mg^{2+} or Ca^{2+} (PBS+).
3. 10 μ/ml DNAse1 in PBS+: Add 500 μl of DNAse1 to 50 ml PBS+.

4. 1% Penicillin/streptomycin/PBS+: Add 5 ml of Penicillin/streptomycin to 50 ml PBS+. This solution can be stored at 4°C before use.

2.4. Preparation of the CDM Prior to Experiment

1. 2% BSA/PBS: Add 1 mg of Bovine Serum Albumin (BSA) to 50 ml PBS+ before use.
2. PBS−.
3. Cells of interest.

2.5. Preparation of Chambers for Live Imaging

1. Imaging chamber (see Fig. 1).
2. 7× Detergent: Prepare a 5–10% solution in tapwater.
3. Acetone.
4. Ethanol.
5. Hot wax (Vaseline: Paraffin: beeswax – 1:1:1): Time-lapse microscopy using cell-derived matrices.
6. 5% FCS/Optimem: Add 2.5 ml FCS to 50 ml Optimem.

2.6. Fixation of CDM (if Required) for Microscopy

1. PBS−.
2. 4% PFA/PBS: Add 4 g of Paraformaldehyde to 100 ml PBS+ and heat in a fume hood to 80°C, add NaOH while stirring and PH to 7.5. Then allow to cool (store at −20°C for long term or at 4°C for short term).

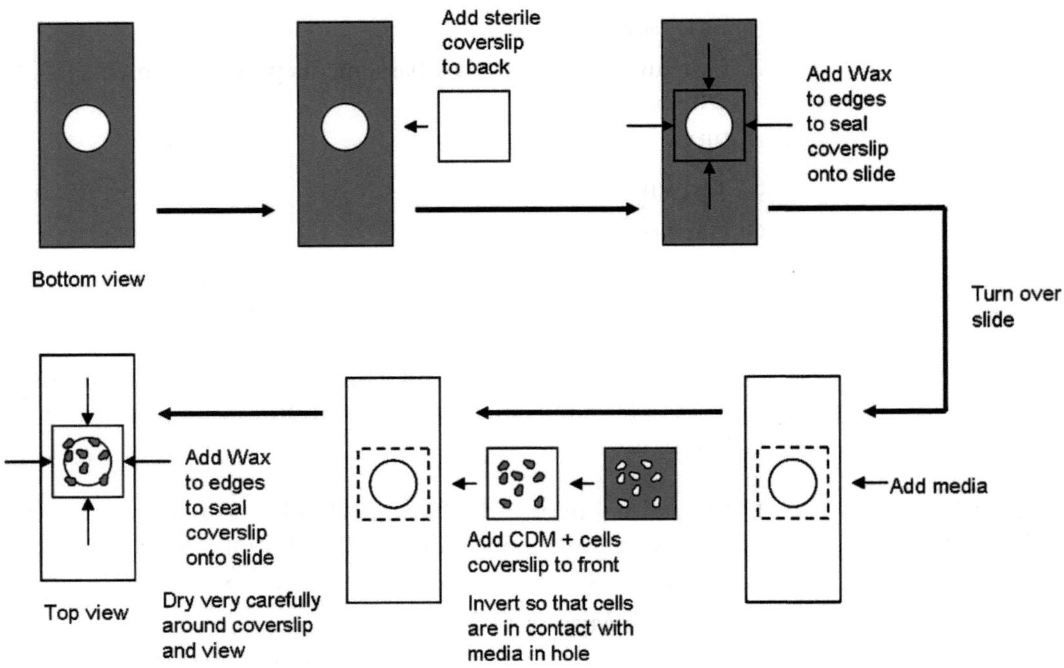

Fig. 1. Imaging Chamber setup. Wash the imaging chamber in ethanol and leave to air dry. Then add a sterile coverslip to one side of the image chamber and seal the edges with wax. Then wash the well with media and add the CDM coverslip plated with cells inverted to the other side of the imaging chamber and seal the edges with wax.

3. 0.2% Triton X-100/PBS: Add 100 μl of Triton X-100 to 50 ml PBS−.
4. 3% BSA/PBS: 1.5 g BSA dissolved in 50 ml PBS−.
5. Microscope slides.
6. Mounting media (e.g. Fluorsave).

3. Methods

3.1. Preparation of Coverslips or Imaging Dishes for Generation of CDM

1. Autoclave coverslips (see Note 2).
2. Place a coverslip (with forceps) into each well of a cell culture dish and wash the coverslips in PBS−.
3. Incubate the coverslips with 0.2% gelatin/PBS− for 1 h at 37°C. Then remove the gelatin and rinse the coverslips three times in PBS−.
4. Incubate the coverslips in 1% glutaraldehyde/PBS− for 30 min at room temperature (leave in the cell culture hood) to permit cross-linking. Then remove the glutaraldehyde and rinse the coverslips three times in PBS−.
5. Transfer the coverslips to new wells in a new cell culture dish (see Note 3) and incubate the coverslips in DMEM for 30 min at 37°C. Then remove the DMEM and rinse the coverslips three times in PBS.

The coverslips can now be stored in PBS− at 4°C or used immediately.

3.2. Plating Fibroblasts

Culture primary human dermal fibroblasts (HDFs) in human dermal fibroblast basal media plus supplements (complete media) in a 75 cm^2 cell culture flasks (see Note 4). HDFs are maintained at 37°C in a water saturated incubator with 5% CO_2. When planning experiments, it is advisable to maintain one stock flask of HDFs and plate out the other flasks for CDM generation (see Note 5).

1. Remove the growth media from the HDFs in a 75 cm^2 cell culture flasks and wash twice in PBS−. Then incubate the HDFs in 2 ml trypsin/EDTA at 37°C until they have detached from the cell culture flask (this can be verified on a light microscope).
2. Centrifuge HDFs to pellet and resuspend in complete media. Perform a cell count using a haemocytometer (add 10 μl cell suspension to the haemocytometer and cover with the coverslip). Count the number of cells in 3 of the squares. (Total number of cells counted/number of squares counted) × 10,000 = number of cells per ml.

3. Centrifuge HDFs again to pellet and resuspend in complete media at a maximum of 50,000 cells per ml. Then plate onto the previously prepared gelatin-coated, glutaraldehyde cross-linked coverslips. Cells should be around 90% confluent (i.e. covering the surface) after they have fully spread following plating.
4. Incubate HDFs overnight at 37°C. Remove media and incubate with complete media plus 35 μg/ml ascorbic acid and 5 μg/ml fibronectin (see Note 6).
5. Change media on HDFs for complete media plus 35 μg/ml ascorbic acid and 5 μg/ml fibronectin every day for 10–14 days. Check the progress of the matrix regularly after initial plating of the fibroblasts using a light microscope (see Note 7).

3.3. Fibroblast Removal and DNAse Treatment

In order to use the CDM to study cells of interest in it, it is necessary to denudate the HDFs to leave the cell-free CDM. The CDM is delicate, so the following steps need to be undertaken with great care in order not to disrupt the integrity of the CDM fibres and structure.

1. Remove the media from the HDFs and wash the cells carefully in PBS−.
2. Incubate with 1–2 ml pre-warmed extraction buffer (20 mM NH_4OH, 0.5% Triton-×100 in PBS−) for 3–5 min at 37°C and check for cell lysis down the microscope (see Note 8).
3. Remove half of the extraction buffer from each well and refill with PBS+. Then incubate for 3–5 min at 37°C and check again for cell lysis.
4. Remove half of the extraction buffer and refill with PBS+. Then incubate again for 3–5 min at 37°C and check for cell lysis.
5. Remove all the PBS in each well and incubate with 2 ml PBS+ containing 10 μ/ml DNAseI for 30 min at 37°C. Then remove the DNAse solution and wash twice with PBS+.
6. Store the CDM-coated coverslips or dishes in PBS+-containing penicillin/streptomycin at 4°C.

   These matrices can be stored for up to 6 months (see Notes 9 and 10).

3.4. Preparation of CDM for Plating Cells of Interest

Heat denature 2% BSA/PBS by heating 2% BSA/PBS at 75°C in a water bath for 3 min and then allowing it to cool.

1. Wash the CDM twice with PBS+ and block with 2 ml heat denatured 2% BSA/PBS for 1 h at 37°C.

2. Remove the BSA and wash the CDM three times with PBS+. Then plate an appropriate number of cells in complete media onto the matrix (see Note 11).
3. Incubate for the appropriate amount of time for cells of interest to adhere and then image for migration or fix for immunofluorescence studies (see Note 12).

3.5. Time-Lapse Microscopy of Cells Within CDM

The following description refers to the use of widefield live cell imaging performed on an inverted Zeiss Axiovert 100 microscope using a Ludl automated XYZ motorised stage.

1. Wash the imaging chambers in 7× detergent or acetone to remove any wax and rinse thoroughly in distilled water. Sterilise the imaging chambers in ethanol before use and air-dry in the tissue culture hood.
2. Place a sterile coverslip on one side of the chamber and seal the edges with hot wax to create a base for the hole in the imaging chamber (see Fig. 1 and Note 13).
3. Rinse the hole of the imaging chamber in 5% FCS/Optimem and re-fill with 5% FCS/Optimem (see Note 14).
4. Place one of the CDM coverslips previously plated with cells of interest inverted onto the imaging chamber and seal the edges with hot wax (see Fig. 1).
5. Place the imaging chamber on the microscope stage with the CDM coverslip on the bottom of the chamber. Acquire images by phase contrast imaging using a 10× phase contrast objective (see Fig. 2 and Notes 15 and 16).
6. Depending upon the acquisition software being used, set either the number of frames to be acquired or the time interval between frames and the total amount of time of the entire experiment (for example 12 h).
7. Time-lapse sequences can be converted into file-type of choice (for example .AVI) using appropriate post-acquisition software (for example ImageJ).
8. If further analysis is required, cells can be tracked in the resultant files in a tracking software (see Note 17). Import the resulting tracking files into analysis software to determine migration velocity and persistence of movement (see Fig. 3).

3.6. Immunostaining CDM for Microscopy Analysis

The matrix is delicate and so all of the washes need to be completed very carefully.

1. Wash coverslips of CDM plus cells of interest in PBS− and fix for 20 min in 4% PFA/PBS at room temperature.
2. Wash the coverslips three times in PBS− and permeabilise for 10 min in 0.2% Triton X-100/PBS.

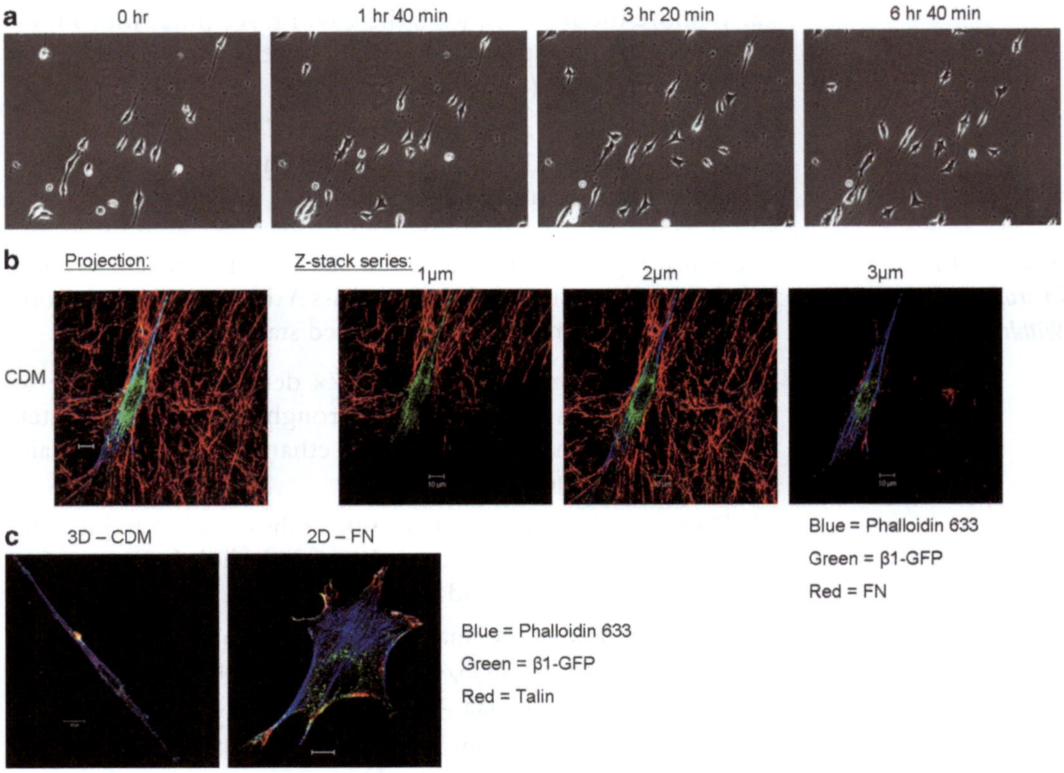

Fig. 2. Structure of fibronectin-rich cell-derived matrix (CDM) and fibroblast morphology in CDM and fibronectin (FN). (**a**) Four representative images from a time-lapse sequence of mouse embryo fibroblasts migrating on CDM, shown under phase contrast. The *black lines* seen in these images are the CDM fibres. (**b**) A confocal z-stack projected in the *Z*-axis and four corresponding stacks from the original Z-stack of a fibroblast expressing β1-GFP in CDM, stained for Phalloidin633 (*blue*) and FN (*red*). (**c**) A confocal image of z-stack projected in the *Z*-axis of a fibroblast expressing β1-GFP in CDM, stained for Phalloidin633 (*blue*) and Talin (*red*). A confocal image of a fibroblast expressing β1-GFP plated on FN, stained for Phalloidin633 (*blue*) and Talin (*red*). Scale bars in (**b**) and (**c**) are 10 μm.

3. Wash the coverslips further two times in PBS– and block for 20–30 min in 3% BSA/PBS at room temperature.
4. Add the primary antibody at an appropriate dilution (diluted in 3% BSA/PBS) to the coverslips (see Note 18) for at least 2 h at room temperature. Then wash three times in PBS– (see Note 19).
5. Add the secondary fluorescent-conjugated antibody at an appropriate dilution (diluted in 3% BSA/PBS) to the coverslips for at least 1 h at room temperature in the dark. Then wash further three times in PBS–.
6. If the cells require further staining repeat this protocol from step 4, if no more staining is required move on to step 7.
7. Wash the coverslip once in distilled water and mount inverted onto microscope slides with mounting media (e.g. Mowiol).

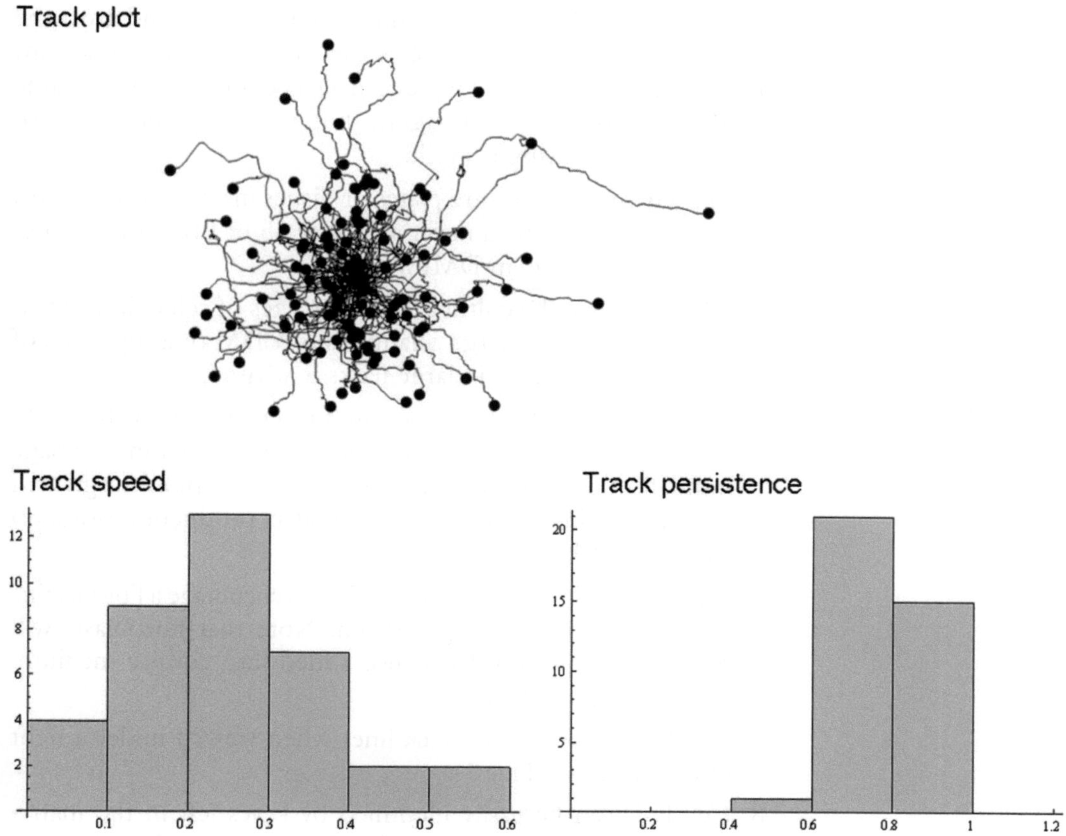

Fig. 3. Analysis of cell migration on CDM. An example of analysis gained from migration experiments. The track plot shows the tracks of each cell plotted originating from an origin and the dot shows where the cell ended. The graphs of migration speed and migration persistence are shown that correspond to the cells from the track plot.

8. Acquire images of the fixed cells using a confocal microscope (see Fig. 2). Z-stacks are preferable to single sections when imaging CDM to allow 3D reconstruction (see Note 20). CDM is on average around 50 μm deep.

4. Notes

1. Different types of fibroblast from different companies can be used to create CDM. Primary fibroblasts, i.e. primary human dermal or foreskin fibroblasts, are generally the best to use for matrix production; however immortalised cells such as NIH3T3 fibroblasts can also be used. If a different fibroblast cell is used to generate CDM wherever HDFs are mentioned in this protocol use these fibroblasts instead.
2. 22 × 22 coverslips are best placed in a well of a 6-well cell culture dish, and 13 mm diameter coverslips are best placed

in a well of a 24-well cell culture dish to prepare and grow the matrix. Other sized coverslips can be used in the same way. Cell culture dishes and plates can also be prepared in the same way as coverslips for imaging, but note that these do not require autoclaving if sterile.

3. In preparing cell culture plates or dishes instead of coverslips, it is not possible to transfer so just wash in PBS– three times and then incubate in DMEM.
4. Fibroblasts can be cultured in other sizes of cell culture flask, but for generating large quantities of CDM (e.g. upwards of 10 plates), culturing in large flasks is advised.
5. When primary fibroblasts are used to create CDM the ECM production slows down over time with increasing passage number of cells. Therefore, it is necessary to monitor growth and cease using the fibroblasts if matrix production slows up to passage 10.
6. The fibronectin is added to the HDFs to encourage a fibronectin-enriched matrix to be generated. Note that fibroblasts will produce fibronectin if it is not added into culture medium, and this step is optional.
7. The CDM appears as black lines when viewed under a light microscope (see Fig. 2).
8. Cell lysis can be easily identified by holes left in the matrix where the fibroblasts used to be. There may still be a few cells around the edges of the coverslips after denudation; if this is the case, repeat the steps until these cells are removed, ensuring that the CDM is handled with great care throughout wash steps to avoid disruption.
9. Check regularly for CDM degradation which can be identified by the CDM breaking apart. The CDM appears as black lines when viewed under a light microscope; when the CDM is degrading, these black lines appear less defined and have larger gaps.
10. Use Parafilm to seal around the edges of the cell culture dishes containing the matrix in order to prevent evaporation and drying out during storage.
11. The appropriate number of cells to plate into the CDM varies depending on the experiment and the cell type.
12. Cells have a more in vivo-like appearance in CDM than on 2D substrates, i.e. fibroblasts are more elongated (see Fig. 2c).
13. Different imaging chambers are available and these can be used instead of the imaging chambers shown; however home-made imaging chambers can be cheaper.
14. Hepes buffered medium such as Optimem is required if there is no CO_2 supply on the microscope in order to buffer the

media. Five per cent FCS was added in order to allow cell survival over the whole time-lapse period; the amount of serum needed to be added to your media will be dependent on the cell type. It is necessary to use phenol red free media if you are imaging using fluorescence as phenol red is excited at 488 nm and emits over a wide spectrum.

15. These imaging chambers can be used on an upright microscope. Invert the chamber so that the CDM coverslip is on the top.
16. For imaging CDM be careful to choose an objective with the correct working distance so that the whole of the matrix can be seen through.
17. Approximately 70–100 cells should be tracked in total for each experimental condition from at least three independent experiments.
18. The antibody can be added to the top of the coverslips or the coverslips can be inverted onto the antibody. If the second method is used, be very careful as the matrix is very delicate.
19. Wash the coverslips in PBS– by picking up the coverslip using forceps and dunking into a beaker of PBS– three times very carefully.
20. Since the matrix is 3D and the cells can align upon many planes in the matrix, the resulting depth of cells in the matrix requires the acquisition of z-stacks to allow imaging of the whole depth.

References

1. Cukierman, E., Pankov, R., Stevens, D. R., and Yamada, K. M. (2001) Taking Cell-Matrix Adhesions to the Third Dimension, in *Science*, pp 1708–1712.
2. Morgan, M. R., Byron, A., Humphries, M. J., and Bass, M. D. (2009) Giving off mixed signals - Distinct functions of alpha5beta1 and alphaVbeta3 integrins in regulating cell behaviour, *IUBMB Life* **61**, 731–738.
3. Zaidel-Bar, R., Ballestrem, C., Kam, Z., and Geiger, B. (2003) Early molecular events in the assembly of matrix adhesions at the leading edge of migrating cells, *J. Cell. Sci.* **116**, 4605–4613.
4. Wozniak, M. A., Modzelewska, K., Kwong, L., and Keely, P. J. (2004) Focal adhesion regulation of cell behavior, *Biochim. Biophys. Acta* **1692**, 103–119.
5. Zamir, E., Katz, M., Posen, Y., Erez, N., Yamada, K. M., Katz, B.-Z., Lin, S., Lin, D. C., Bershadsky, A., Kam, Z., and Geiger, B. (2000) Dynamics and segregation of cell-matrix adhesions in cultured fibroblasts, *Nat. Cell Biol.* **2**, 191–196.
6. Pankov, R., Cukierman, E., Katz, B., Matsumoto, K., Lin, D., Lin, S., Hahn, C., and Yamada, K. (2000) Integrin dynamics and matrix assembly: tensin-dependent translocation of alpha5beta1 integrins promotes early fibronectin fibrillogenesis, *J. Cell Biol.* **148**, 1075–1090.
7. Katz, B.-Z., Zamir, E., Bershadsky, A., Kam, Z., Yamada, K. M., and Geiger, B. (2000) Physical State of the Extracellular Matrix Regulates the Structure and Molecular Composition of Cell-Matrix Adhesions, *Mol. Biol. Cell* **11**, 1047–1060.
8. Yamada, K. M., Pankov, R., and Cukierman, E. (2003) Dimensions and dynamics in integrin function, *Braz. J. Med. Biol. Res.* **36**, 959–966.
9. Bass, M. D., Roach, K. A., Morgan, M. R., Mostafavi-Pour, Z., Schoen, T., Muramatsu,

T., Mayer, U., Ballestrem, C., Spatz, J. P., and Humphries, M. J. (2007) Syndecan-4-dependent Rac1 regulation determines directional migration in response to the extracellular matrix, *The Journal of Cell Biology* **177**, 527–538.

10. Quiros, R. M., Valianou, M., Kwon, Y., Brown, K. M., Godwin, A. K., and Cukierman, E. (2008) Ovarian normal and tumor-associated fibroblasts retain in vivo stromal characteristics in a 3-D matrix-dependent manner, *Gynecol. Oncol.* 110, 99–109.

11. Serebriiskii, I., Castelló-Cros, R., Lamb, A., Golemis, E. A., and Cukierman, E. (2008) Fibroblast-derived 3D matrix differentially regulates the growth and drug-responsiveness of human cancer cells, *Matrix Biol.* 27, 573–585.

Chapter 6

Live Cell Imaging of Neuronal Growth Cone Motility and Guidance In Vitro

Daniel M. Suter

Abstract

The neuronal growth cone, a highly motile structure at the tip of neuronal processes, is an excellent model system for studying directional cell movements. While biochemical and genetic approaches unveiled molecular interactions between ligand, receptor, signaling, and cytoskeleton-associated proteins controlling axonal growth and guidance, in vitro live cell imaging has emerged as a crucial approach for dissecting cellular mechanisms of growth cone motility and guidance. Important insights into these mechanisms have been gained from studies using the large growth cones elaborated by *Aplysia californica* neurons, an outstanding model system for live cell imaging for a number of reasons. Identified neurons can be isolated and imaged at room temperature. *Aplysia* growth cones are five to ten times larger than growth cones from other species, making them suitable for quantitative high-resolution imaging of cytoskeletal protein dynamics and biophysical approaches. Lastly, protein, RNA, fluorescent probes, and small molecules can be microinjected into the neuronal cell body for localization and functional studies. This chapter describes culturing of *Aplysia* bag cell neurons, live cell imaging of neuronal growth cones using differential interference contrast and fluorescent speckle microscopy as well as the restrained bead interaction assay to induce adhesion-mediated growth cone guidance in vitro.

Key words: Neuronal growth cone, *Aplysia californica*, Cell motility, Cytoskeletal dynamics, Live cell imaging, Differential interference contrast, Fluorescent speckle microscopy, Restrained bead interaction

1. Introduction

Neuronal growth cones are highly motile structures at the tip of neuronal processes that detect molecular information in the environment via cell surface receptors during both development and regeneration. Various signaling cascades transduce this information from the plasma membrane to the underlying actin and microtubule cytoskeleton to affect the dynamic properties of

Claire M. Wells and Maddy Parsons (eds.), *Cell Migration: Developmental Methods and Protocols*, Methods in Molecular Biology, vol. 769, DOI 10.1007/978-1-61779-207-6_6, © Springer Science+Business Media, LLC 2011

these structures and ultimately lead to changes in speed and direction of growth cone movement (1–3). In vitro experiments using cultured neurons from various species have contributed significantly to our understanding of the molecular and cellular mechanisms of neuronal growth cone motility and guidance. The benefits of in vitro experiments include 1. working with an identified cell type, 2. the possibility of manipulating specific neurons in a temporarily and spatially defined manner, and 3. the capability of observing cell motility and protein dynamics by live cell imaging without the need for confocal or multi-photon fluorescence microscopy. Commonly used neuronal cell culture systems for growth cone studies include retinal ganglion cells and spinal cord neurons from Xenopus embryos (4, 5), dorsal root ganglion neurons from chicken embryos (6), superior cervical ganglion neurons from rat and mouse embryos (7), hippocampal neurons from rat and mouse embryos (8), and bag cell neurons from adult *Aplysia californica* (9, 10).

Aplysia growth cones are five to ten times larger than typical vertebrate growth cones, making them an excellent model system for high-resolution live cell imaging of protein dynamics, particularly of cytoskeletal proteins through the use of quantitative fluorescent speckle microscopy (FSM) (11–13). Other benefits of this system include the purity of neuronal cultures as well as the fact that neurons can be cultured and imaged without any serum or growth factors at room temperature. The large size of the growth cones and the room temperature compatibility make these cells suitable for biophysical manipulations, e.g., by micropipettes (14) and atomic force microscopy (15). Lastly, the large cell bodies of *Aplysia* neurons (50–100 μm) allow microinjection of various probes, including protein, RNA, fluorescent probes, and small molecules. Thus, *Aplysia* growth cones represent one of the best model systems for studying growth cone motility and guidance in vitro.

To study directional growth cone movements, various approaches of presenting soluble and substrate-bound guidance cues in a gradient or patterned fashion have been developed in recent years. A commonly used method of applying a gradient of a soluble factor is the so-called pipette-gradient assay, which involves pulsed ejections of diffusible small molecules or proteins from micropipette in a defined angle to the advance axis of the growth cone (16, 17). Different techniques have been developed to produce gradients of substrate-bound proteins, including filter application (18), diffuse printing (19), microcontact printing (20), microfluidics (21), and photoimmobilization (22). In vitro growth cone turning responses have also been studied with patterned substrates, using a variety of approaches to create stripe patterns of two different substrates, including microcontact printing (23), microfluidics (24), and silicon masks (25, 26). We have developed an assay referred to as "restrained bead interaction

(RBI)" assay to investigate adhesion-mediated growth cone responses (14). This assay involves microbeads coated with adhesion proteins manipulated with a micropipette in order to study the transition of growth cone adhesion from one substrate to another. The large *Aplysia* growth cones are particularly suitable for this RBI assay. This chapter discusses preparation of cultured *Aplysia* bag cell neurons, live cell imaging of growth cone motility and cytoskeletal dynamics using differential interference contrast (DIC) and fluorescent speckle microscopy (FSM) of actin and microtubule dynamics, as well as setting up the restrained bead interaction assay.

2. Materials

2.1. Aplysia Bag Cell Neuronal Culture

1. Leibovitz's L15 cell culture medium supplemented with artificial seawater (L15-ASW): L15 (powder, containing glutamine, Invitrogen #41300-039), pH 7.9, including 400 mM NaCl, 27 mM $MgSO_4$, 28 mM $MgCl_2$, 9 mM $CaCl_2$, 4 mM L-glutamine, 50 μg/ml gentamicin, 5 mM 4-(2-hydroxyethyl)-1-piperazineethanesulfonic acid (HEPES). Osmolarity should be 950–1,000 mmol/kg. Filter through 0.22 μm sterile filter. Store at 4°C for up to 1 month (see Note 1).
2. Dispase II solution: 10 mg dispase II (Roche # 04942078001) dissolved in 900 μl L15-ASW plus 100 μl sterile H_2O ultrapure. Prepare freshly right before dissection of ganglia. 1 ml of dispase II solution is sufficient for the digestion of 1–3 ganglia.
3. Magnesium chloride solution: 0.5 M $MgCl_2$, prepared in H_2O ultrapure. Store at 4°C.
4. 6 N HCl for acid-cleaning of coverslips. To prepare mix 1 volume H_2O ultrapure with 1 volume of 36% HCl.
5. 60 ml syringe with Luer-lok tip (BD), 18.5 gauge sterile hypodermic needle, 2 in. long.
6. Microdissecting scissors: Vannas Spring Scissor, 8.5 cm long, 7 mm blade (WPI #500086); regular dissecting scissors; two pairs of forceps: Dumont Tweezers #5 – 11 cm long (WPI #500341).
7. Poly-L-lysine (10×): 200 μg/ml Poly-L-lysine (70–150 kDa; Sigma #P-6282) stock solution is prepared in sterile H_2O ultrapure. Store 1 ml aliquots at –20°C.
8. *A. californica*, adult, 100–200 g, from Marinus Scientific or the National Resource for Aplysia at the University of Miami.

9. Coverglass: #1.5, 22×22 mm, acid-cleaned and stored in 100% ethanol.
10. 35 mm Petri dishes.
11. Water bath; laminar flow bench; dissecting microscope (e.g., Leica MZ 9.5); 14°C incubator.

2.2. Live Cell Imaging Using Differential Interference Contrast and Fluorescent Speckle Microscopy

1. Artificial seawater (ASW): 400 mM NaCl, 10 mM KCl, 55 mM $MgCl_2$, 10 mM $CaCl_2$, 15 mM HEPES, pH 7.8. Osmolarity should be 950–1,000 mmol/kg. Filter through 0.45 μm sterile filter. Store at 4°C.
2. Imaging medium: ASW supplemented with 2 mg/ml BSA, 1 mg/ml L-carnosine and 0.25 mM vitamin E. Filter through 0.22 mm sterile filter. Prepare freshly on the day of the experiment.
3. Injection buffer: 100 mM 1,4-Piperazinediethanesulfonic acid (PIPES) pH 7.0, 1 mM $MgCl_2$, 1 mM ethylene glycol tetraacetic acid (EGTA). Filter through 0.22 mm sterile filter. Store at 4°C.
4. G-buffer: 5 mM Tris pH 8.1, 0.2 mM $CaCl_2$, 0.2 mM dithiothreitol, 0.2 mM ATP, 10% (w/v) sucrose. Filter through 0.22 mm sterile filter. Prepare freshly for dilution of G-actin.
5. Rhodamine tubulin (for microinjection): 1 mg/ml rhodamine-labeled bovine tubulin (Cytoskeleton, Inc, old Cat. # TL331M; new Cat. # TL590M) prepared in injection buffer, store at −80°C (see Note 2).
6. Alexa Fluor 488-phalloidin (for microinjection): 20 μM Alexa Fluor 488-phalloidin (Invitrogen # A12379) freshly prepared in injection buffer. Therefore, transfer a small amount (10 μl) of phalloidin dissolved in methanol into microcentrifuge tube, remove the methanol in a speed vacuum concentrator, and reconstitute the phalloidin in injection buffer (see Note 2).
7. Alexa Fluor 568-G-actin (for microinjection): 1 mg/ml Alexa Fluor 568-G-actin (rabbit muscle actin; Invitrogen # A12374) freshly prepared in G-buffer; store stock aliquots of 7–8 mg/ml at −80°C (see Note 2).
8. Imaging chamber: We use a custom-made holder for an imaging chamber made by two coverslips separated by plastic spacers (Fig. 1).
9. Diamond pen; 22×2 mm plastic spacers (cut from 0.020 in. plastic shims, # SHSP-200 SmallParts, Inc.); nylon grommets (# CGN-250-02 SmallParts, Inc.) used as clips to hold top and bottom part of the holder together; high vacuum grease (Dow Corning); Sparkle glass cleaner; Q-tips.

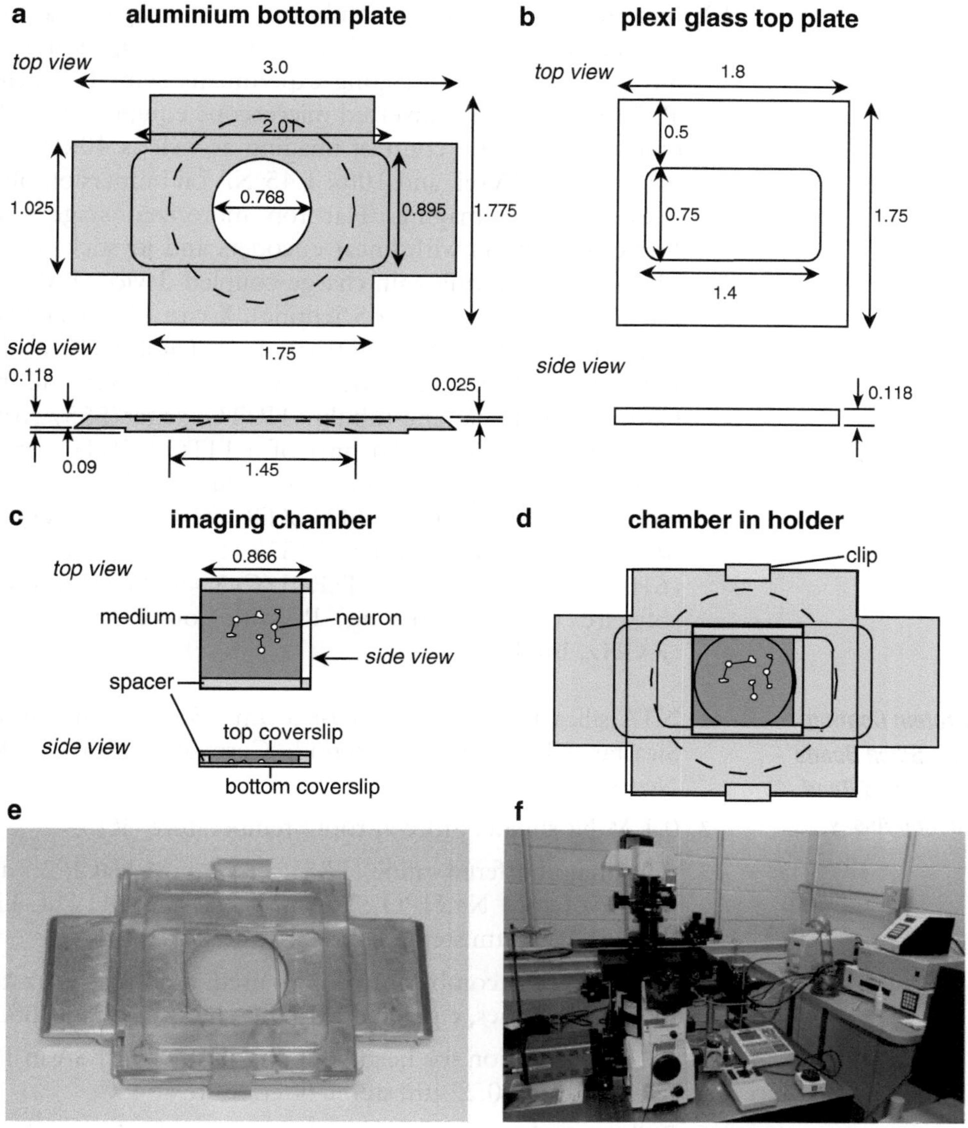

Fig. 1. Imaging chamber used for live cell imaging of neuronal growth cones. (**a**) Schematic drawing of bottom aluminum plate of custom-made holder used for the imaging chamber described in this chapter. Dimensions are in inches. (**b**) Plexiglass top plate of custom-made holder. (**c**) Imaging chamber made by two coverslips separated by two plastic spacers. The bottom coverslip contains the cultured neurons. Left and right sides are open for medium exchange. (**d**) Imaging chamber mounted between bottom and top parts held together by plastic clips. (**e**) Picture of an assembled chamber within the holder. (**f**) Live cell imaging workstation based on an Eclipse TE2000E2 (Nikon) inverted microscope as described in this article.

10. Microinjection needles: borosilicated glass capillaries (1B100F-4, World Precision Instruments). Microloader tips (Eppendorf #930001007).
11. Vertical micropipette puller (Narishige PP830).
12. Micromanipulator: NP-2 patchman (Eppendorf); injection system: FemtoJet (Eppendorf).

13. Imaging set up: Various combinations of imaging hardware and software are possible. The following setup is used for most of our live cell imaging experiments (Fig. 1f): Eclipse TE2000E2 (Nikon) inverted microscope equipped with 10× objective for phase contrast imaging as well as 40× 1.3 NA oil, 60× 1.4 NA oil, and 100× 1.45 NA oil immersion objectives for DIC imaging. Flat top motorized stage H117 ProScan™ (Prior) with linear encoders and joystick control. Electron-multiplying gain charge-coupled device (EMCCD) camera Cascade II (Roper Scientific). X-cite 120 metal halide lamp (EXFO Photonic Solutions, Inc). Lambda 10-3 high-speed filter wheels and shutters (Sutter). Filter sets (all from Chroma) for rhodamine-tubulin: TRITC exciter (555/25 nm) and emitter (HQ610/60 nm) of a FITC/TRITC set; for Alexa Fluor 488-phalloidin: FITC exciter (485/20 nm) and emitter (517/30 nm) of a FITC/TRITC set; for Alexa Fluor 568-G-actin: TxRed exciter (575/15 nm) and emitter (610/40 nm) of a FITC/TxRed/Cy5 set. MetaMorph 7.0 software (Universal Imaging). PC with Core 2 Duo processor (3 GHz, Intel).

2.3. Protein Coating of 5 μm Silica Beads for Restrained Bead Interaction Assay

1. NTA silica beads (diameter of 5 μm); 5% w/v in solution; Sicastar 43-11-503 (Micromod, Germany). Store at 4°C. Never freeze!
2. 0.1 M Ni-sulfate. Store at room temperature (RT).
3. Phosphate buffered saline (PBS): 137.0 mM NaCl, 2.7 mM KCl, 10.1 mM Na_2HPO_4, 1.8 mM KH_2PO_4, pH 7.4. Filter through 0.22 μm sterile filter. Store at 4°C.
4. 6His-tagged recombinant protein used as adhesive substrate for growth cones, e.g., apCAM, 200–400 μg/ml in PBS.
5. Blocking solution for bead coupling: 5 mg/ml BSA in PBS. Filter through 0.22 μm sterile filter. Store at 4°C.
6. Refrigerated bench top centrifuge, e.g., Beckman-Coulter AllegraTM X-22R.
7. End-over-end rotator, e.g., Barnstead/Thermolyne Rotisserie Labquake.

2.4. Restrained Bead Interaction Assay

1. L15-ASW containing 5 mg/ml BSA.
2. Phosphate buffered saline (PBS): 137.0 mM NaCl, 2.7 mM KCl, 10.1 mM Na_2HPO_4, 1.8 mM KH_2PO_4, pH 7.4. Filter through 0.22 μm sterile filter. Store at 4°C.
3. Protein-coated 5 μm silica beads.
4. 5 μl calibrated pipettes (#2-000-001 Drummond Scientific Company).
5. Vertical micropipette puller (Narishige PP830).

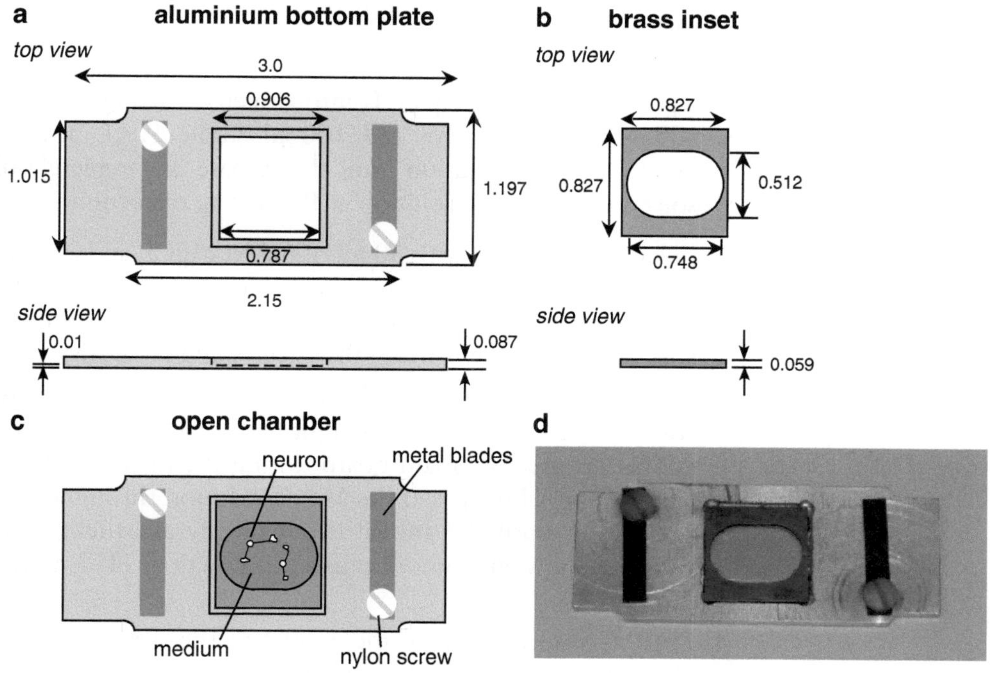

Fig. 2. Open imaging chamber used for RBI growth one guidance assay. (**a**) Schematic drawing of *bottom aluminum plate* of custom-made holder used as open imaging chamber. The chamber is open at the top allowing accessing cultured cells via micropipettes. Dimensions are in inches. (**b**) Brass inset to be mounted on coverslip with cells using high vacuum grease. (**c**) Assembled open chamber. *Brass inset* with coverslip attached is positioned into the *bottom plate*. Metal blades can be used to hold tubing for perfusion of the medium. (**d**) Picture of a fully assembled open chamber.

6. Narishige 3D hydraulic micromanipulator.
7. Custom-made open imaging chamber with aluminum base and brass inset (Fig. 2).

3. Methods

3.1. Aplysia Bag Cell Neuronal Culture

Individual identified *Aplysia* neurons can be isolated and cultured alone or together with other neurons to form small circuits in vitro, e.g., between sensory and motor neurons (27). The following protocol describes culturing bag cell neurons derived from the abdominal ganglion of adult *Aplysia*. The main function of the neurosecretory bag cell neurons is to produce an egg laying pro-hormone and related peptides in response to electrical stimulation. We use bag cell neurons to investigate molecular and cellular mechanisms of growth cone motility and guidance; however, other neurons such as sensory neurons derived from the pleural ganglia can be used for this purpose as well. Video protocols for preparing cultures of bag cell (10) and sensory/motor neurons

(28) have recently been published in the *Journal of Visualized Experiments*.

1. Transfer 10 mg of dispase II into a microcentrifuge tube and add 900 μl L15-ASW and 100 μl sterile H_2O ultrapure. Gently mix the solution until the dispase is completely dissolved and keep the solution at RT during dissection.
2. Fill a 60 ml syringe with 50 ml of 0.5 M $MgCl_2$ and attach an 18.5 gauge needle. Use a styrofoam box cover as dissection area. Sterilize regular dissecting scissors, microdissecting scissors, and two forceps with 70% ethanol and let air dry.
3. Take an adult *Aplysia* from the tank and inject all of the $MgCl_2$ solution into the body cavity right behind the head. Keep the animal on the styrofoam board during the injection. Gently rub the animal to spread the $MgCl_2$ solution throughout the body. Wait until the animal is completely anesthetized and does not show any signs of muscle contractions (1–3 min; see Note 3).
4. Pin the head and tail of the animal to the dissection board using two needles, head facing to the right.
5. Take one forceps to lift up the skin and body wall muscle. Open the body wall with regular dissecting scissors without damaging internal organs. Cut the body wall open along the side toward the head and the tail.
6. Locate the abdominal ganglion toward the mantle and gill. Use Fig. 4–5 in (29) as a reference. Use new forceps to lift up the connecting nerves just rostrally to the ganglion and cut the connecting nerves rostrally and caudally to the abdominal ganglion with microdissecting scissors. Be sure not to damage the ganglion. The two bag cell clusters are attached rostrally to the larger hemiganglia (Fig. 4–7 in (29)). Residual nervous system tissue can be collected if needed. The residual body of the animal is stored at −20°C before incineration.
7. Transfer the ganglion into a microcentrifuge tube with 1 ml of dispase II solution. Digest the tissue for 15–16 h at 22°C using a temperature-controlled water bath (see Note 4).
8. On the following day, bag cell plating is performed in a laminar flow bench. Prepare three 35 mm Petri dishes filled with 4 ml of L15-ASW each for dissecting the bag cell clusters. Transfer the digested abdominal ganglion to one of these dishes using either a 20 μl pipette or sterile forceps. Make sure the medium temperature has been adjusted to RT.
9. Preparing poly-lysine-coated coverslips:
   (a) Use acid-cleaned #1.5 square glass coverslips (22 × 22 mm; see Note 5 regarding acid-cleaning), stored in 100% EtOH. Take the coverglass out of the storage container

using sterile forceps and briefly flame it with a Bunsen burner before placing it onto the edge of an uncovered empty 35 mm Petri dish. Alternatively, air dry the coverglass without flaming. Depending on the number of neurons plated per coverglass, 6–12 is a reasonable number of coverslips to prepare for bag cells derived from one animal.

(b) Prepare enough 20 μg/ml poly-L-lysine solution by diluting 200 μg/ml stock solution ten times with sterile H_2O ultrapure. Once the coverslip is dry, place it inside the Petri dish and coat it with 0.5 ml of 20 μg/ml poly-L-lysine solution. Incubate for at least 20 min at RT (see Note 6).

(c) Wash each coverglass three times with 0.5 ml of sterile H_2O ultrapure. Remove solutions with vacuum. After the last wash add 4 ml of L15-ASW into each dish.

10. Clean microdissecting scissors and two forceps with 70% EtOH. Separate the two bag cell clusters from the rest of the abdominal ganglion while observing them with a dissection microscope. Therefore, first cut the abdominal ganglion into two halves right through the center along the rostral-caudal axis. Then, cut each cluster from the hemiganglion and the residual nerve. Be sure not to damage the clusters with the dissecting tools and to keep the clusters and cells in the medium all the time.

11. Use two forceps to remove the connective tissue sheath from each cluster one at a time. Therefore, hold the cluster with one forceps while pushing against the connective tissue sheath with the second forceps. Try to find an opening where the bag cell neuronal cluster pops out through the connective tissue sheath. Ideally separate the sheath from the cellular cluster as one piece. The bag cell cluster will appear as a "berry-like" structure.

12. Use a forceps to bend a yellow P20 pipette tip that has been briefly (1–2 s) warmed up over a Bunsen burner flame. Be careful not to melt the tip. The tip is bent at a 45° angle to facilitate the positioning of the tip during cell plating. Use the bent tip to transfer the cluster into a new dish with L15-ASW (see Note 7).

13. Triturate the cluster with the yellow tip to remove individual bag cells. Therefore, gently suck the cluster into the tip and push it out. Start out with low shear forces to avoid killing too many neurons and increase the force as needed. As you remove cells from the inside of the cluster, you may have to increase the force by changing the angle between the tip and the coverslip. Healthy cells are bright in the center of the cell

body, while dead cells are transparent and appear dark against a dark background.

14. Pick up three to five healthy cells with the medium and transfer them onto the first poly-lysine-coated coverslip. Avoid picking up dead cells or pieces of extracellular tissue.
15. Repeat steps 13 and 14 and place about 20–40 live neurons into the center of the coverslip. The number of cells depends on the experiment; thus, plate more for microinjections since some cells will not survive the injections. Pay attention to the positioning of the cells because it will be impossible to image cells that are located close to the edge of the coverslip when using oil immersion objectives and the sample holder described here. Remove extracellular debris and dead cells from the coverslips by picking them up with the pipette or gently blowing them away from healthy cells.
16. After all cells from one cluster have been plated, start removing the sheath of the second cluster and proceed with cell separation and cell plating as for the first cluster. The whole plating procedure will take between 1 h 30 min and 2 h 30 min depending on the number of cells. The size of the bag cell clusters varies between animals. A large cluster contains about 300–400 neurons. A good plating typically results in about 200–300 live neurons from one animal.
17. After all healthy neurons have been positioned onto the coverslips, keep the culture dishes for at least 2 h at RT to allow cells to attach. Avoid any potential vibrations during this time. Then, place the dishes into an incubator at 14°C overnight or until further use. Growth cones develop at the tips of cut neuronal processes or emerge from cell bodies within 4–10 h after plating. We use the cells within the first 48 h after plating for most experiments, since growth cones get smaller with time in culture.

3.2. Live Cell Imaging Using Differential Interference Contrast and Fluorescent Speckle Microscopy

3.2.1. DIC Imaging

1. Inspect the cells on the day after cell plating using a tissue culture microscope. Perform a half volume medium change with fresh L15-ASW that has been adjusted to RT. If growth cones have not yet developed during the overnight incubation at 14°C, one can speed up the process by keeping the cultures at RT for 1–2 h.
2. For live cell imaging, place the coverslip with the cells in either custom-made or commercial imaging chamber, for example the Series 20 platform from Warner Instruments. We use a custom-made chamber system originally designed and developed by Paul Forscher and Stephen Smith at Yale University (9). This system is relatively simple and consists of a sandwich chamber made from two coverslips mounted between an aluminum bottom plate and plexiglass top plate

(Fig. 1a, b). This system allows exchange of solutions on the microscope while imaging with both high numerical aperture oil immersion objectives and condenser.

3. To make the imaging chamber, use a diamond pen and a ruler to cut off a 2 mm-wide stripe of a clean square coverslip (22 × 22 mm; Fig. 1c). Cutting off this stripe will facilitate the exchange of solution through the chamber. Fill a 1 ml syringe with high vacuum grease and attach a yellow pipette tip to the syringe for grease application (see Note 8). Apply grease along each of the shorter sides of the cut cover glass. Then, cut two 22 × 2 mm plastic spacers from 0.020 in. yellow plastic shims and press the spacers onto the applied grease using the wooden end of two Q-tips.
4. Apply grease on top of each of the spacers. Using forceps lift up the coverslip/spacer assembly and invert it onto a coverslip with cultured neurons within a Petri dish. Align the inverted coverslip with the one containing the cells and push it down gently using the wooden end of two Q-tips. Make sure the two coverslips stick well together before gently lifting the sandwich out of the medium using forceps (Fig. 1c; see Note 9).
5. Align the two coverslips before positioning the sandwich over the round hole of the bottom part of the sample holder. The slightly lowered inner portion of the aluminum bottom part is covered with a piece of parafilm with a cut out center hole. Align the sandwich well before mounting the plexiglass top part. Use plastic clips cut from nylon grommets on each side to hold the top and bottom parts together (Fig. 1d; see Note 10).
6. Wipe off any excess medium. Clean each cover glass with a Q-tip soaked with Sparkle glass cleaner. Quickly dry off any of the glass cleaner using a dry Q-tip. Clean glass surfaces are essential to high-quality imaging, particularly using differential interference contrast (DIC) optics.
7. Mount the chamber holder onto the microscope stage of an inverted microscope. We use an Eclipse TE2000E2 (Nikon; Fig. 1f). Use the 10× phase objective to locate individual bag cells. Since the number of neurons is relatively low (~20–40) compared to other neuronal culture systems, it is helpful to use a motorized stage with memory function to mark positions of individual cells.
8. After marking the cell body positions, switch the objective to a high-resolution DIC oil immersion objective, e.g., 60×. Set up the microscope for DIC imaging and perform Koehler illumination. Search for an appropriate growth cone for imaging.

9. The optimal camera settings will depend on the type of CCD camera. We use either a Cascade II or a CoolSnap HQ (both from Roper Scientific) and normally select an exposure time of 50 ms for DIC imaging. Time-lapse sequences of growth cone dynamics are acquired in 5–10 s intervals, typically for 10–30 min (Fig. 3a, d). Minimize the light exposure to the amount needed for imaging without negatively impacting growth cone dynamics (see Note 11).
10. Make sure that the chamber is always filled with medium since evaporation will occur over time. Exchange with fresh medium (1–3 ml) at least every 30 min by adding medium into one side of the chamber while removing the medium on the other side with a vacuum line and a yellow pipette tip. Pay attention to possible leaks of solution onto the objective.
11. Using spatial calibrations and image analysis software, analyze image stacks for various parameters of growth cone dynamics, including rates of filopodia and lamellipodia growth (Fig. 3).

3.2.2. Actin and Microtubule FSM Imaging

1. Pull needles for microinjection of cytoskeletal probes. A tip opening of 1 μm is suitable for injections into bag cell neuronal cell bodies. We use borosilicate glass capillaries and perform a double pull with all weights on the vertical puller PP830 using the heat setting "65" for the first pull and "53" for the second pull. Inspect the pulled needles on a microscope for desired tip shape before storing or using them.
2. Prepare cytoskeletal probes for microinjection as described in Subheading 2.2, items 5–7.
3. Clarify cytoskeletal probes by centrifugation at 10,000 × *g* for 30 min at 4°C before microinjection. This reduces the clogging problem when performing microinjections (see Note 12).
4. Prepare your microinjection setup. We use the FemtoJet system for controlled pressure application. Start up the FemtoJet following the instructions of the manufacturer. Turn on micromanipulator if you use a motorized one. We use an NP-2 patchman. Position the microneedle holder in a 40° angle to the stage.
5. Position the Petri dish containing a coverslip with cultured bag cell neurons onto the stage. Search for position of individual neurons using a 10× phase objective and mark the cell positions.
6. Using microloader tips, backfill injection needles with a small amount of cytoskeletal probe (see Note 13).
7. Connect the microneedle to the needle holder. Keep compensation pressure at ~30–40 hPa.

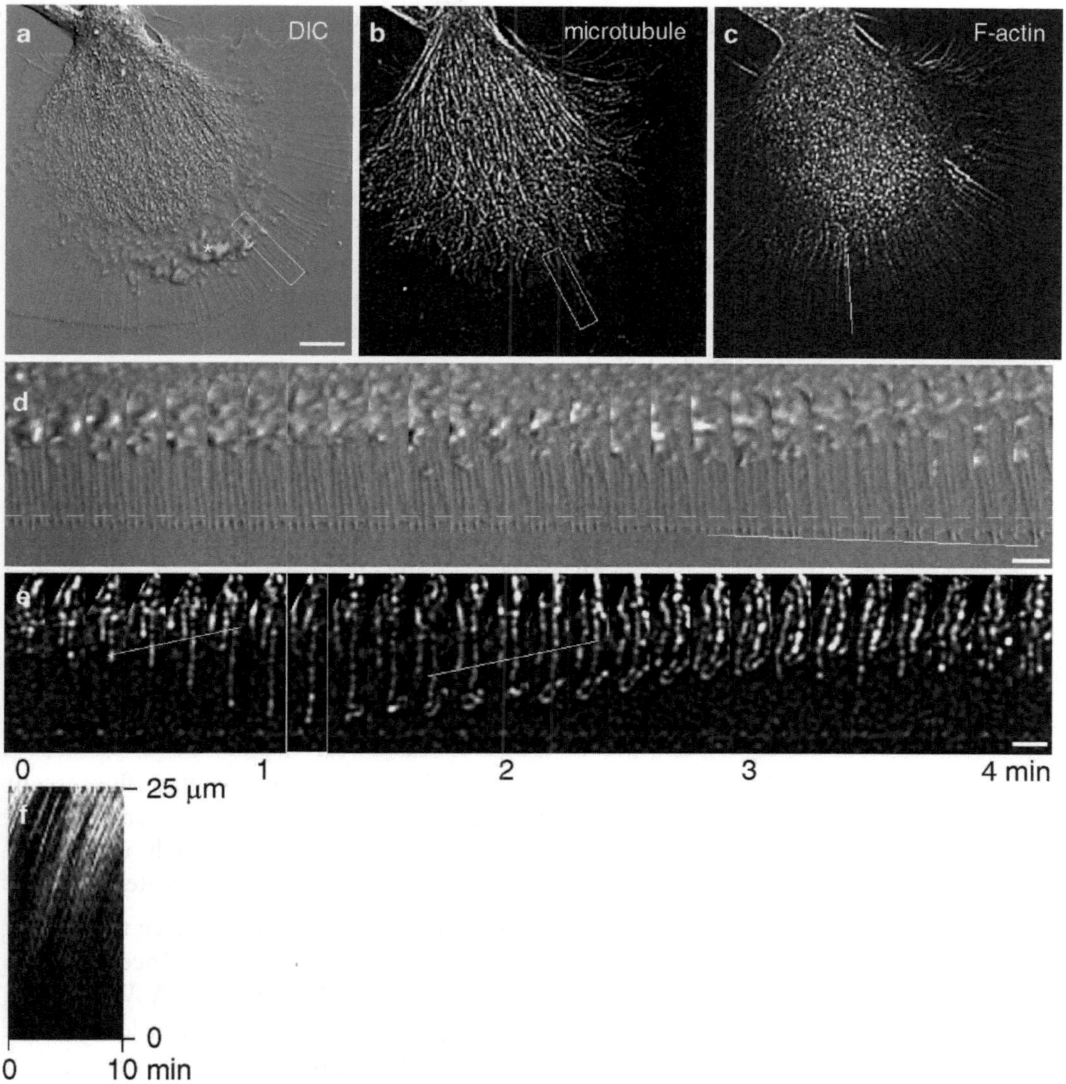

Fig. 3. *Aplysia* bag cell growth cone visualized by DIC and actin/microtubule FSM. (**a**) DIC image of live *Aplysia* bag cell neuronal growth cone on poly-L-lysine. *White box* (5 × 25 μm) demarks region of interest in the peripheral domain used for time-lapse montage shown in (**d**). (**b**) Microtubule FSM image of the same growth cone. Rhodamine-labeled tubulin was injected into the neuronal cell body according to the Subheading 3.2.2, resulting in a "speckled" labeling of individual microtubules. *White box* (5 × 25 μm) demarks region of interest different from the one shown in (**a**) and used for time-lapse montage shown in (**e**). This region contains a dynamic microtubule with fluorescent marks. The corresponding time point shown in image (**b**) is marked in (**e**) with a *white border*. (**c**) Actin FSM image of the same growth cone using Alexa 488 phalloidin labeling. Note that F-actin-labeling with phalloidin highlights actin bundles in the peripheral and central domains, but not highly dynamic actin structures such as ruffles between the central and peripheral domains [indicated by an *asterisk* in (**a**)]. *White line* demarks position of line scan used for kymograph shown in (**f**). (**d**) DIC time-lapse montage of region of interest indicated in (**a**) imaged at 10 s intervals. During the time period of 4 min and 10 s, the leading edge and filopodia slowly advance from the initial position marked by the *dashed line*. The *solid line* demarks the tip position of a filopodium growing at a rate of 1 μm/min. (**e**) Microtubule FSM time-lapse montage of region of interest indicated in (**b**). The position of the leading edge is at the *bottom*. This montage shows a fast growing microtubule that loops around and undergoes retrograde flow. The *two solid white lines* indicate internal reference speckles moving retrogradely. (**f**) Kymograph of F-actin signals scanned along the *white line* shown in (**c**) over a time period of 10 min. The *slope of these lines* indicates the rate of retrograde actin flow. Scale bars: 10 μm in (**a**)–(**c**); 5 μm in (**d**)–(**e**).

8. Quickly lower the injection needle into the medium. Position the tip of the needle in the field of view without touching the surface. Switch to a 40× long working distance phase objective for microinjection and position the cell body of the neuron into the center of the field of view.
9. Check either by phase or fluorescence optics whether there is continuous flow coming out of the pipette. A small "cloud" of fluorescent material should appear in front of the tip. If not the case, increase the compensation pressure or clear out the clog by pressing the "clean" function on the FemtoJet (see Note 12).
10. Position the needle tip adjacent to the cell body using the fine speed setting of the NP-2 micromanipulator. Inject the cell body by moving the needle tip quickly into an area of cytoplasm avoiding the nucleus. A fast movement increases the chance of actually penetrating the plasma membrane. An injection will result in a visible local expansion of cell volume. Leave needle in cell for ~1–3 s and then quickly move needle out since a too large injection volume can kill the cell. Inject an estimated volume of no more than 10% of the cell body. Compensation pressure is usually sufficient for injections; if needed use an increased injection pressure.
11. To make sure the cell body was injected, switch to fluorescent mode using an appropriate fluorescent filter. A clearly injected cell will immediately exhibit fluorescent signals in the cell body that can be seen with the eye piece (see Note 14).
12. After all cell bodies in one dish are injected, place the dish in a light-protected area for the cells to recover. Once all dishes are injected, carefully exchange half of the L15-ASW medium in each dish. Let the injected cells recover for at least 1 h in the dark before starting the imaging.
13. Assemble an imaging chamber as described under steps 3–6 of Subheading 3.2.1 and exchange the L15-ASW with imaging medium.
14. Mark cell positions with 10× phase objective before switching to 60× or 100× oil immersion lenses in order to identify suitable growth cones for imaging.
15. Set up the appropriate illumination and acquisition parameters suitable for the injected fluorescent probes. We use a TRITC exciter (555/25 nm) and emitter (HQ610/60 nm) of an FITC/TRITC set for rhodamine-tubulin and an FITC exciter (485/20 nm) and emitter (517/30 nm) for Alexa Fluor 488-phalloidin. Reduce the fluorescence intensity with neutral density filters (see Note 15). The gain setting and exposure time will depend on the camera used. We typically use 500 ms exposure, gain 3 and additional EM gain on the

Cascade II camera. Time intervals between images are typically 5–10 s. Healthy neurons can be imaged up to 1 h under such conditions. It is recommended to exchange fresh imaging medium every 30 min, or perfuse with medium throughout the experiment. Make sure the sample remains in focus during time-lapse imaging and perform the necessary adjustments.

16. Image processing and data analysis: Raw images will show labeled microtubules and actin bundles; however, only after appropriate image processing one can clearly see the speckles on the filaments (Fig. 3). We use MetaMorph 7.0 software for image processing, quantitative analysis of microtubule and actin dynamics as well as for making movies and montages. To enhance speckles, microtubule and actin images are processed with spatial filters in the following sequence: (1) Low pass 4×4; (2) Laplace 2 edge enhancement; (3) Low pass 3×3. Select clearly labeled microtubules or actin bundles for quantitative speckle analysis (see Note 16). Make a montage and movie of the selected filament and make sure that the appropriate spatial calibration is applied to the montage (Fig. 3e). To measure microtubules assembly rate, measure the distance between a newly added speckle at the microtubule plus end and an internal previously added speckle and divide by the time passed between the two speckle additions. To measure translocation rate, measure the distance an internal speckle moves between two time points and divide by the time. Enter the data into an Excel spreadsheet. For final rate calculation we time-weigh individual rate measurements; thus, rates are multiplied by the time during which they are measured, added up and divided by the total observation time. To measure percentage of time spent in one dynamic behavior, add up all the relevant times during which the behavior is observed and divide the sum by the total time. For actin retrograde flow analysis, use the kymograph function of MetaMorph and draw a line along the lines of speckle movements (Fig. 3f). The slope of this line will directly indicate the rate of actin flow. Calculate average values, standard error of the mean, and significance values in Excel. To get reliable data, it is advised to analyze at least 30 filaments from 10 different growth cones in two to three independent experiments.

3.3. Protein Coating of 5 μm Silica Beads for restrained Bead Interaction Assay

The following protocol describes the preparation of microbeads coated with recombinant Aplysia cell adhesion protein, apCAM. These beads can be used to induce adhesion-mediated neuronal growth as described in Subheading 3.4. Other adhesion molecules and protein coupling methods can be used as well.

1. Transfer 0.1 ml of 5 μm NTA silica beads (5% weight/volume stock solution) into a 1.5 ml microcentrifuge tube (see Note 17).
2. Wash beads 3× with 1 ml sterile H_2O ultrapure. Therefore, resuspend beads with wash solution and sediment the beads by centrifugation at 20,000 × *g*, 4°C for 5 min using a refrigerated table-top centrifuge.
3. Resuspend bead pellet with 0.1 ml of 0.1 M Ni-sulfate and incubate for 30 min at RT by placing the tube on a rotating device, such as an end-over-end rotator, in order to keep the beads in suspension.
4. Transfer 20 μl of Ni-NTA beads (at 5% w/v) into a new tube and wash 2× with 1 ml H_2O ultrapure as described in step 2.
5. Wash beads 2× with 1 ml sterile PBS sterile as described in step 2.
6. Resuspend bead pellet with 100 μl purified 6His-tagged protein, such as apCAM (250–400 μg/ml in PBS), and incubate on rotator overnight at 4°C.
7. Sediment the beads by centrifugation at 20,000 × *g*, 4°C for 5 min. Transfer the supernatant containing free apCAM protein solution into a separate tube.
8. Resuspend bead pellet with 100 μl blocking solution for bead coupling (5 mg/ml BSA in PBS sterile) and incubate for 30 min at RT on the rotator.
9. Spin the bead solution as above, remove free blocking solution and add free apCAM ligand solution back for storage of beads (see Note 18). These apCAM-coupled beads can be used for the RBI assay described in the next section. Beads that were not incubated with apCAM can be used as controls.

3.4. Restrained Bead Interaction Assay to Study Growth Cone Guidance

1. Culture *Aplysia* bag cell neurons on poly-lysine-coated coverslips overnight as described in Subheading 3.1.
2. To prepare apCAM beads for RBI experiments, remove any unbound free ligand from the bead solution. Therefore, pellet the beads at 20,000 × *g*, 4°C for 5 min and transfer free apCAM ligand into a separate tube.
3. Resuspend apCAM bead pellets once with 500 μl sterile PBS and spin at 20,000 × *g*, 4°C for 5 min; then resuspend beads in 100 μl PBS.
4. Prepare a 1:500 dilution of washed beads in an experimental medium, such as L15/ASW 5 mg/ml BSA (in case of DIC imaging) or imaging medium (in case of actin and microtubule FSM imaging) by adding 1 μl beads into 500 μl medium.

50 μl diluted beads are typically sufficient for one open chamber. Spin residual bead stock, remove PBS supernatant, and resuspend beads in apCAM solution for storage.

5. Pull microneedles with short shanks using 5 μl calibrated pipettes. We use a double pull (heating settings 61/54) on the PP830 puller. The tip should be no more than 1 μm wide and closed. If it is open, it can be fire-polished with a microforge.
6. Identify a coverslip containing several neurons elaborating healthy and large growth cones. We use an open chamber system containing an aluminum base plate in slide format and a brass inset to hold the coverglass (Fig. 2a, b). The design shown in Fig. 2 is slightly different from the original one made by Paul Forscher (Yale University) as it uses square 22×22 mm coverslips as opposed to round coverslips (14). Apply vacuum grease generously to the bottom side of the brass inset (Fig. 2b) and mount the brass inset onto the coverslip with cells within the Petri dish filled with medium. Use forceps to lift the brass inset together with the coverslip and medium out of the dish.
7. Align the brass inset well with the coverslip, wipe off excess medium on the bottom side, and apply some vacuum grease along the edges on the bottom side of the coverslip to attach the inset to the aluminum plate. Position the brass inset together with the cells into the aluminum plate (Fig. 2c, d) and wipe off an extra medium. Clean glass coverslip on the bottom side as described in step 6 of Subheading 3.2.1.
8. Exchange L15-ASW with L15-ASW containing 5 mg/ml BSA (if performing DIC imaging only) or with imaging medium (if performing both DIC and FSM imaging). Mount open chamber onto motorized stage and mark cell positions with 10× phase objective.
9. Position the microneedle into the center of the field of view with a relatively flat angle (25° to stage) using a hydraulic micromanipulator (see Note 19). Avoid touching the surface as the needle may break.
10. Switch to 60× oil immersion DIC objective and search for large growth cones with distinct cytoplasmic domains that are suitable for the RBI assay.
11. Add 50 μl of diluted beads into the center of the chamber. Identify a bead in the vicinity of the growth cone. Using the micromanipulator, position the bead close to the leading edge of the growth cone. This can be achieved by picking up the bead with the microneedle or by moving it over the coverslip. One can position the bead at various places along the leading edge of the growth cone; however, orienting the bead–growth

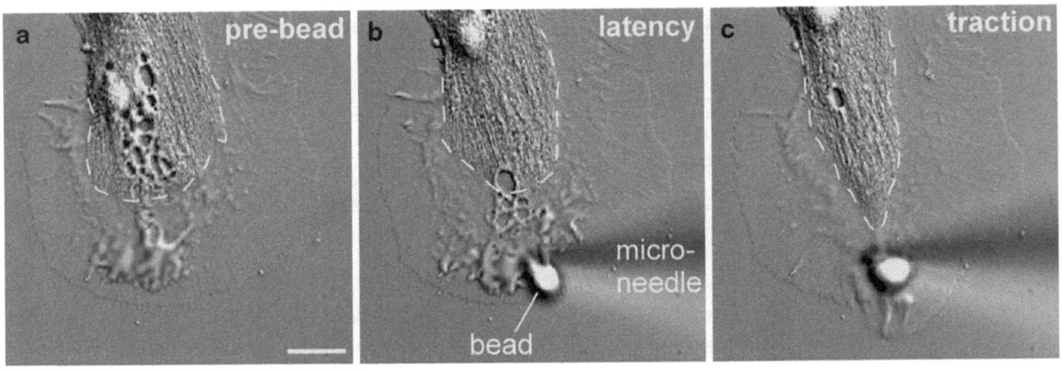

Fig. 4. Restrained bead interaction assay. (**a**) DIC image of an *Aplysia* growth cone on poly-lysine substrate. *Dashed line* indicates the central domain boundary. (**b**) The same growth cone is shown shortly after an apCAM-coated bead is positioned in the peripheral domain close to the leading edge and restrained with a microneedle against retrograde translocation. This image is taken at the beginning of latency phase during which the coupling between apCAM and F-actin is weak. (**c**) As coupling strength increases, the growth cone undergoes rearrangements, such as central domain extension toward the bead and leading edge growth in front of the bead. Scale bar: 10 μm.

cone axis perpendicular to the microneedle provides the best control of the bead during the RBI assay (Fig. 4).

12. Set up the program for time-lapse imaging. Using the microneedle, move the bead onto the leading edge of the growth cone (Fig. 4b). If the protein on the bead binds to cell surface receptors that can couple to the actin flow, the bead will undergo retrograde translocation. Position the microneedle tip behind the bead to block its movement and start the time-lapse program taking images every 10 s (see Note 20). Replace medium every 30 min or set up a continuous perfusion system because of evaporation.
13. During the initial phase of bead restraint, called "latency" phase (Fig. 4b), there are hardly any structural changes that can be observed; during the later phase, termed "interaction" or "traction" phase (Fig. 4c), the central domain including microtubules extend toward the restrained bead (12, 14). This is accompanied by attenuation of retrograde actin flow, tension increase, and protrusion of the leading edge specifically along the growth cone–bead axis (Fig. 4c). These cytosolic and cytoskeletal rearrangements mimic events observed during growth cone interactions with cellular targets. Parameters to analyze are the percentage of successful bead interactions and the latency time. A high percentage of interactions and short latency time indicate a favorable growth condition. This assay can be combined with live cell imaging of fluorescently labeled proteins, such as actin/microtubule FSM (12), as well as immunocytochemistry (14).

4. Notes

1. Use L15-ASW within a month or add additional glutamine after 1 month because of its degradation in aqueous solutions.
2. We have used the cytoskeletal probes mentioned in Subheading 2.2; however, both Cytoskeleton Inc. and Invitrogen cell tubulin, actin, and phalloidin conjugated to different fluorochromes as well. Alternatively, tubulin and actin can be purified and labeled in the lab (30). Stock solutions of labeled tubulin and G-actin are prepared in the buffers described in Subheading 2.2, and small aliquots (1–5 μl) are stored at −80°C after drop-freezing in liquid nitrogen. An aliquot is typically used only once but could be refrozen once and used a second time.
3. Inking and body wall muscle contractions can make the dissection more difficult. Both can be avoided most of the time by gentle handling of the animal and firm injection procedure. Avoid getting ink into the body cavity while dissecting the animal. Make sure to penetrate the needle through the body wall muscle into the cavity without damaging internal organs.
4. The activity of the dispase, the time of digestion, and temperature will determine how effectively the tissue is digested. The goal is to loosen the fibrous connective tissue sheaths without killing many neurons. The optimal amount of enzyme depends on its specific activity and needs to be established for each batch. We use a heating water bath placed in a cold room to maintain 22°C.
5. Acid cleaning of coverslips can be done ahead of time. Incubate coverslips in 6 N HCl for at least 1 h, followed by thorough washing with H_2O ultrapure. Make sure that coverslips do not stick together during the acid-incubation. After extensive washing with water transfer the coverslips with forceps into a new container with 100% EtOH for sterilization and storage at RT.
6. Poly-L-lysine is a non-physiological adhesive substrate that works excellent for plating *Aplysia* neurons. Growth cones are fan-shaped and neurites grow at a rate of 5–10 μm/h on this substrate (Fig. 3). Poly-lysine-coated coverslips can also be stored dry.
7. Keep the bag cell clusters and cells within the medium during the whole dissection procedure. Avoid air bubbles attaching to the cluster as they may lift up the cluster to the medium/air interface which will damage the cells.

8. To fit the yellow pipette tip onto the 1 ml syringe, use a razor blade to cut off the wider end of the tip. Make sure the tip is tightly attached before filling it with grease.
9. The assembly needs to be done with the coverslips fully covered with medium. Avoid getting the cells exposed to air. Make sure to press the coverslips firmly against each other without breaking them. It is okay if the vacuum grease is pushed outside the spacers. This indicates a well-sealed chamber.
10. It is essential that the two coverslips are well aligned with each other as well as with the groove of the aluminum bottom part before the top part is mounted. Uneven positioning will generate local pressure after applying the clips and result in breaking the coverslips.
11. Light intensity can be controlled by the lamp voltage, neutral density, and color filters. For DIC live cell imaging of growth cones, we often use a 590 nm cut off filter to avoid the shorter wavelength light spectrum. Make sure to shut off the illumination between taking images either by using a transillumination shutter or by reducing the lamp voltage.
12. Clogging of the injection needle is one of the biggest challenges in performing microinjections. Samples should be spun before injection. If clogging persists, try the following: (a) spin sample at higher g forces; (b) coat needles with hexamethyldisilazane (HMDS, Pierce Chemical) by placing a few drops of volatile HMDS on the bottom of a needle-holding box and allowing the vapor to coat needles overnight in a fume hood; (c) clear out clog by increasing the pressure of the microinjection system.
13. Filling about 1 cm of the narrow microloader tip provides a sufficient amount for injecting several cells. Make sure not to touch the bottom of the tube where protein aggregates could have accumulated. Gently tap needle to remove any potential air bubbles from the solution in the microneedle. Alternative to pulling injection needles: commercial injection capillaries such as Femtotips (Eppendorf). Alternative to microloader tips: long and thin pulled glass capillaries.
14. The key to successful injections is using minimal pressure to yield a high percentage of injected cells without killing them. As the probability of clogging increases with time, it is important to perform the injections as quickly as possible to avoid the need for continuously increasing the pressure or frequently exchanging the needle.
15. To minimize phototoxicity use at least one neutral density filter and a minimally open field diaphragm in the fluorescent light path.

16. Select clearly labeled filaments for quantitative speckle analysis. For example, an ideal microtubule has clear speckles against background, does not cross over with other microtubules or ruffling structures containing a lot of free tubulin, and preferably has distinct growth/collapse phases.
17. This protocol describes non-covalent coupling of recombinant 6His-tagged protein to Ni-NTA silica beads. There are many other ways of coupling proteins to different types of microbeads in oriented and non-oriented manner, covalently and non-covalently.
18. Always store the beads (1% w/v) in ligand solution (such apCAM in PBS) at 4°C. Otherwise the density of the protein on the beads will gradually decrease over time since the 6His-Ni-NTA binding has an on- and off-rate.
19. We found that a non-motorized, hydraulic micromanipulator works best for well-controlled bead manipulations without any vibrations.
20. The RBI assay is vibration-sensitive. It is critical to minimize any vibrations by using an optical table, detaching shutters from the microscope body, and/or controlling the transillumination via the lamp voltage instead of a shutter.

Acknowledgments

I would like to thank Aih Cheun Lee for contributing the images for Figs. 3 and 4. I am also grateful to Aih Cheun Lee and Yingpei He for excellent suggestions on this chapter. Work in the Suter lab is supported by grants from the NIH (R01 NS049233) and the Bindley Bioscience Center at Purdue University to D.M.S.

References

1. Bashaw, G. J., and Klein, R. (2010) Signaling from axon guidance receptors, *Cold Spring Harb Perspect Biol* **2**, a001941.
2. Huber, A. B., Kolodkin, A. L., Ginty, D. D., and Cloutier, J. F. (2003) Signaling at the growth cone: ligand-receptor complexes and the control of axon growth and guidance, *Annu Rev Neurosci* **26**, 509–563.
3. Lowery, L. A., and Van Vactor, D. (2009) The trip of the tip: understanding the growth cone machinery, *Nat Rev Mol Cell Biol* **10**, 332–343.
4. Gomez, T. M., Harrigan, D., Henley, J., and Robles, E. (2003) Working with Xenopus spinal neurons in live cell culture, *Methods Cell Biol* **71**, 129–156.
5. Leung, K. M., and Holt, C. E. (2008) Live visualization of protein synthesis in axonal growth cones by microinjection of photoconvertible Kaede into Xenopus embryos, *Nat Protoc* **3**, 1318–1327.
6. Gallo, G., Lefcort, F. B., and Letourneau, P. C. (1997) The trkA receptor mediates growth cone turning toward a localized source of nerve growth factor, *J Neurosci* **17**, 5445–5454.
7. Lewis, A. K., and Bridgman, P. C. (1992) Nerve growth cone lamellipodia contain two populations of actin filaments that differ in organization and polarity, *J. Cell Biol.* **119**, 1219–1243.

8. Goslin, K., and Banker, G. (1989) Experimental observations on the development of polarity by hippocampal neurons in culture, *J Cell Biol* **108**, 1507–1516.
9. Forscher, P., Kaczmarek, L. K., Buchanan, J. A., and Smith, S. J. (1987) Cyclic AMP induces changes in distribution and transport of organelles within growth cones of Aplysia bag cell neurons, *J Neurosci* 7, 3600–3611.
10. Lee, A. C., Decourt, B., and Suter, D. (2008) Neuronal cell cultures from aplysia for high-resolution imaging of growth cones, *J Vis Exp.*
11. Waterman-Storer, C. M., Desai, A., Bulinski, J. C., and Salmon, E. D. (1998) Fluorescent speckle microscopy, a method to visualize the dynamics of protein assemblies in living cells, *Curr Biol* **8**, 1227–1230.
12. Lee, A. C., and Suter, D. M. (2008) Quantitative analysis of microtubule dynamics during adhesion-mediated growth cone guidance, *Dev Neurobiol* **68**, 1363–1377.
13. Schaefer, A. W., Kabir, N., and Forscher, P. (2002) Filopodia and actin arcs guide the assembly and transport of two populations of microtubules with unique dynamic parameters in neuronal growth cones, *J Cell Biol* **158**, 139–152.
14. Suter, D. M., Errante, L. D., Belotserkovsky, V., and Forscher, P. (1998) The Ig superfamily cell adhesion molecule, apCAM, mediates growth cone steering by substrate-cytoskeletal coupling, *J Cell Biol* **141**, 227–240.
15. Xiong, Y., Lee, A. C., Suter, D. M., and Lee, G. U. (2009) Topography and nanomechanics of live neuronal growth cones analyzed by atomic force microscopy, *Biophys J* **96**, 5060–5072.
16. Lohof, A. M., Quillan, M., Dan, Y., and Poo, M. M. (1992) Asymmetric modulation of cytosolic cAMP activity induces growth cone turning, *J Neurosci* **12**, 1253–1261.
17. Zheng, J. Q., Felder, M., Connor, J. A., and Poo, M. M. (1994) Turning of nerve growth cones induced by neurotransmitters, *Nature* **368**, 140–144.
18. Halfter, W. (1996) The behavior of optic axons on substrate gradients of retinal basal lamina proteins and merosin, *J Neurosci* **16**, 4389–4401.
19. Mai, J., Fok, L., Gao, H., Zhang, X., and Poo, M. M. (2009) Axon initiation and growth cone turning on bound protein gradients, *J Neurosci* **29**, 7450–7458.
20. von Philipsborn, A. C., Lang, S., Loeschinger, J., Bernard, A., David, C., Lehnert, D., Bonhoeffer, F., and Bastmeyer, M. (2006) Growth cone navigation in substrate-bound ephrin gradients, *Development* **133**, 2487–2495.
21. Dertinger, S. K., Jiang, X., Li, Z., Murthy, V. N., and Whitesides, G. M. (2002) Gradients of substrate-bound laminin orient axonal specification of neurons, *Proc Natl Acad Sci USA* **99**, 12542–12547.
22. Adams, D. N., Kao, E. Y., Hypolite, C. L., Distefano, M. D., Hu, W. S., and Letourneau, P. C. (2005) Growth cones turn and migrate up an immobilized gradient of the laminin IKVAV peptide, *J Neurobiol* **62**, 134–147.
23. von Philipsborn, A. C., Lang, S., Bernard, A., Loeschinger, J., David, C., Lehnert, D., Bastmeyer, M., and Bonhoeffer, F. (2006) Microcontact printing of axon guidance molecules for generation of graded patterns, *Nat Protoc* **1**, 1322–1328.
24. Romanova, E. V., Fosser, K. A., Rubakhin, S. S., Nuzzo, R. G., and Sweedler, J. V. (2004) Engineering the morphology and electrophysiological parameters of cultured neurons by microfluidic surface patterning, *FASEB J* **18**, 1267–1269.
25. Turney, S. G., and Bridgman, P. C. (2005) Laminin stimulates and guides axonal outgrowth via growth cone myosin II activity, *Nat Neurosci* **8**, 717–719.
26. Walter, J., Kern-Veits, B., Huf, J., Stolze, B., and Bonhoeffer, F. (1987) Recognition of position-specific properties of tectal cell membranes by retinal axons in vitro, *Development* **101**, 685–696.
27. Martin, K. C., Casadio, A., Zhu, H., Yaping, E., Rose, J. C., Chen, M., Bailey, C. H., and Kandel, E. R. (1997) Synapse-specific, long-term facilitation of aplysia sensory to motor synapses: a function for local protein synthesis in memory storage, *Cell* **91**, 927–938.
28. Zhao, Y., Wang, D. O., and Martin, K. C. (2009) Preparation of Aplysia sensory-motor neuronal cell cultures, *J Vis Exp.*
29. Kandel, E. R. (1976) *Cellular basis of behavior*, W. H. Freeman and Company, San Francisco.
30. Waterman-Storer, C. (2002) Fluorescent speckle microscopy (FSM) of microtubules and actin in living cells, *Curr Protoc Cell Biol* Chapter 4, Unit 4 10.

Chapter 7

Boyden Chamber

Marco Falasca, Claudio Raimondi, and Tania Maffucci

Abstract

The Boyden chamber, initially designed to study leukocyte chemotaxis, has become one of the most used tools to assess cell motility and invasion. The classical Boyden chamber consists of two compartments separated by a membrane representing a physical barrier that cells can overcome only by active migration. Since its initial introduction, a number of different Boyden chamber devices have been developed. The Boyden chamber can be adapted to study tumour cells' invasive properties by coating the membrane with different extracellular matrix proteins. The method described in this chapter is intended specifically for measuring the migration or invasion of human endothelial and cancer cells.

Key words: Cell migration, Cell invasion, Chemotaxis, Chemotactic agent, Endothelial cells, Cancer cells

1. Introduction

The ability of cells to migrate and to move towards a chemoattractant is critical for many physiological conditions such as embryogenesis, wound healing, and during immune system response (1). Cell migration and invasion also play a very important role in the dissemination of cancer cells during the metastatic spread (2, 3). In this process malignant cells are able to escape from the primary tumour through the circulatory or lymphatic system, invading distant healthy tissues to initiate a new malignant lesion called metastasis. The mechanisms involved in the regulation of cell motility are the results of an intricate network of signals activated by different growth factors, chemoattractants, and extracellular matrix components in vivo (4).

The Boyden chamber is a powerful tool to study this process in a simplified condition in vitro. It consists of two chambers separated by a microporous membrane (Fig. 1a). Cells are seeded in

Claire M. Wells and Maddy Parsons (eds.), *Cell Migration: Developmental Methods and Protocols*,
Methods in Molecular Biology, vol. 769, DOI 10.1007/978-1-61779-207-6_7, © Springer Science+Business Media, LLC 2011

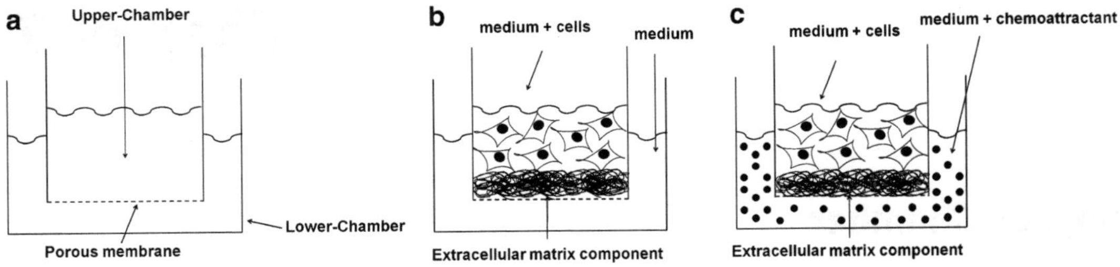

Fig. 1. Schematic representation of Transwell® chambers. (**a**) Transwell® chambers consist of a *lower* and an *upper chamber* separated by a porous membrane. (**b**) When testing migration induced by an extracellular matrix component only, the porous membrane is coated with the matrix and the lower chamber is filled with medium (usually serum-free medium). Cells are plated in the *upper chamber*. (**c**) When testing migration induced by extracellular matrix component and in the presence of a chemoattractant, the porous membrane is coated with the matrix and the lower chamber is filled with a medium containing the chemoattractant. Cells are plated in the *upper chamber.*

the upper chamber and let to migrate toward chemotactic agents present in the bottom chamber media (such as growth factors) for an appropriate incubation time (Fig. 1c). Alternatively, Boyden chamber can be used to study migration induced by activation of signalling pathways by a specific extracellular matrix component that is used to coat the porous membrane (Fig. 1b). Since the cells migrate from the upper to the lower compartment through a porous membrane, the Boyden chambers are also called transwell® chambers. Once the cells pass through the membrane, they adhere to it and can be fixed and stained for quantification by manual counting.

2. Materials

2.1. Materials Needed Two Days Before the Experiment

1. *Plastics.* Six well plate, sterile pipettes.
2. *Solutions.* Phosphate buffered saline solution without Ca^{2+} and Mg^{2+} (PBS), Trypsin solution (PBS containing 20 mg/ml trypsin and 1 mM EDTA).
3. *Media.* Usually cancer cells grow in DMEM or RPMI supplemented with 10% FBS and penicillin/streptomycin; human umbilical vein endothelial cells (HUVEC) can be grown in specific endothelial media, for instance EBM supplemented with 10% FBS and SingleQuots (5).

2.2. Materials Needed the Day Before the Experiment

1. *Plastics.* 24 well plate, sterile pipettes.
2. *Media.* Serum deprivation is usually needed to maximally stimulate cell migration. The length of starvation and the starvation media depend on cell type and must be optimised in preliminary experiments. Most (but not all) cancer cells can be starved overnight in the same media used to culture them but in the absence of serum. Alternatively, if complete

serum deprivation induces apoptosis/death of the tested cell line, 0.5 or 1% FBS can be added to the starvation media. HUVEC can be starved overnight in EBM or M199 supplemented with 0.5% FBS.

3. *Transwell® inserts.* Polycarbonate inserts exist in different diameters and in different pore sizes, and the appropriate insert must be chosen based on the cell line used (Note 1).
4. *Coating solutions.* Although some cells are able to migrate even in the absence of coating, usually Transwell® inserts need to be coated with a specific extracellular matrix component to allow migration. The appropriate coating solution depends on the cell type and the type of migration to be tested, and it must be optimised in preliminary experiments. In this respect it is important to underline that extracellular matrix components are not inactive compounds. On the contrary, they can activate several signalling molecules (in particular integrins) which in turn can activate signalling pathways including signals involved in cell migration (6, 7). Indeed, as discussed before, Transwell® can be used to determine migration induced by the extracellular matrix component alone, in the absence of an additional chemoattractant in the lower chamber (4). If this is the case, the choice of the coating compound depends on the integrin signalling to be studied. On the other hand, when the extracellular matrix component is added only as a coating solution and the aim of the experiment is to determine migration induced by a specific chemoattractant, it is preferable to choose as a coating compound an extracellular matrix component that induces a limited migration in order to better appreciate migration induced by the chemoattractant. Examples of coating solutions include (but are not limited to) fibronectin (10 μg/ml in PBS), collagen IV (10–100 μg/ml in PBS), and gelatin (10–100 μg/ml in PBS).

2.3. Materials Needed on the Day of the Experiment

1. *Plastics.* 24 well plate, sterile pipettes.
2. *Blocking media.* Starvation media supplemented with 0.5–1% BSA.
3. *Resuspending media.* Cells to be plated on Transwell® inserts must be detached and resuspended in the appropriate media, according to the experiment to be performed. Usually cells are resuspended in blocking media. When the aim of the experiment is to test the effect of a particular inhibitor on migration, cells are resuspended in blocking media in the presence or absence of the inhibitor (to compare migration of untreated cells versus cells treated with the inhibitor).
4. *Fixing/staining solutions.* Paraformaldehyde (4% w/v solution in PBS); Cristal Violet (0.1% w/v solution in water).

3. Methods

The following method specifically refers to the investigation of the effect of two distinct inhibitors on migration induced by a specific chemoattractant. A very similar protocol can be used to compare the effect of migration in cells expressing a control scrambled siRNA versus migration of cells transfected with two distinct siRNAs targeting a protein of interest. The procedure can be easily adapted to different experimental investigations.

3.1. Day 1

On day 1 the appropriate number of cells is seeded in the chosen plate. In our experience, cells detached from two nearly confluent wells of a six well plate are an appropriate number to perform the experiment. Therefore on day 1 we normally plate cells on a six well plate.

1. Detach cells from petri dish or flask by washing with 5 ml of PBS and incubating with trypsin solution for few minutes. Stop trypsin with media, collect and pellet cells (440 × *g*, 5 min). Resuspend cells in culturing media.
2. Plate cells in a six well plate. The number of wells that must be plated depends on the experimental design. For instance, in the specific method described here (investigation of the effect of two distinct inhibitors on migration induced by a specific chemoattractant), a full six well plate must be plated. Indeed, on day 3, cells will be pre-treated with the inhibitors before being detached and plated on the Transwell® inserts (i.e. two wells will stay untreated, two will be pre-treated with inhibitor 1 and two will be pre-treated with inhibitor 2 before being detached and plated on the Transwell® inserts, as shown in the scheme below).

Untreated	Pre-treated with inhibitor 1	Pre-treated with inhibitor 2
Untreated	Pre-treated with inhibitor 1	Pre-treated with inhibitor 2

3.2. Day 2 (Evening)

On day 2 cells are serum deprived and Transwell® inserts are coated with the appropriate coating solution. The number of Transwell® inserts to be coated depends on the experimental conditions. At least duplicate inserts are needed for each experimental condition. In our example, 12 inserts are required (untreated cells ± chemoattractant, treated with inhibitor 1 ± chemoattractant, treated with inhibitor 2 ± chemoattractant).

3.2.1. Cells

1. Wash six well plate with PBS and add starvation media. Incubate overnight.

3.2.2. Inserts

Inserts are coated overnight by positioning them on a 24 well plate (one insert per well: the insert represents the upper chamber, the well the lower chamber) containing an appropriate volume of coating solution. Specifically, add 500 μl of coating solution to the appropriate number of wells of a 24 well plate.

1. Add Transwell® inserts on each coated well of the 24 well plate. Handle the inserts with clean forceps and do not touch it directly. Make sure that there are no bubbles between the lower surface of the membrane and the coating solution since an uneven coating may result in non homogeneous migration and may compromise the final result. Seal the 24 well plate with parafilm and incubate overnight at 4°C. Alternatively, coating can be performed for 1–2 h at 37°C on Day 3 (see below).

3.3. Day 3 (Morning)

On day 3 migration assay is performed. The first step consists of pre-treating the cells with the inhibitors to be tested, detaching them from the six well plate, counting them, and plating them on the inserts. Then cells are left to migrate for the appropriate time (optimised on the specific cell line tested and chemoattractant used) and inserts are then fixed, stained, and analysed.

3.3.1. Inserts

1. Add 500 μl of blocking solution to the appropriate number of wells of a 24 well plate.
   (a) Transfer Transwell inserts from the coating solution to wells containing the blocking solution. Incubate for at least 1 h at 37°C.

3.3.2. Preparation of Cells

1. In our example (investigation of the effect of two distinct inhibitors on migration induced by a specific chemoattractant), incubate cells in the six well plate with starvation media in the presence or absence of the tested inhibitor(s) for 15–30 min (as required) before detaching (step 2).
2. Detach cells from six well plate as in step 1, day 1. In this step it is important to stop trypsin action with blocking media (i.e. media with no serum), as the addition of serum can activate signalling pathways involved in cell migration and therefore affect the final result.
3. Resuspend cell pellets in appropriate media (blocking media with or without inhibitor(s)). In our experience, for a pellet of endothelial cells detached from two wells of a six well plate, 800 μl of media is an appropriate volume.

4. Count resuspended cells using the haemocytometer. The number of cells to be plated on the Transwell® inserts must be optimised based on cell types, coating, and chemoattractant. Preliminary experiments should be performed to optimise these conditions in order to choose the right experimental conditions. Typically, an average of ~100 cells/field is a reasonable number of cells to be counted in a Transwell® after migration. Higher numbers can be difficult to be counted (high possibility of errors in counting), lower numbers can be non-significant. In our experience, to obtain this number of migrated cells, 15,000–20,000 HUVEC cells must be plated (this is valid for cells plated on 10 mm Transwell insert, 8 μm pore coated with gelatin, and for migration induced by fibroblast growth factor or vascular endothelial growth factor, refs. 5, 8). However, if HUVEC are plated on collagen, a lower number of cells must be plated since the collagen-induced basal migration is higher than migration induced by gelatin. In the case of lysophosphatidic acid-induced migration of ovarian cancer cells SKOV-3 and cervical cancer cells HeLa 30,000 cells/insert is an appropriate number (9). A volume of 100–150 μl of cell suspension has to be plated on each insert (in the case of a 10 mm Transwell® insert, 8 μm pore). A lower volume is not sufficient to cover the Transwell® and a higher volume does not allow cells to attach on the membrane (i.e. cells float on top of the insert). Once cells are resuspended and cell number is counted, the most accurate way to proceed is to adjust the volume of the pellet so that it contains the exact number of cells in 150 μl (method 1). However, this is not always possible, in particular when a limited amount of compounds to be tested (for instance inhibitors) is available (it must be remembered that cells are resuspended in media containing the inhibitor). In this case, transfer the precise volume of cell suspension corresponding to the required number of cells into a sterile eppendorf tube and adjust the final volume to 150 μl with blocking media ± inhibitor(s) (method 2). In this latter case, it is important to resuspend the cells carefully during this operation, and it is advisable to have two separate eppendorf tubes for duplicate Transwell® in order to minimise errors. Attention must be paid to this part because errors in pipetting and non-homogeneous resuspension can highly affect the final result. When possible, it is advisable to use method 1 since method 2 requires experienced operator to avoid errors.

3.3.3. Start Migration

1. Transfer the inserts from the wells containing the blocking solutions to wells containing the different "migration media,"

according to the experimental conditions to be tested. In our example the inserts scheme will be as below:

	1	2	3	4	5	6
A	Blocking media no inhibitor	Blocking media no inhibitor + chemoattractant	Blocking media plus inhibitor 1	Blocking media plus inhibitor 1 + chemoattractant	Blocking media plus inhibitor 2	Blocking media plus inhibitor 2 + chemoattractant
B	Blocking media no inhibitor	Blocking media no inhibitor + chemoattractant	Blocking media plus inhibitor 1	Blocking media plus inhibitor 1 + chemoattractant	Blocking media plus inhibitor 2	Blocking media plus inhibitor 2 + chemoattractant

Inserts on columns 1, 3, and 5 will determine the basal migration in the absence of chemoattractant and in the absence or presence of the tested inhibitors; inserts on columns 2, 4, and 6 will determine migration in the presence of chemoattractant in the absence or presence of the inhibitors. Put the Transwell inserts into the wells and plate cells carefully on top of the inserts. Cells need to be resuspended very well and plated homogeneously on the inserts (Note 2).

2. Leave in the incubator for the appropriate time (in our experience, 4 h is sufficient for migration of most cancer and endothelial cells).

3.3.4. Stop Migration

1. Wash the inserts once by transferring them on wells of a 24 well plate containing 500 μl of PBS.
2. Transfer inserts into empty wells and gently remove non-migrated cells using cotton swab (Note 3).
3. Transfer inserts into wells containing 500 μl of paraformaldehyde. Leave it for 30 min to 1 h.
4. Wash inserts once by transferring them on wells of a 24 well plate containing 500 μl of PBS.
5. Transfer inserts into wells containing 500 μl of crystal violet solution. Leave it for 30 min to 1 h.
6. Wash extensively by repeatedly transferring the inserts into wells of a 24 well plate containing 500 μl of PBS. Washing is critical because cells need to be clearly defined and inserts must be clear of any debris or crystals from the staining solution. On the other hand, too many washes can lead to a faint staining of the cells which makes the counting step difficult and more prone to error.

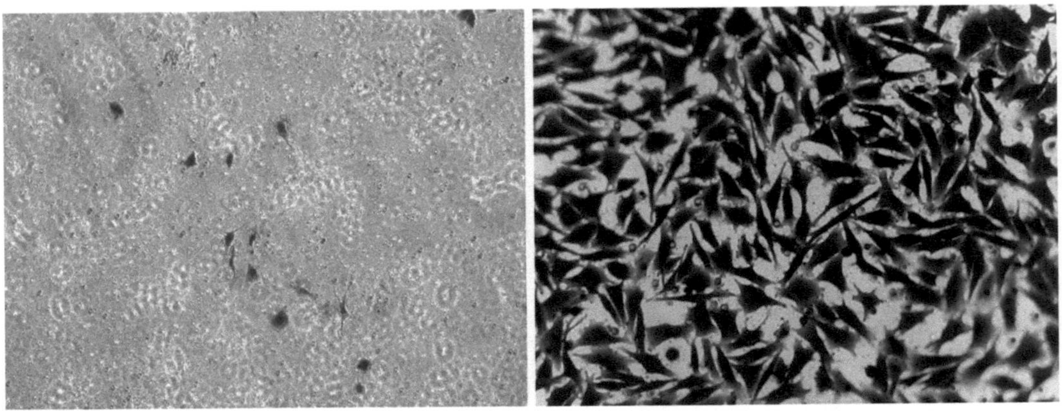

Fig. 2. Representative phase-contrast images of migrated MDA-MB-231 plated on Transwell® chambers uncoated (*left panel*) or coated with fibronectin (*right panel*) upon staining with crystal violet.

Typical images of breast cancer cells in unstimulated (left) and stimulated (right) conditions are shown in Fig. 2. Cells are shown as large violet/purple bodies (in the case of big cells, it is possible to distinguish the cell shape). The black spots are the membrane pores.

7. Count the number of cells/field using a 10× magnification. Since inserts do not possess grids, counting must be performed very carefully. A complete homogeneous distribution of the cells within the insert is difficult to achieve (especially to less experienced operators) and therefore it is advisable to count as many fields as possible (ideally cells in the whole insert). Attention must be paid to avoid counting cells in a field more than once and not to overlap fields (some guide can be provided by the edge of the insert).

4. Notes

1. The size of the membrane porous is a critical choice for the analysis. Pores do not have to be too large (to avoid that cells can pass by gravity alone) or too tight (to avoid that cells squeeze through the pore of the membrane). A right dimension has to be found according to the cell type:

   12–8 μm: epithelial and endothelial cells.

   5 μm: leukocytes, monocytes, dendritic cells, lymphoblast.

   3 μm: neutrophils and eosinophils.

2. Seeding of the cells on the transwell insert is of crucial importance. A non-homogeneous seeding will lead to non-homogeneous migration and will affect the quantification of the

assays since it will result in over-confluent fields and empty fields within the same insert.

3. Step 2 of the Subheading 3.3.4 is important to remove the non-migrating cells from the chamber before fixing the cells. This step has to be performed gently in order to prevent breaking the membrane, but it has to be strong enough to remove all the cells in the upper chamber.

References

1. Ridley AJ, Schwartz MA, Burridge K, Firtel RA, Ginsberg MH, Borisy G, Parsons JT, Horwitz AR. (2003) Cell migration: integrating signals from front to back. *Science* **302**:1704–9
2. Kassis J, Lauffenburger DA, Turner T, Wells A. (2001) Tumor invasion as dysregulated cell motility. *Semin Cancer Biol.* **11**:105–17
3. Sala G, Dituri F, Raimondi C, Previdi S, Maffucci T, Mazzoletti M, Rossi C, Iezzi M, Lattanzio R, Piantelli M, Iacobelli S, Broggini M, Falasca M. (2008) Phospholipase Cgamma1 is required for metastasis development and progression. *Cancer Res.* **68**:10187–96.
4. Howard J (2009) Mechanical signaling in networks of motor and cytoskeletal proteins. *Annu Rev Biophys.* **38**:217–34.
5. Maffucci T, Raimondi C, Abu-Hayyeh S, Dominguez V, Sala G, Zachary I, Falasca M. (2009) A phosphoinositide 3-kinase/phospholipase Cgamma1 pathway regulates fibroblast growth factor-induced capillary tube formation. *PLoS One.* **4**:e8285.
6. Albini A, Iwamoto Y, Kleinman HK, Martin GR, Aaronson SA, Kozlowski JM, McEwan RN. (1987) A rapid in vitro assay for quantitating the invasive potential of tumor cells. *Cancer Res.* **47**:3239–45.
7. Albini A, Benelli R. (2007) The chemoinvasion assay: a method to assess tumor and endothelial cell invasion and its modulation. *Nat Protoc.* **2**:504–11.
8. Maffucci T, Piccolo E, Cumashi A, Iezzi M, Riley AM, Saiardi A, Godage HY, Rossi C, Broggini M, Iacobelli S, Potter BV, Innocenti P, Falasca M. (2005) Inhibition of the phosphatidylinositol 3-kinase/Akt pathway by inositol pentakisphosphate results in antiangiogenic and antitumor effects. *Cancer Res.* **65**:8339–49.
9. Maffucci T, Cooke FT, Foster FM, Traer CJ, Fry MJ, Falasca M. (2005) Class II phosphoinositide 3-kinase defines a novel signaling pathway in cell migration. *J Cell Biol.* **169**:789–99.

Chapter 8

Transwell® Invasion Assays

John Marshall

Abstract

The need to identify inhibitors of cancer invasion has driven the development of quantitative in vitro invasion assays. The most common assays used are based on the original Boyden assay system. Today commercially available plastic inserts for multi-well plates, which possess a cell-permeable membrane, as typified by Transwell® Permeable Supports, permit accurate repeatable invasion assays. When placed in the well of a multi-well tissue culture plate these inserts create a two-chamber system separated by the cell-permeable membrane. To create an invasion assay the pores in the membrane are blocked with a gel composed of extracellular matrix that is meant to mimic the typical matrices that tumour cells encounter during the invasion process in vivo. By placing the cells on one side of the gel and a chemoattractant on the other side of the gel, invasion is determined by counting those cells that have traversed the cell-permeable membrane having invaded towards the higher concentration of chemoattractant. In this chapter, in addition to protocols for performing Transwell invasion assays, there is consideration of the limitations of current assay designs with regard to available matrices and the absence of tumour microenvironment cells.

Key words: Fluoroblok, Invasion, Transwell, Matrigel

1. Introduction

The principal reason for death from solid cancer is the ability of malignant cells to translocate from their site of origin to different tissues and establish secondary sites of growth – a process termed metastasis. However, this initial ability to invade, i.e. to pass forcibly through the dense surrounding extracellular-matrix of basement membranes and stromal compartments, is a major reason why metastasis is the principal reason for death from cancer (1–3). The requirement to develop effective strategies to inhibit cancer cell invasion has required the development of assays that permit quantitative and reproducible measurement of the invasive propensity of cancer cells. Such assays help to delineate the molecular basis of

Claire M. Wells and Maddy Parsons (eds.), *Cell Migration: Developmental Methods and Protocols*, Methods in Molecular Biology, vol. 769, DOI 10.1007/978-1-61779-207-6_8, © Springer Science+Business Media, LLC 2011

cancer cell invasion and thus identify those key targets for which therapeutic interventions may be developed and then to use such assays again to select the most effective inhibitors. However, as cancer cell invasion in vivo is such a complex process, involving dynamic interactions with cells of the microenvironment including angiogenic blood vessels, fibroblasts, and inflammatory cells (4–7), it is not possible to replicate, simultaneously, all of its complexities in simple in vitro assays. However it is possible to construct assays that investigate different aspects of the invasion process. Based on the original Boyden Assay (8) commercially available plastic inserts for multi-well plates, which possess a cell-permeable membrane, as typified by Transwell® Permeable Supports, permit accurate repeatable invasion assays. When placed in the well of a multi-well tissue culture plate these inserts create a two-chamber system separated by the cell-permeable membrane (see Fig. 1a). The 10 μm thick membrane has evenly distributed pores of pre-determined diameter (Fig. 1b). Pore size is chosen

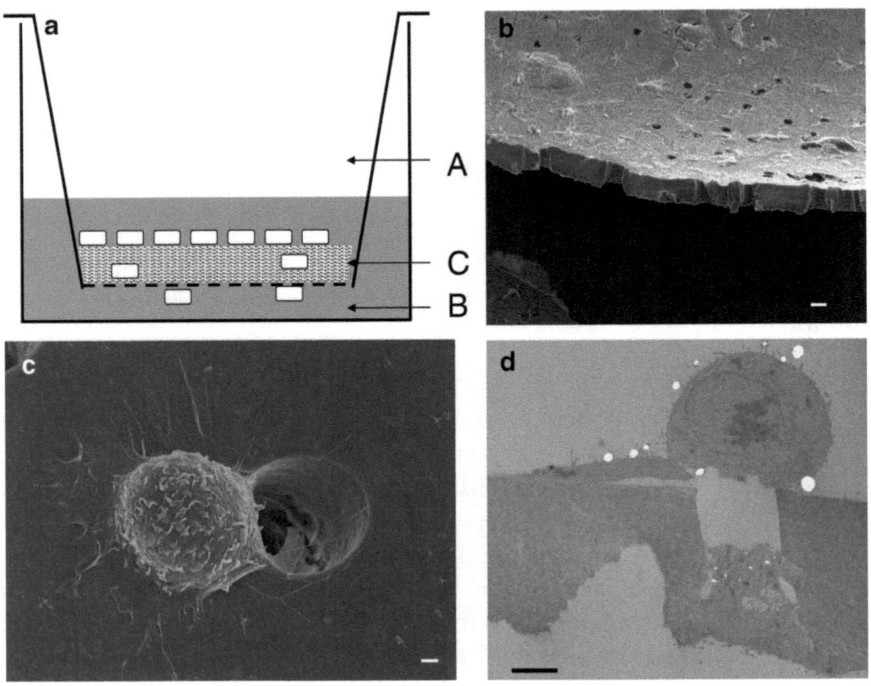

Fig. 1. (**a**) Diagram of a Transwell invasion chamber. An insert (A) with a semi-permeable membrane is suspended in a well (B). ECM gels (C) are formed in situ on the membrane. Medium with a chemotactic agent is placed in B and tumour cells are added and they invade towards the gradient. (**b**) Freeze fracture SEM image of an 8 μm pore-sized polycarbonate membrane. Squamous carcinoma cells (SCCs) were added in serum-free medium and allowed to migrate toward 10% FCS in the lower chamber (B). Magnification bar = 5 μm. (**c**) SEM image of an SCC cell in the process of migrating down through an 8 μm pore. Magnification bar = 1 μm. (**d**) TEM image of an SCC cell in the process of migrating down through an 8 μm pore. Note that the invasive pseudopod has split from the main body of the cell during processing. Note also that Magnification bar = 5 μm.

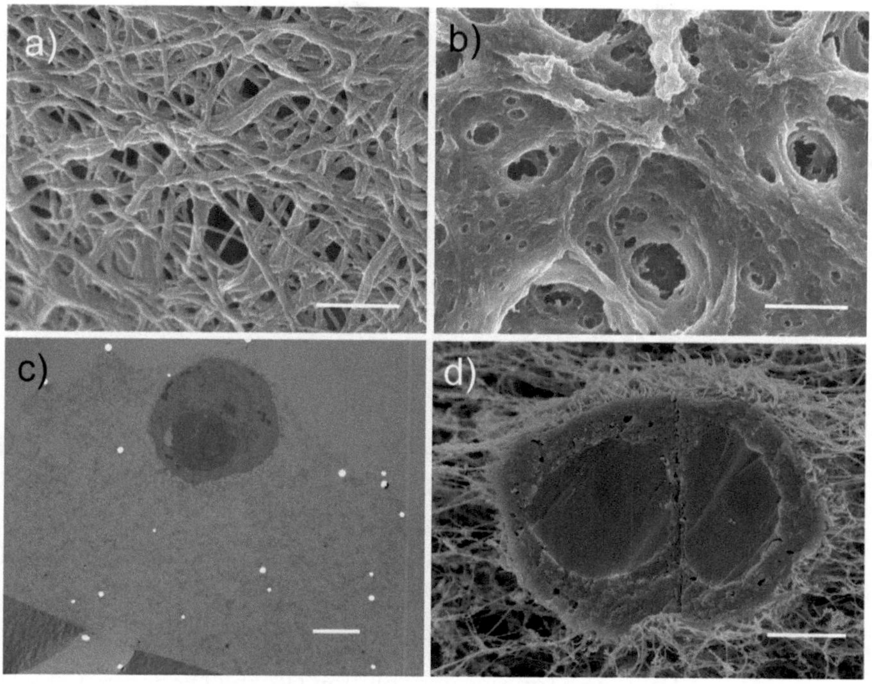

Fig. 2. (**a**) SEM image of a 100% type I collagen (2.5 mg/ml approximately) gel on a Transwell membrane. Note that inter-fibril spaces are much smaller than 1 μm. Magnification bar = 1 μm. (**b**) 100% Matrigel (3.5 mg/ml approximately) gel on a Transwell membrane. Note that the structure is not fibrillar but more gel-like. Magnification bar = 1 μm. (**c**) TEM picture of an SCC cell invading towards an 8 μm-diameter pore. The membrane was coated with 70 μl of a 50:50 type I collagen:Matrigel gel. Note the cell "sinks" into the gel which is more than 2 cell diameters thick. Magnification bar = 10 μm (**d**) Freeze fracture SEM image of two SCC cells invading a 50:50 type I collagen:Matrigel gel. Note that pores in the collagen are much too small to allow for protease-independent passage of cells. Magnification bar = 5 μm.

to be much smaller than the diameter of the target cell (Fig. 1c, d). To create an invasion assay the pores in the membrane are blocked with a gel composed of extracellular matrix that is meant to mimic the typical matrices that tumour cells encounter during the invasion process in vivo. (see Fig. 2a, b for images of collagen type 1 versus Matrigel). By placing the cells on one side of the gel and a chemoattractant on the other side of the gel, an invasive tumour cell will degrade the underlying adjacent matrix and migrate through the ECM gel (the combination of matrix degradation with, usually, directional motility = invasion). Quantitative invasion is determined by counting those cells that have traversed the cell-permeable membrane having invaded towards the higher concentration of the chemoattractant. The following methods are routinely used in our laboratory.

2. Materials

2.1. Transwell Assays with Unstained Cells

1. Matrigel® (Becton Dickinson, Oxford, UK; Cat No. 354234).
2. Acid-extracted rat-tail collagen type I.
3. 10× strength DMEM (Dulbecco's Modification of Eagle's Medium; Sigma Aldrich D2429) supplemented with $NaHCO_3$ at 3.7 g/L.
4. 1× DMEM (see Note 1).
5. Bovine foetal calf serum.
6. Transwell Permeable supports, 8 μm pore-sized, 6.5 mm diameter (Costar, Corning Inc., Cat No. 3422).
7. Casy Cell Counter and Analyser system (CASY-TTC, Roche Diagnostics, West Sussex, UK; Cat No 05651727001).
8. Trypsin (0.25%)/EDTA (0.04%).
9. 1× strength growth media e.g. DMEM.
10. CASY ton isotonic solution (Roche Diagnostics, 1 × 10 L, Cat No 05651808001).
11. CASY cups (Roche Diagnostics, 1 × 500 cups with lids, Cat No. 05650623001).
12. Millipore 0.2 μm Steritop filter unit.
13. Cell Recovery Solution (Becton Dickinson).

2.2. Transwells Assays with Stained Cells

1. Crystal Violet.
2. Millipore 0.45 μm syringe filter, 13 mm diameter.
3. 20 ml plastic disposable syringe.
4. Suitable plate spectrophotometer e.g. Molecular Devices SpectraMax 250 Microplate Reader (BC-MDSMX250) with 600 nm filter.
5. 96-well flat-bottomed Microplate.
6. Poly-L-lysine.
7. 1× Phosphate buffered saline.

2.3. Transwell Assay Using Fluorochrome Labelled Cells

1. Calcein-AM (Molecular Probes; cat no C3100).
2. Celltracker Green CMFDA (Molecular Probes; Cat No. C2925).
3. Suitable fluorescence plate reader e.g. FLx800 Fluorescence Microplate Reader (Biotek UK, Bedfordshire, UK).
4. Nunc 96 Microwell black plates, Cat No 237108 (Thermo Scientific, UK).

3. Methods

3.1. Transwell Matrigel Invasion Assay with Unstained Cells

1. Thaw the stock 10 ml Matrigel® overnight at 4°C by placing inside a refrigerator or cold room. Once thawed, always keep Matrigel chilled on wet ice. Thaw small aliquots under cold running water (see Notes 2–4).
2. Dilute Matrigel 1:1 with chilled serum-free growth medium just prior to coating (equivalent to approximately 5–6 mg protein/ml). To generate a homogenous mixture pipette up and down slowly, ensuring that no bubbles are introduced into the mixture.
3. 70 μl of the chilled diluted Matrigel is placed directly onto the centre (see Note 5) of the 24-well, 6.5 mm diameter, 8 μm pore-size Transwell insert (see Note 6).
4. Place plates in a humidified incubator at 37°C for 60 min to allow gelling.
5. Meanwhile, tumour cells are detached with trypsin/EDTA (0.25%/0.04%). After detachment trypsin is inactivated in 10% (v/v) serum-containing medium. Cells are then washed twice in serum-free growth medium by centrifugation (580 *g*, 3 min).
6. 50,000 cells (see Note 7) in 200 μl in serum-free growth medium are added to the upper chamber of replicate (3–4 wells minimum) and 500 μl of medium containing a chemoattractant added to the lower chamber (see Note 8).
7. The plate is incubated for 24–72 h at 37°C depending on the particular tumour cell line. Thus highly invasive cell lines will require a shorter incubation time in order to achieve sufficient invasion for quantification. Although the optimal time should be determined empirically routinely we assay new lines initially at 72 h and then adjust subsequently depending on invasive propensity.
8. Place 9 ml of 0.2 μm filtered isotonic solution (CASY ton) into CASY cups (plastic measuring tubes, one per insert) (see Note 9). Lid and tap the tubes and allow the tubes to sit for an hour so that any bubbles can rise to the top.
9. Take plate from the incubator. Manually, using a micropipette, remove and discard the medium in the upper chamber. Transfer 500 μl of medium in the lower chamber to a CASY cup.
10. The insert is gently washed to remove serum by dipping in PBS in a spare well then drained of excess fluid.
11. The insert is replaced in its original well and 500 μl of trypsin/EDTA (0.25%/0.04%) added to the lower chamber.

12. The plate is incubated for 1 h at 37°C (see Note 10).
13. Residual cells still attached to the membrane are removed by repeatedly pipetting the trypsin/EDTA solution across the underside of the membrane.
14. This detached suspension of cells is added to the appropriate CASY cup, giving a final volume of 10 ml (see Note 11).
15. The CASY cup is re-lidded. After harvesting the whole plate(s), immediately before measuring each sample, it is gently inverted two to three times but without generating bubbles to achieve an homogenous suspension (see Note 12).
16. Open and place the CASY cup on the CASY TTC (see Note 13) sample port. The CASY will determine the average counts/ml from three consecutive (200 μl or 400 μl) samples. Onboard software is adjustable for the dilution factor (×10) so that the total cell number is calculated.
17. Relative invasion can be determined using the equation: (Mean number of cells invaded/50,000) × 100. Standard deviation of invasion can be determined using the equation: (Standard deviation of number of cells invaded/Mean number of cells invaded) × Relative Invasion. Using these calculations it is then possible to compare treatment groups with control treatments (see Notes 14 and 15).

3.2. Transwell Collagen Invasion Assay with Unstained Cells

1. Into an ice-chilled tube is placed (1) 700 μl of acid-extracted rat-tail collagen, (2) 100 μl of 10× serum-free DMEM containing phenol red (Sigma-Aldrich Cat No. D2429), (3) 100 μl of fetal calf serum (FCS), and (4) 100 μl of 1× DMEM. Keep collagen solutions on wet-ice until used (see Notes 16 and 17).
2. Mix by pipetting, avoiding the generation of bubbles.
3. Add 10 μl of 0.1 M NaOH and re-mix. Keep adding similar volumes of 0.1 M NaOH until after mixing the colour remains a near neutral pH i.e. pink-orange.
4. Add 70 μl to the centre of the 6.5 mm diameter Transwell membrane.
5. Place in an humidified incubator 37°C for 60 min to allow gelling.
6. Follow the same procedure as described in Subheading 3.1 from step 5.

3.3. Transwell Invasion Assay with Stained Cells

1. Follow the procedures in Subheading 3.1 or 3.2 until removal of the plate after 24–72 h.
2. Manually remove and discard the medium from upper and lower chambers. Residual ECM gel in the upper chamber can also be pipetted off.

3. Place 500 μl of alcohol (100% methanol, 70% ethanol, 100% industrial methylated spirits are all fine) in the lower chamber to fix the cells for 15 min at ambient temperature.
4. Remove the insert to a spare well to air-dry.
5. Pick up the insert with your gloved fingers and remove all of the cells and the remaining gel material on the upper membrane surface with a dampened cotton bud. More than one cotton bud is usually required (see Note 18).
6. Prepare 0.5% (w/v) crystal violet in distilled water. Using a syringe and 0.45 μm filter, filter the solution. When stored at 4°C this solution is OK for 1 week.
7. Place 250 μl of 0.5% crystal violet into the lower chamber for 30′ at ambient temperature.
8. Drain off the excess dye and rinse in running tap water (dip inserts individually in a beaker of water) until no more dye is released. Allow to air-dry completely (a warming oven can speed this process) (see Note 19).
9. Elute dye in the well with 250 μl of 33% (v/v) acetic acid in distilled water in the lower chamber. Incubate at ambient temperature for 10 min shaking plate intermittently to release all the dye.
10. Drain and discard the insert. Transfer 100 μl of eluate to 96-well Microplates and measure optical density (OD) at 600 nm on a plate spectrophotometer.
11. Relative invasion can be determined by plotting mean OD. To transform OD values into cell numbers a standard curve must be established as follows.
12. Place 100 μl Poly-L-lysine (0.1 mg/ml PBS) for 30 min into quadruplicate wells of a clear 96-well Microplate.
13. Discard solution by flicking plate over a sink. Rinse wells with PBS and discard washings.
14. Add 250, 500, 1,000, 2,000, 4,000, 8,000, 16,000, 32,000 of cells in 100 μl of serum-free medium. Place at 37°C for 1 h.
15. Discard media and fix in 100 μl alcohol (100% methanol, 70% ethanol, or 100% industrial methylated spirits) for 10 min at ambient temperature.
16. Discard fixative and air-dry.
17. Add 100 μl of freshly prepared and filtered 0.5% crystal violet in PBS for 30 min.
18. Wash as above and elute dye in 100 μl of 33% acetic acid in water.
19. Plot OD versus cell number for a standard curve, subtracting background OD (i.e. from the no cell wells).

3.4. The Future of Cancer Cell Invasion Assays

It is now better understood that assays must get closer to the natural tumour microenvironment if the data from the in vitro systems are to be truly representative of the in vivo behaviour. I predict that commercially available invasion assays that incorporate stromal fibroblasts and endothelial cells will eventually become a reality. In fact some companies already supply this mixture of cells for measuring angiogenesis (TCS cellworks, Buckingham, UK; Angiokit Cat No ZHA-1000) and thus it is a short step to generate a similar system in a Transwell. Under these systems it will be necessary to prelabel the tumour cells to discriminate from the microenvironmental cells. Although very expensive now one hopes that in the future costs will reduce. Until then, the "simple" reductionist assays described above have much to offer but users must be aware of their limitations.

4. Notes

1. Use media appropriate for the cells under study. The 10× DMEM can be used for making collagen gels whichever media is used.
2. Essentially most invasion assays that incorporate an ECM equivalent "gelled" in the laboratory are attempting to mimic either the basement membrane or stromal collagen milieu. Commercial products for the basement membrane are all based on the secreted matrix of a single sarcoma cell line. Thus Matrigel® (aka EHS extract and Cultrex®) (9) is purified from Englebreth-Holm Swarm (EHS) sarcoma murine tumours and contains a matrix rich in collagen type IV, laminin, and perlecan. As this combination of glycoproteins is similar to that found in basement membranes (10), this has led to Matrigel® being considered a basement membrane substitute. However, users should be aware that, in addition to having a composition that differs significantly from natural basement membranes, the principal laminin chain is laminin $\alpha 1$ which is not found in all basement membranes (11, 12). In addition Matrigel gels lack the structural integrity of naturally occurring insoluble basement membranes (10) and thus extrapolation from data derived from its use as a BM substitute should be done so with these caveats in mind (13). Similarly, when modelling stromal collagens, naturally occurring type I collagen fibrils possess important inter-strand crosslinks that are removed by pepsin-extraction, but not acid-extraction methods (14, 15). The protease dependency for invasion through gels of pepsin-extracted collagen differs significantly from that through acid-extracted collagen gels or

naturally occurring basement membranes (15, 16). Thus, a source of plasminogen (from which plasmin is generated by the action of a plasminogen activator released by tumour cells e.g. uPA) is often required. As serum contains plasminogen, if it is the chemoattractant, it is usually not necessary to add exogenous plasminogen. Therefore when using commercial collagen preparations, choose the product that best suits your scientific question.

3. Matrigel arrives frozen and is stored at –20°C till used. Growth factor reduced Matrigel (BD Cat No. 354230) is also available. Note that this product still possesses some growth factors. Also, plates pre-coated with Matrigel (Becton Dickinson, Biocoat Matrigel® 354480) are also available. We prefer to coat our own to ensure sufficient thickness of coating (see Fig. 2c, d). Thus 70 μl of Matrigel gives a thickness of at least 70 μm (Fig. 2c).

4. Avoid using refrigerator doors for thawing as the temperature of their contents can rise rapidly on repeated opening resulting in premature gelling of the Matrigel. If rapid thawing is required holding the Matrigel by the bottleneck and thawing under running cold water with swirling is also possible. Take care not to hold the sides of the container as warmth from fingers can generate a point for gelling. Once thawed always keep Matrigel chilled on wet ice. Upon thawing we usually sterilely aliquot the Matrigel into 0.5 ml or 1.0 ml volumes and re-freeze at –20°C; this avoids accidental gelling before use.

5. Do not drain the mixture down the side of the well as the gel may not set evenly across the membrane. This can result in false high-invasion rates as tumour cells traverse the poorly coated parts of the membrane. A 70 μl coating of 4 mg/ml Matrigel creates a gel whose thickness, by TEM, is at least 2–3 cell diameters (Fig. 2c), a significant physical barrier for invasion.

6. Transwell® supports come in various formats to fit 6-, 12-, or 24-well plates and 10 cm Petri-dishes. Pore-sizes include 0.4, 3.0, 5.0, 8.0, and 12.0 μm (Fig. 1b, c). For tumour invasion assays we use the 24-well, 8 μm-pore-sized, polycarbonate plates (Costar, Corning Inc., Cat No. 3422). The diameter of the insert membrane is 6.5 mm, equivalent to 0.33 cm^2, and the density of the pores is $10^5/cm^2$ (see Fig. 1). The average diameter of cell lines in our lab is approximately 12–20 μm so the 8 μm pore size affords the best balance of allowing directed translocation through the membrane while minimising random motility through. For small diameter tumour cell lines (10 μm) or inflammatory cells use the 5 μm pore-sized membranes. In order that the assay validly be called an invasion

assay these pores must be covered with a suitable ECM (see Fig. 2c). Note that robot-ready Transwell® plates are now available, so it is possible to adapt the assays for high-throughput assays.

7. The cell number should be determined empirically. Test from 10,000–100,000 cells and choose the number that gives maximal percentage invasion compared with original number added (input).
8. In our experience it is always necessary to generate a gradient or tumour cells will not invade significantly (data not shown). Normal growth media supplemented with 10% foetal calf-serum is a useful standard chemoattractant. This provides both a gradient and also a source of plasminogen from which tumour cells can generate the protease plasmin, enhancing their invasive capacity (17). However, others have shown that gradients do not last for the whole 72-h period, and for large molecular weight proteins, perhaps the first 24–36 is the extent of the gradient (18).
9. Other isotonic solutions can be used including PBS.
10. The requirement for the extended incubation is to allow cells that still have cellular projections within the pores of the 10-μm thick membrane to withdraw these processes. See Fig. 8.1d.
11. Check that all the cells have been detached by checking a few wells on an inverted phase-contrast microscope. If a significant number remain, it is usually only necessary to continue squirting the trypsin-EDTA across the underside of the membrane until they are removed. If this still does not remove cells this may suggest your trypsin is of poor activity and you should repeat the assay with fresh active trypsin solution.
12. Micro-bubbles can be counted as "particles" on the CASY and give a false count. Large bubbles will block the machine.
13. The CASY® cell analysers permit very accurate cell counting of low (100 cells/ml) numbers of cells as well as many additional analyses (https://www.roche-applied-science.com/cfm/prodsrch-index.jsp?SearchTerm=CASY%20cups&showAll=true). For this reason we use the CASY® for analysing migration and invasion assays. Note that it is not possible to use a standard haemocytometer in place of the CASY as the total number of cells harvested per 6.5 mm diameter Transwell® is too low (usually between 1,000 and 30,000 cells). An advantage of the CASY is that ALL the invaded cells are counted rather than a sample, important if the sampling method is not random. However, as not all labs possess a CASY one of the alternative methods described above will almost certainly suit.

14. If significant amounts of proliferation take place during the invasion assay time-frame then the absolute invasion levels calculated will be affected. Thus the shorter the invasion assay time, the less likely that proliferation will impact on the interpretation of the data. However, although it is likely that the tumour cells in your assay will have proliferated if the assay was for 48 or 72 h, providing the amount of proliferation was similar in each treatment group, this still permits direct comparison of relative invasion data between treatment groups. Thus we usually plate the same number of cells (as in the invasion assay) into replicate wells of 24-well plates and give identical treatments (e.g. antibodies or drugs) as is being assessed in the assay. At the time of harvesting the invasion assay we also trypsin/EDTA detach and count the cells in these control wells thereby determining whether differences in proliferation are likely to affect interpretation of the invasion results. If there is a significant difference in proliferation then invasion should be expressed as a percentage of the total number of cells in each well at the end of the assay. Thus, although time-consuming, it is necessary and possible to harvest the cells from the upper chamber. Note it is likely that for highly invasive cells the Matrigel mostly will be degraded or gel fragments remain. Thus the task is to harvest cells in these fragments and those on the exposed membrane. Cells within Matrigel can be released with Cell Recovery Solution (Becton-Dickinson) which can depolymerise Matrigel at 2–8°C. Before trypsinising cells (Subheading 3.1, step 9), transfer the medium and any gel fragments from the upper chamber, using a 1 ml micro-pipette, into a chilled centrifuge tube (e.g. Falcon 2095) containing 500 μl of cold Cell Recovery Solution. Rinse the upper chamber manually with 200 μl of cold PBS and add to the tube. Add 250 μl of cold Cell Recovery Solution to the upper and lower chambers. After 30 min at 4°C remove the upper chamber contents to the tube. Check microscopically to see if cells remain attached to the membrane. If so these cells should be trypsinised at the same time as the lower chamber cells by adding 100 μl of trypsin/EDTA then added to the centrifuge tube. Spin down the cells (250 g; 1,200 rpm for 3′ in bench-top centrifuge) and count. If gel fragments remain, resuspend the pellet in 500 μl of fresh Cell Recovery Solution for an additional 30 min at 4°C, then centrifuge the released cells. Resuspend in 1 ml of serum-free, and assay 100 μl in 9.9 ml of CASYton. The invaded cells (in the lower chamber) can be assessed as described.
15. What constitutes the control varies depending on the experiment. For example, if a potential chemoattractant is being

tested, control wells have nil chemoattractant. If the ability of an inhibitor is being tested, all wells would have a chemoattractant (10% FCS is a cheap one) and the controls are wells without the inhibitor.

16. Adjust total collagen volume depending on the number of wells for coating. In some assays we utilise 50:50 collagen type I: Matrigel gels, i.e. 35 μl of collagen and 35 μl of Matrigel. Note that collagen gel porosity decreases and gel stiffness increases with increasing collagen concentration. In the collagen gel described, collagen is at approximately 2.5 mg/ml and the pore size (see Fig. 2a) is 0.5 μm or less. Thus it is essential that cells degrade the collagen for invasion to occur. Note that if tumour stromal fibroblasts were present (as in a tumour) they would contract the gel further, decreasing pore size and further increasing stiffness, similar to what occurs in most developing cancers.

17. We use the Becton Dickinson acid-extracted rat-tail collagen type I (Cat No. 354236). This arrives as a chilled liquid dissolved at or about 3.6 mg/ml of 20 mM acetic acid, but absolute concentration varies from batch to batch. This is stored at 4°C till used. High-concentration, acid-extracted rat-tail collagen (8–11 mg/ml) is also available (BD Cat No. 354249) if higher concentration collagen gels are required. A pepsin-extracted bovine dermal collagen type 1 called PureCol (formerly called Vitrogen; available from Advanced Biomedix, SD, USA) is available as a 3 mg/ml solution on 10 mM HCl.

18. The membrane is quite tough but can be split, so be careful. It is possible to ensure all cells have been removed by viewing on a simple inverted microscope.

19. It is possible to terminate the assay here and count cells manually under a microscope. Carefully, with a scalpel, cut out the filter from the Transwell. Place membrane, lower-surface upward, on a microscope slide (Fisher Scientific, Cat No 12-544-15) and mount in DPX mountant (Sigma Aldrich Cat No 44581-100ML) and a glass coverslip (VWR International, Cover glass superslips No1 thickness; cat No 631–0433). Count multiple random high-power fields.

5. Additional Notes

1. A variety of fluorescence-based methods are also used for labelling cells. For post-labelling cells (i.e. the plates are removed after the 24–72 h incubation) Calcein-AM (Molecular Probes; Cat No C3100MP; 4 μM for 1 h at 37°C in serum-free medium) is often used. If the cells are then

detached they can be placed in black 96-well plates and read on a "top-reading" fluorescence plate reader at 485 nm excitation/520 nm emission. This is an endpoint assay. If the cells are pre-labelled (e.g. Celltracker Green CMFDA; Molecular Probes; Cat No.C2925; 5-25 μm in growth medium for 15–45 min) then it is possible to do kinetic analysis by the use of BD Falcon™ 8 μm pore, individual FluoroBlok™ 24-well inserts. These are designed so the membranes of the inserts are impervious to light in the range 490–700 nm. Thus with a bottom-reading fluorescence plate reader the only fluorescence is that associated with cells. Thus the plate can be read repeatedly over time (provided the fluorescence plate reader can read 24-well plates) and get multiple readings during the period of invasion. Detailed protocols are available in Becton-Dickinson Technical Bulletins #442 and #441.

2. Another commonly used invasion assay allows cells to invade upwards. Briefly, in a laminar flow cabinet, the 24-well plate containing the Matrigel-coated Transwells is inverted and the base-plate removed. Tumour cells in 50–100 μl medium are plated onto the exposed underside of the membrane. The base-plate is replaced and the plate is incubated at 37°C until the cells have attached (approximately 2–3 h). The 24-well plate is then placed the right way up and 200 μl of medium containing a chemoattractant added to the upper chamber and 500 μl serum-free medium to the lower chamber. The plate is incubated at 37°C for 24–72 h. At the end of the assay the cells are labelled with 4 μM Calcein-AM (see above) and invasion assessed on an inverted confocal microscope, taking fluorescent images every 10 μm from the upper side of the membrane. Relative invasion is determined from at least three sites per well and the pixel counts associate therein, using image analysis (e.g. Image J freeware). One important advantage of this method is that it is not necessary for cells to completely traverse the Matrigel as invasion is measured within the gel. See Ref. (19) for more details.

3. It is possible that the absolute level of invasion for individual cell lines may vary from experiment to experiment. This can be for many reasons including that different batches of Matrigel and collagen vary in absolute composition, the density at which you harvest your cells can affect their adhesion molecule (especially integrin) expression levels and that continued growth in culture can lead to changes in cell phenotype. Thus to minimise these changes between repeat experiments try to use identical parameters: consider batch buying the same lot of a pre-tested matrix protein, always use cells at a similar passage number, always harvest your cells at the same density, and always using the same method. For example, trypsin/EDTA detachment of epithelial-derived

cells results in cleavage of surface E-cadherin whereas EDTA-detachment will not, changing the ability of individual cells to move. The latter method will also often result in cell clumping, particularly if the cells are epithelially derived, method making determining cell number difficult.

4. Good positive control cell lines that invade well are HT1080 (fibrosarcoma) and MDA MB231 (breast carcinoma).

Acknowledgments

I wish to thank Ms Anne Weston, Electron Microscopy Unit, London Research Institute, Cancer Research UK who performed all electron microscopy processing and imaging.

References

1. Dorudi S and Hart IR (1993) Mechanisms underlying invasion and metastasis Curr Opin Oncol 5:130–135
2. Hanahan D and Weinberg RA (2000) The Hallmarks of Cancer. J Cell 100:57–70
3. Gupta GP and Massague J (2006) Cancer Metastasis: Building a framework. Cell 127: 679–695
4. Orimo A and Weinberg RA (2006) Stromal fibroblasts in cancer: a novel tumour-promoting cell type cell Cycle 5:1597–1601
5. Petras K and Ostman A (2010) Hallmarks of cancer: interactions with the tumour stroma Exp Cell Res 316:1324–1331
6. Mantovani A, Allavena P, Sica A and Balkwill F (2008) Cancer-related inflammation Nature 454:436–444
7. Colotta F, Allavena P, Sica A, Garlanda C and Mantovani A. (2009) Cancer-related inflammation, the seventh hallmark of cancer: links to genetic instability. Carcinogenesis 30:1073–1081
8. Boyden S (1962) The chemotactic effect of mixtures of antibody and antigen on polymorphonuclear leukocytes J Exp Med 115: 453–466
9. Kleinmann HK and Martin GR (2005) Matrigel: Basement membrane matrix with biological activity Seminars Cancer Biol. 15:378–386
10. Kalluri R (2003) Basement membranes: structure, assembly and role in tumour angiogenesis Nat Rev Cancer 3:422–433
11. Timpl R, Rohde H, Robey PG, Rennard SI, Foidart JM, Martin GR (1979) Laminin- a glycoprotein from basement membranes J Biol Chem 254:9933–9937
12. Virtanen I, Gullberg D, Rissanen J, Kivilaakso E, Kiviluoto T, Laitinen LA, Lehto V-P and Ekblom P (2000) Laminin α1-chain shows a restricted distribution in epithelial basement membranes of fetal and adult tissues Exp Cell res 257:298–309
13. Sodek, KL, Brown TJ and Ringuette MJ. (2008) Collagen I but not Matrigel matrices provide MMP-dependent barrier to ovarian cancer cell penetration. BMC Cancer 8:223
14. Reiser K, McCormick RJ, Rucker RB (1992) Enzymatic and n on-enzymatic cross-linking of collagen and elastin FASEB J 6:2439–2449
15. Sabeh F, Shimizu-Hirota R and Weiss SJ (2010) Protease-dependent versus independent cancer cell invasion programs: three-dimensional amoeboid movement revisited. J Cell Biol 185:11–19
16. Ota I, Li X-Y, Hu Y and Weiss SJ (2009) Induction of a MT1-MMP and MT2-MMP-dependent basement membrane transmigration program in cancer cells by Snail1 PNAS 106:20318–20323
17. Thomas GJ, Poomswat S, Lewis MP, Hart IR, Speight PM and Marshall JF (2001) αvβ6 integrin upregulates matrix metalloproteinase 9 and promotes migration of normal oral keratinocytes J Invest Dermatol 116:898–904
18. Hooper S, Marshall JF, Sahai E. (2006) Tumor cell migration in three dimensions. Methods Enzymol.;406:625–43
19. Lamb RF, Hennigan RF, Turnbull K, Katsankis KD, MacKenzie ED, Birnie GD and Ozanne BW (1997) AP-1 mediated invasion requires increased expression of the hyaluronan receptor CD44 Mol Cell Biol 17:963–976

Chapter 9

Imaging Podosome Dynamics and Matrix Degradation

Taylor W. Starnes, Christa L. Cortesio, and Anna Huttenlocher

Abstract

Invasive cell migration is critical for leukocyte trafficking into tissues. Podosomes are matrix-degrading adhesive structures that are formed by macrophages and are necessary for macrophage migration and invasion. Here, we describe methods for imaging and quantifying podosomes in primary human macrophages and in THP-1 cells, a monocyte cell line that can be differentiated to a macrophage-like state. Moreover, we outline detailed methods for live imaging of podosomes, which are highly dynamic, and for the quantification of rates of podosome turnover. Finally, we discuss methods for the quantitative analysis of matrix degradation on fluorescent-gelatin-coated cover slips.

Key words: Podosomes, Podosome turnover, Podosome dynamics, Macrophage, THP-1, Live imaging, Gelatin degradation, Lifeact, Invadopodia

1. Introduction

Cell migration and tissue invasion are important functions in both physiological and pathological processes. Some highly motile and invasive cells form podosomes and invadopodia, which are structures uniquely adapted to these functions. Podosomes are the characteristic adhesive structures of monocyte-lineage cells but are also formed in Src-transformed fibroblasts (1) and smooth muscle cells (2, 3), while invadopodia are protrusive structures primarily formed by invasive cancer cells (Fig. 1) (4). Both podosomes and invadopodia share the ability to protrude into extracellular matrix and degrade it with proteolytic enzymes (reviewed in ref. 4). With half-lives of generally less than 5 min, podosomes are able to rapidly alter their organization, which is presumably necessary for migrating leukocytes to respond to inflammatory stimuli (4, 5).

Claire M. Wells and Maddy Parsons (eds.), *Cell Migration: Developmental Methods and Protocols*,
Methods in Molecular Biology, vol. 769, DOI 10.1007/978-1-61779-207-6_9, © Springer Science+Business Media, LLC 2011

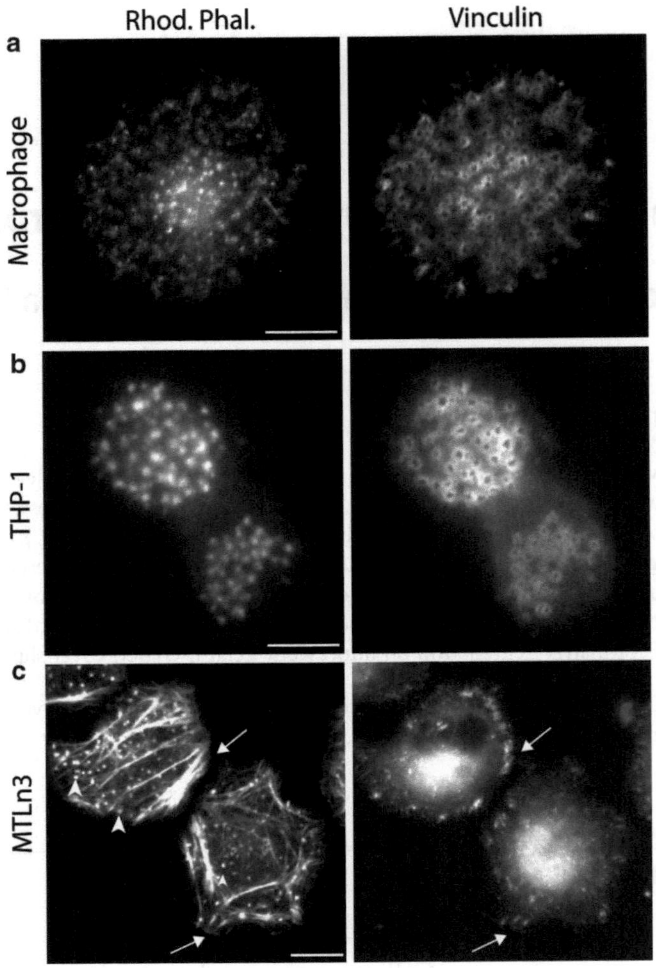

Fig. 1. Comparison of podosomes, invadopodia, and focal adhesions. (**a**) Primary human macrophages differentiated with MCSF for 7 days and (**b**) THP-1 cells differentiated for 48 h after PMA treatment were stained with rhodamine phalloidin and an anti-vinculin antibody. Note the prominent ring structures in the vinculin images that are characteristic of podosomes. (**c**) MTLn3 rat mammary carcinoma cells stained with rhodamine phalloidin and an anti-vinculin antibody. Note the localization of vinculin to focal adhesions, which are formed at the termini of actin stress fibers (*arrows*), and the absence of vinculin rings in areas corresponding to the punctuate invadopodia (*arrowheads*). Scale bar = 10 μm for all panels.

Podosomes are easily distinguished from other types of adhesions due to their characteristic dot and ring morphology (Fig. 1a, b). In addition to the morphology of individual podosomes, another striking feature of podosomes is the ability to form highly organized rings and rosettes containing tens to hundreds of individual podosomes (4, 6). The dot at the core of the podosome is predominantly composed of F-actin and F-actin regulatory proteins. Wiskott–Aldrich Syndrome Protein (WASP)/N-WASP, cortactin, and Arp2/3 are several components of podosome cores that are shared with invadopodia (Table 1 and Fig. 2) (7). The podosome ring structure is formed by integrins

Table 1
Components common to podosomes and invadopodia

Component	Function	References
Actin	Cytoskeletal	a
Arp 2/3	Actin-nucleation	b, c
WASP/N-WASP	Actin-binding	d, e
Cortactin	Actin-binding	f, g
Src kinases	Signaling	h–j
Matrix metalloproteases	ECM degradation	k–m

[a]Tarone, G., D. Cirillo, F.G. Giancotti, P.M. Comoglio, and P.C. Marchisio. 1985. Rous sarcoma virus-transformed fibroblasts adhere primarily at discrete protrusions of the ventral membrane called podosomes. *Exp Cell Res.* 159:141–57
[b]Linder, S., H. Higgs, K. Hufner, K. Schwarz, U. Pannicke, and M. Aepfelbacher. 2000. The polarization defect of Wiskott Aldrich syndrome macrophages is linked to dislocalization of the Arp2/3 complex. *J Immunol.* 165:221–5
[c]Kaverina, I., T.E. Stradal, and M. Gimona. 2003. Podosome formation in cultured A7r5 vascular smooth muscle cells requires Arp2/3-dependent de-novo actin polymerization at discrete microdomains. *J Cell Sci.* 116:4915–24
[d]Linder, S., D. Nelson, M. Weiss, and M. Aepfelbacher. 1999. Wiskott-Aldrich syndrome protein regulates podosomes in primary human macrophages. *Proc Natl Acad Sci USA.* 96:9648–53
[e]Lorenz, M., H. Yamaguchi, Y. Wang, R.H. Singer, and J. Condeelis. 2004. Imaging sites of N-wasp activity in lamellipodia and invadopodia of carcinoma cells. *Curr Biol.* 14:697–703
[f]Hiura, K., S.S. Lim, S.P. Little, S. Lin, and M. Sato. 1995. Differentiation dependent expression of tensin and cortactin in chicken osteoclasts. *Cell Motil Cytoskeleton.* 30:272–84
[g]Bowden, E.T., M. Barth, D. Thomas, R.I. Glazer, and S.C. Mueller. 1999. An invasion-related complex of cortactin, paxillin and PKCmu associates with invadopodia at sites of extracellular matrix degradation. *Oncogene.* 18:4440–9
[h]Gavazzi, I., M.V. Nermut, and P.C. Marchisio. 1989. Ultrastructure and gold-immunolabelling of cell-substratum adhesions (podosomes) in RSV-transformed BHK cells. *J Cell Sci.* 94 (Pt 1):85–99
[i]Cougoule, C., S. Carreno, J. Castandet, A. Labrousse, C. Astarie-Dequeker, R. Poincloux, V. Le Cabec, and I. Maridonneau-Parini. 2005. Activation of the lysosome-associated p61Hck isoform triggers the biogenesis of podosomes. *Traffic.* 6:682–94
[j]Chan, K.T., C.L. Cortesio, and A. Huttenlocher. 2009. FAK alters invadopodia and focal adhesion composition and dynamics to regulate breast cancer invasion. *J Cell Biol.* 185:357–70
[k]Samanna, V., T. Ma, T.W. Mak, M. Rogers, and M.A. Chellaiah. 2007. Actin polymerization modulates CD44 surface expression, MMP-9 activation, and osteoclast function. *J Cell Physiol.* 213:710–20
[l]Kelly, T., S.C. Mueller, Y. Yeh, and W.T. Chen. 1994. Invadopodia promote proteolysis of a wide variety of extracellular matrix proteins. *J Cell Physiol.* 158:299–308
[m]Nakahara, H., L. Howard, E.W. Thompson, H. Sato, M. Seiki, Y. Yeh, and W.T. Chen. 1997. Transmembrane/cytoplasmic domain-mediated membrane type 1-matrix metalloprotease docking to invadopodia is required for cell invasion. *Proc Natl Acad Sci USA.* 94:7959–64

and integrin-associated proteins, including vinculin, talin, and paxillin (Fig. 2) (8). Interestingly, these and many other proteins are shared by both podosomes and focal adhesions (Fig. 1c and Table 2). The presence of these focal adhesion-associated proteins in podosomes may help explain the observation that podosomes play a more significant role in substratum adhesion than invadopodia (4, 9).

In response to inflammatory stimuli, the effector functions of podosomes are needed at multiple stages as monocytes traffic from the blood and as macrophages and dendritic cells migrate through tissues. The processes of endothelial transmigration,

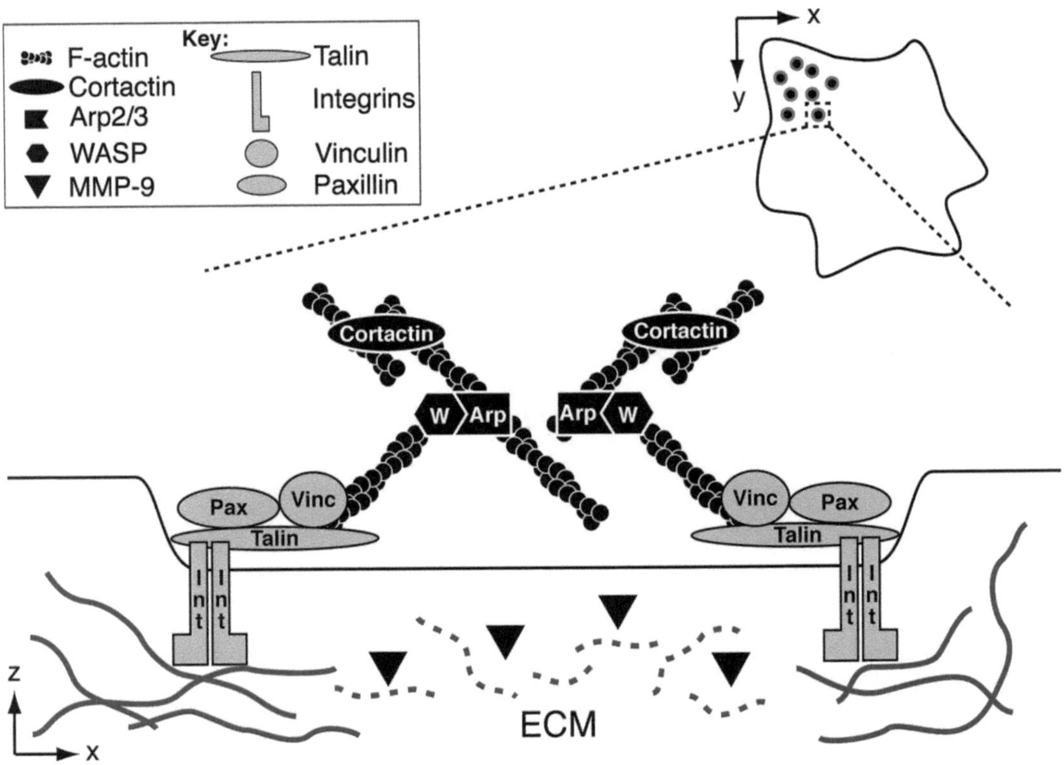

Fig. 2. Schematic of podosome structure. Pictured on the *top right* is a cell forming a cluster of podosomes as viewed from above. The gross podosome structure is depicted as *black dots* and *gray rings*. Below is a magnified cross section of an individual podosome. The colors of the proteins in the cross section correspond to their location in the podosome core (*black*) or ring (*gray*). Matrix metalloproteinase 9 (MMP-9) is secreted from podosome cores and is capable of degrading the underlying extracellular matrix (ECM). Degraded matrix is depicted as *dashed line*.

invasion through the basement membrane, and chemotaxis to sites of inflammation all appear to require podosomes. The initial steps of endothelial transmigration, classically described as rolling, activation, and arrest, are well defined (reviewed in ref. 10). During the process of endothelial transmigration, an early step in the inflammatory response, evidence suggests an important role for podosomes. When lymphocytes and monocytes are added to a monolayer of endothelial cells, podosome-like structures create invaginations in endothelial cells that coalesce into pores, thereby facilitating transcellular diapedesis (11). Reaching a site of inflammation also requires invasive migration through the vascular basement membrane and connective tissue. Macrophages from mice lacking Hck, an Src-family kinase in myeloid cells, form fewer podosomes and fail to accumulate in the peritoneal cavity after thioglycollate-induced peritonitis, indicating that podosomes may be playing a role in macrophage invasion (12). Furthermore, podosomes are known to form near the leading edge of chemotaxing cells (8, 13). The loss or mutation of proteins that regulate

Table 2
Components common to podosomes and focal adhesions

Component	Function	References
Talin	Cytoskeletal	a, b
Paxillin	Cytoskeletal/signaling	c, d
Zyxin	Cytoskeletal/signaling	e, f
Integrins	Cytoskeletal/signaling	g
Src kinases	Signaling	h, i, j
FAK/Pyk2	Cytoskeletal/signaling	k, l
Vinculin	Cytoskeletal	m, n
Tensin	Cytoskeletal/signaling	o, p

[a]Burridge, K., and Connell, L. (1983). Talin: a cytoskeletal component concentrated in adhesion plaques and other sites of actin membrane interaction. *Cell Motil.* 3, 405–417

[b]Marchisio, P. C., Bergui, L., Corbascio, G. C., Cremona, O., D'Urso, N., Schena, M., Tesio, L., and Caligaris-Cappio, F. (1988). Vinculin, talin, and integrins are localized at specific adhesion sites of malignant B lymphocytes. *Blood* 72, 830–833

[c]Turner, C. E., Glenney, J. R., Jr., and Burridge, K. (1990). Paxillin: a new vinculin-binding protein present in focal adhesions. *J. Cell Biol.* 111, 1059–1068

[d]DeFife, K.M., C.R. Jenney, E. Colton, and J.M. Anderson. 1999. Cytoskeletal and adhesive structural polarizations accompany IL-13-induced human macrophage fusion. *J Histochem Cytochem.* 47:65–74

[e]Crawford, A. W., and Beckerle, M. C. (1991). Purification and characterization of zyxin, an 82,000-dalton component of adherens junctions. *J. Biol. Chem.* 266, 5847–5853

[f]Kaverina, I., Stradal, T. E., and Gimona, M. (2003). Podosome formation in cultured A7r5 vascular smooth muscle cells requires Arp2/3-dependent de-novo actin polymerization at discrete microdomains. *J. Cell Sci.* 116, 4915–4924

[g]Block, M.R., C. Badowski, A. Millon-Fremillon, D. Bouvard, A.P. Bouin, E. Faurobert, D. Gerber-Scokaert, E. Planus, and C. Albiges-Rizo. 2008. Podosome-type adhesions and focal adhesions, so alike yet so different. *Eur J Cell Biol.* 87:491–506

[h]Rohrschneider, L. R. (1980). Adhesion plaques of Rous sarcoma virus-transformed cells contain the src gene product. *Proc. Natl. Acad. Sci. USA.* 77, 3514–3518

[i]Gavazzi, I., Nermut, M. V., and Marchisio, P. C. (1989). Ultrastructure and gold-immunolabelling of cell-substratum adhesions (podosomes) in RSV-transformed BHK cells. *J. Cell Sci.* 94 (Pt 1), 85–99

[j]Cougoule, C., S. Carreno, J. Castandet, A. Labrousse, C. Astarie-Dequeker, R. Poincloux, V. Le Cabec, and I. Maridonneau-Parini. 2005. Activation of the lysosome-associated p61Hck isoform triggers the biogenesis of podosomes. *Traffic.* 6:682–94

[k]Schaller, M.D., C.A. Borgman, B.S. Cobb, R.R. Vines, A.B. Reynolds, and J.T. Parsons. 1992. pp125FAK a structurally distinctive protein-tyrosine kinase associated with focal adhesions. *Proc Natl Acad Sci USA.* 89:5192–6

[l]Duong, L.T., P.T. Lakkakorpi, I. Nakamura, M. Machwate, R.M. Nagy, and G.A. Rodan. 1998. PYK2 in osteoclasts is an adhesion kinase, localized in the sealing zone, activated by ligation of alpha(v)beta3 integrin, and phosphorylated by src kinase. *J Clin Invest.* 102:881–92

[m]Tarone, G., D. Cirillo, F.G. Giancotti, P.M. Comoglio, and P.C. Marchisio. 1985. Rous sarcoma virus-transformed fibroblasts adhere primarily at discrete protrusions of the ventral membrane called podosomes. *Exp Cell Res.* 159:141–57

[n]Bloch, R. J., and Geiger, B. (1980). The localization of acetylcholine receptor clusters in areas of cell-substrate contact in cultures of rat myotubes. *Cell* 21, 25–35

[o]Hiura, K., S.S. Lim, S.P. Little, S. Lin, and M. Sato. 1995. Differentiation dependent expression of tensin and cortactin in chicken osteoclasts. *Cell Motil Cytoskeleton.* 30:272–84

[p]Bockholt, S. M., Otey, C. A., Glenney, J. R., Jr., and Burridge, K. (1992). Localization of a 215-kDa tyrosine phosphorylated protein that cross-reacts with tensin antibodies. *Exp. Cell Res.* 203, 39–46

podosome formation, such as WASP (13, 14), verprolins (15), and ARF6 (16), also results in dysfunctional chemotaxis. While the evidence that podosomes play a role in these functions is not yet definitive, diseases in which podosome function is altered provide additional support for their physiological importance.

Diseases that result in aberrant podosome function highlight the essential role of podosomes for in vivo immune responses. Macrophages from patients with disease-associated mutations in WASP lack podosomes and have revealed a key role for WASP in both podosome formation and dynamics (6, 17). The classic presentation of Wiskott–Aldrich Syndrome is the triad of thrombocytopenia, eczema, and immune deficiency. Additionally, these patients are predisposed to autoimmunity (17). Because of the key role of WASP in regulating podosome dynamics (7), matrix degradation (14), and chemotaxis (13, 15), altered podosome function is likely an important component of the immune deficiency experienced by these patients. Macrophages from patients with the autoinflammatory disease, pyogenic sterile arthritis, pyoderma gangrenosum, and acne (PAPA) syndrome, also exhibit impaired podosome formation and decreased invasive potential (18). While the associations of disease with altered podosome function are correlative, there seems to be a substantial link between podosomes and appropriately regulated immune responses. There is, therefore, considerable interest in understanding the mechanisms that regulate podosome formation and turnover. Here we present methods for studying the function of podosomes – live imaging of podosome dynamics and quantitative analysis of podosome-mediated gelatin degradation.

2. Materials

2.1. Blood Collection and Differentiation of Monocyte-Derived Macrophages

1. Phlebotomy Equipment: 30 mL syringes, alcohol prep pads, tourniquets, 2×2 in. sterile gauze pads, bandages, and a sharps container. We find that a Becton Dickinson Vacutainer® Safety-Lok™ Blood Collection Set with the following specifications is best for our needs: 21G ¾in. needle with 12 in. tubing and female luer. However, many functional equivalents are available.
2. Heparin sodium from porcine intestinal mucosa 1,000 USP units/mL (Abraxis Pharmaceutical Products) is stored at room temperature.
3. Phosphate buffered saline (without Ca^{2+} or Mg^{2+}). All PBS used in this protocol is without Ca^{2+} or Mg^{2+}.
4. Lymphoprep™ (Axis-Shield) is stored at room temperature. It must be kept in the dark due to light sensitivity.

5. 10 cm tissue culture-treated plates.
6. Macrophage medium. Based on RPMI-1640 plus the following additional components: 10% (v/v) fetal bovine serum (FBS), 0.15% (w/v) $NaHCO_3$, 0.45% (w/v) glucose, 10 mM HEPES, 1 mM sodium pyruvate, 100 IU/mL of penicillin, 100 μg/mL of streptomycin, and 0.05 mM β-mercaptoethanol. The FBS used for macrophage medium is not heat inactivated.
7. Recombinant human macrophage colony stimulating factor (rhM-CSF) (R & D Systems) is diluted with PBS to 20 μg/mL and stored in 31 μL aliquots at −20°C. After thawing on the day of blood collection, it is stored at 4°C during the subsequent week when it will be used for the differentiation of macrophages.

2.2. Culture and Differentiation of THP-1

1. THP-1 cells. This monocyte cell line is available from the ATCC (ATCC# TIB-202).
2. THP-1 medium. This medium is based on RPMI-1640 with the following additions: 10% heat-inactivated fetal bovine serum, 0.05 mM β-mercaptoethanol, 100 IU/mL of penicillin, and 100 μg/mL of streptomycin.
3. Phorbol 12-myristate 13-acetate (PMA). A stock solution is made by dissolving the powder/film to a concentration of 2 mg/mL in 100% ethanol. The stock is stored in small aliquots at −20°C for several months.
4. THP-1 Imaging Medium. Ham's F-12 medium (without phenol red) with 10% heat-inactivated fetal bovine serum, 10 mM HEPES, and 0.05 mM β-mercaptoethanol.

2.3. Lentiviral Transduction of THP-1

1. Lentiviral particles encoding shRNA targeted against a gene of interest.
2. 4 mg/mL polybrene.

2.4. Nucleofection of Primary Macrophages

1. Nucleofector® II Device (Lonza).
2. Human Macrophage Nucleofector® Kit (Lonza). Prior to the use of the kit, mix the supplement with the main vial of nucleofector solution. Store the nucleofector solution at 4°C. This kit includes the cuvettes and transfer pipettes required for the procedure, as well as an aliquot of the pMaxGFP® vector at 0.5 μg/μL.
3. Lifeact-mRuby at 1 μg/μL.
4. pEGFP-vinculin at 1 μg/μL.
5. 1× TRED (50 mL): 0.25% trypsin and 0.02% (w/v) EDTA in PBS (without Ca^{2+} or Mg^{2+}).
6. Gel-loading pipette tips.
7. Glass-bottom 35 mm tissue culture dishes (see Notes 1 and 2).

8. 10 μg/mL fibrinogen. 10 mg/mL fibrinogen stock solution is prepared by solubilizing powdered human fibrinogen in ddH_2O. It is stored in single use aliquots at −80°C. For coating glass-bottom dishes the stock solution is diluted to 10 μg/mL in PBS. 250 μL is needed per dish. 10 μg/mL fibronectin may be substituted. We purify fibronectin from human plasma as previously described (19).

2.5. Live Imaging of Podosomes

1. Macrophage Imaging Medium. Made with Ham's F-12 medium (without phenol red) and the following additions: 10% fetal bovine serum (FBS), 0.15% (w/v) $NaHCO_3$, 0.45% (w/v) glucose, 10 mM HEPES, 1 mM sodium pyruvate, and 0.05 mM β-mercaptoethanol (see Note 3).
2. Inverted epifluorescence or confocal microscope with a 60–100× objective, filter wheels with excitation and emission filters appropriate for imaging EGFP and mRuby/RFP, a camera, and a shutter. The filters, shutter, and camera should be computer-controlled. The microscope should be contained in an environmental chamber, which can be maintained at 37°C by a temperature control unit.
3. Microscope control and image editing software. We use Metamorph® software (Molecular Devices) for integrated microscope control and image analysis. However, other software capable of performing similar functions is also appropriate.

2.7. Production of Gelatin-Coated Cover Slips

1. 70% (v/v) ethanol (EtOH).
2. 50 μg/mL poly-l-lysine (PLL): 1 part PBS and 1 part 0.01% (100 μg/mL) PLL. Make 150 μL per cover slip on the day of use.
3. 0.5% glutaraldehyde: Dilute 25% glutaraldehyde stock solution 1:50 in PBS on the day of use. Make 150 μL per cover slip.
4. 200 μg/mL green gelatin: Make a 2 mg/mL stock by dissolving Oregon Green® 488-conjugated gelatin from pig skin (Invitrogen) in PBS. Make a 200 μg/mL working solution by diluting the stock 1:10 in PBS, 50 μL per cover slip. Extra stock solution can be stored at −20°C for future use.
5. 5 mg/mL $NaBH_4$: Make 150 μL per cover slip by dissolving powdered $NaBH_4$ in PBS at the time specified in the protocol. Allow the tube to vent after the solution is made.

2.8. Imaging Podosome-Mediated Gelatin Degradation

1. Acid washed 10 mm cover slips (see Note 2).
2. Fixative: 4% paraformaldehyde and 0.25 mg/mL saponin in PBS. This should be made fresh on the day of use, 500 μL per cover slip.
3. Quenching solution: 150 mM glycine in PBS, 500 μL per cover slip.

4. Blocking buffer: 10% goat serum (alternatively, heat-inactivated FBS) and 0.25 mg/mL saponin in PBS. This should be made on the day of use, 600 μL per cover slip.
5. 200 units/mL rhodamine phalloidin.
6. DAPI: Dissolve in ddH_2O to make a 20 mg/mL stock and store in small aliquots at −80°C. Dilute stock 1:2,000 in ddH_2O for staining, 100 μL per cover slip.
7. Accutase cell detachment solution (Innovative Cell Technologies) is needed if THP-1 cells are being used. Store aliquots at −20°C. Once thawed, the solution is stable at 4°C for several weeks.
8. Mounting medium (see Note 4).

3. Methods

3.1. Blood Collection and Differentiation of Monocyte-Derived Macrophages

This section will describe the methods used to obtain blood, isolate mononuclear cells, and differentiate monocytes into macrophages. As a general guideline, 30 mL of blood will yield one 80–100% confluent 10 cm plate of macrophages ($1–3 \times 10^6$ cells). We often collect 60 mL to ensure that enough cells are available for multiple nucleofection conditions. This protocol assumes that proper IRB approval for human blood collection has been obtained.

1. Assemble all supplies needed for blood collection before beginning the procedure.
2. Using a fine gauge needle and 1 mL syringe, add heparin to the syringe being used for the draw so that the final concentration will be 10 U/mL after the desired volume of blood is collected (i.e., use 300 μL of 1,000 U/mL heparin for a 30 mL collection). Express any remaining air from the syringe and attach the Blood Collection Set. Alternatively, heparinized vacutainer® tubes and holders may be used, but this requires multiple tube changes for the volumes being collected.
3. Collect 30 mL of venous blood according to standard practices.
4. Working in a sterile tissue culture hood, empty the blood from the syringe into a sterile flask or beaker. Dilute the blood 1:1 in PBS (without Ca^{2+} or Mg^{2+}).
5. Add 15 mL of Lymphoprep™ to each of two 50 mL conical tubes. Gently overlay 30 mL of the diluted blood on the Lymphoprep™. Do not allow the layers to mix.
6. All of the subsequently described spins for this prep require a swinging bucket centrifuge and are performed at 25°C without

the brake. Spin the samples for 30 min at 500 × *g*. After this first spin, the following four layers will be present from top to bottom: a straw-colored plasma layer, the white "buffy coat" containing leukocytes, the clear lymphoprep, and the red blood cells.

7. Collect the buffy coats from both tubes as completely as possible and pool them in a fresh 50 mL conical tube. Add PBS to a final volume of 50 mL and spin at 500 × *g* for 15 min. Gently aspirate and discard the supernatant, which may be opaque after this first spin. The pellets are relatively soft after this and the subsequent spins. Take care not to disturb them.
8. Resuspend the pellet in 25 mL of room temperature PBS and spin at 330 × *g* for 12 min.
9. Resuspend the pellet in 25 mL of macrophage medium prewarmed to 37°C and spin at 330 × *g* for 12 min.
10. After this final spin, aspirate the supernatant and resuspend the pellet in 10 mL of macrophage medium. Add to a 10 cm tissue culture-treated dish and incubate at 37°C with 5% CO_2 for approximately 4 h to allow the monocytes to adhere to the plate (see Note 5).
11. Remove the lymphocyte fraction from the prep by swirling the plate *very gently* to resuspend any non-adherent cells that have settled at the bottom of the dish, and aspirate the medium. The monocytes are only lightly adherent at this point, and it is important not to disturb them. Adding and removing medium from a single location on the plate will further minimize disruption.
12. Add 10 mL of warmed macrophage medium containing 20 ng/mL rhM-CSF. On days 3 and 5 after isolation, the medium should again be aspirated from the macrophages and replaced with 10 mL of fresh medium plus rhM-CSF. The macrophages will be fully differentiated and ready for use on days 6–7 after the initial isolation.

3.2. Culture and Differentiation of THP-1

THP-1 cells are an accepted model system for the study of macrophage function (20, 21). Here, we describe the culture requirements of THP-1 cells and the use of PMA to differentiate them to a macrophage-like state. After differentiation, they can be used in the place of primary macrophages for live imaging or gelatin degradation with some modifications of the described protocols. If THP-1 cells are used for nucleofection and live imaging, then they should be differentiated after nucleofection.

1. THP-1 are maintained in THP-1 medium at 37°C with 5% CO_2. Cultures are maintained according to ATCC recommendations, keeping cells within the range of 2×10^5 to

1×10^6 cells/mL. They are generally subcultured to a density of 2×10^5 after reaching $7–9 \times 10^5$ cells/mL.

2. In order to differentiate the cells to a macrophage-like state, spin down an aliquot of cells for 4 min at $500 \times g$. Resuspend the cells at a density of 5×10^5 cells/mL in THP-1 medium plus 500 nM PMA. Add the cells to a 10 cm suspension plate (smaller if necessary) and return to a 37°C with 5% CO_2 incubator for 3 h. Although a suspension plate is preferred, a tissue culture-treated plate will work, but lifting the cells requires more pipetting. This 3 h treatment with PMA initiates the differentiation process, but it is necessary to wait 48–72 h after treatment before using the cells for imaging, as differentiation continues in the absence of PMA.
3. After 3 h, the cells are lightly adherent. Lift them by pipetting over the entire plate. Pellet the cells by spinning for 4 min at $500 \times g$. Wash once in a volume of THP-1 medium at least as large as the volume used for differentiation. Pellet the cells as described.
4. Resuspend the cells at a concentration that is appropriate for the intended use in THP-1 medium without PMA. This will be described in subsequent steps of the protocol.

3.3. Lentiviral Transduction of THP-1

Manipulation of endogenous gene expression is a valuable tool for studying many cellular processes. shRNA-mediated knockdowns have been used to study the role of several proteins in podosome function (14, 22). Lentiviruses are a convenient vehicle for shRNA delivery because of their high transduction efficiency (23). Ready-to-use lentiviral particles are available from multiple companies. Alternatively, several systems are available for producing shRNA-encoding viruses in the lab. Here, we describe an infection protocol that we have used to successfully knock down target genes in THP-1 cells.

It is important to note that while lentiviral vector systems are generally replication incompetent, viruses with tropism for human cells can infect lab workers. Caution should be taken during the construction of lentiviruses to ensure that genes with significant potential to damage humans, such as oncogenes, will not be expressed. Furthermore, personal protective equipment should be used when handling lentiviruses, and proper decontamination procedures should always be followed after handling them (safety precautions reviewed in ref. 24).

1. On the day before infection, split a culture of THP-1 cells to a density of 4×10^5 cells/mL. 10^6 cells will be needed for each knockdown condition.
2. On the day of infection, spin down aliquots of 10^6 THP-1 cells for 4 min at $500 \times g$. At a minimum, the following samples and controls should be used:

(a) Lentivirus targeting the gene(s) of interest. Typically, approximately 3 shRNA targets are tested for each gene in order to identify an optimal knockdown.

(b) Lentivirus encoding a scrambled or off-target control.

(c) An uninfected sample.

3. Mix the viral particles to be used for each infection in 4 mL of THP-1 medium containing 4 μg/mL polybrene.
4. Resuspend the pelletted THP-1 cells in the appropriate virus-containing medium. Add the cells to a six well plate.
5. Spin the cells on a microplate carrier in a 32°C swinging bucket centrifuge for 1.5 h at 400 × *g*.
6. Return the cells to a 37°C, 5% CO_2 incubator for 3 days to maximize infection prior to beginning selection.
7. Common selection mechanisms for lentiviral systems are puromycin resistance and fluorescent protein (i.e., GFP) expression. If a negative selection system, such as puromycin resistance, is being used, transduced cells should be selected for 7–14 days in selection medium. When all of the uninfected control cells have died and the transduced cells are growing rapidly in the selection medium, selection is complete. The cells may be kept in the selection medium indefinitely, but we have found that fully selected lines reliably maintain their knockdowns when cultured in standard THP-1 medium. If your system uses a positive selection marker, such as GFP expression, then transduced cells can be separated from non-transduced cells using fluorescence-activated cell sorting (FACS).
8. The knockdown should be confirmed with western blot analysis. Knockdown lines can then be used in the following protocols to study podosome function.

3.4. Nucleofection of Primary Macrophages

Nucleofection is a proprietary transfection method that utilizes solutions and electrical impulses that have been developed for individual cell types. It is useful for expressing constructs in cells that are otherwise difficult to transfect, such as terminally differentiated primary macrophages (25). In this protocol a Lifeact-mRuby fusion is nucleofected into macrophages to monitor the podosome cores. Lifeact is a 17aa actin probe derived from the yeast protein, Abp140, which is reported to have minimal effects on actin dynamics due to a relatively low binding affinity (26). EGFP-vinculin is used to study the behavior of the podosome rings. It is also possible to nucleofect THP-1 cells for use in live imaging. We use a THP-1 nucleofection protocol described by Schnoor et al. before treating the cells with PMA to induce differentiation (27). After a 3 h PMA treatment, they can be added to glass bottom dishes and imaged after 2 days of differentiation.

1. Before preparing cells for nucleofection, it is best to have all of the necessary reagents at hand due to the toxicity of the nucleofector solution if exposure is prolonged.
2. Approximately 1 h before nucleofecting the cells, coat the glass-bottom dishes being used with 10 μg/mL fibrinogen. Alternatively, 10 μg/mL fibronectin may be used. Add 250 μL of 10 μg/mL fibrinogen to the glass area of the dish only. Incubate at 37°C for 1 h, and wash the plates twice with 2 mL of PBS. The second wash may remain in the plate until the macrophages are ready to be added.
3. At least 1 h prior to the actual nucleofection, place 10 mL of macrophage medium in a dish at 37°C with 5% CO_2 to charge.
4. In 500 μL eppi tubes, mix 100 μL of Human Macrophage Nucleofector Solution with each of the following plasmids (see Note 6):
   (a) 0.5 μg of Lifeact-mRuby plus 0.5 μg of pEGFP-vinculin
   (b) 1 μg of pMaxGFP
   (c) No DNA
5. These solutions should reach room temperature before use in nucleofection. Leave the stock vial of nucleofector solution at room temperature until all nucleofections are completed in case extra is needed (see Note 7).
6. Turn on the Nucleofector device and enter program code Y-010.
7. Aspirate medium from the plate of macrophages and wash 2×5 mL with warm PBS. Add 2 mL of warm TRED and return the plate to the incubator for 5–7 min. Remove plate from the incubator and pipette vigorously over the entire plate using a P1000 to lift the cells (see Note 8). Confirm that all cells have been lifted under the microscope, and transfer the cells to a 15 mL conical containing 6 mL of warm macrophage medium. Add 2 mL of macrophage medium to the plate to wash off any remaining cells, and add this to the conical (final volume ≈ 10 mL).
8. Perform a rough count of the cells. Expect to have $1–3 \times 10^6$ macrophages in total. Pellet the cells for 4 min at 500 × *g*, preferably in a swinging bucket centrifuge.
9. Thoroughly resuspend the cells in a volume of PBS that would result in a concentration of $\sim 1 \times 10^6$ cells/mL according to the previous count.
10. Perform an accurate count and aliquot 5×10^5 cells into each of four eppi tubes (1 extra in case of nucleofection problems), and spin down for 4 min at 500 × *g*.
11. Completely aspirate the supernatant and resuspend in 100 μL of the appropriate nucleofector solution/DNA mixture.

Resuspend gently, being certain to avoid the formation of bubbles. After adding nucleofector solution to the cells, work as quickly as possible until it has been neutralized post-nucleofection.

12. Add the contents of each of these tubes to nucleofector cuvettes using gel-loading pipette tips. Again, be sure not to introduce bubbles. If the solution is not spread evenly across the bottom of the cuvette, tap it lightly on the work surface to redistribute the liquid.
13. Nucleofect all of the samples, using the Y-010 program.
14. As soon as possible after nucleofection, add 1 mL of the charged macrophage medium to each cuvette, allowing any debris to float to the surface. Use the transfer pipettes supplied with the nucleofector kit to remove all but the topmost 50–100 μL from the cuvette to reduce collection of debris. This volume should contain enough cells to split between two glass-bottom dishes. Add cells to the glass-bottom dishes so that the medium is pooled in the center of the dish, concentrating the cells in the area to be imaged.
15. Allow the cells to adhere for approximately 4 h. Gently aspirate the medium, wash the plate with 2 mL of macrophage medium to remove debris, and add 2 mL of fresh medium, this time filling the entire dish.
16. Allow cells to recover and begin expressing the transfected constructs for at least 12 h prior to imaging. Proceed with live imaging as described in Subheading 3.5.

3.5. Live Imaging of Podosomes

Podosomes are highly dynamic structures with half-lives generally less than 5 min (4, 5). Manipulations that alter podosome stability also perturb normal podosome function, demonstrating the important link between the two. For example, inhibition of calpains, intracellular proteases that cleave many cytoskeletal proteins, results in stabilized podosomes due to decreased disassembly (28). Increased podosome stability correlated with decreased random migration and transmigration across an endothelial monolayer (28). Conversely, cells that expressed a phospho-mimetic WASP mutant formed more dynamic podosomes and degraded gelatin to a greater extent than their wild-type counterparts (14). The following methods describe the techniques required for live fluorescence imaging of podosomes and the subsequent analysis of their turnover (Fig. 3). The method for quantification of podosome turnover was developed by Webb and Horwitz, and modified by our lab to quantify the turnover of multiple adhesion types (29, 30). The rate constants of podosome assembly and disassembly are determined as functions of the change in fluorescence intensity over time. It is important to note that the method

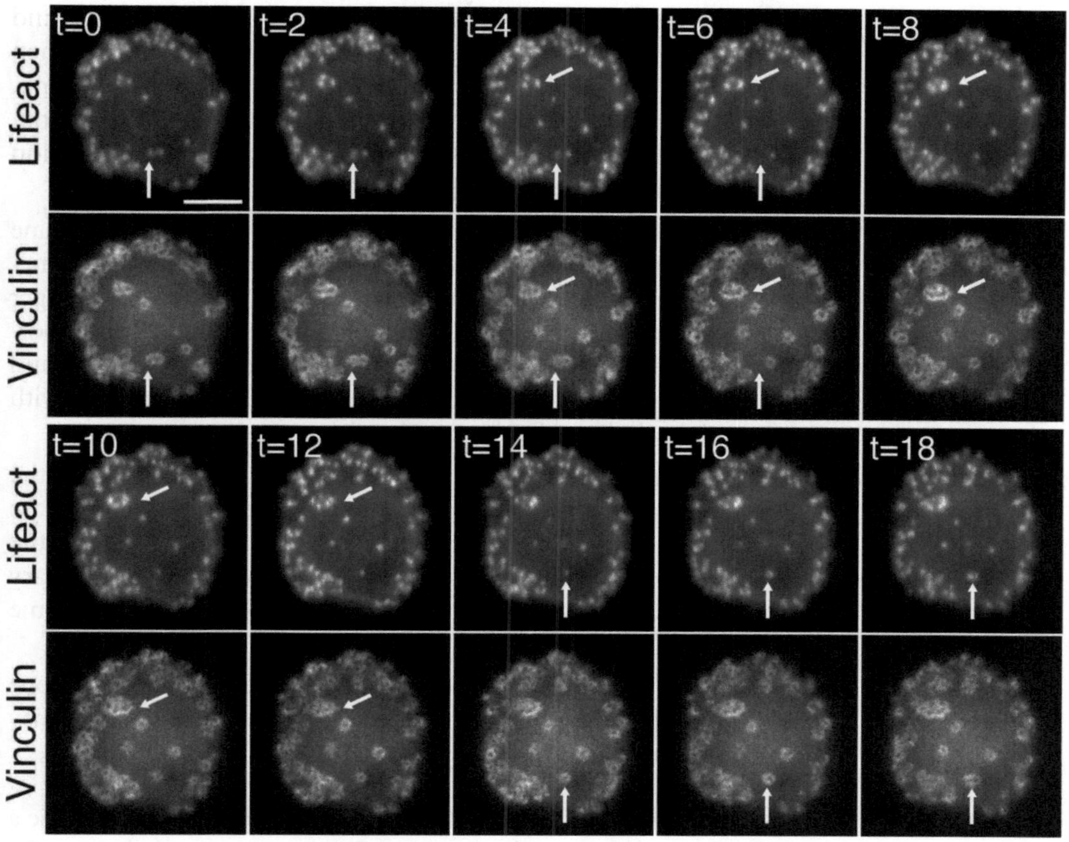

Fig. 3. Live fluorescent imaging of podosomes in primary human macrophages. Primary human macrophages were differentiated for 7 days with MCSF. On day 7, the macrophages were nucleofected using 0.5 μg of Lifeact-mRuby and 0.5 μg of pEGFP-vinculin and plated on fibrinogen-coated glass-bottom dishes. Time-lapse images were acquired using the following settings: binning = 3 × 3, gain = 2, Lifeact exposure = 15 ms, and vinculin exposure = 15 ms. Images were taken at 1 min intervals for 20 min. Lifeact and vinculin images are shown at 2-min intervals. *Arrows* indicate assembling or disassembling podosomes. Scale bar = 10 μm.

can only be used to measure turnover of the podosome cores, as the rings are typically too close together to isolate.

1. At least 2 h (preferably overnight) before the beginning of imaging, turn on the temperature control unit. This will allow all components of the microscope and stage to reach 37°C, which improves the maintenance of focus during time-lapse movies.
2. 30–60 min before imaging a plate, replace the macrophage medium with Macrophage Imaging Medium. If THP-1 cells are being used for imaging, use THP-1 Imaging Medium for this step. Return cells to the incubator to equilibrate.
3. Just before imaging, overlay the medium with 1–2 mL of mineral oil to prevent evaporation. Place a drop of microscope oil on the bottom of the dish and on the 100× objective.

4. Find a cell(s) with strong expression of Lifeact-mRuby and pEGFP-vinculin. If possible, use neutral density filters when searching for the cells of interest to avoid photobleaching. While it is important to find a cell with strong expression, extremely strong expression may actually decrease resolution and should be avoided.
5. With the filters that will be used during the time lapse, determine the optimal exposure time. In order to minimize photobleaching, choose an exposure time that is no more than 20–50% longer than needed to obtain a clear image (see Note 9). It is important to strike a good balance between image resolution and exposure time so that multiple images may be acquired with minimal photobleaching and toxicity.
6. Acquire a multi-dimensional time lapse. Take an image of the red and green fluorescence channels, as well as a DIC or phase contrast image every 15–30 s for 15–60 min. Because of the rapid movement and turnover of podosomes, it is necessary to take images at short intervals. In an active cell, podosome organization can completely change in 1–5 min.
7. Additional movies may be made in parallel if your microscope is equipped with an automated stage. Sequential movies may also be made.

3.6. Quantification of Podosome Turnover

1. After acquiring a time lapse of macrophage podosomes, use a filter to decrease background fluorescence in the image. We use a filter based on the algorithm described by Zamir et al. that works well for reducing diffuse signal around punctate structures (31). Alternatively, most image processing software has the ability to subtract background fluorescence. Applying these filters should increase the resolution of the podosome cores and facilitate their tracking.
2. Scan through the movie to find a podosome that appears and disappears completely during the course of the movie. Make note of the frames at which the podosome is first visible and at which it has completely disappeared. The time at which the podosome initially appears is $t=0$. Because the rate of formation and dissolution is based on fluorescence intensity, it is important to find a podosome that does not contact other podosomes during its lifetime. If the podosomes are moving actively, no other podosome should move into the path taken by the podosome being analyzed.
3. Create a region that encompasses the area occupied by the podosome of interest during its entire lifetime, but which will not contain other podosomes. Transfer this region to all images in the time lapse that contain the podosome of interest.

4. Determine the integrated intensity of the region for all relevant points and log this data to a spreadsheet. Record the corresponding times, which are determined by the interval at which the time lapse was taken. Integrated intensity is the sum of the intensity of all the pixels in the region. It should increase and decrease as the podosome assembles and disassembles, respectively. At this point, plotting intensity vs. time would result in a period of exponential assembly and disassembly.
5. To make this data linear, generate a semi-logarithmic plot by taking natural logs of the intensity using the following guidelines. To determine the rate of assembly, take the natural log of the product of the intensity (I) at each time point from $t=0$ (I_0) until the intensity has reached a plateau (I_{Max}) divided by the intensity at $t=0$.

$$\text{Assembly:}\Delta I = \text{Ln}\frac{[I_{0\rightarrow \text{Max}}]}{[I_0]}.$$

The change in intensity during disassembly is calculated in a similar manner. Here I_{Max} is the intensity of the frame just prior to the beginning of disassembly and I_{End} is the intensity of the frame in which the podosome is no longer visible.

$$\text{Disassembly:}\Delta I = \text{Ln}\frac{[I_{\text{Max}}]}{[I_{\text{Max}\rightarrow \text{End}}]}.$$

6. Create a scatter plot for both assembly and disassembly by plotting the change in intensity vs. time. Fit a linear trend line to the data and take the slope of this line, which is the rate constant of assembly or disassembly. As the intensity value has arbitrary units, the calculated rate constant has units of time^{-1}. In order to determine if this model can accurately be applied to the podosome of interest, determine the R^2 value. An $R^2 > 0.7$ is generally acceptable for this analysis.
7. Because podosomes in different parts of the cell may have different rates of assembly, sample podosomes from multiple locations within the cell to determine accurate average rates. Where possible, we determine the assembly/disassembly rates of 10–25 podosomes per cell for 5–10 cells in each experimental condition.

3.7. Production of Gelatin-Coated Cover Slips

The matrix-degrading ability of podosomes is unique in comparison to other types of adhesions. Macrophage podosomes are able to degrade matrix through the secretion of matrix metalloproteinases (32). Podosomes in Src-transformed fibroblasts and osteoclasts have also demonstrated the ability to degrade matrix through the secretion of lysosomal contents (33, 34). While this function has not been directly demonstrated in macrophages, lysosomes colocalize with podosome rosettes in macrophages, and the lysosome-associated protein p61Hck plays a role in podosome formation (12, 35).

Gelatin-coated cover slips are a useful tool for studying the degradative ability of podosomes, as they provide a qualitative result and are amenable to quantification (Fig. 4). The following protocol describes a method adapted from Artym et al. for coating cover slips with fluorescent gelatin (36). We have used Oregon Green 488-conjugated gelatin, but gelatin conjugated to other fluorophores may be substituted. Additionally, we demonstrate the techniques for imaging and quantifying podosome-mediated gelatin degradation.

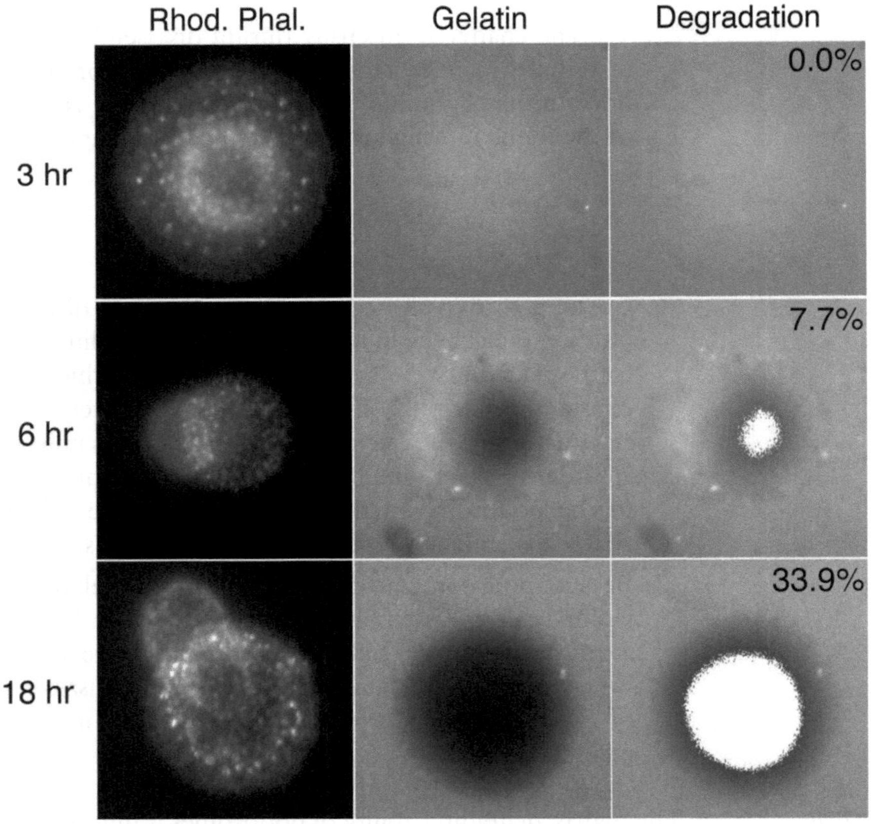

Fig. 4. Gelatin degradation by primary macrophages. Cover slips were coated with 200 μg/mL Oregon Green-488 gelatin. 3×10^4 primary human macrophages, which had been differentiated for 7 days with MCSF, were added to the cover slips. Podosomes were stained with rhodamine phalloidin. To determine the percentage of cell area degraded, a degradation coefficient of 0.3 was used for each image. The *white shading* in the *right-most column* indicates the area that was thresholded as degradation. The percentage of cell area degraded is indicated. Images were selected to demonstrate various degrees of degradation. Scale bar = 10 μm.

1. Sterilize 24 acid-washed cover slips by dunking in 70% EtOH, and place them on parafilm. Remove residual EtOH by washing each cover slip twice with 1 mL PBS (see Note 10).
2. Add 150 μL of 50 μg/mL poly-l-lysine to each cover slip, allowing the liquid to form a dome over the cover slip. Incubate for 10 min at room temperature, aspirate the PLL, and wash once with 1 mL of PBS. For this and all subsequent washes in this protocol, leave ~100 μL of PBS in a dome on the cover slip, and aspirate it just prior to adding the next wash or reagent.
3. Add 150 μL of 0.5% glutaraldehyde. Incubate for 10 min at room temperature. Aspirate the glutaraldehyde and wash each cover slip five times with 1 mL of PBS.
4. Line the bottom of a dark box with a sheet of parafilm. Place several 50 μL drops of the 200 μg/mL gelatin solution on the parafilm, being careful to avoid bubbles.
5. Aspirate the previous wash from a few cover slips at a time. Lift them from the parafilm using fine tweezers and invert them onto the gelatin drops. The glutaraldehyde-treated face should be in contact with the gelatin, and there should be no air bubbles. After all of the cover slips have been transferred in this manner, cover the box and incubate at room temperature for 15 min.
6. During this incubation, prepare the 5 mg/mL $NaBH_4$.
7. Prepare a fresh sheet of parafilm. Lift the cover slips off of the gelatin solution and place them gelatin side up on the fresh sheet of parafilm.
8. Place 150 μL of $NaBH_4$ on the cover slips and incubate at room temperature for 10 min. $NaBH_4$ is a reducing agent that quenches any remaining glutaraldehyde. Aspirate the $NaBH_4$, and transfer the cover slips to a 24 well plate for use or storage. Wash the cover slips twice with 1 mL of PBS.
9. Add 1 mL of PBS to each well to keep the cover slips hydrated until use. The cover slips can be stored at 4°C for 2–4 weeks if they are protected from light.

3.8. Imaging Podosome-Mediated Gelatin Degradation

1. Draw blood and differentiate macrophages as described in Subheading 3.1. On day 6–7 following the blood draw, the macrophages will be ready for use. (If THP-1 cells are being used differentiate them as described in Subheading 3.2.)
2. Lift and count the cells as described in step 7 of Subheading 3.4. Pellet the cells at 500 × *g* for 5 min. Resuspend the cells in a volume of macrophage medium that will result in a concentration of ~1×10^5 cells/mL based on the first count. (If THP-1 cells are being used, lift the cells by washing once

with PBS and adding 2 mL of accutase. Incubate the dish at room temperature for approximately 5–10 min. Cells that remain adherent can be lifted by pipetting. Resuspend the THP-1 cells as described in this step for primary macrophages, but use the THP-1 medium. Proceed with step 3).

3. Perform a final count of the macrophages and dilute them to 6×10^4 cells/mL.
4. Remove a gelatin-coated cover slip from the 24 well plate and place it in a new 24 well plate. Add 3×10^4 macrophages (500 μL) to the cover slip.
5. Incubate at 37°C with 5% CO_2 for 3–24 h, depending on the amount of degradation desired. A time course may be taken by fixing cover slips at chosen time intervals.
6. To fix the cover slips, aspirate the medium from each well, wash once with 500 μL of PBS, and add 500 μL of Fixative to the wells. Allow the cells to fix for 10 min at room temperature.
7. Aspirate the Fixative, and 500 μL of Quenching Solution for 10 min at room temperature.
8. Aspirate the Quenching Solution, and add 500 μL of Blocking Solution for 30 min at room temperature. If you are taking a time course, place the cover slips at 4°C until all of the cover slips are ready for staining.
9. Without aspirating the Blocking Solution, carefully lift cover slips out of their well and place them in a parafilm-lined, dark, plastic box. Add 100 μL of rhodamine phalloidin diluted 1:1,000 in blocking buffer to the top of the cover slips, allowing it to form a dome (see Note 11). Incubate for 45 min at room temperature.
10. Aspirate the rhodamine phalloidin, and add 100 μL of DAPI solution. Incubate for 5 min at room temperature.
11. Wash three times with 1 mL of PBS. Leave ~100 μL of the PBS on the cover slips for 5–10 min between each wash.
12. Spot 7 μL of mounting medium (MM) on a clean glass slide, being careful not to introduce bubbles. Aspirate the final wash and invert the cover slip onto the MM. Allow the MM to dry overnight at room temperature in a dark place.
13. Seal the edges of the cover slip with clear nail polish. Allow the nail polish to dry completely before use. After the nail polish has dried, cover slips should be stored at −20°C when not in use.
14. Using a microscope with a 100× objective, take pictures using filters appropriate for rhodamine phalloidin, GFP, DAPI, and DIC/phase contrast. In determining the percent of cell area degraded, the DIC or rhodamine phalloidin image will be used to define the cell area.

3.9. Quantification of Gelatin Degradation

The quantification of podosome-mediated gelatin degradation raises several technical challenges. The areas of degradation are generally irregular in size and shape. Partially degraded areas may retain some fluorescence, creating a gradient of degradation rather than a sharply defined area of degradation. Additionally, even when images are acquired at the same exposure, the fluorescence intensity of the gelatin may vary between cover slips or even regions of the same cover slip. Therefore, we have developed a protocol for defining degraded areas based on the fluorescence intensity of undegraded gelatin within each individual image. The intensity of undegraded gelatin is used as the basis for defining a threshold that encompasses areas of degraded gelatin. We believe that this method reduces the error produced by manually assigning areas of degradation.

1. Open the DIC or rhodamine phalloidin image so that the edge of the plasma membrane can easily be identified.
2. Determine the cell border. Manually trace a region around the plasma membrane, and transfer this region to the corresponding area of the gelatin image.
3. Define the average intensity of undegraded gelatin as a reference. Place a 100 × 100 pixel region in an area of the gelatin image that contains no cells or areas of degradation. If there is variable intensity across the image, more than one reference region may be required. Determine the average intensity of the reference region (in arbitrary units of intensity). We will define this value as $\text{Int}_{\text{undeg}}$.
4. Determine the value of the lowest intensity represented in the gelatin image. We will define the lowest intensity of the whole image as Int_{lo}.
5. Determine the high threshold value for your image. In order to determine the areas of degradation in the image, it is necessary to calculate the value for the high threshold for the image ($\text{thresh}_{\text{hi}}$). The value for $\text{thresh}_{\text{hi}}$ will be defined by the following equation:

$$\text{thresh}_{\text{hi}} = \text{Int}_{\text{lo}} + \left(\left(\text{Int}_{\text{undeg}} - \text{Int}_{\text{lo}}\right) \cdot x\right),$$

   where x is the "degradation coefficient" (see Note 12). We find that a degradation coefficient of 0.3–0.6 is a good starting point for our analyses, but it needs optimization based on experimental conditions (see Note 13).
6. Apply a threshold to the image so that areas with intensities less than $\text{thresh}_{\text{hi}}$ are selected. This threshold should highlight degraded areas in the image.
7. Determine the area of the cell that was traced in steps 2 and 3. Determine the area of the thresholded region that falls within

the borders of the cell. The percent of cell area degraded is as follows:

$$\text{Percent degraded} = \frac{\text{Area threshholded}}{\text{Total cell area}}.$$

4. Notes

1. Glass-bottom dishes: The dishes may be purchased commercially, but they are relatively easy to make in house. Drill a 16–18 mm hole in the center of the bottom half of a standard 35 mm tissue culture dish. Invert the dish so that the outside surface is facing up. Place a thin bead of Norland Optical Adhesive 68 around the hole. Place an acid-washed 22 mm cover slip (see Note 2) on the bead of glue and press down lightly. This glue is cured by UV light. Approximately 1 h in a UV transilluminator or overnight in the UV light of a tissue culture hood is sufficient to cure the glue. Sterilize by rinsing with 70% ethanol and exposing to a UV light source.
2. Acid-washed cover slips: In a 500 mL beaker on ice (this solution gets hot), make 200 mL of 3.6 M H_2SO_4. When the solution has cooled to room temperature, add the desired amount of cover slips and stir to ensure that all cover slips are exposed to the acid. Cover and incubate overnight. The next morning, pour as much of the acid as possible into a waste beaker and neutralize appropriately. Neutralize the beaker of cover slips by adding 200 mL of 0.1 M NaOH. Adjust the pH of the cover slips to between 6 and 8. Discard this solution. Rinse the cover slips two times with ddH_2O. Dry the cover slips in a single layer on Whatman paper. These can be stored for several months.
3. There are two especially important components of the imaging medium that deserve mention. It is necessary to use medium without phenol red in order to reduce background fluorescence. The addition of HEPES maintains the neutral pH of the medium in an environment with atmospheric levels of CO_2.
4. We make our own mounting medium, which utilizes phenylenediamine as an anti-fading agent. This recipe makes approximately 25 mL. Weigh 6.0 g of ultrapure glycerol into a 50 mL conical tube. Weigh 2.4 g of polyvinyl alcohol and add it to the tube of glycerol. Stir this mixture with a glass rod before adding 6 mL of sterile ddH_2O. Incubate 4 h to overnight at room temperature. Make 12 mL of 0.1 M Tris base pH 8.5 and add 10 mL of this solution to the mixture. Place the beaker in a 50°C water bath and stir for 10 min. After 6 min of stirring, weigh 30 mg of phenylenediamine and add

it to the remaining 2 mL of 0.1 M Tris. It is very important to work quickly after the phenylenediamine is in solution, as it oxidizes rapidly. Add the phenylenediamine solution to the 50 mL conical and stir with a glass rod until a uniform solution has been created. Cool for 3 min on ice and spin the mounting medium down in a 4°C swinging bucket centrifuge for 15 min at 2000×*g*. Make 0.5–1 mL aliquots and store at −80°C. When using the mounting medium, it is important to thaw it on ice and return it to −80°C as soon as possible to slow oxidation. When fresh, the mounting medium is pale brown, and it should be discarded if it becomes a deeper brown.

5. When placed in a tissue culture-treated dish, the lymphocytes remain in suspension while the monocytes and platelets adhere. During the course of the three medium changes over the subsequent 5 days, the vast majority of the remaining lymphocytes are removed. During this time, many platelets will be removed, but they will continue to detach over the course of the 7 days of differentiation. By the time the cells are used for experiments, they are a relatively pure population of macrophages.
6. The pMax-GFP nucleofection is included so that the nucleofection procedure may be optimized in reference to other published protocols. These tend to define transfection efficiency by the percentage of pMax-GFP expressing cells. The "No DNA" nucleofection is included as a control for the toxicity of the nucleofection solution. Additionally, macrophages are auto-fluorescent at a wavelength that is visible in the GFP channel, and it is good to familiarize yourself with this background in order to evaluate real vs. auto-fluorescent signal.
7. Despite the precautionary measures that we recommend in this protocol, we still find that nucleofections occasionally do not work. In these cases, the nucleofector device will report an error, and cells in that sample will no longer be viable. For this reason we recommend having extra cells, DNA, and nucleofector solution ready until all of the desired conditions are completed.
8. Macrophages are strongly adherent and difficult to lift. In order to increase viability, keeping the cells in TRED for the minimum amount of time is important. After incubating the cells at 37°C for 5–7 min, we find that the cells can be lifted efficiently by pipetting with a P1000. Generally, pipetting 10–15 times across the plate while holding in one orientation, turning the plate 90°, and repeating the washing will lift all of the macrophages.
9. In order to prevent photobleaching of the samples, it is necessary to minimize the exposure times used for acquisition.

However, it is prudent to use an exposure time that is longer than the absolute minimum, knowing that some photobleaching will occur. We typically use the following exposure settings: exposure time = 10–100 ms, gain = 1, and pixel binning = 2. In order to decrease exposure time, gain may be increased to 2, and binning may be increased to 3 or 4. However, both of these changes will lead to decreased resolution. Additionally, the use of neutral density filters in the excitation path can substantially decrease the rate of photobleaching. If a confocal microscope is being used, minimize the laser power as much as possible.

10. For all procedures requiring the manipulation of cover slips the choice of forceps is important. We find that very fine-tipped forceps help substantially when lifting cover slips from parafilm or the bottom of 24 well plates.
11. The cover slips are transferred to parafilm at this step to minimize the volume required for staining and minimize the use of expensive reagents. The cover slips may be stained in the plate, but it requires a substantially greater volume to completely coat the cover slip.
12. To determine $thresh_{hi}$, we first subtract the lowest intensity represented in the image (Int_{lo}) from the average intensity of undegraded gelatin (Int_{undeg}) to normalize the differences that will be compared across a series of images. The degradation coefficient is the fraction of this difference which will be added to Int_{lo}, thereby determining the value of $thresh_{hi}$. The values of the degradation coefficient may range from 0 to 1.
13. The degradation coefficient used in this calculation must be determined by the user. When we quantify gelatin degradation in this manner, we try several values on a subset of our images to determine which value of the degradation coefficient best represents the actual degradation. After a value is chosen for the degradation coefficient, it remains constant throughout the experiment and replicates of the experiment. We suggest testing values in the range of 0.3–0.6 because this is the range that has worked for us under several conditions.

References

1. Tarone, G., Cirillo, D., Giancotti, F., Comoglio, P., and Marchisio, P. (1985) Rous sarcoma virus-transformed fibroblasts adhere primarily at discrete protrusions of the ventral membrane called podosomes., *Exp Cell Res 159*, 141–157.
2. Hai, C., Hahne, P., Harrington, E., and Gimona, M. (2002) Conventional protein kinase C mediates phorbol-dibutyrate-induced cytoskeletal remodeling in a7r5 smooth muscle cells., *Exp Cell Res 280*, 64–74.
3. Lener, T., Burgstaller, G., Crimaldi, L., Lach, S., and Gimona, M. (2006) Matrix-degrading podosomes in smooth muscle cells., *Eur J Cell Biol 85*, 183–189.
4. Linder, S. (2007) The matrix corroded: podosomes and invadopodia in extracellular matrix degradation., *Trends Cell Biol 17*, 107–117.
5. Evans, J., Correia, I., Krasavina, O., Watson, N., and Matsudaira, P. (2003) Macrophage podosomes assemble at the leading lamella by

growth and fragmentation., *J Cell Biol 161*, 697–705.

6. Linder, S., Nelson, D., Weiss, M., and Aepfelbacher, M. (1999) Wiskott-Aldrich syndrome protein regulates podosomes in primary human macrophages., *Proc Natl Acad Sci USA 96*, 9648–9653.
7. Calle, Y., Antón, I., Thrasher, A., and Jones, G. (2008) WASP and WIP regulate podosomes in migrating leukocytes., *J Microsc 231*, 494–505.
8. Linder, S., and Aepfelbacher, M. (2003) Podosomes: adhesion hot-spots of invasive cells., *Trends Cell Biol 13*, 376–385.
9. Buccione, R., Orth, J., and McNiven, M. (2004) Foot and mouth: podosomes, invadopodia and circular dorsal ruffles., *Nat Rev Mol Cell Biol 5*, 647–657.
10. Ley, K., Laudanna, C., Cybulsky, M., and Nourshargh, S. (2007) Getting to the site of inflammation: the leukocyte adhesion cascade updated., *Nat Rev Immunol 7*, 678–689.
11. Carman, C., Sage, P., Sciuto, T., de la Fuente, M., Geha, R., Ochs, H., Dvorak, H., Dvorak, A., and Springer, T. (2007) Transcellular diapedesis is initiated by invasive podosomes., *Immunity 26*, 784–797.
12. Cougoule, C., Le Cabec, V., Poincloux, R., Al Saati, T., Mège, J., Tabouret, G., Lowell, C., Laviolette-Malirat, N., and Maridonneau-Parini, I. (2010) Three-dimensional migration of macrophages requires Hck for podosome organization and extracellular matrix proteolysis., *Blood 115*, 1444–1452.
13. Jones, G., Zicha, D., Dunn, G., Blundell, M., and Thrasher, A. (2002) Restoration of podosomes and chemotaxis in Wiskott-Aldrich syndrome macrophages following induced expression of WASp., *Int J Biochem Cell Biol 34*, 806–815.
14. Dovas, A., Gevrey, J., Grossi, A., Park, H., Abou-Kheir, W., and Cox, D. (2009) Regulation of podosome dynamics by WASp phosphorylation: implication in matrix degradation and chemotaxis in macrophages., *J Cell Sci 122*, 3873–3882.
15. Tsuboi, S. (2006) A complex of Wiskott-Aldrich syndrome protein with mammalian verprolins plays an important role in monocyte chemotaxis., *J Immunol 176*, 6576–6585.
16. Svensson, H., West, M., Mollahan, P., Prescott, A., Zaru, R., and Watts, C. (2008) A role for ARF6 in dendritic cell podosome formation and migration., *Eur J Immunol 38*, 818–828.
17. Ochs, H., and Thrasher, A. (2006) The Wiskott-Aldrich syndrome., *J Allergy Clin Immunol 117*, 725–738; quiz 739.
18. Cortesio, C., Cooper, K., Wernimont, S., Kastner, D., and Huttenlocher, A. (2010) Impaired podosome formation and invasive migration of macrophages from patients with a PSTPIP1 mutation and PAPA syndrome., *Arthritis Rheum.*
19. Ruoslahti, E., Hayman, E., Pierschbacher, M., and Engvall, E. (1982) Fibronectin: purification, immunochemical properties, and biological activities., *Methods Enzymol 82 Pt A*, 803–831.
20. Tsuboi, S., Takada, H., Hara, T., Mochizuki, N., Funyu, T., Saitoh, H., Terayama, Y., Yamaya, K., Ohyama, C., Nonoyama, S., and Ochs, H. (2009) FBP17 Mediates a Common Molecular Step in the Formation of Podosomes and Phagocytic Cups in Macrophages., *J Biol Chem 284*, 8548–8556.
21. Dostert, C., Pétrilli, V., Van Bruggen, R., Steele, C., Mossman, B., and Tschopp, J. (2008) Innate immune activation through Nalp3 inflammasome sensing of asbestos and silica., *Science 320*, 674–677.
22. Carrithers, M., Chatterjee, G., Carrithers, L., Offoha, R., Iheagwara, U., Rahner, C., Graham, M., and Waxman, S. (2009) Regulation of podosome formation in macrophages by a splice variant of the sodium channel SCN8A., *J Biol Chem 284*, 8114–8126.
23. Reiner, N. (2009) Methods in molecular biology. Macrophages and dendritic cells. Methods and protocols. Preface., *Methods Mol Biol 531*, v–vi.
24. Mosier, D. E. (2004) Introduction for "Safety Considerations for Retroviral Vectors: A Short Review", pp 68–75, Applied Biological Safety Association, Applied Biosafety.
25. Zhang, X., Edwards, J., and Mosser, D. (2009) The expression of exogenous genes in macrophages: obstacles and opportunities., *Methods Mol Biol 531*, 123–143.
26. Riedl, J., Crevenna, A., Kessenbrock, K., Yu, J., Neukirchen, D., Bista, M., Bradke, F., Jenne, D., Holak, T., Werb, Z., Sixt, M., and Wedlich-Soldner, R. (2008) Lifeact: a versatile marker to visualize F-actin., *Nat Methods 5*, 605–607.
27. Schnoor, M., Buers, I., Sietmann, A., Brodde, M., Hofnagel, O., Robenek, H., and Lorkowski, S. (2009) Efficient non-viral transfection of THP-1 cells., *J Immunol Methods 344*, 109–115.
28. Calle, Y., Carragher, N., Thrasher, A., and Jones, G. (2006) Inhibition of calpain stabilises podosomes and impairs dendritic cell motility., *J Cell Sci 119*, 2375–2385.

29. Webb, D., Donais, K., Whitmore, L., Thomas, S., Turner, C., Parsons, J., and Horwitz, A. (2004) FAK-Src signalling through paxillin, ERK and MLCK regulates adhesion disassembly., *Nat Cell Biol 6*, 154–161.
30. Chan, K., Cortesio, C., and Huttenlocher, A. (2007) Integrins in cell migration., *Methods Enzymol 426*, 47–67.
31. Zamir, E., Katz, B., Aota, S., Yamada, K., Geiger, B., and Kam, Z. (1999) Molecular diversity of cell-matrix adhesions., *J Cell Sci 112 (Pt 11)*, 1655–1669.
32. Yamaguchi, H., Pixley, F., and Condeelis, J. (2006) Invadopodia and podosomes in tumor invasion., *Eur J Cell Biol 85*, 213–218.
33. Tu, C., Ortega-Cava, C., Chen, G., Fernandes, N., Cavallo-Medved, D., Sloane, B., Band, V., and Band, H. (2008) Lysosomal cathepsin B participates in the podosome-mediated extracellular matrix degradation and invasion via secreted lysosomes in v-Src fibroblasts., *Cancer Res 68*, 9147–9156.
34. Mulari, M., Zhao, H., Lakkakorpi, P., and Väänänen, H. (2003) Osteoclast ruffled border has distinct subdomains for secretion and degraded matrix uptake., *Traffic 4*, 113–125.
35. Cougoule, C., Carréno, S., Castandet, J., Labrousse, A., Astarie-Dequeker, C., Poincloux, R., Le Cabec, V., and Maridonneau-Parini, I. (2005) Activation of the lysosome-associated p61Hck isoform triggers the biogenesis of podosomes., *Traffic 6*, 682–694.
36. Artym, V., Zhang, Y., Seillier-Moiseiwitsch, F., Yamada, K., and Mueller, S. (2006) Dynamic interactions of cortactin and membrane type 1 matrix metalloproteinase at invadopodia: defining the stages of invadopodia formation and function., *Cancer Res 66*, 3034–3043.

Chapter 10

Endothelial Cell Migration Under Flow

Beata Wojciak-Stothard

Abstract

The endothelial cells lining blood vessels are continuously exposed to fluid shear stress generated by pulsatile flow of blood. In order to minimise forces acting on their surface, endothelial cells adapt to shear stress by alignment and migration within the direction of flow. Failure to adapt to shear stress results in endothelial damage contributing to generation of atherosclerotic plaques or abnormal vessel repair. This chapter describes methods of generating laminar flow in vitro and studying endothelial cell alignment and motility under flow.

Key words: Endothelial, Flow, Shear stress, Flow chambers, Migration, Alignment

1. Introduction

1.1. Shear Stress and Endothelial Cells

Endothelial cells are continuously subjected to mechanical forces imposed by circulating blood, which include laminar and turbulent (nonlaminar) frictional wall shear stress, circumferential distension, and blood pressure. Shear stress, the tangential component of haemodynamic forces, induces extensive changes in endothelial cell behaviour and has been implicated in vasculogenesis, reendothelialization of vascular grafts, atherosclerosis, and angiogenesis (1, 2). Normal, physiological levels of shear stress are required for endothelial production of nitric oxide, expression of genes encoding coagulation molecules, growth factors, and adhesion molecules (3).

In order to minimise forces acting on their surface, endothelial cells adapt to shear stress by alignment and migration within the direction of the flow. Failure to adapt to shear stress results in endothelial damage contributing to generation of atherosclerotic plaques or abnormal vessel repair (4, 5). Endothelial cells in large

Claire M. Wells and Maddy Parsons (eds.), *Cell Migration: Developmental Methods and Protocols*,
Methods in Molecular Biology, vol. 769, DOI 10.1007/978-1-61779-207-6_10,

vessels are subjected to high laminar flow and are well aligned, but in branching points, where shear stress is low or turbulent, endothelial cells lack alignment (Fig. 1). These areas are prone to the formation of atherosclerotic plaques (4). Shear stress acting on endothelial cells is higher in arterial vessels (~15–30 dyn/cm^2 (~1.5–3 N/m^2), compared with shear stress in venous vessels (~1 dyn/cm^2) (6). Shear stress in narrow, occluded arteries increases enormously, for example in an artery with 50% concentric stenosis shear stress may increase to 2,500 dyn/cm^2 (7).

Realignment of endothelial cells under flow is essentially a two-step process involving rapid reorganization of the cell cytoskeleton (8, 9). In the first stage, the cells round up and lose their original orientation (minutes) and then re-spread and migrate within the direction of flow (hours) (Fig. 2). The first step is mediated predominantly by small GTPase RhoA, while the second step requires sequential, coordinated activation of Rho GTPases Rac1, Cdc42, and RhoA (8–10). Cell rounding is less pronounced in confluent endothelial cells, where cells can remodel their shape without losing junctional integrity (11, 12).

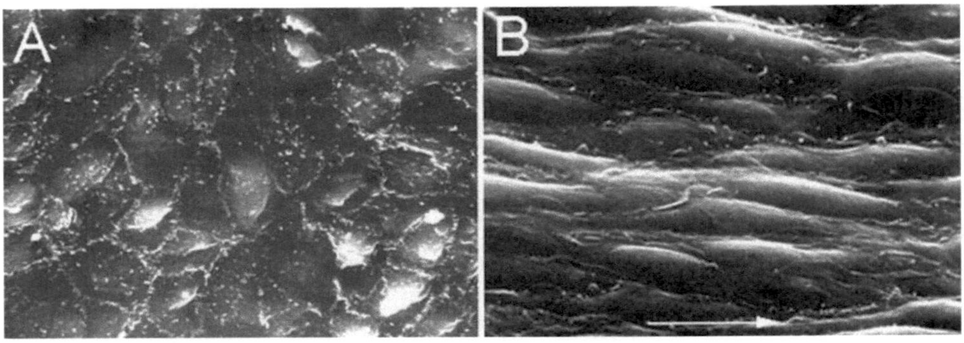

Fig. 1. Differences in endothelial cell alignment between cells from the area of aortic branching (**a**) and main aorta (*rabbit*) (**b**). Images from the Pathology of Atherosclerosis by N. Woolf (20).

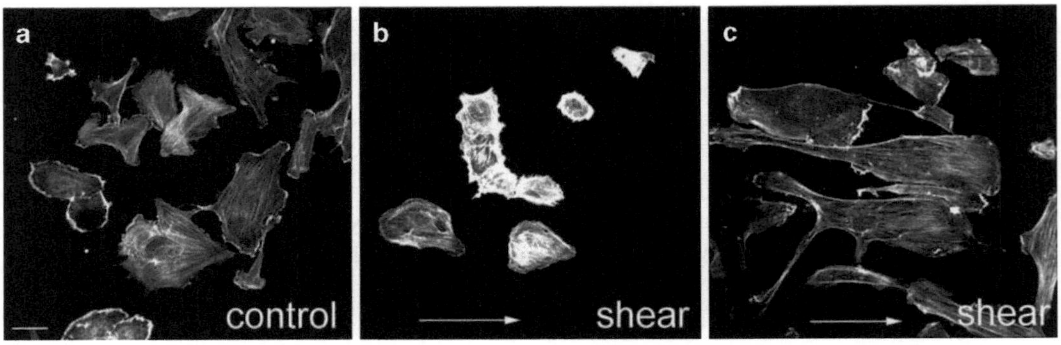

Fig. 2. Alignment of HUVECs under shear stress in vitro, parallel plate flow chamber. The cells were left untreated (**a**) or were subjected to shear stress of 3 dyn/cm^2 for 15 min (**b**) or 4 h (**c**). Bar = 20 μm. The *arrow* in (**c**) indicates the direction of flow. The image (**c**) was previously published in (9).

1.2. Parallel Plate Flow Apparatus

The parallel plate apparatus represents a convenient and inexpensive way of analysing cells under flow. The chambers are transparent and portable allowing direct observation of cells (4, 9). The design of the parallel plate flow system described here was based on the one described by Braddock et al. (4, 13).

A glass plate or a slide with growing cells is inserted into the flow chamber. A peristaltic pump creates laminar flow of media over the endothelial cells in a closed sterile system with controlled temperature and CO_2 levels (Fig. 3). In this system cells can be cultured for several days. Flow chambers can be set in series to allow exposure of several slides to shear at the same time. The peristaltic pump capacity, chamber dimensions, and the internal diameter of capillary tubing determine the value of shear stress, which usually falls between 1 and 80 dyn/cm^2 (4). The parallel plate flow chamber produces plane Poiseuille flow. The flow rate through the assembled system is calibrated by collecting the volume of fluid discharged over a given time.

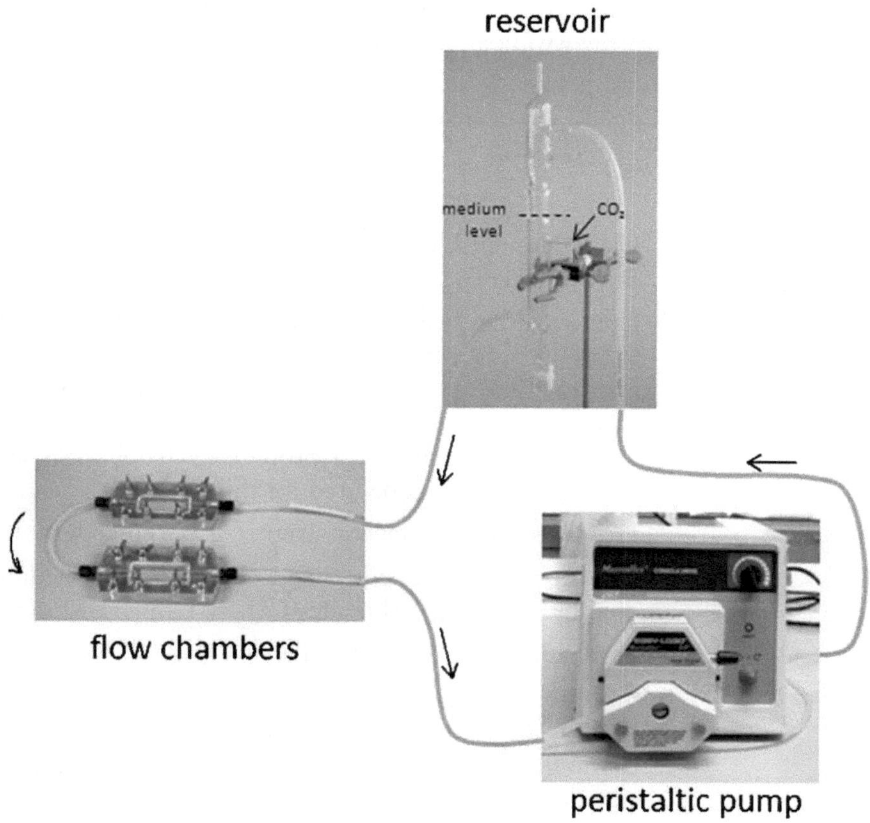

Fig. 3. Schematic diagram of a parallel plate flow chamber system.

For a laminar flow system, wall shear stress (ω) is defined as:

$$\omega = 3\mu Q/2a^2 b$$

where ω is the wall shear stress (dyn/cm^2); μ is the viscosity at 37°C (poise), in our system $\mu = 0.007$ poise; Q is the volumetric flow rate (ml/s), determined by the rate of peristaltic pump flow and the diameter of tubing; a is the half channel height (cm), in our system $a = 0.0075$ cm; b is the channel width (cm), in our system $b = 1.3$ cm.

1.3. Cone-and-Plate Apparatus (Not Shown Here)

In this device cells are cultured on a stationary plate while a rotating cone placed above the cell layer produces a stable, three-dimensional flow (14, 15). It has been shown that fluid shear stress of 0–200 dyn/cm^2 generated in this apparatus is laminar and predominantly oriented in a circular direction (14). The commercially available automated rheological in vitro system based on the cone-and-plate viscosimeter allows exposure of several cone plate chambers at the same time. The system allows computerized control of shear stress and may be coupled to other devices such as a microscope, video camera, or impedance spectroscopy (http://www.mos-technologies.de).

2. Materials

1. Cells: Human Umbilical Vein Endothelial Cells (HUVECs; Biowhittaker).
2. Culture dishes. Cells are grown in polystyrene slide flaskettes 9 cm^2 (Nunc cat. No 170920) (for small chambers). To obtain larger numbers of cells for biochemical analysis, cells may be grown on 80/100 mm glass plates (outer glass plates used for the assembly of protein electrophoresis apparatus mini Protean II can be used for this purpose).
3. Human serum fibronectin (Sigma, 1 mg/ml solution).
4. Culture media. Specialised endothelial growth medium buffered by sodium bicarbonate/CO_2 system (Biowhittaker).
5. Peristaltic pump: Masterflex Easyload L/S 7518-00/7543-20.
6. Silicone tubing: (sizes): L/S® 16 (Cole Parmer, cat no 96400-16, inside diameter 3.1 mm); L/S® 14 (Cole Parmer cat no 06485-14, inside diameter 1.6 mm); fine tubing – outside diameter 1.6 mm, inside diameter 1.2 mm.
7. Flow chambers (custom-made) consist of: (a) a stainless steel base plate with eight bolts and a machined depression to hold a slide; (b) machined Perspex with inlet and outlet ports and a gasket-retaining groove; (c) Neoprene gasket; (d) silicone

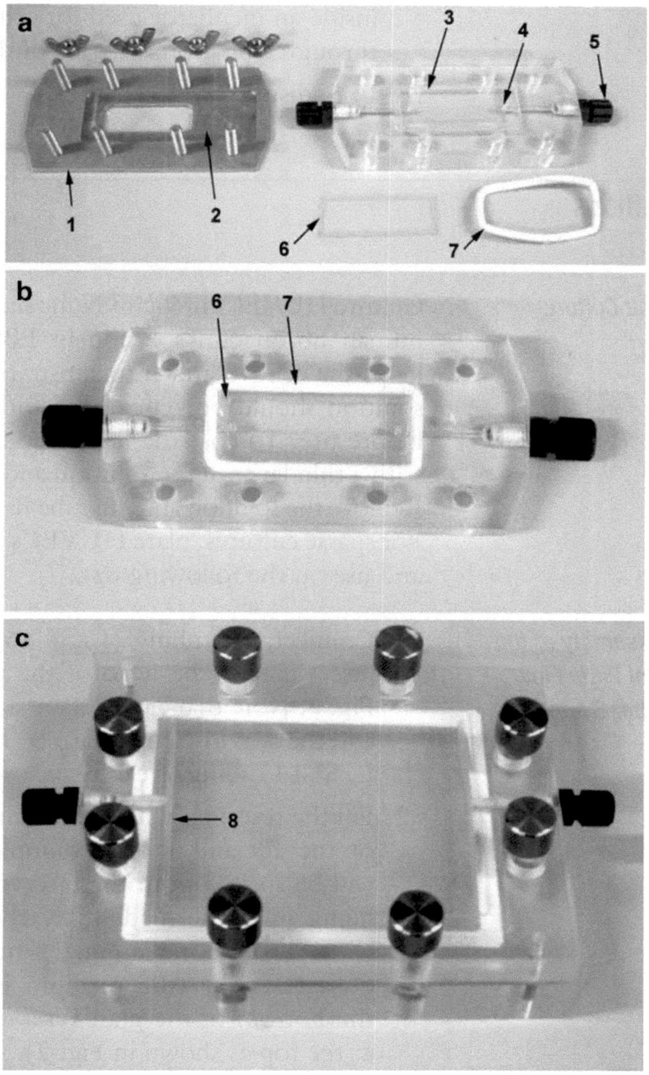

Fig. 4. Parallel plate flow chambers. (**a**) and (**b**) show individual components of a small flow chamber while (**c**) shows a large flow chamber suitable for the analysis of larger numbers of cells. 1, Stainless steel base with eight bolts and winged nuts; 2, machined depression to hold a slide; 3, gasket-retaining groove; 4, machined inlet/outlet for media flow; 5, inlet/outlet port for tubing; 6, silicone gasket; 7, neoprene gasket; 8, a groove in Perspex allowing an even distribution of media flow within the large chamber.

gasket thickness 0.5 mm (e) eight winged nuts. Flow chambers are shown in Fig. 4.

8. Glass media reservoir with inlets and outlets (custom-made), plastic tap.
9. Temperature and CO_2 control. The flow system is placed inside a Perspex box supplied with a heater and a temperature sensor. CO_2 from a compressed air/CO_2 cylinder is delivered to the medium via appropriate tubing. Alternatively, the flow system (tubing, media reservoir, and chambers) may be placed

inside an incubator used for cell culture with tubing inserted through a port on the back of the incubator.

10. PBS, distilled water, beakers.

3. Methods

3.1. Cell Culture

Culture HUVECs in 9-cm^2 Nunc slide flasks coated with 10 μg/ml human fibronectin solution in PBS (see Note 1). Fibronectin helps the cells to attach to substratum (see Note 2). Fibronectin solution should be kept in the fridge and pre-warmed to 37°C before use. To coat the flask, cover the growth area of the flask with a thin layer of the solution and incubate at 37°C for 15 min. Remove the solution and rinse the flask with PBS before cell plating. For sparse cultures, plate HUVECs at a density of 2×10^4 cells/ml and use on the following day.

3.2. Assembly of the Parallel Plate Flow Chamber System

1. Connect the tubing (L/S® 16) to the media reservoir and insert it into the head of the peristaltic pump as shown in Fig. 3. Note that at this stage flow chambers are not inserted, and the two lose ends of L/S® 16 tubing are connected by an L/S® 14 tubing, creating a closed circuit.
2. Fill the reservoir with culture medium using one of the inlets on the side and start the pump on high setting. The medium will be sucked from the reservoir into the tubing. Continue filling until the medium reaches the desired level (marked with a broken line in Fig. 2) and starts to circulate within the glass reservoir (the medium is sucked into the tubing at the bottom of the glass reservoir and returns via the outlet at the top as shown in Fig. 2).
3. Start CO_2 supply at a rate that produces a steady stream of bubbles in the reservoir and allow the system to equilibrate for about 15–30 min. Equilibration time is also needed if the apparatus is placed inside an incubator with controlled temperature and CO_2 levels (see Note 3).
4. Stop the pump. Place all the parts of the flow chamber (metal base with bolts, winged nuts, machined Perspex, Neoprene gasket, silicone gasket as shown in Fig. 4), sterile syringe, thin tubing, and a flaskette with cells in the laminar flow cabinet (see Notes 4 and 5).
5. Assemble the flow chamber like a sandwich. Place the metal plate at the bottom with bolts facing upwards. Detach the bottom of the slide flaskette and place it "cell side" up in the machined area of the metal plate. Drop a small amount of medium onto the cells to prevent them from drying out during the assembly. Place the Neoprene gasket in the groove of

the Perspex block and the silicone gasket on the inside, so that it does not obstruct the inlet and outlet holes (Fig. 4b). Turn the Perspex block "gasket side" down and place it on the metal base so that the metal bolts go through the holes. Fasten the two parts together with winged nuts (start with the opposing nuts). Insert two thin tubes (1.6 mm outer diameter) into the chamber's inlet and outlet ports.

6. Using 1 ml syringe with a needle fill the chamber very slowly with pre-warmed medium through the inlet tubing (the flow rate should not be higher than 1 ml/min). The chamber should be held vertically so that the medium level rises from the bottom to the top and all air bubbles leave the chamber through the outlet at the top. If there is any remaining air bubbles trapped on the chamber wall, tap gently to dislodge the bubbles. Allow some of the media leak through the outlet tubing.
7. Insert the chamber's tubing into the flow system (make sure you know the flow direction) and re-start the pump on a very low setting (the flow should not exceed 1 ml/min). Hold the chamber vertically with the inlet tubing at the bottom so that all the air bubbles can leave through the outlet at the top. It is important that there are no air bubbles trapped within the chamber as they may damage cells when travelling through the chamber. Rest the plate horizontally and let the medium circulate slowly. If needed, the next chamber may be assembled and inserted downstream of the existing chamber. The medium in the top reservoir may be replenished, as required.
8. To induce alignment of cells, shear stress should be increased stepwise, every 10 min (see Note 6). However, flow may also be changed rapidly to simulate sudden changes of shear stress or transition from stationary to shear conditions.

3.3. Disassembly and Cleaning Procedure

1. Stop the pump. Disconnect the chambers from the main tubing and close the media circuit by joining the loose ends of the tubing.
2. Remove the medium by opening the tap at the bottom of the glass reservoir. Start the pump again on high flow setting and let the remaining medium leak out.
3. Fill in the glass reservoir with water containing detergent (1% washing up liquid), switch the pump on high flow, and let the fluid circulate for at least 15 min. Let the fluid out, disconnect glass container, and wash it first in tap water and then in distilled water several times.
4. Insert one of the loose ends of the tubing into a large (2 l) beaker filled with distilled water and the other end into an empty beaker. Start the pump so that water is sucked from the beaker and is spewed out from the other end of the tubing. Use at least 2 l of water to flush the system.

5. Disassemble and wash all parts of the flow chambers with detergent and rinse thoroughly with water. Leave them to dry. The tubing should be sterilised by autoclaving (15 min at 120°C), while the glass reservoir and flow chambers may be autoclaved or sterilised with 70% ethanol.

3.4. Analysis of the Cell Shape and Migration

Flow chambers are transparent and may be placed under the microscope for time-lapse recording. The images are analysed with image analysis software to calculate parameters of cell shape and migration. Cell orientation (alignment) is an angle between the long axis of the cell and the chosen direction (usually taken as a vertical line). Cell elongation is taken as a proportion of the long axis of the cell to the shortest axis of the cell and varies between 0 and 1 (perfectly round cell) (16).

1. Place the assembled chamber on a heated stage of a microscope. Select a field of view with 15–20 evenly spaced cells.
2. Start filming. As endothelial cells do not move very quickly, it is sufficient to take frames every 10 min and analyse cell movement for at least 6–24 h.
3. Image analysis. Select the area of interest manually (by drawing around the cell perimeter) or automatically. Phase contrast microscopy provides images with varying levels of darkness/brightness within each cell, which often precludes automated selection of the cell as a whole. Fluorescent staining of cells (for instance with Fluorescent Cell Tracker dyes, Invitrogen) often enables a faster, automated analysis. Free image analysis software ImageJ may be used to calculate cell area, elongation, orientation, to visualise cell trajectories and calculate the speed of movement. Other commercially available software packages include Image Pro Plus, Volocity 3 (Improvision), and Kinetic Imaging. Figure 5 shows examples of such an analysis. The cells are digitally recorded by video microscopy using a time-lapse interval of 10 min over a 5–24 h period. Cells are tracked with Kinetic Imaging software, and the trajectories are statistically analyzed using Mathematica software (Wolfram Research Inc.) (17). Cell displacement is evaluated by calculating the percentage of the total cells analyzed that had migrated a distance of 50 μm or more from the starting point during 5 h in each experiment. To visualize the directionality of cell movement in response to shear stress, statistical analysis of directional data is used and displayed as direction plots (18).

A detailed protocol for cell tracking and analysis of cell motility with the use of Kinetic Imaging software can be found in another chapter of the Methods in Enzymology (19).

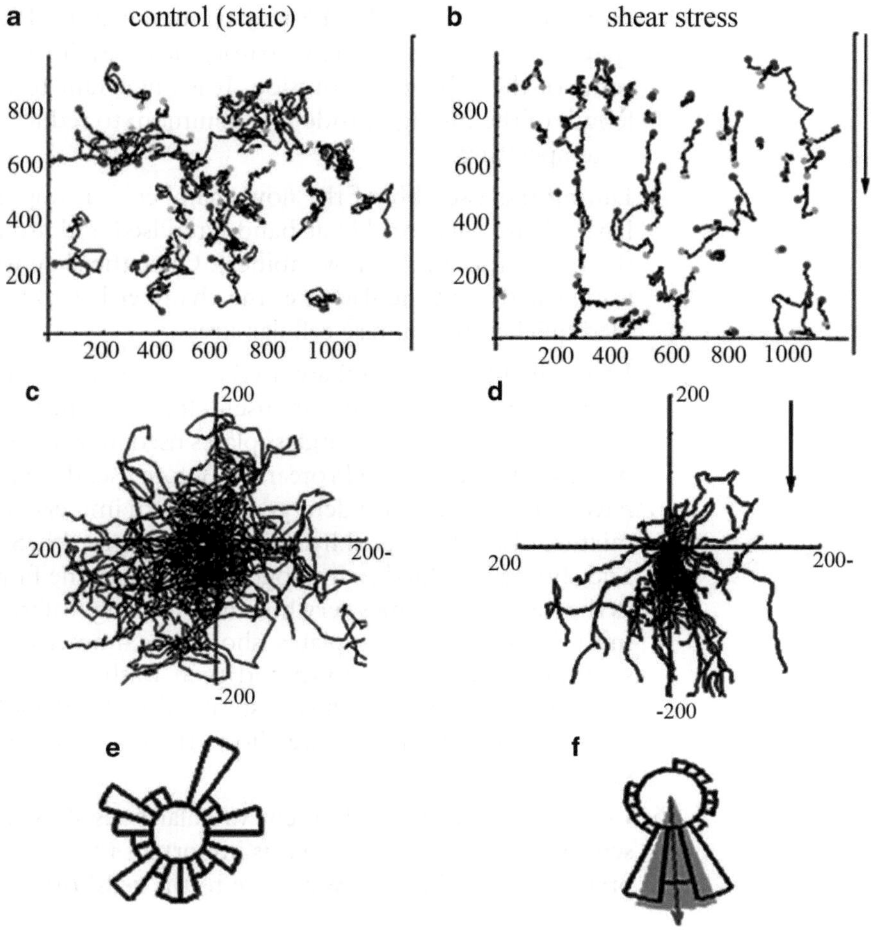

Fig. 5. Cell trajectories during a 12-h experiment are shown in static conditions (**a**) or under shear stress, 3 dyn/cm^2 (**b**). The starting point of each trajectory is plotted at the intersection of the *X* and *Y* axes (**c**, **d**). Circular histograms (**e** and **f**) show the proportion of cells migrating into each of 20 equal segments, when each cell had migrated 50 μm from its starting point. The arrow in (**f**) indicates the mean direction of the cell population when this is significant, and the marks the 95% confidence interval of statistical significance, as calculated using a Rayleigh test (9).

4. Notes

1. It is best to use HUVECs between passages 2 and 6. Early passage cells are well attached, show cobblestone morphology when confluent, and show no signs of overlapping. Cells at later passages (>8) tend to elongate and overlap in culture and detach more easily upon shear stress stimulation.
2. Bovine serum fibronectin can be used instead of human serum fibronectin as it is cheaper and equally effective in supporting adhesion of HUVECs.
3. CO_2 supply is not necessary for short-term experiments (minutes–few hours) and media w/o sodium bicarbonate

buffered by 10 mM HEPES may be used instead. If the flow system is placed inside an incubator, the peristaltic pump and some of the tubing stay outside. It is important to keep the length of the tubing outside to a minimum to reduce any loss of temperature.

4. During the assembly of the flow chamber, it is important to have all the parts readily at hand (sterilised and arranged in the right order in the flow cabinet). Once the slide with cells is detached from the flaskette, the chamber has to be assembled quickly to minimise cellular stress.
5. Large chambers (Fig. 4c) are used to expose high numbers of the cells to shear stress and are useful for biochemical analysis. We use 80/100 mm outer glass plates used in protein electrophoresis apparatus mini Protean II. Before use, the glass plates have to be cleaned with detergent and then immersed in a 3:1 mixture of sulphuric acid:hydrogen peroxide for 1 h. Sulphuric acid should be added to hydrogen peroxide in the fume hood as the mixture becomes very hot and bubbling (safety goggles must be worn). Glass plates should be immersed in this solution until it cools down and then washed several times with distilled water and finally sterilised with ethanol before use. This cleaning procedure allows the plates to be re-used several times.
6. Endothelial cells from different vascular beds show different sensitivity to shear stress, so it is important to use the appropriate levels of shear stress relative to the cells' origin.

References

1. Chien, S., Li, S., and Shyy, Y. J. (1998) Effects of mechanical forces on signal transduction and gene expression in endothelial cells. *Hypertension* **31**, 162–9.
2. Urbich, C., Dernbach, E., Reissner, A., Vasa, M., Zeiher, A. M., and Dimmeler, S. (2002) Shear stress-induced endothelial cell migration involves integrin signaling via the fibronectin receptor subunits alpha(5) and beta(1). *Arterioscler Thromb Vasc Biol* **22**, 69–75.
3. Azuma, N., Akasaka, N., Kito, H., Ikeda, M., Gahtan, V., Sasajima, T., and Sumpio, B. E. (2001) Role of p38 MAP kinase in endothelial cell alignment induced by fluid shear stress. *Am J Physiol Heart Circ Physiol* **280**, H189–97.
4. Braddock, M., Schwachtgen, J. L., Houston, P., Dickson, M. C., Lee, M. J., and Campbell, C. J. (1998) Fluid Shear Stress Modulation of Gene Expression in Endothelial Cells. *News Physiol Sci* **13**, 241–246.
5. Albuquerque, M. L., Waters, C. M., Savla, U., Schnaper, H. W., and Flozak, A. S. (2000) Shear stress enhances human endothelial cell wound closure in vitro. *Am J Physiol Heart Circ Physiol* **279**, H293–302.
6. Morawietz, H., Talanow, R., Szibor, M., Rueckschloss, U., Schubert, A., Bartling, B., Darmer, D., and Holtz, J. (2000) Regulation of the endothelin system by shear stress in human endothelial cells. *J Physiol* **525** Pt 3, 761–70.
7. MacIsaac, A. I., Thomas, J. D., and Topol, E. J. (1993) Toward the quiescent coronary plaque. *J Am Coll Cardiol* **22**, 1228–41.
8. Tzima, E. (2006) Role of small GTPases in endothelial cytoskeletal dynamics and the shear stress response. *Circ Res* **98**, 176–85.
9. Wojciak-Stothard, B., and Ridley, A. J. (2003) Shear stress-induced endothelial cell polarization is mediated by Rho and Rac but not Cdc42 or PI 3-kinases. *J Cell Biol* **161**, 429–39.

10. Tzima, E., del Pozo, M. A., Shattil, S. J., Chien, S., and Schwartz, M. A. (2001) Activation of integrins in endothelial cells by fluid shear stress mediates Rho-dependent cytoskeletal alignment. *Embo J* **20**, 4639–47.
11. Seebach, J., Dieterich, P., Luo, F., Schillers, H., Vestweber, D., Oberleithner, H., Galla, H. J., and Schnittler, H. J. (2000) Endothelial barrier function under laminar fluid shear stress. *Lab Invest* **80**, 1819–31.
12. Seebach, J., Donnert, G., Kronstein, R., Werth, S., Wojciak-Stothard, B., Falzarano, D., Mrowietz, C., Hell, S. W., and Schnittler, H. J. (2007) Regulation of endothelial barrier function during flow-induced conversion to an arterial phenotype. *Cardiovasc Res* **75**, 596–607.
13. Schwachtgen, J. L., Houston, P., Campbell, C., Sukhatme, V., and Braddock, M. (1998) Fluid shear stress activation of egr-1 transcription in cultured human endothelial and epithelial cells is mediated via the extracellular signal-related kinase 1/2 mitogen-activated protein kinase pathway. *J Clin Invest* **101**, 2540–9.
14. Schnittler, H. J., Franke, R. P., Akbay, U., Mrowietz, C., and Drenckhahn, D. (1993) Improved in vitro rheological system for studying the effect of fluid shear stress on cultured cells. *Am J Physiol* **265**, C289–98.
15. Buschmann, M. H., Dieterich, P., Adams, N. A., and Schnittler, H. J. (2005) Analysis of flow in a cone-and-plate apparatus with respect to spatial and temporal effects on endothelial cells. *Biotechnol Bioeng* **89**, 493–502.
16. Dunn, G. A., and Brown, A. F. (1986) Alignment of fibroblasts on grooved surfaces described by a simple geometric transformation. *J Cell Sci* **83**, 313–40.
17. Allen, W. E., Zicha, D., Ridley, A. J., and Jones, G. E. (1998) A role for Cdc42 in macrophage chemotaxis. *J Cell Biol* **141**, 1147–57.
18. Zicha, D., Dunn, G., and Jones, G. (1997) Analyzing chemotaxis using the Dunn direct-viewing chamber. *Methods Mol Biol* **75**, 449–57.
19. Wells, C. M., and Ridley, A. J. (2005) Analysis of cell migration using the Dunn chemotaxis chamber and time-lapse microscopy. *Methods Mol Biol* **294**, 31–41.
20. Woolf N. The Pathology of Atherosclerosis, Elsevier (1982), p. 28.

Chapter 11

In Vitro Analysis of Chemotactic Leukocyte Migration in 3D Environments

Michael Sixt and Tim Lämmermann

Abstract

Cell migration on two-dimensional (2D) substrates follows entirely different rules than cell migration in three-dimensional (3D) environments. This is especially relevant for leukocytes that are able to migrate in the absence of adhesion receptors within the confined geometry of artificial 3D extracellular matrix scaffolds and within the interstitial space in vivo. Here, we describe in detail a simple and economical protocol to visualize dendritic cell migration in 3D collagen scaffolds along chemotactic gradients. This method can be adapted to other cell types and may serve as a physiologically relevant paradigm for the directed locomotion of most amoeboid cells.

Key words: 3D, Interstitial migration, Chemotaxis, Chemokine gradient, Cell motility, Collagen, Extracellular matrix, Interstitium, Connective tissue

1. Introduction

Stromal cells, together with their secreted extracellular matrix, constitute the structural and mechanical backbone of most organs. Collagen type I has been widely used to re-create three-dimensional (3D) networks for the in vitro study of organogenesis, vessel development, fibril assembly, fibroblast mechanics, cell polarity, tumor metastasis, and leukocyte migration (1–7). Although other 3D in vitro systems exist (with Matrigel and fibrin gels as the next most common), 3D collagen I gels most closely resemble the in vivo interstitial matrix in connective tissues as collagen I is the major constituent of most interstitial tissues (4). Importantly, it is becoming increasingly evident that cell migration analysis in 3D

Claire M. Wells and Maddy Parsons (eds.), *Cell Migration: Developmental Methods and Protocols*,
Methods in Molecular Biology, vol. 769, DOI 10.1007/978-1-61779-207-6_11, © Springer Science+Business Media, LLC 2011

systems reflects physiological cell behavior much better than results obtained from 2D in vitro systems (e.g., cell culture dish) (2, 5, 8–10). The physical nature of collagen gels (density, pore size, stiffness, and bundle thickness) is highly dependent on the initial collagen source and extraction process (4). Most commonly, 3D in vitro collagen models are based on rat-tail or bovine dermis collagen, which differ in their chemical cross-linking, extraction protocol, and fiber assembly, leading to discrepancies in experimental outcome (4, 11).

Migrating cells are characterized by their polarized appearance with a leading and a trailing edge. Cells either undergo spontaneous polarization or polarize in response to gradients of chemokines, cytokines, growth factors, or extracellular matrix fragments. In connective tissues, directional migration toward a chemotactic source has been observed for neutrophils, monocytes (toward sites of inflammation and tissue injury) (12), dendritic cells, and tumor cells (toward lymphatic vessels) (2, 13, 14). Cells chemotaxing in fibrillar 3D networks face different challenges than in 2D systems. Apart from polarizing, they also have to maintain integrity while navigating through a geometrically complex porous environment (7, 15). While the 3D chemotaxis assays can be studied in collagen gel-coated transwells, this experimental system is restricted to end-point analysis and it is not trivial to differentiate between a general cell motility response to a chemokine (chemokinesis) and true directional migration (chemotaxis). Time-lapse video microscopy overcomes these limitations and allows direct analysis of parameters such as directionality (chemotactic index) and cell velocity. Simple cell culture bright-field microscopes equipped with a heating chamber and triggered by software are sufficient to record low magnification movies and determine chemotactic parameters (2, 8). For more detailed analysis of moving cells in the context of their fibrillar environment, both polarized cells and collagen fibers can be visualized in high magnification with different kinds of microscopy (Table 3) (2, 4, 9, 16).

We provide here a simple method for visualizing chemotaxis of cells in 3D bovine collagen gels by time-lapse video bright-field-microscopy and analysis of cell velocity and directionality. This experimental setup can also be extended to the study of polarized cells and cell–collagen matrix interactions with confocal microscopy. Our primary intention here is to describe a method that can be quickly established in almost every laboratory without the acquisition of extremely specialized or costly materials, reagents, or instruments. 3D assay systems that include other tissue parameters such as lymphatic flow in connective tissues have been detailed elsewhere (14, 17).

2. Materials

2.1. Cell Culture

1. Murine long bones (femurs).
2. One pair of standard forceps (curved or straight, with serrated tips).
3. One pair of scissors (straight, small).
4. Petri dishes (sterile, 100 × 15 mm).
5. Tissue-culture dishes (sterile, 60 × 15 mm).
6. R10 medium: Heat-inactivate fetal calf serum (FCS) or fetal bovine serum (FBS) in a 56°C water bath for 30 min and then sterile-filter it (0.22 μm). R10 medium is RPMI-1640 supplemented with 2 mM L-glutamine, 100 U/mL penicillin, 100 μg/mL streptomycin, and heat-inactivated, filtered FCS or FBS (10%) (see Note 1).
7. Plastic syringes (sterile, 20 mL).
8. Needles (sterile, 23G 1″ or 25G 1″).
9. LPS solution: Lipopolysaccharides from *Escherichia coli* 0127:B8 (suitable for cell culture, γ-irradiated, Sigma, St. Louis, MO). Prepare a 1 mg/mL solution in sterile phosphate-buffered saline (PBS) and freeze aliquots at −20°C (see Note 2).
10. GM-CSF solution: Recombinant murine granulocyte macrophage-colony stimulating factor (GM-CSF). Prepare a 50 μg/mL solution in sterile water and freeze aliquots at −80°C.
11. Polypropylene conical centrifuge tube (50 mL).

2.2. Paraffin Mix

1. Paraffin: paraffin wax (mp > 60°C) or paraffin pellets.
2. Petroleum jelly (e.g., Vaseline).
3. Laboratory hot plate.

2.3. Standard Migration Chamber

1. Microscopy glass slides (microslides, 75 × 25 mm). As an alternative you can use glass-bottom culture dishes (35 mm, 7–14 mm glass, No. 1.5 thickness) (MatTek, Ashland, MA).
2. One paint brush (brush size: 6–12, flat or round brush, any bristle material will work).
3. Microscope cover glasses (18 × 18 mm, No. 1.5 thickness).

2.4. Collagen Network and Chemokine Gradient

1. Bovine collagen solution: Purecol® (3 mg/mL) or Nutragen® (6 mg/mL) (both Advanced Biomatrix, CA). Both collagen preparations are soluble atelo-collagen in 0.01 N HCl; therefore, the pH is ca. 2.0 (see Note 3), store at 4°C.
2. Sodium bicarbonate solution ($NaHCO_3$, 7.5%, Sigma, St. Louis, MO), sterile-filtered, cell culture tested, store at 4°C.

3. Minimal essential medium (Eagle) (10×, Sigma), with Earle's salts, without L-glutamine and sodium bicarbonate, liquid, sterile-filtered, cell culture tested, store at 4°C.
4. Plastic microslide box (to hold 75×25 mm microslides, as used for histology samples).
5. CCL19 solution: Recombinant murine CCL19 (MIP-3β). Prepare a 25 μg/mL solution in sterile PBS and 1 mg/mL bovine serum albumin (BSA) and store aliquots at –20°C.

2.5. Bright-Field Video Microscopy

For low magnification bright-field movies, any standard bright-field microscopy setup with a heated microscopy stage, 10× or 20× objective, camera, and software-triggered recording will be suitable.

Our experimental setup comprised an inverted Axiovert 40 (Zeiss) cell culture microscope, equipped with custom-built climate chambers (5% CO_2, 37°C, humidified) and PAL cameras (Prosilica) triggered by custom-made software (SVS Vistek). For parallel recording of several experiments, we ran one experiment at one of three microscopes simultaneously. Other microscopes are available which allow several experiments to run simultaneously by recording multiple stage positions at defined time intervals.

2.6. Analysis of Dendritic Cell Chemotaxis with ImageJ

1. ImageJ free software (http://rsbweb.nih.gov/ij/, National Institutes of Health, Bethesda, MD).
2. Image J plugins: Manual Tracking, Chemotaxis, and Migration Tool (http://rsbweb.nih.gov/ij/plugins).

2.7. Preparing Fluorescent Collagen

1. Alexa Fluor 647 carboxylic acid, succinimidyl ester (A-20006, Invitrogen).
2. Dimethyl sulfoxide (DMSO).
3. Dialysis solution: Add 1.15 mL glacial acetic acid (17.4 N) to 999 mL distilled water to create glacial acetic acid of 0.02 N and pH 3.9.
4. Dialysis tubing (molecular weight cut off (MWCO) 3,500 Da).
5. Two dialysis clamps.

3. Methods

The whole procedure is comprised of four principal steps: (1) Construction of a migration chamber (which can be done on the day before the experiment), (2) casting a collagen gel matrix containing cells (which takes ca. 45–60 min), (3) application of a chemotactic gradient and subsequent time-lapse video microscopy, and (4) analysis of the imaging data with ImageJ software.

This method is not limited to bright-field microscopy only, but can also be used to visualize chemotactic cells in the context of their fibrillar environment with different kinds of confocal microscopy at high magnification and resolution. Table 3 summarizes the common techniques used to visualize collagen networks and we describe in Subheading 3.10 how to generate fluorescent collagen gels. Although this method is potentially applicable to any chemotactic cell type, we will describe a protocol for the generation of dendritic cells from murine bone marrow (18). Dendritic cell migration toward the chemokine CCL19 shows very robust directional migration over several hours and can be used as a positive control to test other 3D chemotaxis setups (2, 8).

3.1. Generation of Primary Dendritic Cells from Murine Bone Marrow

1. Euthanize the mouse according to your animal study protocol and local regulations.
2. Dissect the mouse and carefully remove the legs from the pelvic bone without destroying the head of the femur. The femur and tibia can be separated by cutting the knee ligaments with scissors. Once separated, rub the muscles gently off the femoral bone with utility wipes. Place the muscle-free femurs in a Petri dish with 70% ethanol and continue your work in a biological safety cabinet.
3. Immerse the femurs in 70% ethanol for 1 min and let them air-dry in a new Petri dish. With sterile forceps and scissors (e.g., incubated in 70% ethanol before), pick up the femur in the middle of the shaft with the forceps and cut off both ends with the scissors.
4. Hold the femur (with the forceps) over a centrifuge tube and flush out the red marrow with sterile PBS, using a syringe and a needle. The marrow will appear as a red strand in the tube and the cavity of the femur will turn from red to white color.
5. Centrifuge the resultant cell suspension for 5 min at $300 \times g$ (4–20°C).
6. Discard supernatant and take the cell pellet up in R10 medium, then adjust the cell concentration to 2.5×10^6 cells/mL.
7. For each Petri dish, use 1 mL bone marrow-cell suspension to 9 mL R10 medium, then add 4 μL GM-CSF solution to a final concentration of 20 ng/mL GM-CSF. Incubate the cells at 37°C, 5% CO_2, 95% H_2O. This is the start day of the dendritic cell culture (day 0).
8. Three days later (day 3), add 10 mL R10 medium and 4 μL GM-CSF solution.
9. Three days later (day 6), carefully remove 10 mL of the medium and add again 10 mL R10 medium and 4 μL GM-CSF solution. Until days 8–9, the culture is highly enriched for immature dendritic cells.

10. At days 8–9 of culture, collect the medium and non-adherent cells in suspension in a centrifuge tube and centrifuge the cells for 5 min at 300 × *g* (4–20°C).
11. Discard the supernatant and take up the cell pellet of one Petri dish (100 × 15 mm) in 10 mL R10 medium. Transfer this cell suspension to one tissue-culture dish (60 × 15 mm) and add 4 μL GM-CSF solution. The adhesive surface of the tissue-culture dish will further separate adherent immature dendritic cells and remnant macrophages from non-adherent mature dendritic cells.
12. To induce maturation of dendritic cells, add 2 μL LPS solution to the tissue-culture dish. After 24–36 h of incubation at 37°C, 5% CO_2, 95% H_2O, mature dendritic cells have a characteristic morphology (Fig. 3a, see Notes 1 and 2). Do not use dendritic cells at time points later than 36 h after LPS stimulation, as they then undergo apoptosis.

3.2. Preparation of the "Paraffin Mix"

1. Heat paraffin wax or pellets at 60–80°C/140–178°F in a small beaker (100 mL size) on a hot plate until the paraffin is fluid.
2. Prepare the paraffin mix by adding petroleum jelly to the fluid paraffin in a 1:3 – 1:5 ratio by volume (see Note 4).

3.3. Preparation of the "Standard Migration Chamber"

1. Take a glass slide (or glass-bottomed microscopy dish, see Note 5) (Fig. 1a).
2. Take the brush, dip it into the paraffin mix, and paint a "squared U" (ca. 20 × 20 mm) with the paraffin in the center of the glass slide (Fig. 1b). The paraffin will quickly become solid on the colder glass slide. The paraffin line should be ca. 3–5 mm in width. Repeat this step 2–3 times until the paraffin line is approximately 1-mm thick in height.
3. Take the coverslip (18 × 18 mm) and lay it on top of the "squared U"-paraffin line (Fig. 1c).
4. Fix the coverslip with your fingers and brush a layer of paraffin mix over the border of the coverslip. The coverslip should be sealed between the paraffin lines providing a migration chamber that can be filled from its open top (Fig. 1d).

3.4. Preparation of the Cell Suspension

1. After 24–36 h of stimulation of dendritic cells with LPS, take off the supernatant from the 6-cm cell culture dish (collect the non-adherent mature dendritic cells) and transfer it to a 50-mL plastic tube. You might have to pool cells from several dishes.
2. Centrifuge the cell suspension for 5 min at 300 × *g* (4–20°C).
3. Discard the supernatant and take the cell pellet up in 1 mL R10 medium.

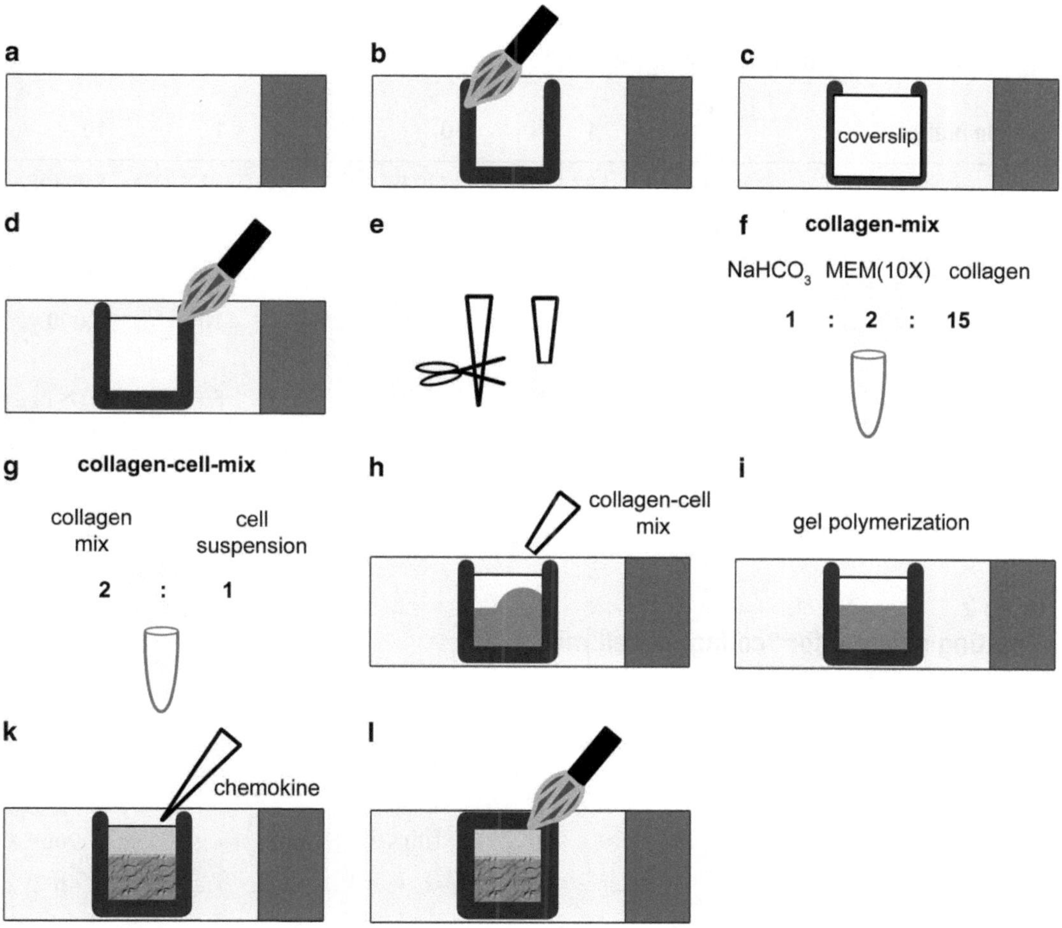

Fig. 1. General scheme (for detailed instructions, please see Subheadings 3.2–3.8). (**a–d**), building the migration chamber; (**e–i**), casting the 3D collagen gel; (**k** and **l**), application of the chemokine gradient.

4. Count the cell number and then adjust the cell concentration to 3×10^6 cells/mL with R10 medium (see Note 6).
5. Keep the cell suspension at 37°C, 5% CO_2 (before continuing with it in Subheading 3.6) (see Note 7).

3.5. Preparation of the "Collagen Mix"

1. Take a 1.5-mL reaction tube.
2. Cut off one-third of some pipette tips (Fig. 1e, see Note 8).
3. For a standard collagen gel of 1.7 mg/mL, prepare a collagen mix of sodium bicarbonate, MEM (10×), and Purecol® (3 mg/mL) in a 1:2:15 ratio (Fig. 1f, see Note 9). Calculate 100 µL of collagen mix for one standard migration chamber, although less is actually needed (Tables 1 and 2, see Note 10).

3.6. Preparation of the "Collagen–Cell Mix"

1. Take a 1.5-mL reaction tube.
2. Cut off one-third of some pipette tips.

Table 1
Pipetting scheme for the "collagen mix"

Sample number		1	10		1	10
Collagen[a] (μL)	75 (P)	83.3 (P)	833 (P)	75 (N)	83.3 (N)	833 (N)
MEM (10×) (μL)	10	11.1	111	10	11.1	111
$NaHCO_3$ (7.5%) (μL)	5	5.6	56	5	5.6	56
Final volume of "Collagen mix" (μL)	90	100	1,000	90	100	1,000
Collagen conc. in "Collagen mix" (mg/mL)	2.5 (P)	2.5 (P)	2.5 (P)	5 (N)	5 (N)	5 (N)

[a]P = Purecol, N = Nutragen

Table 2
Pipetting scheme for "collagen–cell mix"

Sample number	1	10	1	10	1	10
Collagen conc. in "Collagen mix" (mg/mL)	2.5 (P)	2.5 (P)	2.5 (P)	2.5 (P)	5 (N)	5 (N)
"Collagen mix" (μL)	90	900	100	1,000	90	900
Cell suspension (μL)	50	50	50	500	50	50
R10 medium (μL)	10	10	–	–	10	10
Final volume of "Collagen–cell mix" (μL)	150	1,500	150	1,500	150	1,500
Final Collagen conc. in "Collagen–cell mix" (mg/mL)	1.5	1.5	1.7	1.7	3	3

P = Purecol, N = Nutragen

3. For a standard collagen gel of 1.7 mg/mL, prepare a collagen–cell mix of collagen mix and cell suspension (3 mg/mL) in a 2:1 ratio (Fig. 1g) (see Note 11). Make up 150 μL of collagen–cell mix for one standard migration chamber, although less is actually needed (Table 2).

3.7. Polymerization of the Collagen Network (Filled with Cells)

1. Cut off one-third of some pipette tips (as described in Subheading 3.4) and use in the following steps.
2. Place the migration chamber (from Subheading 3.3) upright in a slide holder box.

3. Fill two-thirds of the migration chamber by pipetting the collagen–cell mixture from the upper open side of the chamber and let the mixture drop into the chamber by gravity (Fig. 1h).
4. If the collagen–gel mix does not immediately form an even horizontal lining, then take the migration chamber out of the slide holder, tap it several times gently on the bench, and then put it back into the slide holder (see Note 12).
5. Put the slide holder (with migration chamber(s)) in an incubator at 37°C, 5% CO_2 for at least 30 min for gel polymerization (Fig. 1i, see Note 13).

3.8. Application of the Chemokine Gradient and Time-Lapse Video Microscopy

1. Dilute the CCL19 stock concentration in R10 medium to a concentration of 50–500 ng/mL. You will require ca. 50 μL for one standard migration chamber.
2. Take the slide holder (with the migration chamber(s)) out of the incubator.
3. Fill the empty, upper one-third of the migration chamber with the CCL19 dilution (50–500 ng/mL) laying it on top of the collagen–cell mixture (Fig. 1k). The diluted chemokine will diffuse quickly into the gel forming a chemotactic gradient within 5–30 min.
4. Seal the open upper end of the migration chamber with the paraffin mix. The migration chamber represents a closed system (Fig. 1l, see Note 14).
5. Start the time-lapse video microscopy soon after applying the chemokine solution. With high chemokine concentrations, the first chemotactic response of dendritic cells can be seen within few minutes at the collagen gel–medium interface.
6. The resultant CCL19 gradient is reasonably stable over time and the dendritic cells directionally migrate toward the chemotactic source for a number of hours. For standard chemotaxis analysis, we employed time-lapse video microscopy with an inverted bright-field microscope, 10× objective, one frame per minute over 3–4 h (see Notes 15 and 16).

3.9. Analysis of Dendritic Cell Chemotaxis with ImageJ

This section outlines the minimal requirements for analyzing 3D collagen chemotaxis with the free online software ImageJ that is accessible to everyone. This method allows manual tracking of individual chemotactic cells and is sufficient to determine the basic chemotactic parameters (velocity and directionality) (see Note 17). This analysis method requires two additional plugins (Manual Tracking, Chemotaxis, and Migration Tool) that can also be downloaded from the ImageJ homepage (http://rsbweb.nih.gov/ij/plugins). Several data types of the initial time-lapse image sequence can be analyzed; we will describe the analysis of a sequence of .tiff files.

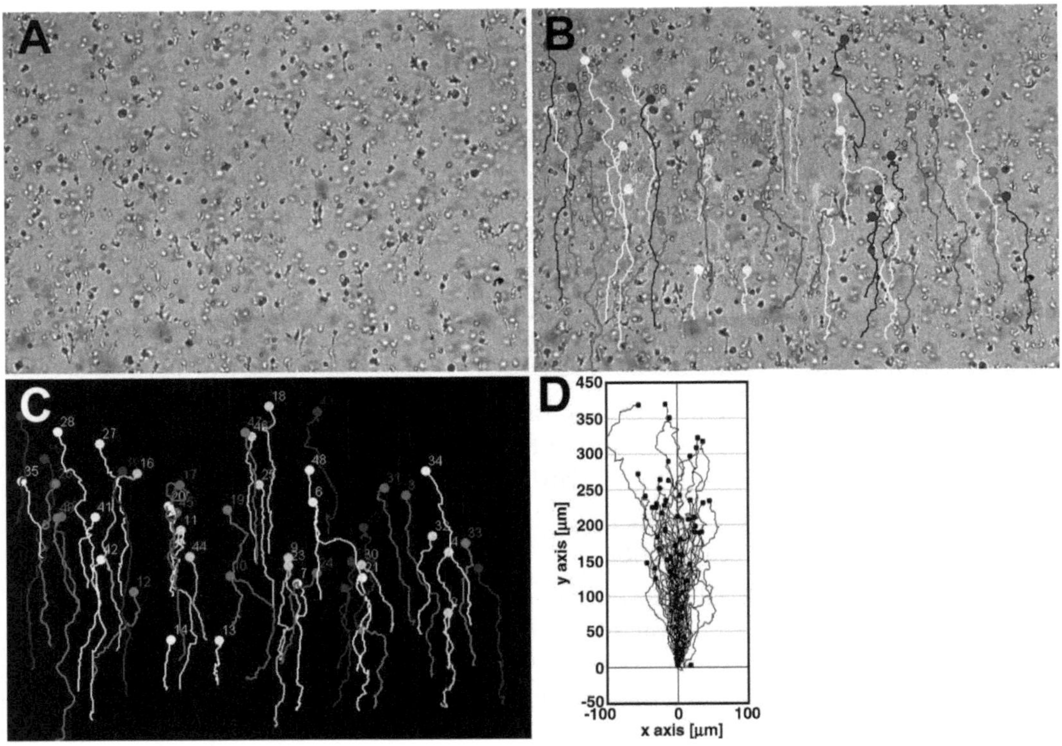

Fig. 2. Chemotaxis analysis with ImageJ (for detailed instructions on how to use ImageJ, please *see* Subheading 3.9). (**a**), original image stack; (**b**), image stack overlaid with cell tracks (overlay *dots* and *lines*); (**c**), cell tracks only (*dots* and *lines*); (**d**) chemotaxis plot graph.

1. Open ImageJ and import your time-lapse image sequence (File – Import – Image Sequence).
2. Under "sequence options," choose the details of your data set to be analyzed, which includes number of images, starting image, and image increment. This will then display the image sequence as image stack (Fig. 2a) that can be animated (Image – Stacks – Start Animation).
3. Open the Manual Tracking plugin. The commands (4 and 5) are performed in the Manual Tracking plugin window.
4. Select "add track," choose one chemotaxing cell and follow its track by clicking into the image stack (one click per frame). The track will either finish automatically when you reached the last frame or you click "end track" before (e.g., when a cell moves out of the field of view). Then choose the next cell and continue as before. All tracks will automatically be numbered. The coordinates for every single click are the basis for the resulting track and calculated parameters. They are automatically recorded as a .txt file which must be stored at the end of the tracking. Saved .txt files can be loaded at any time with "Load Previous Track File."

5. If you want to display the resultant tracks of several cells, you can choose between different options. Figure 2b shows "overlay dots and lines," and Fig. 2c shows "dots and lines" of the image stack.
6. Open the Chemotaxis and Migration Tool plugin. All the following commands are operated in this box.
7. Select "Import data" and choose your saved .txt file. Under "number of slices," select "Use slice range from … to …," type the first and last number of your image stack, and then select "add dataset."
8. Check one "selected dataset." You can analyze up to four data sets at once.
9. Under "settings," set your X/Y calibration and time interval. All further parameters will be calculated by the numbers and units that are selected here.
10. Then "Apply settings."
11. When choosing "Show Info," you will immediately get different calculated average parameters that also include directionality (value 0-1, whereby 1 corresponds to chemotaxis along a straight line toward the chemotactic source) and velocity (speed).
12. For obtaining graphic data, we prefer to present chemotaxis data as a plot graph (Fig. 2d). Therefore, select "plot feature," then check "open in new window" and select "plot graph."
13. To adjust the size of the graph, go to "set axis scaling" and select "manual" for your preferred axis lengths. Then select "plot graph" again. This graph can be saved as .jpg or .tiff file when choosing "save as" in the ImageJ toolbar.

3.10. Labeling Collagen with a Fluorescent Dye

To study moving cells in their fibrillar environment, collagen networks can be visualized by several technical means (see Note 18, Fig. 3b–f, Table 3). While differential interference contrast (DIC) and confocal reflection microscopy make use of the physical properties of unlabeled collagen fibers, confocal laser scanning and spinning-disc microscopy on fluorescent collagen gels provide some other advantages. The pros and cons of the three methods are listed in Table 3. To generate fluorescent collagen, any fluorescent reactive dye can be coupled to pure collagen. Here, we describe collagen labeling with Alexa Fluor 647 carboxylic acid, adapted from the suggested manufacturer protocol.

1. Dissolve 5 mg of Alexa Fluor 647 dye into 0.5 mL DMSO. As the reactive compounds of the dye are not stable in solution, dissolve the dye immediately before starting the labeling solution.
2. Place 10 mL of Purecol® in a small beaker with a stir bar and slowly stir it on a magnetic plate. Slowly add 100 μL of the

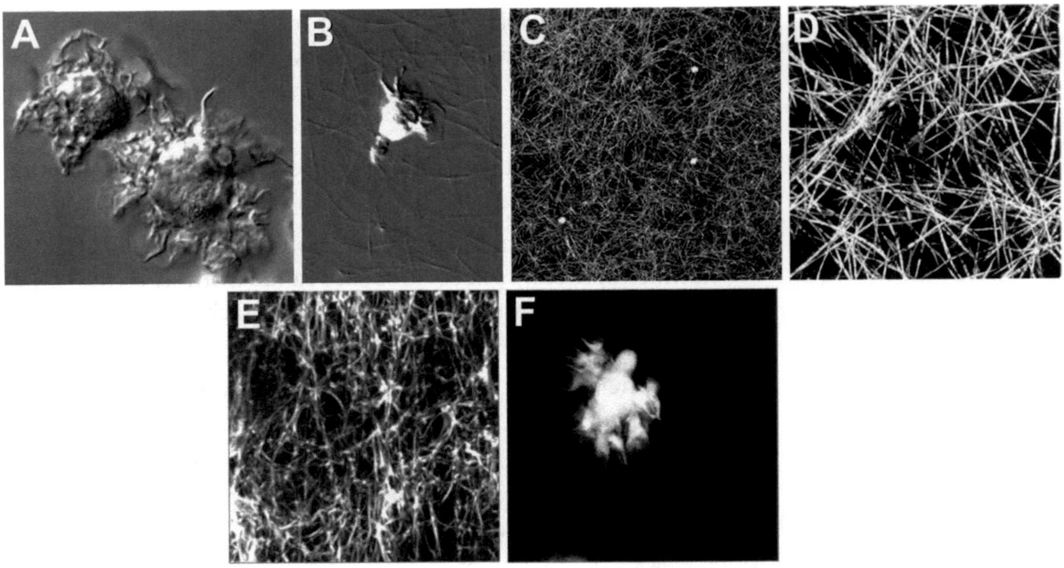

Fig. 3. Different ways of visualizing cells and the collagen network. (**a**) 24–36 h after LPS stimulation, mature dendritic cells are non-adherent and show prominent protrusions ("veils"). (**b**) Cells and collagen fibers are visualized by differential interference contrast (DIC) microscopy. (**c** and **d**) Merged confocal stacks of fibrillar collagen networks obtained by reflection microscopy. Cells might also give a signal by reflection (*small round circles* in **c**). (**e** and **f**) Merged confocal stacks of fluorescent collagen networks. If collagen (**e**) and cells (**f**) are differentially fluorescently labeled, they can clearly be distinguished and separated. We courteously thank Dr. Caren Petri Aronin for kindly providing Fig. 3e and f.

Table 3
Pros and cons for different ways of visualizing the collagen network

	Pros	Cons
Differential interference contrast (DIC) (Bright-field microscopy)	– no collagen labeling required – requires only DIC objective	– collagen fibers only visualized in one plane of focus – topographically inaccurate
Reflection (confocal microscopy)	– no collagen labeling required – requires only laser scanning confocal microscope	– can produce reflection of the cells as well – requires high laser power – not suitable for spinning-disc confocal microscopy
Fluorescence (confocal microscopy)	– suitable for spinning-disc confocal microscopy – requires only low laser power	– requires fluorescent collagen – varying degrees of photobleaching of the fluorescent collagen over time

Alexa Fluor 647 reactive dye to the stirred Purecol®. Then, continue stirring in the dark for 48 h at 4°C. After labeling, free dye molecules have to be removed by dialysis.

3. Prepare 3.5 in. of dialysis tubing, pre-soak it for 20 min in distilled water, and then rinse it thoroughly with clean distilled

water. Set one clamp 1 in. apart from the bottom side of the tubing.

4. Fill the 10 mL acidic, dye-labeled collagen solution into the tubing and place the second clamp at the other side to form a not too tightly packed, closed dialysis bag in the middle of the tubing.
5. Place the dialysis bag in a large beaker with 1 L acetic acid and slowly stir for 1 week at 4°C in the dark. Try to prevent the dialysis bag touching the stir bar by fixing it at the upper part of the beaker. Keep the Alexa Fluor 647-labeled collagen solution at 4°C until use.
6. To generate fluorescent collagen gels, we recommend mixing Alexa Fluor 647-labeled collagen with unlabeled collagen in a 1:20 ratio for the collagen–mix (see Subheading 3.5).

4. Notes

1. Dendritic cells with a high migratory potential are morphologically characterized by (1) loss of adhesion to the tissue-cell culture plastic 24–36 h after LPS stimulation and (2) prominent cell protrusions ("veils") all over the cell body. We and other investigators have observed that this phenomenon is highly dependent on the batch or lot of FCS/FBS used in the R10 medium. Since the composition of serum varies between manufacturers and between batches, we recommend an initial assessment of the effect of different sera on DC maturation. This can be achieved by flow cytometric assessment of surface activation markers (such as MHCII, CD86, CD80, and CD40), although the presence of such markers does not completely guarantee that the mature dendritic cells will have a high migratory potential.
2. Incubating immature dendritic cells at days 8–9 of culture with LPS is only one way to acquire migratory DCs. Alternatively, other stimuli such as 500 U/mL TNF-α can be used (19).
3. We have used Purecol® and Nutragen® (2, 8), but bovine collagen preparations are also provided by other manufacturers.
4. Petroleum jelly will make the paraffin more fluid and easier to brush. You may wish to test the paraffin mixture by brushing it onto a test glass slide.
5. Commercially available glass-bottomed microscopy dishes (e.g., MatTek) can be used as an alternative to glass slides. These dishes already have some spacing between the glass-bottom and cell culture plastic, and you can skip step 2. This approach is probably less time consuming, but more expensive. If you are doing confocal microscopy or differential interference

contrast microscopy (requiring imaging through a No. 1.5 glass slide), then it is helpful to assess which setup approach will be otimal for your specific microscope (upright or inverted light path) and microscopy stage.

6. Start with a cell suspension at a concentration of 3×10^6 cells/mL (this will provide approximately 100,000 cells per one standard migration chamber). This cell number is optimal for bright-field imaging at low magnification (10× to 20× objectives) when cells of different focal planes are visualized simultaneously (see Note 17). In contrast to dendritic cells (with a cell diameter >10 μm), collagen gels with smaller cells (granulocytes and lymphocytes) will require higher cell concentrations (5–6×10^6 cells/mL) to ensure sufficient cell numbers in the field of view during microscopy (2). For confocal microscopy, you may also want to increase the cell concentration when you are only imaging in single planes of focus.
7. Bone marrow-derived dendritic cells are pretty robust and can also be kept at 4°C (or on ice) for up to 1–2 h without losing migratory potential.
8. A larger tip opening reduces the likelihood of air bubbles produced while pipetting the viscous collagen. Depending on the number of assays you are preparing, you will either choose 20–200-μL or 200–1,000-μL tips.
9. As the extracted bovine collagen is only soluble after acidification, the addition of sodium bicarbonate is required to bring the pH to physiological levels (4). When pipetting collagen, add MEM (10×) first, followed by sodium bicarbonate (the color of the mix will change from yellow to orange-red, indicating pH change). The collagen mix will stay soluble at room temperature and only polymerize at 37°C.
10. Tables 1 and 2 give pipetting schemes for collagen gels of different concentrations resulting in different pore-sized collagen networks. For higher collagen concentrations, you will have to use Nutragen® (stock concentration: 6 mg/mL) instead of Purecol® (stock concentration: 3 mg/mL). The table also accounts for various numbers of samples to be prepared.
11. If using irreversible small chemical inhibitors (2), these can be added at this step of the collagen gel procedure in order to have the inhibitor continuously present in the migration assay.
12. The even horizontal lining of the collagen–gel mix will later define the border between polymerized collagen gel and soluble chemokine solution. It very much depends on the initial thickness of the paraffin line when constructing the migration chamber (see Subheading 3.2, step 2). If the paraffin lining is too thin, then capillary forces will counteract the gravity-induced

fall of the viscous collagen–gel mix. If the paraffin lining is too thick, cells will be visualized in too many planes of focus by bright-field microscopy, which may cause problems during the final migration analysis (see Note 17). Please also bear in mind that the final microscopic field of view is only a small portion of the entire migration chamber.

13. In contrast to rat-tail collagen that polymerizes within few minutes, the polymerization of bovine collagen gels takes approximately 30 min at 37°C. As cells in the collagen–cell mix can still "fall" by gravity, it is therefore essential to keep the migration chamber in an upright position during gel polymerization. This ensures that cells will be distributed in all different planes of focus of the collagen gel. Otherwise, 30 min is sufficient time for the cells to "fall" to the bottom glass side of a 1-mm-thick migration chamber. Occasionally, the collagen–gel mixture does not fully polymerize within 30 min and takes up to 1 h. This problem can be resolved by increasing the pH of the sodium bicarbonate solution (pH 7.5–7.9) (as happens when leaving an aliquot of sodium bicarbonate open to the air).

14. The 3D collagen gel system described here has been successfully used for visualizing defined chemotactic responses of murine dendritic cells, neutrophils, and activated lymphoblasts. As this setup is a closed system, the gas exchange with the surrounding might be limited and some cell types that are very sensitive to oxygen supply or nutrient flow (e.g., naïve T cells) might not migrate under these conditions (20).

15. In principle, this 3D chemotaxis assay is applicable to every possible chemotactic cell type. Random cell migration in 3D collagen gels has been shown in numerous studies with cells of low-adhesive (e.g., leukocytes) to highly adhesive (e.g., fibroblasts and endothelial cells. This assay might be useful for the study of their specific chemotactic responses to cytokines, chemokines, or growth factors. However, the experimental conditions (e.g., chemokine concentration and duration of migration) might vary depending on the relevant cell type–chemokine combination and will require individual testing. Parameters such as chemokine diffusion, receptor desensitization, cellular activation, and adhesiveness are only a few of the factors that might affect the quality of the chemotactic response. The 3D chemotaxis system described here does not allow exact manipulation of the chemokine gradient, but more complicated setups have been described which allow more precise determination of such parameters (21).

16. The image acquisition time should be chosen according to the expected speed of the cell. As examples, we suggest for fast migrating cells (e.g., neutrophils and lymphoblasts,

10–15 μm/min) one frame per every 15–30 s, for dendritic cells (3–5 μm/min) one frame per 1–2 min, and fibroblasts (1–5 μm/h) one frame per 5 min.

17. When acquired with bright-field microscopy, time-lapse video sequences of 3D collagen gels will finally result in a series of 2D images with cells in different planes of focus. Although the manual tracking is actually 2D tracking and underestimates cell movement in the *z*-axis, this method gives a good representation of the actual cell velocities and directionality as the chemotactic gradient defines migration in the *X–Y* plane. More sophisticated, commercially available image analysis software and custom-made solutions (P. Friedl, personal communication) allow automated tracking or analysis of migration in *X–Y–Z* dimensions.

18. For more detailed information on experimental setups and requirements of differential interference microscopy (2), reflection microscopy (4), and spinning disc confocal microscopy (9), we refer to published data.

19. We would like to thank Dr. Caren Petri Aronin (NIAID, NIH, Bethesda, USA) and Prof. Matthias Gunzer (University of Magdeburg, Germany) for providing protocols and sharing technical expertise on the described methods. We would also like to thank Dr. Menna Clatworthy (NIAID, NIH, Bethesda, USA) for critical reading of the manuscript.

References

1. Cukierman, E., Pankov, R., Stevens, D. R., and Yamada, K. M. (2001) Taking cell-matrix adhesions to the third dimension, *Science (New York, N.Y* **294**, 1708–1712.
2. Lämmermann, T., Bader, B. L., Monkley, S. J., Worbs, T., Wedlich-Soldner, R., Hirsch, K., Keller, M., Forster, R., Critchley, D. R., Fassler, R., and Sixt, M. (2008) Rapid leukocyte migration by integrin-independent flowing and squeezing, *Nature* **453**, 51–55.
3. Martin-Belmonte, F., and Mostov, K. (2008) Regulation of cell polarity during epithelial morphogenesis, *Current opinion in cell biology* **20**, 227–234.
4. Wolf, K., Alexander, S., Schacht, V., Coussens, L. M., von Andrian, U. H., van Rheenen, J., Deryugina, E., and Friedl, P. (2009) Collagen-based cell migration models in vitro and in vivo, *Seminars in cell & developmental biology* **20**, 931–941.
5. Wolf, K., Mazo, I., Leung, H., Engelke, K., von Andrian, U. H., Deryugina, E. I., Strongin, A. Y., Brocker, E. B., and Friedl, P. (2003) Compensation mechanism in tumor cell migration: mesenchymal-amoeboid transition after blocking of pericellular proteolysis, *The Journal of cell biology* **160**, 267–277.
6. Yamada, K. M., and Cukierman, E. (2007) Modeling tissue morphogenesis and cancer in 3D, *Cell* **130**, 601–610.
7. Reichardt, P., Gunzer, F., and Gunzer, M. (2007) Analyzing the physicodynamics of immune cells in a three-dimensional collagen matrix, *Methods in molecular biology (Clifton, NJ)* **380**, 253–269.
8. Lämmermann, T., Renkawitz, J., Wu, X., Hirsch, K., Brakebusch, C., and Sixt, M. (2009) Cdc42-dependent leading edge coordination is essential for interstitial dendritic cell migration, *Blood* **113**, 5703–5710.
9. Lämmermann, T., and Sixt, M. (2009) Mechanical modes of "amoeboid" cell migration, *Current opinion in cell biology* **21**, 636–644.
10. Schmidt, S., and Friedl, P. (2010) Interstitial cell migration: integrin-dependent and alternative adhesion mechanisms, *Cell and tissue research* **339**, 83–92.

11. Sabeh, F., Shimizu-Hirota, R., and Weiss, S. J. (2009) Protease-dependent versus -independent cancer cell invasion programs: three-dimensional amoeboid movement revisited, *The Journal of cell biology 185*, 11–19.
12. Peters, N. C., Egen, J. G., Secundino, N., Debrabant, A., Kimblin, N., Kamhawi, S., Lawyer, P., Fay, M. P., Germain, R. N., and Sacks, D. (2008) In vivo imaging reveals an essential role for neutrophils in leishmaniasis transmitted by sand flies, *Science (New York, N.Y* **321**, 970–974.
13. Pflicke, H., and Sixt, M. (2009) Preformed portals facilitate dendritic cell entry into afferent lymphatic vessels, *The Journal of experimental medicine* **206**, 2925–2935.
14. Shields, J. D., Fleury, M. E., Yong, C., Tomei, A. A., Randolph, G. J., and Swartz, M. A. (2007) Autologous chemotaxis as a mechanism of tumor cell homing to lymphatics via interstitial flow and autocrine CCR7 signaling, *Cancer cell* **11**, 526–538.
15. Quast, T., Tappertzhofen, B., Schild, C., Grell, J., Czeloth, N., Forster, R., Alon, R., Fraemohs, L., Dreck, K., Weber, C., Lämmermann, T., Sixt, M., and Kolanus, W. (2009) Cytohesin-1 controls the activation of RhoA and modulates integrin-dependent adhesion and migration of dendritic cells, *Blood* **113**, 5801–5810.
16. Brightman, A. O., Rajwa, B. P., Sturgis, J. E., McCallister, M. E., Robinson, J. P., and Voytik-Harbin, S. L. (2000) Time-lapse confocal reflection microscopy of collagen fibrillogenesis and extracellular matrix assembly in vitro, *Biopolymers* **54**, 222–234.
17. Bonvin, C., Overney, J., Shieh, A. C., Dixon, J. B., and Swartz, M. A. (2010) A multichamber fluidic device for 3D cultures under interstitial flow with live imaging: development, characterization, and applications, *Biotechnology and bioengineering* **105**, 982–991.
18. Lutz, M. B., Kukutsch, N., Ogilvie, A. L., Rossner, S., Koch, F., Romani, N., and Schuler, G. (1999) An advanced culture method for generating large quantities of highly pure dendritic cells from mouse bone marrow, *Journal of immunological methods* **223**, 77–92.
19. Menges, M., Rossner, S., Voigtlander, C., Schindler, H., Kukutsch, N. A., Bogdan, C., Erb, K., Schuler, G., and Lutz, M. B. (2002) Repetitive injections of dendritic cells matured with tumor necrosis factor alpha induce antigen-specific protection of mice from autoimmunity, *The Journal of experimental medicine* **195**, 15–21.
20. Huang, J. H., Cardenas-Navia, L. I., Caldwell, C. C., Plumb, T. J., Radu, C. G., Rocha, P. N., Wilder, T., Bromberg, J. S., Cronstein, B. N., Sitkovsky, M., Dewhirst, M. W., and Dustin, M. L. (2007) Require-ments for T lymphocyte migration in explanted lymph nodes, *J Immunol* **178**, 7747–7755.
21. Haessler, U., Kalinin, Y., Swartz, M. A., and Wu, M. (2009) An agarose-based microfluidic platform with a gradient buffer for 3D chemotaxis studies, *Biomedical microdevices* **11**, 827–835.

Chapter 12

Quantification of Transendothelial Migration Using Three-Dimensional Confocal Microscopy

Robert J. Cain, Bárbara Borda d'Água, and Anne J. Ridley

Abstract

Migration of cells across endothelial barriers, termed transendothelial migration (TEM), is an important cellular process that underpins the pathology of many disease states including chronic inflammation and cancer metastasis. While this process can be modeled in vitro using cultured cells, many model systems are unable to provide detailed visual information of cell morphologies and distribution of proteins such as junctional markers, as well as quantitative data on the rate of TEM. Improvements in imaging techniques have made microscopy-based assays an invaluable tool for studying this type of detailed cell movement in physiological processes. In this chapter, we describe a confocal microscopy-based method that can be used to assess TEM of both leukocytes and cancer cells across endothelial barriers in response to a chemotactic gradient, as well as providing information on their migration into a subendothelial extracellular matrix, designed to mimic that found in vivo.

Key words: Endothelium, Transmigration, Cancer, Leukocyte, ECM, HUVEC, Quantification, Confocal, Imaging

1. Introduction

Transendothelial migration (TEM) is the process by which leukocytes exit the bloodstream, traversing the walls of blood vessels in order to infiltrate the underlying tissues (1–3). Under normal conditions the number of transmigrating cells is low except for constitutive trafficking through lymphoid tissues, however, it increases dramatically during inflammation and contributes to the pathogenesis of inflammatory diseases such as rheumatoid arthritis as well as noninflammatory diseases such as atherosclerosis and cancer (4, 5). In cancer, the TEM process is subverted to permit cancer cells to breach barriers within the body, leading to tumor invasion and metastasis.

Claire M. Wells and Maddy Parsons (eds.), *Cell Migration: Developmental Methods and Protocols*,
Methods in Molecular Biology, vol. 769, DOI 10.1007/978-1-61779-207-6_12, © Springer Science+Business Media, LLC 2011

During TEM from blood vessels, cells not only need to overcome the endothelial barrier, but also to interact with and negotiate a stromal environment rich in networks of extracellular matrix (ECM) components that include type I collagen, laminin, and fibronectin (6–9). In cancer, this process involves cell adhesion, migration, and often proteolytic matrix degradation, although the mechanism of migration varies depending on the cancer cell type and conditions (8–11). For leukocytes, TEM involves capture from the fast-moving bloodstream via transient interactions between ligands such as selectins and their receptors, which are expressed on the leukocyte and endothelial cell surfaces and are upregulated by tissue inflammation. This is then followed by leukocyte integrin activation leading to firm adhesion and migration, culminating in transit across the endothelium (2, 12–14). Some leukocyte types such as T-cells have been shown to cross the endothelial barrier using either a paracellular route between cell–cell junctions, or a transcellular route directly through the endothelial cell body by formation of a transcellular channel.

TEM can be modeled in vitro using co-cultures of endothelial cells with leukocytes or cancer cells (15–17). The rate of TEM is often quantified using Boyden chambers or transwells (18, 19). Although useful, these methods have limitations, including variable quantification of TEM rate and a lack of visual information on the process. Cell morphology during TEM can be assessed microscopically using confocal and timelapse microscopy-based techniques with cells seeded on glass coverslips, however, this does not allow the study of the complete extravasation process as cells are not able to migrate away from the endothelial monolayer after TEM, nor interact with the ECM. Use of 3D gels of polymerized ECM proteins that resemble the composition of the stromal environment can provide cultured cells with similar determinants to those that control cell migration in vivo. In particular, matrix density, inter-fiber pore size, and flexibility are crucial for determining cell shape and speed during migration in 3D systems (8).

Here, we describe an in vitro confocal microscopy-based assay that utilizes a thick layer of ECM and primary human endothelial cells, which can be used to quantify TEM of a variety of cell types, providing both consistent qualitative and quantitative data, as well as spatial and temporal information on TEM. It also has the advantage of utilizing the two-compartment Boyden chamber, to allow establishment of chemotactic gradients within the assay system.

2. Materials

2.1. Cell Culture and Staining

1. Pooled primary human umbilical vein endothelial cells (HUVECs) (Lonza) (see Note 1).

2. Endothelial cell basal medium-2 (EBM-2) (Clonetics®, Lonza) supplemented with growth factors (ascorbic acid, R3-IGF-1, heparin, rhFGF-B, hydrocortisone, GA-1000, rhEGF and VEGF, and 2% fetal bovine serum. Supplements are provided as part of a bullet kit, purchased with the basal medium). Medium plus supplements is referred to as EGM-2 (stored at 4°C).
3. Human plasma fibronectin diluted in PBS to 10 μg/ml; stored at −20°C.
4. MDA-MB-231 breast cancer cells (ATCC).
5. PC3 prostate cancer cells (ATCC).
6. THP-1 monocyte-like cell line (ATCC).
7. Dulbecco's Modified Eagle's medium (DMEM) containing 4.5 g/l of glucose, L-glutamine, 25 mM HEPES and supplemented with 10% fetal calf serum (FCS), 100 U/ml penicillin, 100 μg/ml streptomycin, and 1 mM sodium pyruvate. Medium is stored at 4°C.
8. Roswell Park Memorial Institute 1640 medium (RPMI-1640) containing L-glutamine, 25mM HEPES and supplemented with 10% FCS, 100 U/ml penicillin, and 100 μg/ml streptomycin (store at 4°C).
9. Trypsin (store at 4°C).
10. 1× Nonenzymatic Cell Dissociation Solution (Sigma-Aldrich).
11. Phosphate-buffered saline (PBS).
12. Cell tracker orange (CTO) (Molecular Probes, Invitrogen); store at −20°C (see Note 2).
13. Cell tracker green (CTG) (Molecular Probes); store at −20°C.
14. Carboxyfluorescein diacetate succinimidyl ester (CFSE) (Molecular Probes); store at −20°C.
15. Chemokine ligand 2 (CCL2)/monocyte chemotactic protein-1 (MCP-1, store at −20°C).

2.2. Collagen Matrix Preparation

1. Buffer 1: 200 mM Na_2HPO_4, 1.3 M NaCl, and 90 mM NaOH. Adjust to final volume with distilled H_2O. Adjust to pH 7 if required. Addition of buffer 1 to liquid collagen catalyzes matrix polymerization.
2. Pure Col™ bovine native collagen I gel (INAMED). Store at 4°C to avoid spontaneous polymerization. Prepare working solution on ice by diluting nine parts of Pure Col™ with one part cold buffer 1 (see Note 3).
3. EGM-2 medium as above.
4. FCS.
5. ThinCert™ (Boyden chamber) multiwell plates with 0.8-μm pore inserts.
6. Hydrophobic barrier marker pen.

2.3. Confocal Microscopy

1. Microscope coverslips (25-mm diameter).
2. Glass slides (Super Premium, 1.0–1.2-mm thickness).
3. Phosphate-buffered saline.
4. Paraformaldehyde: prepare 4% (w/v) solution by dissolving 4 g of paraformaldehyde in 50 ml of distilled deionized water and 1 ml of 1 M NaOH. Stir the mixture gently on a heating block (around 65°C) until paraformaldehyde is dissolved. Add 10 ml of 10× PBS and allow to cool to room temperature. Adjust pH to 7.4 with 1 M HCl (around 1 ml is needed) and adjust the final volume to 100 ml with distilled deionized water. Filter using 0.2-μm membrane filters and store aliquoted at −20°C (see Note 4).
5. Permeabilization solution: 0.1% (v/v) Triton X-100 in PBS.
6. Antibody dilution buffer: 3% (w/v) BSA in PBS.
7. Mouse anti-human PECAM-1 antibody (DakoCytomation, clone SC70A).
8. AlexaFluor-546-conjugated-goat anti-mouse IgG.
9. Disposable scalpels.
10. Nitrile O-rings metric – 1.6-mm cross-section, internal diameter 14.1 mm (Altec Fifty, ORN -0141-16).
11. Microscope grease (Glisseal®).

3. Methods

The goal of this protocol is to provide information on how TEM of different cell types can be analyzed in one assay to provide both visual and quantitative information. The system includes a chemotactic gradient as well as a subendothelial matrix, so that TEM across and subsequent migration away from the endothelium can be measured. Below two different uses of the system are described. The first uses interaction of the cancer cell lines, PC3 and MDA-MB-231, with endothelial cells to illustrate how morphological information can be obtained. In the second TEM of the monocytic cell line THP-1 across the endothelium in response to a gradient of MCP-1 is used to illustrate how quantitative data can be obtained.

3.1. HUVEC Cell Culture

For optimal proliferation and endothelial phenotype, HUVECs are cultured on a fibronectin matrix. Coat the bottom of culture vessels with 10 μg/ml human fibronectin (see Note 5) in PBS for 30 min at 37°C (see Note 6). Thaw frozen cell aliquots in a water bath and add to 15 ml of EGM-2 medium. One cell aliquot of 500,000 cells is suitable for a 75-cm^2 flask, and should reach confluence in 2–3 days grown at 37°C and 5% CO_2 (see Note 7).

Replace medium after 24 h to stimulate cell growth. HUVECs should be passaged at 70–80% confluency, and used for experiments only between passages 1–4 (see Note 8). Passage cells by removing the medium from the flask, and washing cells three times with 10 ml PBS. Trypsinize cells for 3 min (see Note 9) with 1 ml of trypsin, and collect cells with 9 ml of medium to inactivate the trypsin. It may be necessary to tap the bottom of the flask to loosen cells. Centrifuge the cells at 900×*g* for 3 min, and aspirate to remove trypsin, before addition of fresh medium and seeding in fresh culture vessels. Do not dilute cells more than 1:4 during passage, as this leads to unhealthy and clumped cell growth. If the cells do not require splitting, change the medium no less than every 3 days, in order to maintain a healthy culture. Occasionally growth can be slow following defrosting, in which case the medium should be changed more frequently, to stimulate them.

3.2. MDA-MB-231 and PC3 Cell Culture

Thaw cells in a water bath from frozen aliquots and then add to 10 ml warm DMEM or RPMI containing 10% FCS, respectively. Centrifuge cells at 900×*g* for 4 min, then aspirate the supernatant to remove DMSO or other cryopreservatives. Resuspend cells in 10 ml of fresh DMEM or RPMI and seed into a 75-cm^2 flask. Maintain cells at 37°C and 5% CO_2, and replace medium after 24 h. When cells are 80% confluent, passage using trypsin as described for HUVECs above, and reseed into culture flasks or for use in experiments. Split cells at between 1:2 and 1:5 and change the medium every 2–3 days. To improve experimental consistency, discard cells after 1 month and defrost a fresh cell aliquot.

3.3. THP-1 Cell Culture

Thaw cells in a water bath from frozen aliquots and then add to 10 ml warm RPMI containing 10% FCS. Centrifuge cells at 900×*g* for 3 min and aspirate the supernatant to remove cryopreservatives, before reseeding in culture flasks at the desired density. THP-1 is a suspension cell line, and should be cultured at a density of between 500,000 and 2,000,000 cells per ml (see Note 10).

3.4. Collagen Matrix Preparation

1. Prepare the desired volume of collagen I by diluting on ice nine parts of Pure Col™ with one part of cold buffer 1. The phosphates and pH of the buffer aid polymerization of the collagen matrix. One sample requires approximately 250 μl of collagen matrix, but be sure to make excess to what is required as the Pure Col™ solution is very viscous and difficult to handle.
2. Mix buffer carefully to ensure complete homogenization of the solution and to avoid bubble formation. In order to analyze cancer cell invasion into the matrix, the final concentration of collagen I in the gel should be between 1.5 and 2 mg/ml, however in order to keep the endothelial monolayer stable, the ideal concentration should be 2 mg/ml; to obtain the desired concentration, dilute with chilled EBM-2 medium (see Note 11).

3. At this point, chemoattractants can be added to the gel matrix. In our experiments with cancer cells, FCS was tested as a chemoattractant, and added to the matrix at either 2 or 10%, for MDA-MB-231 or PC3 cells, respectively. THP-1 cells do not require a chemoattractant to be added to the gel itself. Keep the matrix solution on ice while preparing the rest of the material for making the gel.
4. Polymerizing the liquid matrix inside the upper ThinCert™ Boyden chamber is problematic, as the liquid will naturally form a meniscus, resulting in a gel that is thicker at the edges, and thinner in the center. To minimize or avoid this completely, a hydrophobic barrier pen, which has ink that is charged such that it repels water, can be used to prevent a meniscus from forming. Use the pen to draw a circle around the inside of the upper insert of the Boyden chamber plate, a few millimeters above the membrane. It is important to keep a steady hand, and to draw this as even as possible. The pen draws a line of a few millimeters thickness (Fig. 1).
5. Add collagen solution to the chamber, until around halfway up the hydrophobic barrier. Depending on the relative position of this circle add a volume of 250–300 μl of collagen I solution to the inserts (see Note 12). Allow matrix to polymerize by placing the insert back into the culture plate and incubating at 37°C (see Note 13). Polymerization time will vary according to the chosen collagen I concentration. A concentration of 2 mg/ml will take around 2–3 h to polymerize. The gel will become opaque when fully polymerized.
6. After gels have polymerized, equilibrate with warm EBM-2 medium for at least 6 h to equilibrate the gel before addition of HUVECs. Washing and aspiration of medium should be done very carefully using micropipettes, especially after adding the cells, as strong washing may disrupt the gel, or cause cells to detach from the gel surface.
7. At this point, cells that are to be stained with fluorescent dyes for confocal visualization should be prepared. Stain HUVECs with 5 μM CTO diluted in EGM-2 medium. Incubate at 37°C for 45 min, and then replace stain with normal EGM-2 medium. As these dyes are retained in cells for several days, this can even be performed the previous day, without loss of staining fluorescence. When antibodies are to be used to analyze endothelial junctions, HUVECs can be left unstained (as Fig. 2) in order to free up laser channels on the confocal microscope.
8. Add 1.5 ml of EGM-2 to the bottom of the Boyden chamber and 200,000 or 150,000 freshly detached HUVECs in 500 μl of EGM-2 to the upper chamber to form a confluent monolayer within 24 or 48 h, respectively (see Note 14). It is extremely important that HUVEC monolayers are completely confluent,

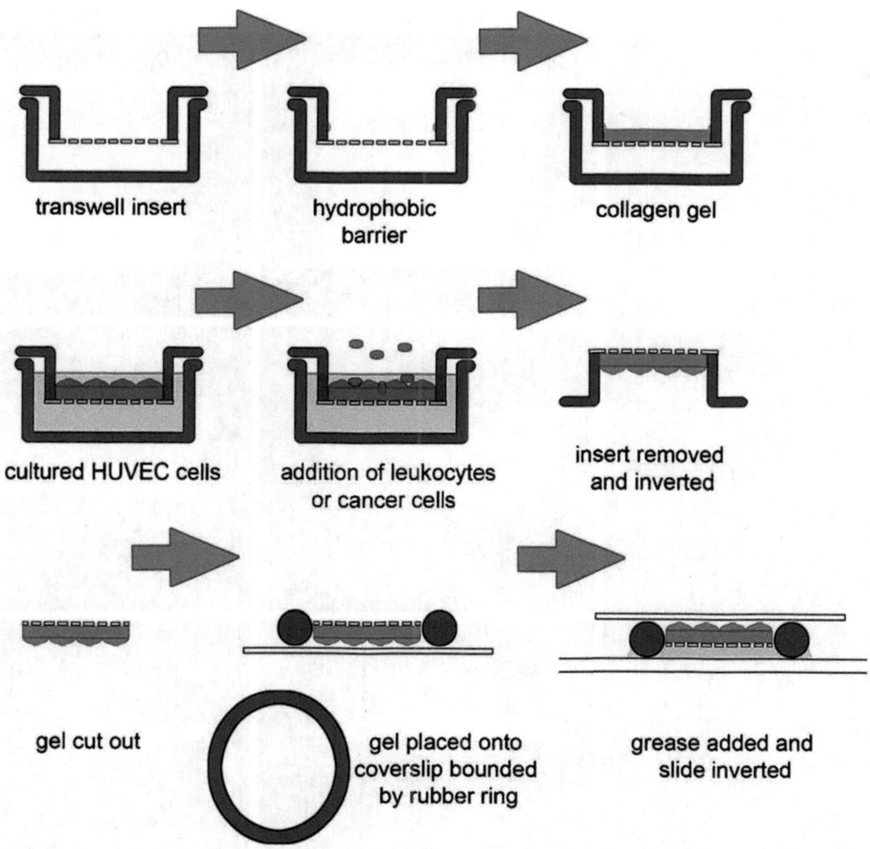

Fig. 1. Schematic illustrating preparation of TEM samples for confocal analysis. *Top row*: Boyden chambers are used to create a co-culture system to study migration of cancer cells or leukocytes across endothelial monolayers into a subendothelial matrix. The matrix is polymerized in the top chamber. In order to ensure no meniscus is formed, a hydrophobic barrier is applied to the sides of the upper insert. *Middle row*: HUVECs are cultured on top of the matrix, and chemoattractant is added to the bottom chamber. Cancer cells or leukocytes are added to the top chamber, and allowed to interact with the endothelium. Samples are subsequently fixed, and the insert removed. *Bottom row*: The gel is removed from the insert with a scalpel and placed facedown onto a glass coverslip. A nitrile o-ring is placed around the gel which is fixed in place with microscope grease. The whole sample is turned over onto a microscope slide, which is then ready for imaging.

with intact intercellular junctional complexes, in order to provide an efficient physical barrier to leukocyte or cancer cell migration (Fig. 1).

9. For cancer cell TEM studies, once a confluent endothelial monolayer has been formed, remove medium from the top chamber, and replace with 400 μl of fresh EGM-2. At this stage stain MDA-MB-231 or PC3 cells, attached on flasks or plates, using 10 μM CFSE diluted in PBS. Incubate for 20 min at 37°C and then wash twice with PBS to remove the excess dye. Detach these using the nonenzymatic dissociation buffer for 10 min. Collect the cells using the assay medium, in this case either DMEM or RPMI containing 0.1% FCS. It may be necessary to tap the bottom of the flask/ plate to

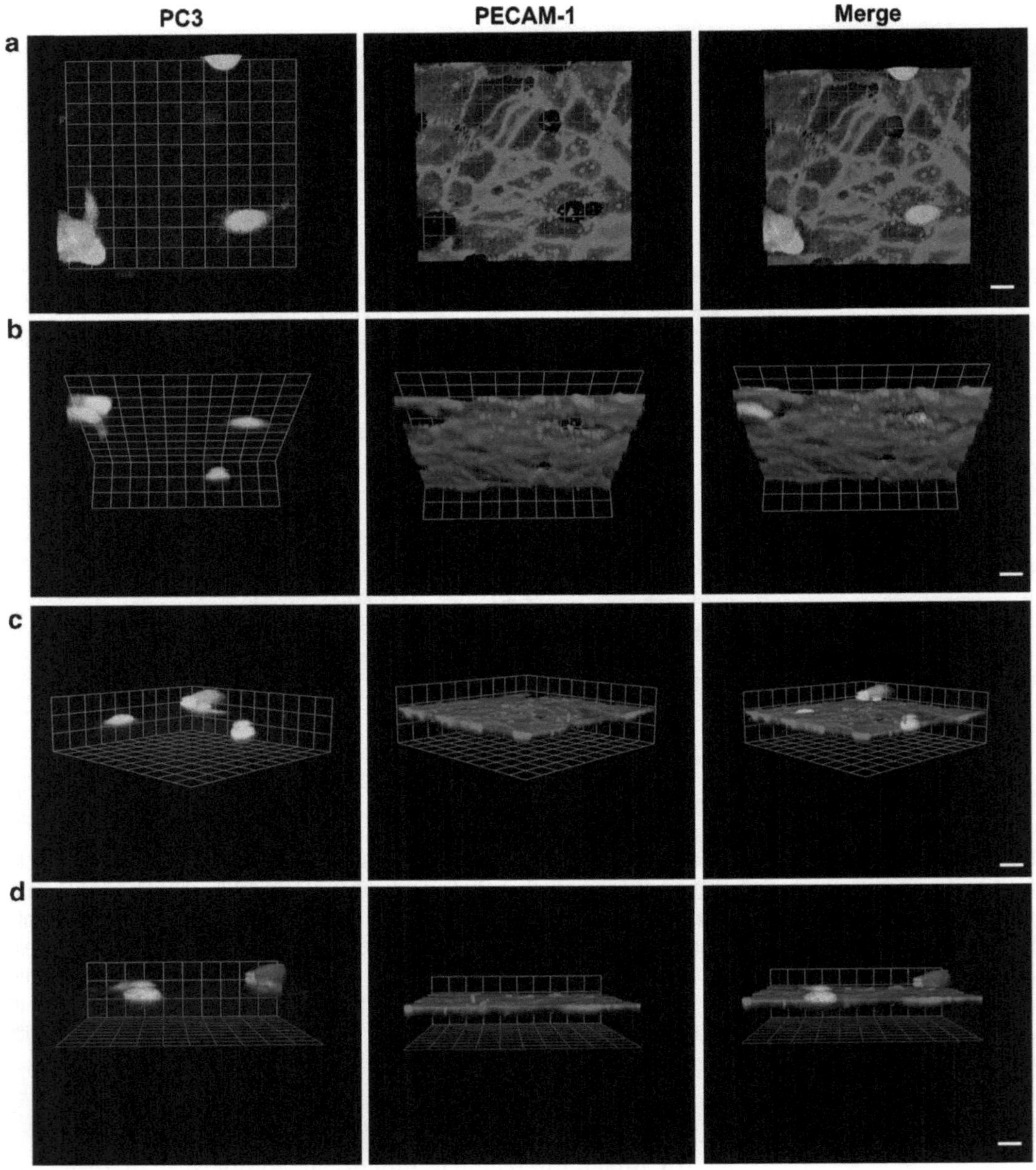

Fig. 2. Three-dimensional confocal reconstruction of PC3 cell TEM using Volocity. CFSE-labeled PC3 cells were added to HUVECs, and TEM allowed to proceed toward a serum gradient in the lower chamber for 2 h before fixation. Fixed samples were stained with mouse anti-PECAM-1 antibodies followed by Alexafluor-546 conjugated anti-mouse antibodies to visualize endothelial junctions, and z-stacks collected. Three-dimensional reconstructions were produced using Volocity software, as described in the text. Different orientations [plan view (**a**), beneath (**b**), isometric (**c**), and side view (**d**)] are shown to illustrate aspects of the TEM process. Scale bar = 20 μm.

loosen cells. Calculate the concentration of cells per ml using a CASY counter or similar. While quantifying, leave the cells in suspension in the assay medium at 37°C. Add 50,000 stained MDA-MB-231 or PC3 cells in 100 μl of DMEM or RPMI containing 0.1% FCS to the HUVECs (Fig. 1).

10. In order to create a serum gradient to act as a chemoattractant for cancer cells, fill the bottom of the Boyden chamber with EGM-2 containing either 2 or 10% FCS (depending on the cell type being used; see above). Samples are fixed at different timepoints to assess progression of TEM. To evaluate adhesion of cancer cells fix the samples after 15 min. To analyze intercalation within the endothelial monolayer fix samples within 1–2 h and to detect complete TEM fix after 4–6 h. The best timepoint to visualize cells that have migrated away from the monolayer and are invading the matrix will depend on the cell type being used. While MDA-MB-231 cells fixed after 24–72 h show good levels of matrix invasion, PC3 cells do not invade the ECM at all in this model, and remain intercalated in the endothelial monolayer.
11. For leukocyte TEM studies, stain THP-1 cells with 1 μM CTG, diluted in RPMI for 45 min, before washing cells in normal RPMI. Stained THP-1 cells can be kept in culture for 24 h prior to TEM experiments without loss of fluorescence.
12. Add 50,000 stained THP-1 cells in 500 μl fresh EGM-2 to the insert of the Boyden chamber containing the gel, and 1.5 ml of EGM-2 containing 10 μg/ml CCL2/MCP-1 as chemoattractant to the bottom chamber (Fig. 1).
13. Cells should start to transmigrate after 10–15 min, and show good migration into the collagen matrix after 45–60 min.

3.5. Preparation of Samples for Confocal Microscopy

1. In order to analyze the interaction of cancer cells or leukocytes with the HUVECs, first aspirate the medium from both chambers carefully with a micropipette, and wash carefully with PBS to remove loosely bound or nonadhered cells. If the aim of the experiment is to analyze cancer or leukocyte adhesion to the monolayer, three quick but careful PBS washes should be performed to avoid losing adhered cells. However, if the aim is to analyze cells that have been interacting for longer timepoints, three longer washes of 5 min are required (Fig. 1).
2. Fix samples with PFA, adding 500 μl to the top chamber and 1.5 ml to the bottom. Incubate at room temperature for 30 min, before aspirating and performing three 5 min washes with PBS.
3. Permeabilize the cells by incubating with 0.1% Triton X-100 (v/v) in PBS for 10 min at room temperature (both top and bottom chambers).
4. Incubate samples with 3% BSA/PBS (antibody dilution buffer) for 1 h to reduce subsequent background fluorescence.
5. Remove the inserts from the Boyden chamber and blot excess liquid with paper towel. Wrap the bottom of the chamber with parafilm, in order to prevent liquid from leaking out of

the top chamber (see Note 15). Incubate sample with 250 μl of the desired primary antibody diluted in 3% BSA, for 1 h at room temperature (see Note 16). In this example (Figs. 2 and 3), PECAM-1 was used to visualize endothelial cell–cell junctions, used at a 1:100 dilution (see Note 17).

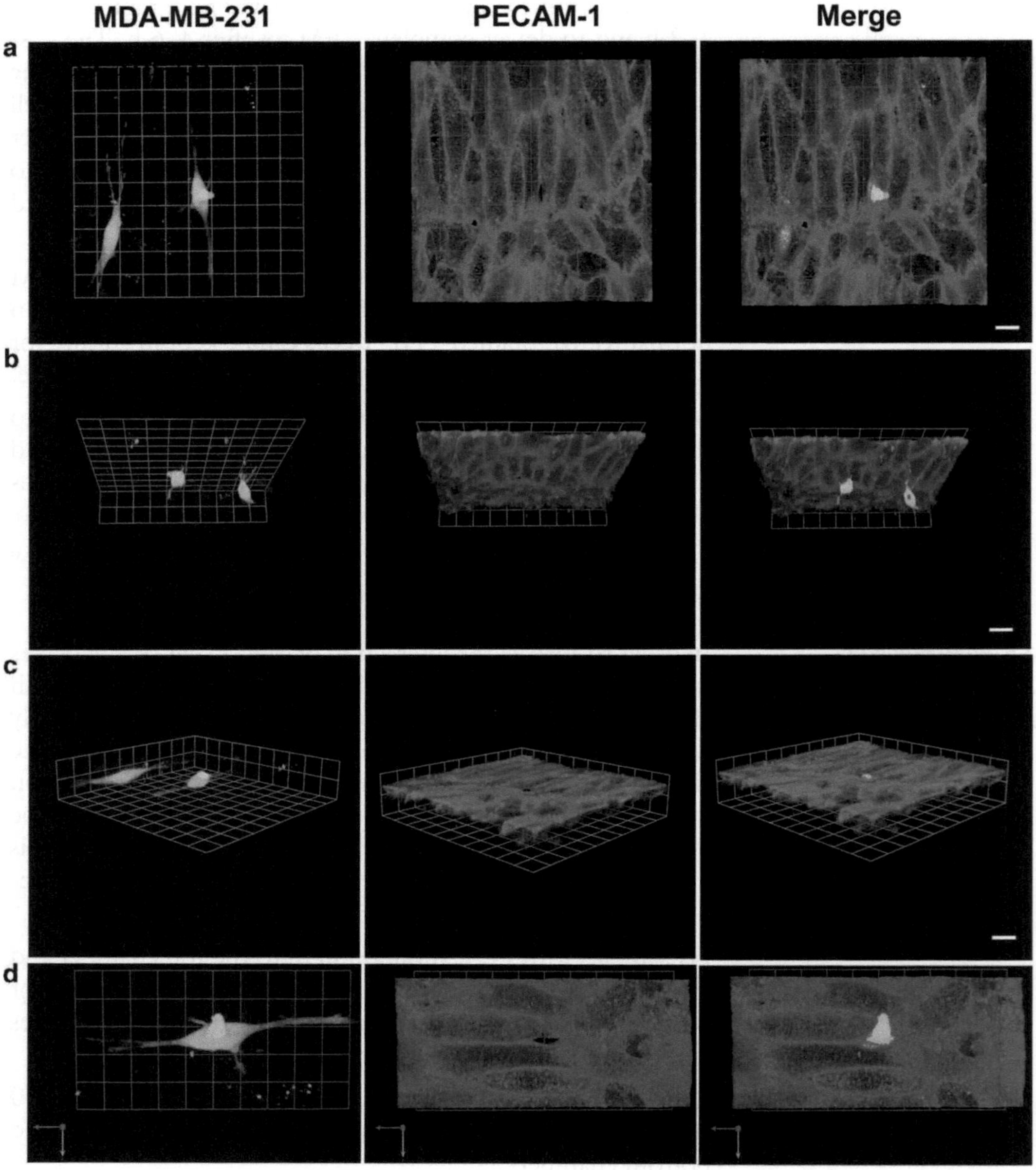

Fig. 3. Three-dimensional confocal reconstruction of MDA-MB-231 cell TEM. CFSE-labeled MDA-MB-231 cells were added to HUVECs, and TEM allowed to proceed toward a serum gradient in the lower chamber for 4 h before fixation. Fixed samples were stained with PECAM-1 antibodies followed by Alexafluor-546 conjugated anti-mouse antibodies to visualize endothelial junctions and z-stacks collected. Three-dimensional reconstructions were produced using Volocity software, as described in the text. Different orientations [plan view (**a**), beneath (**b**), and isometric (**c**)] are shown to illustrate aspects of the TEM process. Panel (**d**) shows a magnified section of the gel, illustrating a single TEM event, where details of cell morphology, including filopodia are visible. Scale bar = 20 μm.

6. Wash sample three times for 5 min with PBS and incubate with an appropriate secondary antibody diluted in 3% BSA for 1 h, followed by subsequent PBS washes (see Note 17).
7. Prepare the microscope slide for the sample, by taking a rubber o-ring, and lubricating the edge with the microscope grease. Place the ring on top of a 25-mm diameter glass coverslip; the grease should prevent the ring from moving and provide a seal (Fig. 1).
8. Remove the gel from the insert carefully using a scalpel blade, making small cuts at the edge of the filter (Fig. 1). Use the large end of a 1-ml micropipette tip to press the sample down evenly through the filter onto the glass coverslip, inside the rubber o-ring. The face of the gel containing the endothelial cells should now be in contact with the coverslip (see Note 18). Using the scalpel as a trowel, fill the gap between the sample and the edge of the o-ring with more microscope grease. This prevents the sample from moving and drying out during confocal analysis.
9. Invert the coverslip onto the glass slide (Fig. 1). The sample is now ready for microscopy.

3.6. Acquisition of z-Stack Data by Confocal Microscopy

1. Data shown in this chapter were collected using a Zeiss LSM510 confocal microscope although any confocal microscope that is capable of collecting z-stack data would be suitable. High-resolution images suitable for analyzing cell morphology (Figs. 2 and 3) are acquired with a 63× oil immersion objective (Plan-Apochromat 63×/1.40 Oil DIC M27), taking a stack depth of 20–30 μm and one slice every 0.41 μm. To improve resolution, each slice should be averaged from eight or more confocal scan passes.
2. For TEM quantification, low-resolution stacks are collected (Fig. 4) with a 20× objective (EC Plan-Neofluor 20×/0.50 M27), taking a stack depth of 150 μm and one slice every 1 μm. Each slice is an average of four confocal scan passes (see Note 19).

3.7. Three-Dimensional Reconstruction of Confocal Data with Volocity

1. Although there are many different image analysis packages available on the market that are suitable for manipulating confocal microscope-acquired z-stack data, we have found Volocity (www.improvision.com) is both easy to use and provides high-resolution images. Here, we explain how to use the package to produce three-dimensional images. Figures 2 and 3 show example results from PC3 cell and MDA-MB-231 cell TEM, respectively. In each case, several different orientations of the three-dimensional stack are shown.
2. Open the confocal stacks in Volocity (see Note 20) and select "3D visualization mode."

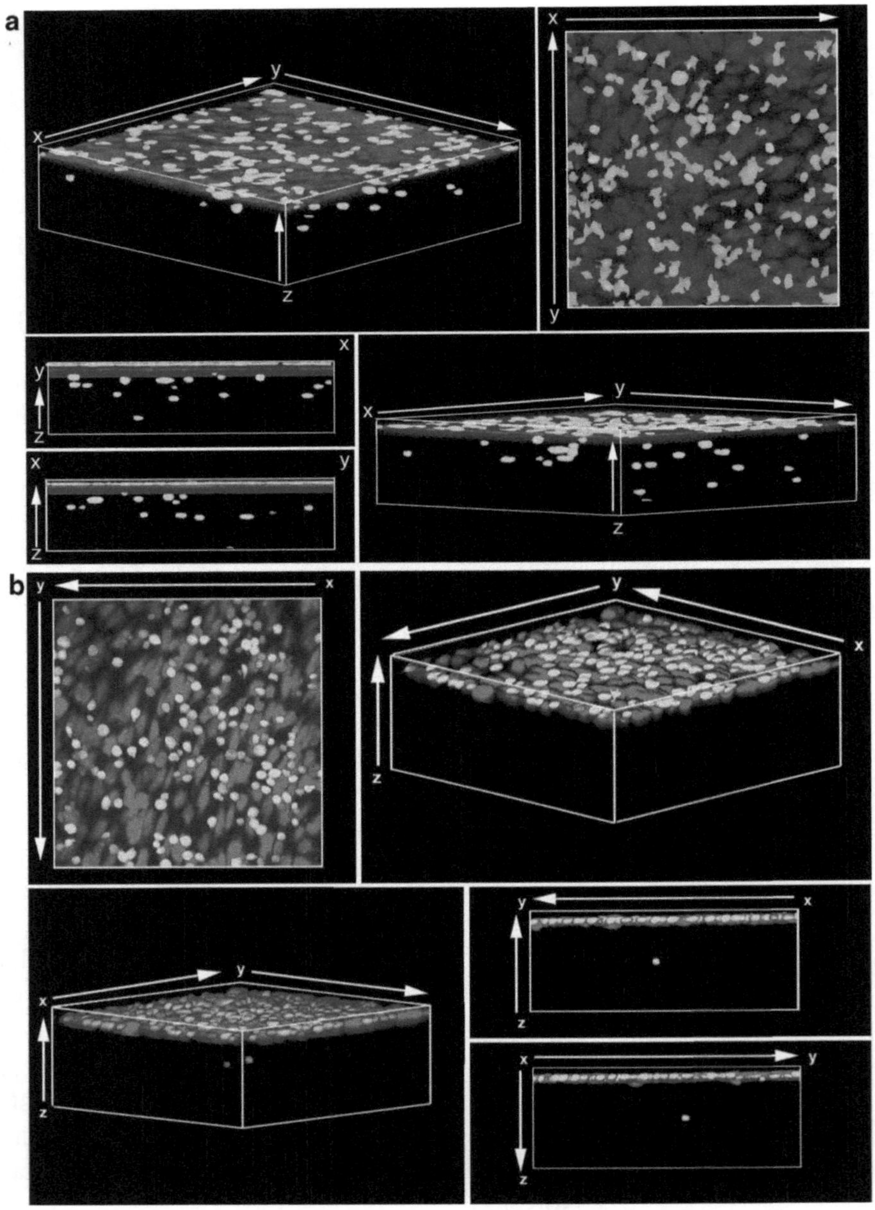

Fig. 4. Three-dimensional confocal reconstruction of THP-1 cell TEM with and without wortmannin treatment. CTG-labeled THP-1 cells (*light gray*), either treated with 100 nM wortmannin or left untreated, were added to CTO-labeled HUVECs (*dark gray*), and TEM allowed to proceed toward a gradient of MCP-1 in the lower chamber for 45 min. Cells were then fixed, z-stacks subsequently collected, and three-dimensional reconstructions produced. Various projections of untreated THP-1 TEM (**a**) and wortmannin-treated TEM (**b**) showing the depth of cell penetration into the collagen matrix.

3. Create a library and drag the .lsm file into the left-hand-panel of the software window (Fig. 5a). Analyze the images using the 3D opaque mode (Fig. 5b) and adjust the background. In the worked example, the floor was removed, scale bar added, and the orientation axis removed. All these modifications can

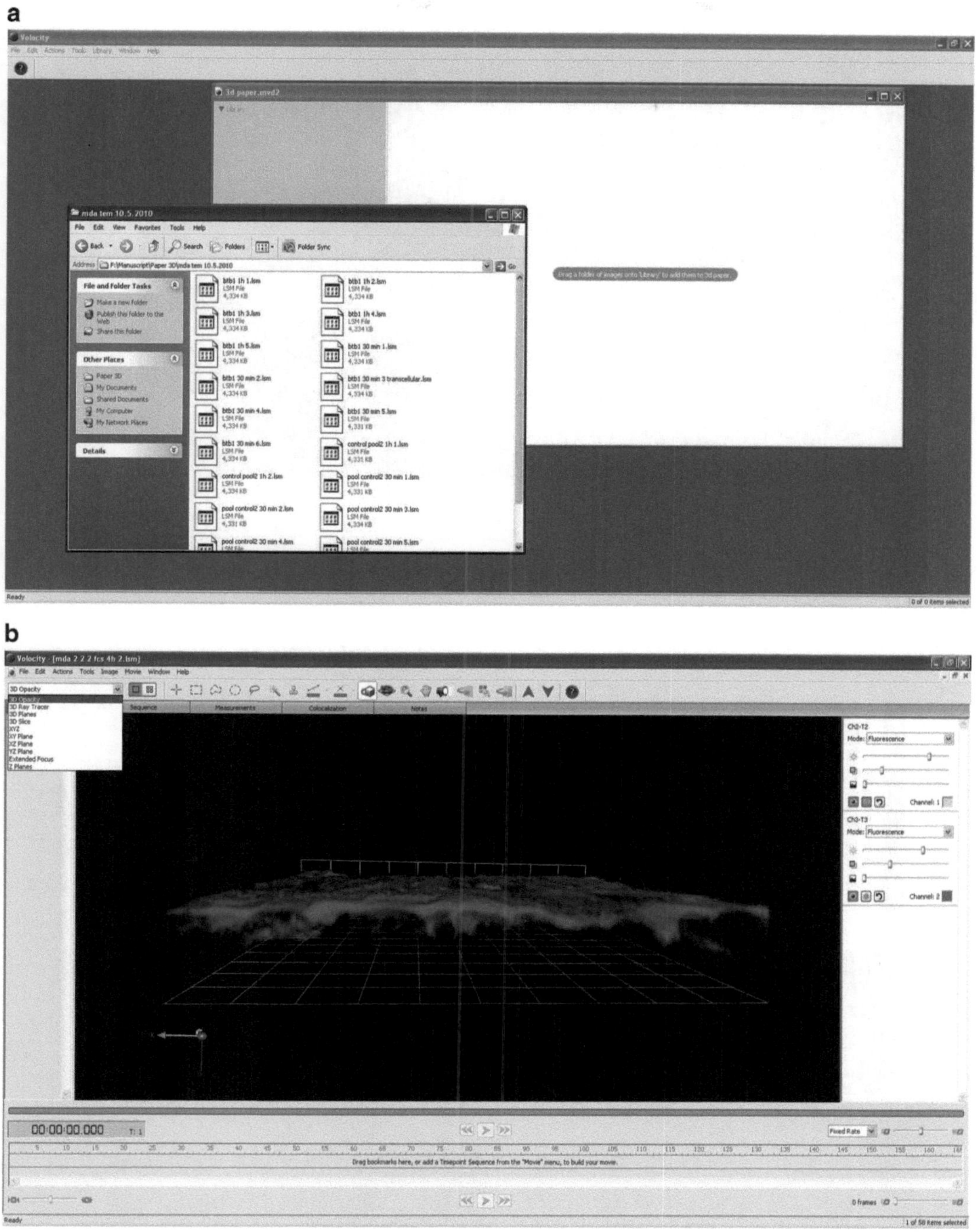

Fig. 5. Volocity screenshots illustrating the individual steps used to turn confocal microscope z-stacks into three-dimensional objects.

be performed by using the top menu bar on the main image window, that was opened for the selected lsm file, by selecting the option "Image" and then "Display." Adjust the brightness, density, and blackness level (Fig. 5c) for each channel,

c

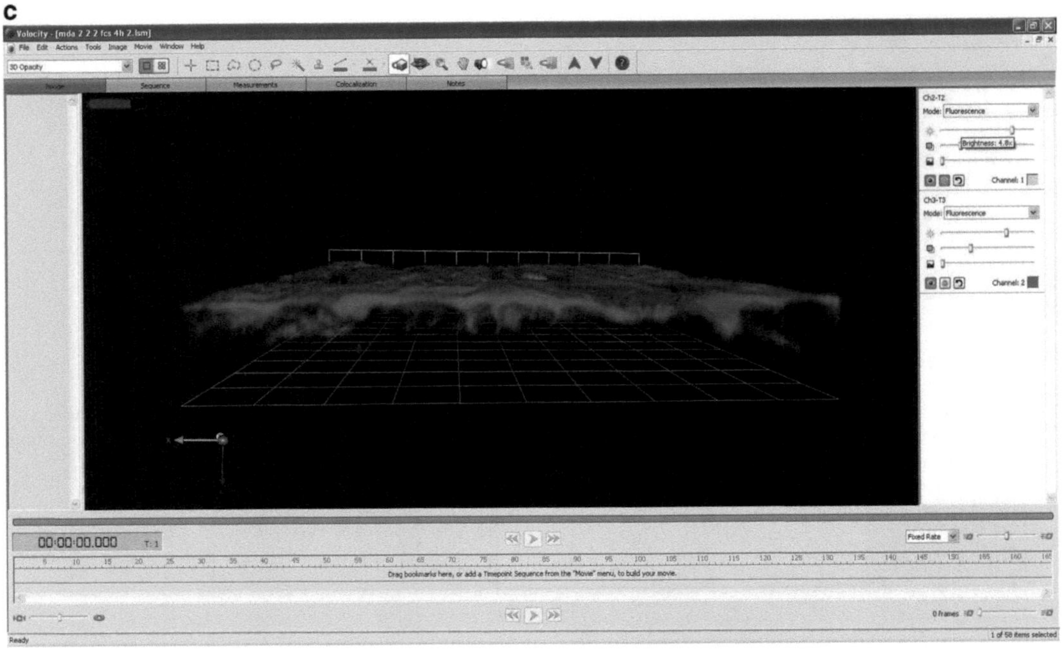

d

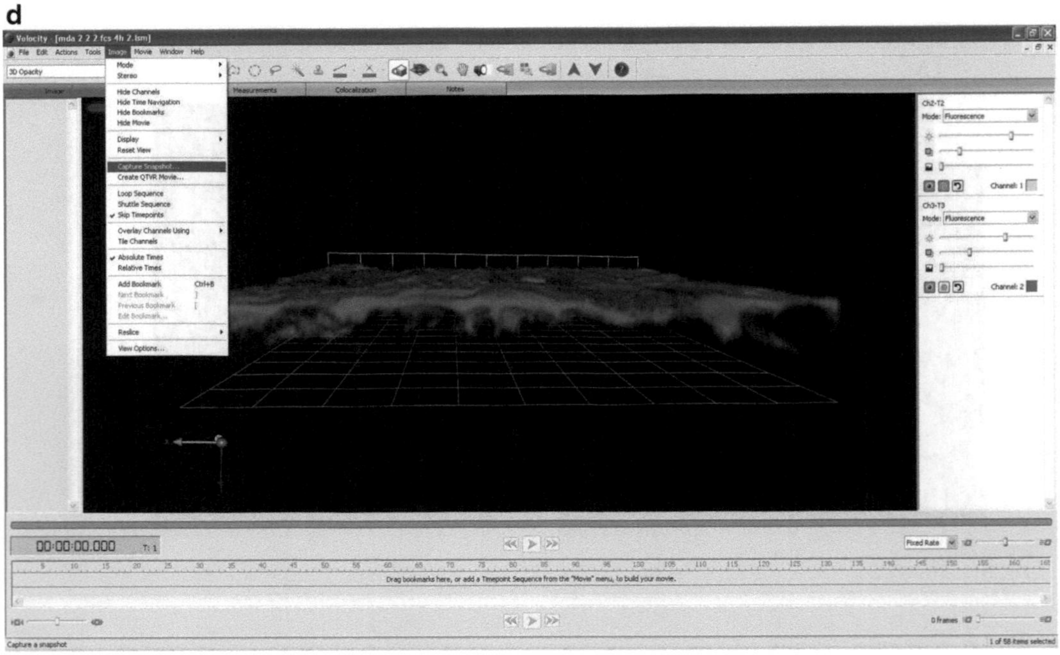

Fig. 5. (continued)

by using the left panel of the image window. For example in Fig. 5, the channel ch2-T2 was adjusted with 4.8×, 22%, 0%; ch3-T3 was adjusted with 3.8×, 21%, 0%; and ch3-T4 3.7×, 25%, 0%. The aim of this is to produce an image that accurately represents what was observed when acquiring the images on the microscope.

e

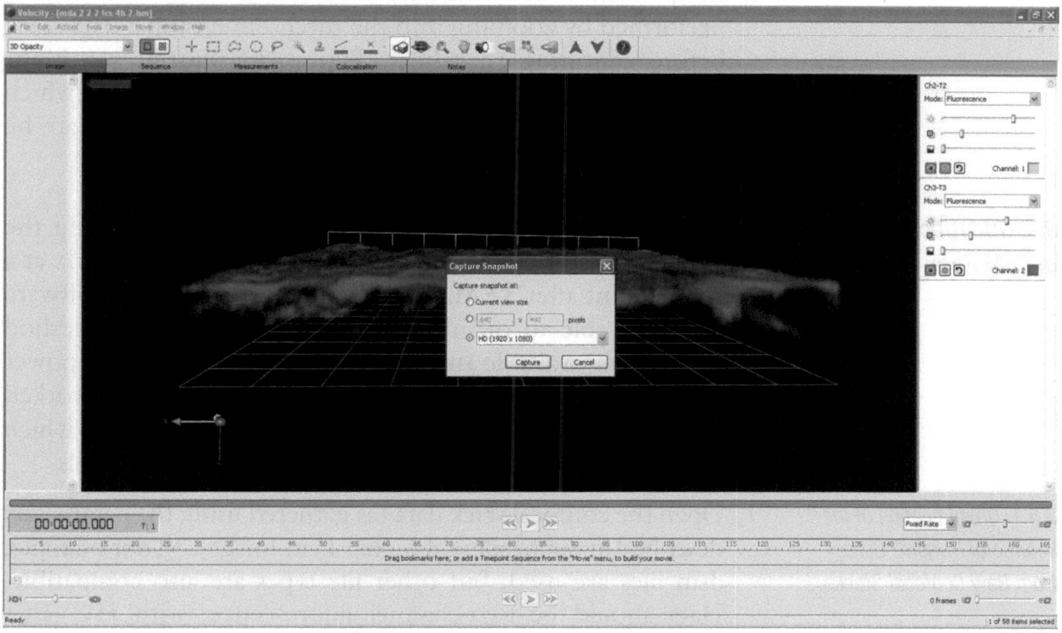

f

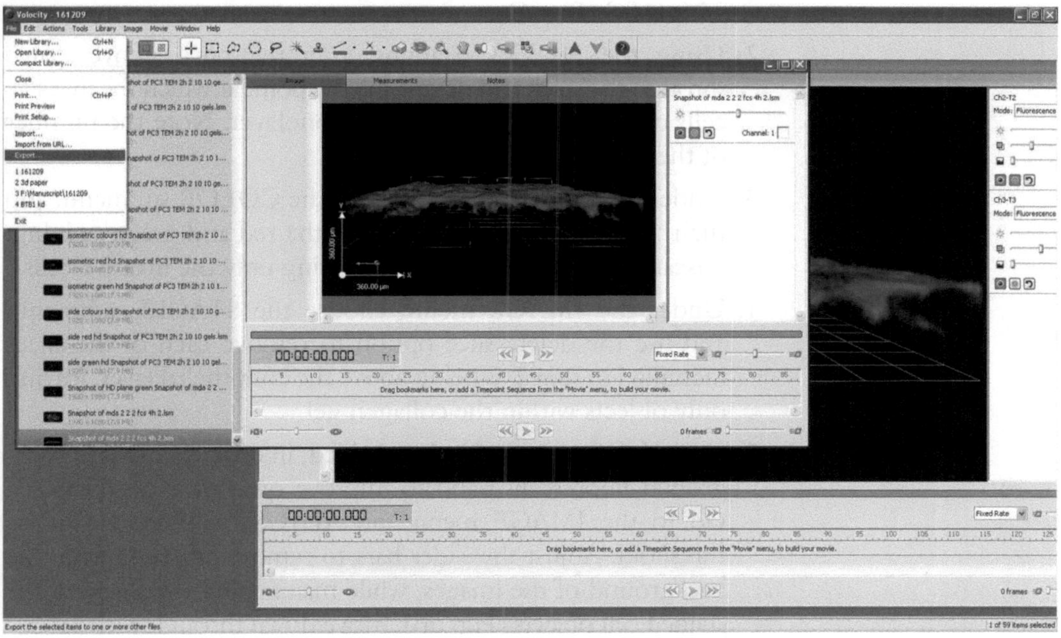

Fig. 5. (continued)

4. After selecting the desired image orientation, capture a high-resolution snapshot of the data and export it in the preferred format (Fig. 5d–f). In the worked example data are exported as a TIFF.

5. Magnification of specific features of the sample can also be performed by zooming into the three-dimensional data stack before capturing snapshots. Figure 3d illustrates a high-magnification image of MDA-MB-231 cell TEM, in which filopodia extending from the transmigrating cell can be clearly seen.

3.8. Quantification of TEM Rate

Quantification of TEM in this assay is based on measuring the volume of stained objects that are within the collagen matrix at a given timepoint after TEM. Although many commercial software packages are able to quantify this, here we describe a method using the freely available ImageJ package. To illustrate the power of the method to quantify TEM differences, we are using a worked example in which THP-1 cells treated with wortmannin, which inhibits leukocyte TEM (1), are compared to untreated cells.

3.8.1. Quantification of the Volume of Invasive Cells from Raw Confocal Stacks

1. Open the confocal stack data (as gathered using the parameters described in Subheading 3.6, step 1) in ImageJ. In our example, .lsm files are used, but other file types are also compatible. Data should be two-channel fluorescence, with HUVECs stained using CTO, and the invasive cell of interest, THP-1, with CTG (Fig. 4).
2. Using the HUVEC staining as a reference, move down through the stack and determine the point at which the invasive cells dip below the endothelial monolayer. Note the number of the slice (Fig. 6a).
3. Under the IMAGE menu, choose the COLOR submenu, and then "Split Channels" to separate the red and green staining. Discard the HUVEC channel, leaving only the invasive cells.
4. Under the IMAGE menu, choose the STACKS submenu, and use the delete slice option to remove each of the slices identified in step 2 as being above the endothelial layer, leaving only objects within the collagen gel.
5. Stack data must next be thresholded, in order to exclude background fluorescence and to improve quantification reliability. Go to the IMAGE menu and ADJUST submenu. Choose threshold. Adjust the slider bars to remove any haze from the background of the images, while maintaining the shape of the stained cell objects (Fig. 6b). Move down through each slice of the stack and perform the same operation. Errors in thresholding do not tend to lead to large errors in volume quantification; Fig. 6b shows a graph of thresholding stringency plotted against quantified volume. The plateau seen illustrates the robustness of the system as data thresholded beyond 60 points is somewhat resistant to variation in the final quantified volume.
6. The data are now ready for volume quantification. Under the PLUGIN menu, choose the 3D object counter command; this

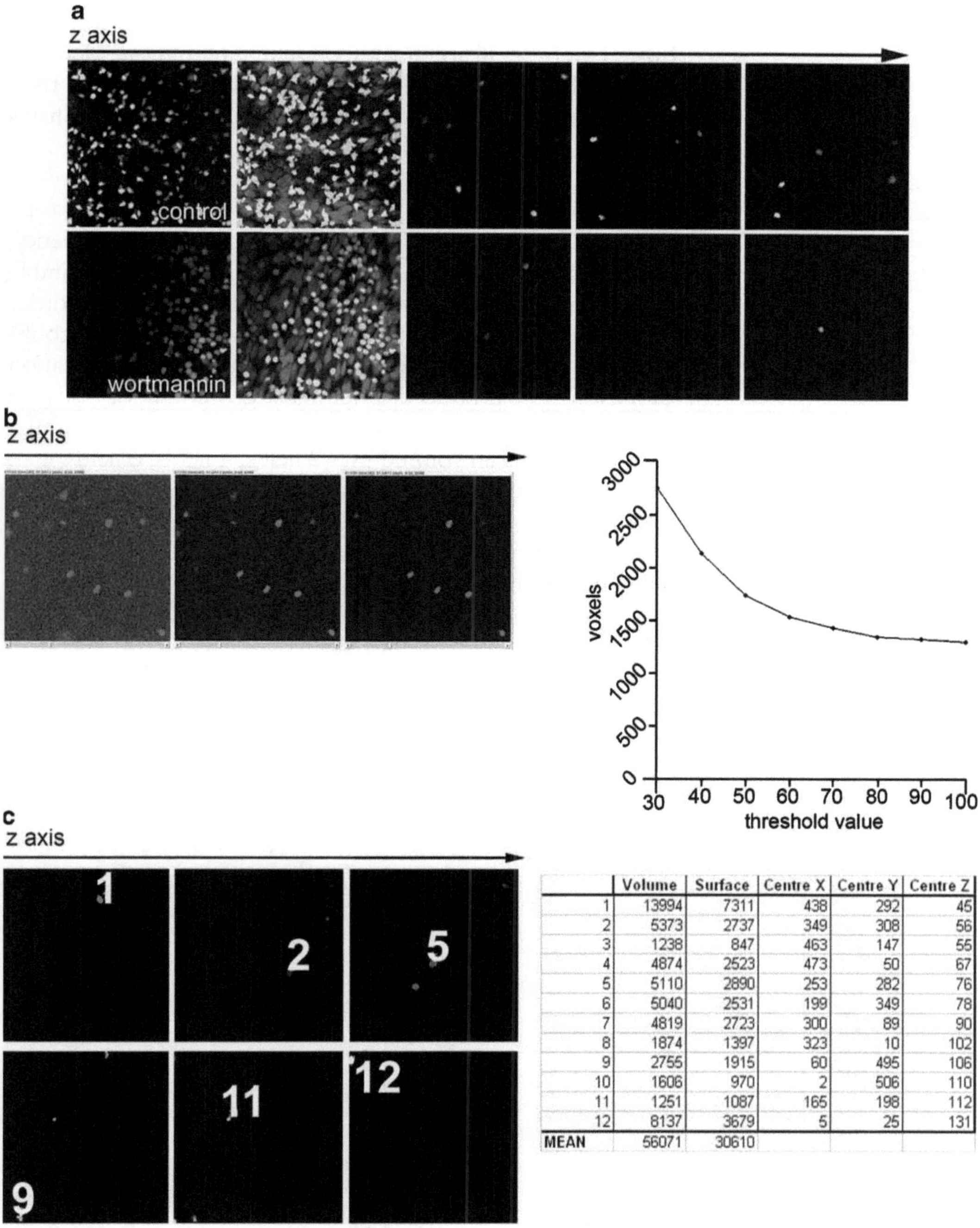

	Volume	Surface	Centre X	Centre Y	Centre Z
1	13994	7311	438	292	45
2	5373	2737	349	308	56
3	1238	847	463	147	55
4	4874	2523	473	50	67
5	5110	2890	253	282	76
6	5040	2531	199	349	78
7	4819	2723	300	89	90
8	1874	1397	323	10	102
9	2755	1915	60	495	106
10	1606	970	2	506	110
11	1251	1087	165	198	112
12	8137	3679	5	25	131
MEAN	56071	30610			

Fig. 6. Quantification of THP-1 TEM using ImageJ software. (**a**) Schematic showing various slices from confocal stacks of untreated and wortmannin-treated THP-1 cells in TEM assays, illustrating how to identify which cells have entered the collagen matrix. (**b**) Illustration of the effect of image thresholding (*left*) on subsequent cell TEM (*right*). (**c**) Illustration of the output file from ImageJ's 3D object counter plug-in, which identifies objects in confocal stacks (*left*) and quantifies their volume (*right*).

plug-in may need to be downloaded and installed from the ImageJ website, if not already present. The plug-in identifies stained objects within the collagen slice and quantifies their volume in voxols, surface area and positional co-ordinates (Fig. 6c).

3.8.2. Quantifying a Representative Sample of the Gel

The data retrieved from one stack represent one field of the total gel volume (Fig. 7a). In order to fully quantify TEM efficiency, several of these fields must be counted, however it is not desirable to count all fields as this is time-consuming and laborious. In order to quantify the minimum number of fields, while still being confident that the whole gel is represented, cumulative quantification curves must first be produced. Briefly, a large number of fields are first quantified, and then the mean volumes plotted in a cumulative fashion (e.g., mean volume after 2 fields, 4 fields, 6 fields, etc.; Fig. 7b). At a certain point the data will plateau, which will be the mean value for the whole sample. In our example of THP-1 TEM (Fig. 7b), the mean stabilized after only 5–6 fields, so this was taken as a representative sample of the whole gel.

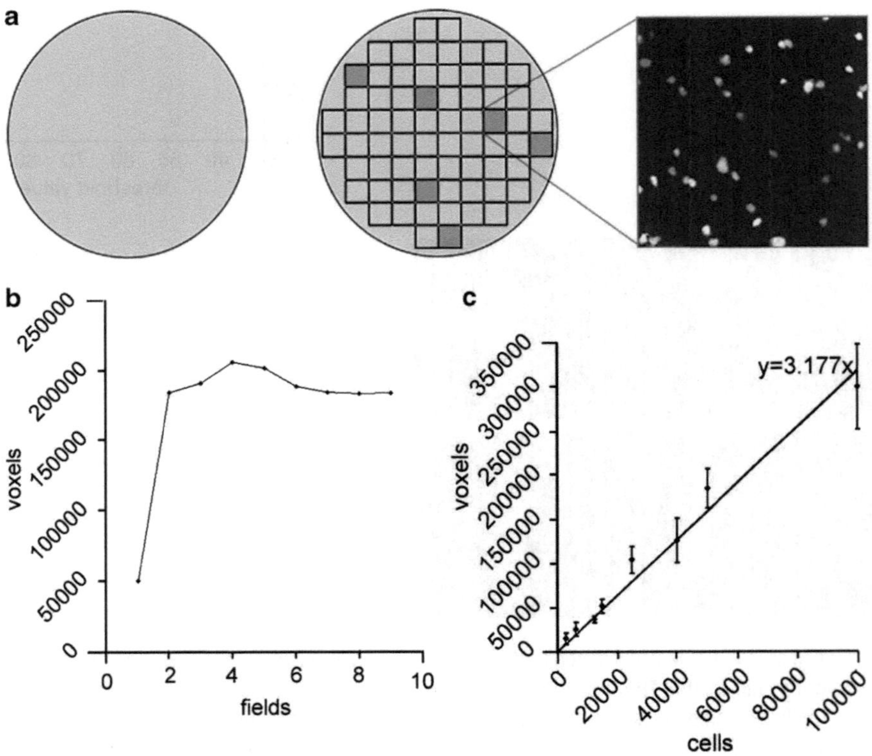

Fig. 7. Identification of field sample size required for representative quantification of TEM. (**a**) Illustration of how gel samples are split into separate visual fields on the microscope. (**b**) Cumulative cell volume quantification graph, illustrating how the quantified volume of objects with the sample stabilizes on the mean value when an increasing number of fields are counted. (**c**) Standard curve produced when correlating known numbers of cells with their quantified volume from the confocal analysis.

3.8.3. Selection of Fields

TEM events are not evenly distributed across the whole of the sample. Figure 8 shows several different patterns of field selection (a), and the resulting quantified TEM values for each (b). In our experience, TEM events are slightly more likely to occur in the center of the sample than at the edges. Once the number of fields to be counted has been established (as above), care should be taken to distribute these evenly throughout the sample, using a pattern such as "even" or "center+edge" (Fig. 7a).

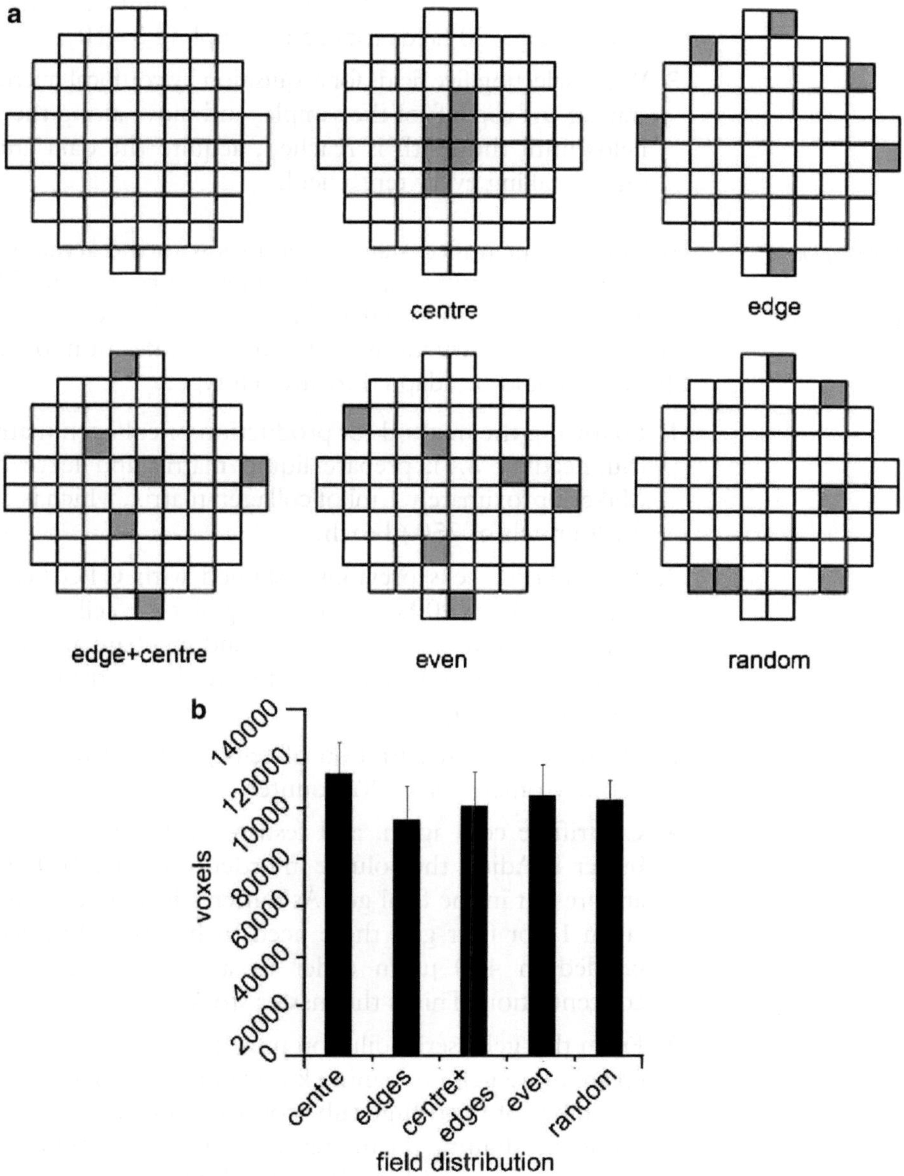

Fig. 8. Illustration of the effect of field distribution on TEM quantification. (**a**) Different potential field distributions within the sample. (**b**) Quantified cell volumes from the same sample using different field distribution patterns.

1. Establish how many fields there are in the sample area. An easy way to do this is to focus the microscope at the top left-hand-corner of the sample, and then move across in rows, using features in the microscope field of view to establish when one is moving completely into a fresh field, taking care to note how many fields have been counted. Ignore partial fields and exclude them from the study. At the end of each row, move the stage down to the next row using the same mechanism, and repeat until the entire sample has been visualized.
2. Divide the total number of fields by the number you wish to sample (e.g., 50 fields total, and sampling 5 = 10).
3. When selecting five fields for acquisition by confocal microscopy, start at the top left of the sample, and move along the row of fields until the tenth is reached, acquire the data and then repeat, taking every tenth field.

3.8.4. Calculating the Number of Cells from Quantified Object Volumes

To correlate quantified stained volumes with the actual number of cells that have transmigrated, a standard curve of four different known cell numbers plotted against volume should be produced. The following instructions relate to quantification of THP-1 TEM, but could be adapted to any cell type.

1. Following the method for production of collagen matrix gels (Subheading 3.4), prepare liquid matrix and leave on ice. Make approximately 1 ml of collagen matrix, which is enough for four gels of 250 μl each.
2. Take THP-1 cells previously stained with CTG (as above) and centrifuge (900 × *g*, 3 min) to pellet the cells. Resuspend in 3.7% PFA in order to fix cells, and incubate at room temperature for 10 min. Wash out fixative by centrifuging again, and resuspending in PBS.
3. Quantify the concentration of cells per ml of buffer using a hemocytometer or CASY counter.
4. Centrifuge cells again, and resuspend in a small volume of buffer 1. Adjust the volume in order to ensure 100,000 cells are present in the final gel. As buffer 1 is diluted 1:10 in collagen I, for four gels there need to be 400,000 cells resuspended in 400 μl in order to achieve the final desired concentration. This is the master stock. Leave on ice.
5. From this gel a serial dilution must be performed in order to create more gels containing known numbers of cells. Prechill a number of microfuge tubes on ice, and label appropriately for serially diluting the master (e.g., 50,000, 25,000, 12,500). Mix master stock well, and serially dilute into chilled microfuge tubes with fresh collagen/buffer 1 solution.

6. Pipette dilutions into Boyden chamber inserts, and incubate at 37°C as described (Subheading 3.4) to polymerize the solution.
7. Once the collagen is set, process for confocal microscope analysis as described (Subheading 3.5) and quantify the volume of CTG-stained objects in each gel.
8. To produce the standard curve, quantify several fields for each serial dilution and calculate the mean. Perform the assay subsequently until three independent mean values exist for each serial dilution (Fig. 7c).
9. Plot voxol values against serially diluted cell numbers to produce a standard curve that will allow you to calculate cell numbers from voxols during actual TEM experiments (Fig. 7c).

3.8.5. Using the Standard Curve to Quantify TEM

In our example of TEM THP-1 treated with wortmannin or untreated was measured using the system described above. Three-dimensional reconstruction (Fig. 4) showed that TEM was reduced after wortmannin treatment, as fewer cells were present in the collagen matrix.

1. To quantify this difference, count voxols present in the gel over five fields as described above.
2. Use the standard curve produced (Fig. 7c) to determine cell number from voxols. The slope of the curve can be used to produce a formula to make this transition. The data clearly show a fourfold decrease in TEM of THP-1 cells after wortmannin treatment (Fig. 9).

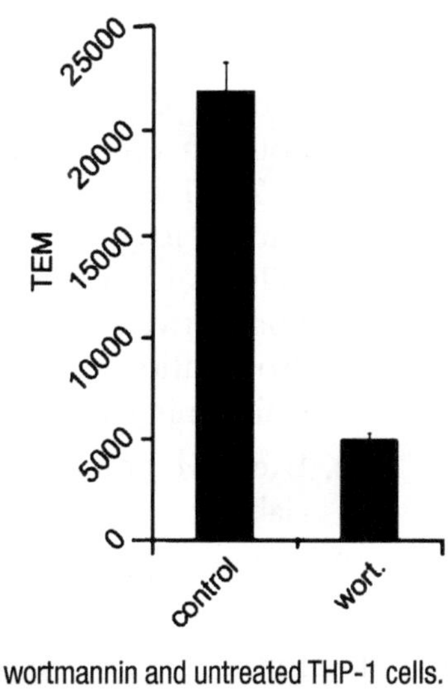

Fig. 9. Quantification of wortmannin and untreated THP-1 cells. Data were taken from the stacks used to produce the images in Fig. 4, and TEM of THP-1 cells quantified utilizing the standard curve (Fig. 7) to translate quantified volumes into numbers of cells.

4. Notes

1. HUVECs can be difficult to culture, and there can be variations between batches. To avoid problems of consistency in experiments, try to use cells from one supplier, and minimize the number of different cell batches used as far as is possible.
2. Similar cell dyes are available from other companies, but we have found that those from Molecular Probes produce good quality consistent results, with low levels of dye leaching out of the cells.
3. Solution can be kept at room temperature for up to 3 months. Chill on ice before use. Keeping on ice for long periods can cause the salts to precipitate.
4. Thaw aliquots as required. PFA is stable at 4°C for few days, but deteriorates over time. Avoid repeated freeze/thawing.
5. Fibronectin is purchased lyophilized, and must be dissolved in PBS and aliquoted. Make aliquots of fibronectin in advance, and freeze as 50× stocks. Upon defrosting, dilute to working concentration in PBS, and store at 4°C. Although fibronectin can be reused between experiments if kept sterile, it is not advisable to keep the working dilution at 4°C for more than a few weeks.
6. Ensure that enough volume of fibronectin solution (e.g., 4 ml of fibronectin solution is enough to coat a 75-cm^2 flask) is used to cover the bottom of the culture vessel adequately – if insufficient liquid is added it can retract from the edges of the flask during incubation, leaving holes in the coating.
7. Do not centrifuge the cells to remove DMSO as this reduces viability.
8. Beyond passage 4, HUVECs begin to loose their endothelial phenotype, and become larger and more spread. We have noted that the presence of some tight junctional markers such as ZO-1 is reduced. Culturing cells to full confluence also leads to irregular cell morphology when assessed microscopically. As HUVECs are a primary cell type, there can be variation between different batches of cells (e.g., morphology, proliferation rate). The effects of this can be minimized by maintaining as consistent a culturing regime as possible.
9. Excessive volume or exposure to trypsin reduces HUVEC viability.
10. Cells will grow poorly below 500,000 cells/ml, and will start to show abnormal phenotypes if grown above 2,000,000 cells/ml.
11. Efficient migration of cancer cells in collagen gels is dependent on optimal collagen I concentration, as very low or very high concentrations will lead to decreased migration. Similarly, HUVECs will not form proper monolayers on top of gels

with a concentration below 1.5 mg/ml, and tend to sink into the matrix and form tubules instead.

12. It should be possible to see the straightened meniscus when holding the insert up to the light and tilting slightly. Add more collagen if necessary in order to produce as flat a surface as possible.
13. Do not polymerize in the cell culture incubator, as CO_2 and increased humidity inhibit the polymerization reaction.
14. Our laboratory has found that HUVECs require at least 24 h to form mature adherens junctions.
15. Removing the gel in this way allows a smaller volume of buffer to be used for subsequent antibody incubations, as only the top chamber of the Boyden chamber needs to be filled.
16. It is very important not to let the gel dry out at any point, as this will distort subsequent sample morphology.
17. We have used different antibodies to visualize cell–cell junctions, including VE-cadherin (1:100; BD Biosciences), β-catenin (1:500; Sigma-Aldrich) as well as fluorescently labeled phalloidin to visualize filamentous (F)-actin. Secondary antibodies were all from the Molecular Probes range of fluorescently labeled antibody conjugates, which give excellent staining and minimal background. The choice of fluorophore-labeled antibodies will depend on the confocal microscope available, and its complement of lasers. Examples shown in this chapter are all possible with two channel fluorescence.
18. Care should be taken to avoid getting air bubbles between the gel and the coverslip. If this happens, they can be released by gently lifting the edge of the gel. Take care, as the gel will be fragile.
19. The Zeiss confocal microscope software saves stack data as .lsm files. These can be opened by a number of different commercially available and free image analysis software packages.
20. Zeiss software saves stack data as .lsm files, but other microscope data types such as Leica LIF files are also supported by Volocity.

Acknowledgments

This work was funded by the Association for International Cancer Research, Cancer Research UK, the Biotechnology and Biological Sciences Research Council, and the Bettencourt-Schueller Foundation (grants to AJR). BBA was funded by Fundação para a Ciência e Tecnologia (Portugal) and the Graduate Program in Areas of Basic and Applied Biology (GABBA – University of Porto, Portugal).

References

1. Engelhardt, B., and Wolburg, H. (2004) Mini-review: Transendothelial migration of leukocytes: through the front door or around the side of the house?, *Eur J Immunol* **34**, 2955–2963.
2. Millan, J., and Ridley, A. J. (2005) Rho GTPases and leucocyte-induced endothelial remodelling, *Biochem J* **385**, 329–337.
3. Nourshargh, S., and Marelli-Berg, F. M. (2005) Transmigration through venular walls: a key regulator of leukocyte phenotype and function, *Trends Immunol* **26**, 157–165.
4. Blankenberg, S., Barbaux, S., and Tiret, L. (2003) Adhesion molecules and atherosclerosis, *Atherosclerosis* **170**, 191–203.
5. Tarrant, T. K., and Patel, D. D. (2006) Chemokines and leukocyte trafficking in rheumatoid arthritis, *Pathophysiology* **13**, 1–14.
6. Hooper, S., Marshall, J. F., and Sahai, E. (2006) Tumor cell migration in three dimensions, *Methods Enzymol* **406**, 625–643.
7. Sabeh, F., Shimizu-Hirota, R., and Weiss, S. J. (2009) Protease-dependent versus -independent cancer cell invasion programs: three-dimensional amoeboid movement revisited, *J Cell Biol* **185**, 11–19.
8. Sahai, E. (2007) Illuminating the metastatic process, *Nat Rev Cancer* 7, 737–749.
9. Vega, F. M., and Ridley, A. J. (2008) Rho GTPases in cancer cell biology, *FEBS Lett* **582**, 2093–2101.
10. Chiang, A. C., and Massague, J. (2008) Molecular basis of metastasis, *N Engl J Med* **359**, 2814–2823.
11. Nguyen, D. X., Bos, P. D., and Massague, J. (2009) Metastasis: from dissemination to organ-specific colonization, *Nat Rev Cancer* **9**, 274–284.
12. Carman, C. V., Sage, P. T., Sciuto, T. E., de la Fuente, M. A., Geha, R. S., Ochs, H. D., Dvorak, H. F., Dvorak, A. M., and Springer, T. A. (2007) Transcellular diapedesis is initiated by invasive podosomes, *Immunity* **26**, 784–797.
13. Muller, W. A. (2009) Mechanisms of transendothelial migration of leukocytes, *Circ Res* **105**, 223–230.
14. Milan, J., Charalambous, C., Elhag, R., Chen, T. C., Li, W., Guan, S., Hofman, F. M., and Zidovetzki, R. (2006) Multiple signaling pathways are involved in endothelin-1-induced brain endothelial cell migration, *Am J Physiol Cell Physiol* **291**, C155–164.
15. Cernuda-Morollon, E., Gharbi, S., and Millan, J. Discriminating between the paracellular and transcellular routes of diapedesis, *Methods Mol Biol* **616**, 69–82.
16. Mierke, C. T., Zitterbart, D. P., Kollmannsberger, P., Raupach, C., Schlotzer-Schrehardt, U., Goecke, T. W., Behrens, J., and Fabry, B. (2008) Breakdown of the endothelial barrier function in tumor cell transmigration, *Biophys J* **94**, 2832–2846.
17. Khuon, S., Liang, L., Dettman, R. W., Sporn, P. H., Wysolmerski, R. B., and Chew, T. L. (2010) Myosin light chain kinase mediates transcellular intravasation of breast cancer cells through the underlying endothelial cells: a three-dimensional FRET study, *J Cell Sci* **123**, 431–440.
18. Roth, S. J., Carr, M. W., Rose, S. S., and Springer, T. A. (1995) Characterization of transendothelial chemotaxis of T lymphocytes, *J Immunol Methods* **188**, 97–116.
19. Ding, Z., Xiong, K., and Issekutz, T. B. (2000) Regulation of chemokine-induced transendothelial migration of T lymphocytes by endothelial activation: differential effects on naive and memory T cells, *J Leukoc Biol* **67**, 825–833.

Chapter 13

Chemotaxis of Slow Migrating Mammalian Cells Analysed by Video Microscopy

Roman Zantl and Elias Horn

Abstract

We present a microfabricated chamber designed for visualising and quantifying the chemotaxis of slow-migrating adherent mammalian cells such as cancer and endothelial cells. Most of the existing solutions for the investigation of chemotaxis are limited to fast migrating cells such as leukocytes or *Dictyostelium discoideum*. Here, we describe the details of an assay using the μ-Slide Chemotaxis to investigate the chemotactic response of human umbilical vein endothelial cells to a gradient of human vascular endothelial growth factor 165. In combination with phase contrast video microscopy and cell tracking, the trajectories of all single cells migrating in temporally stable gradients are derived. The resulting migration data are displayed and analysed in detail by several different parameters for quantifying chemotaxis. We found that with this tool the potential of chemoattractants to migration of mammalian cells as well as the impact of inhibitors to chemotaxis and migration can be evaluated.

Key words: Chemotaxis, Migration, Endothelial cells, Cell tracking, VEGF, Video microscopy, Adherent cells

1. Introduction

Human umbilical vein endothelial cells (HUVEC) migrate with an average speed of roughly 30 μm/h (1) similarly to the velocity of cancer cells such as HT-1080 with roughly 50 μm/h (2). Compared to rather fast migrating cells like T-lymphocytes with a speed of 360 μm/h (3) and *Dictyostelium discoideum* with 450 μm/h (4), they are referred to as slow migrating cells. Though chemotaxis of slow migrating cells is of great importance in many physiological and pathophysiological processes including wound healing (5), angiogenesis (6), and metastasis (7), only few data of characteristic chemotaxis parameters are published.

Claire M. Wells and Maddy Parsons (eds.), *Cell Migration: Developmental Methods and Protocols*, Methods in Molecular Biology, vol. 769, DOI 10.1007/978-1-61779-207-6_13, © Springer Science+Business Media, LLC 2011

This might be due to the leak of chemotaxis assays specially designed for investigating slow migrating cells. Fast migrating cells often move very straight in the direction of the gradient and their trajectories can be described as nearly straight lines. In contrast, the chemotactic movement of slow migrating cells rather resembles arbitrary diffusion with a slight average displacement in the direction to the chemoattractant source. In combination with rather slow migration speed this behaviour makes observation times in the range of 12–24 h necessary in order to measure chemotaxis parameters sufficiently reproducible and reliable. Most of the known assays based on the ideas of Zigmond (8) and Boyden (9) provide gradients that are only over shorter time periods temporally stable and therefore, are suited for fast migrating but in most cases not for slow migrating cells. The µ-Slide Chemotaxis (Fig. 1) developed exactly for this application was already successfully applied for chemotaxis studies of slow migrating cells (10, 11) also including chemotaxis inhibition (12). Based on the well-known Zigmond chamber an observation area with a height of 70 µm and a volume of 140 nl connects two large reservoirs each having a volume of 40 µl. The resulting small cross section of the observation area restricts strongly the exchange of molecules between the reservoirs and is therefore maintaining concentration gradients nearly constant over 2 days (unpublished data).

In this chapter, we describe in detail the preparation of the slide with HUVEC cells in a vascular endothelial growth factor (VEGF)-gradient. This assay can be adapted to other slow migrating adherent cells as well as adherent fast migrating cells (see Note 1). Then, details about image capture and video microscopy are

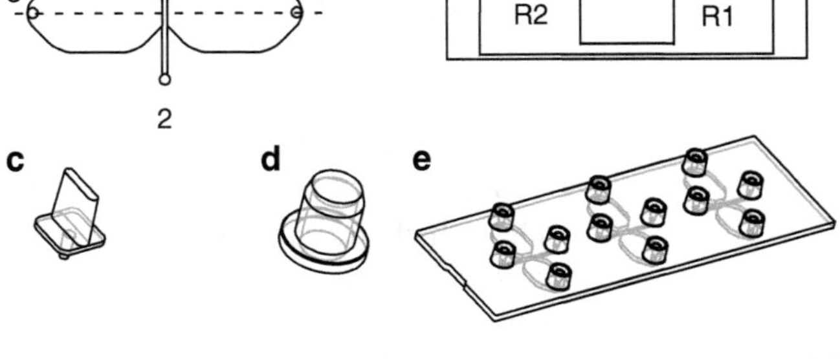

Fig. 1. Schematic drawing of the used µ-Slide. (**a**) Top view on one experimental chamber with reservoirs R1 and R2, and filling ports 1–4. (**b**) Cross section along the dashed line between filling ports 3 and 4. R1 and R2 are connected via an observation area of low height (70 µm). The concentration difference between R1 and R2 leads to a gradient across the observation area. (**c**) Plugs for air and liquid tight closure. (**d**) Caps for sterile covering. (**e**) The µ-Slide Chemotaxis with three experimental chambers has standard microscopy slide format.

presented, as well as finally a way for data analysis and interpretation. We focus on a maximum number of cells for chemotaxis evaluation thus using low resolution video microscopy.

2. Materials

2.1. Cell Culture

1. HUVECs (Lonza, Verviers, Belgium) from passages three to five.
2. *Full medium*: Endothelial Cell Growth Medium (PromoCell GmbH, Heidelberg, Germany) with supplement mix and 10% foetal calf serum (FCS).
3. Solution of trypsin (0.5% w/v) and ethylenediamine tetraacetic acid (EDTA, 0.68 mM).
4. 75-cm^2 cell culture flasks with tissue culture surface for primary cell culture.
5. Dulbecco's PBS, 1× without Ca^{2+} and Mg^{2+}.
6. Neubauer improved counting chamber (depth 0.1 mm, 0.0025 mm^2).

2.2. Chemotaxis Experiment

1. μ-Slide Chemotaxis with collagen IV coating (#80302 ibidi, Martinsried, Germany) (Fig. 1e) (see Note 2).
2. *Starving medium*: Endothelial Cell Growth medium without any supplements.
3. *Full medium*.
4. *Chemoattractant solution*: 2 μg of $VEGF_{165}$ powder (PeproTech GmbH, Hamburg, Germany) is dissolved first in 20 μl ultra pure water, and then further diluted with 180 μl of PBS to a concentration of 10 μg/ml. 10 μl aliquots are stored at −20°C. After thawing 1 μl is diluted with 199 μl *starving medium* in a reaction cap to a final concentration of 50 ng/ml.
5. *Balanced chemoattractant solution*: this solution is made by diluting the *chemoattractant solution* with *starving medium* to a final concentration of 11.25 ng/ml (see Note 3).
6. A simple humidifying chamber is made by using a 100-mm Petri dish with two pieces of "kimwipes" each wetted with 5 ml ultrapure water. In between the two paper pieces a gap of 5 cm is left open for placing the slide.
7. 100 mm Petri dish for handling of the sterile plugs and caps (Fig. 1c, d).
8. Sterile blunt forceps.
9. Pipet tips T-200C from Axygen (VWR, Darmstadt, Germany) (see Note 4).

2.3. Video Microscopy Equipment

1. Nikon Ti-E Eclipse (Nikon, Düsseldorf, Germany), an inverted microscope providing phase contrast, time lapse data acquisition, a motorised stage, and automated illumination control (see Note 5).
2. Nikon Plan Fluor 4× 0.13 PhL DL microscope objective (Nikon, Düsseldorf, Germany) providing phase contrast (see Note 6).
3. Stage top incubator for controlling the environment of the sample to 37°C and 5% CO_2 (ibidi, Martinsried, Germany) (see Note 7).

2.4. Data Analysis

1. Computer (1 GB RAM) with Windows XP or higher operation system.
2. ImageJ bundled with 32-bit Java 1.6.0_10 (free download http://rsbweb.nih.gov/ij/) (see Note 8).
3. Manual tracking plugin for ImageJ (free download from http://rsbweb.nih.gov/ij/plugins/track/track.html) needs to be copied into the folder plugins in the ImageJ installation folder.
4. Chemotaxis and Migration Tool 2.0 (ibidi, free download from http://www.ibidi.de/applications/ap_chemo.html) (see Note 9).
5. Documentations for the Manual Tracking Plugin (http://rsbweb.nih.gov/ij/plugins/track/Manual20Tracking%20plugin.pdf) and the Chemotaxis and Migration Tool (http://www.ibidi.de/applications/ap_chemo.html) are helpful for getting started using the software.

3. Methods

Air bubbles are one of the major problems in all microfluidic systems. Once trapped, they are hard to remove and might influence the integrity of the diffusive gradient. Therefore, it is necessary that exactly the proposed pipet tips are used, and that the filling protocol is followed down to the last detail (see Note 10). Air bubbles emerge by gassing out from liquids and plastic materials when temperature is raised. At least 24 h before the experimental preparation is started it is crucial for equilibration to place both, the slide and the media in a cell incubator or cabinet dryer at a temperature between 37°C and 50°C. All needed media are put in tubes with slightly loose caps. Slides, plugs, and caps can be left in their sterile packaging (see Note 11).

The gradient will start to form immediately after filling in the *chemoattractant solution*. Roughly after 12 h it will be at its maximum, a steady state remaining for at least 48 h without major changes. We found that in case of experiments with HUVEC it is optimal to start the video microscopy measurement right after slide preparation and running it for 18 h. This way, three experiments on one slide can be done every day.

Motility is an important parameter strongly influencing certain chemotaxis parameters like, for example the displacement of the centre of mass. The motility of cells depends on multiple parameters and can also be altered by different chemoattractant concentrations (13). Preparing a negative control containing only *starving medium* (−/−) and a positive control with *balanced chemoattractant solution* of homogeneous concentration (+/+) we are able to measure directly the motility of cells independently from chemotaxis.

3.1. Preparation of Cell Suspension

1. HUVEC cells are split when approaching confluence with trypsin/EDTA in a ratio of 1:2 for amplification in culture flasks.
2. For preparing cells for the experiment HUVEC are grown to 80–90% confluence, trypsinised, resuspended in *full medium*, counted, and centrifuged in a reaction cap. The supernatant is removed with a pipet and the cells are resuspended in *full medium* to a final concentration of 3×10^6 cells/ml.

3.2. Preparation of the Chemotaxis Chamber

3.2.1. Preparation of the Chemotaxis Experiment with Gradient (+/−)

1. All materials required are placed in the sterile workbench: μ-Slide Chemotaxis with caps and plugs, cell suspension with 3×10^6 cell/ml in equilibrated *full medium*, sterile forceps, sterile 100 mm Petri dish (for caps and plugs), and the humidifying chamber.
2. The μ-Slide is put into the humidifying chamber and the sterile caps and plugs are put in the sterile Petri dish (see Note 12).
3. Filling ports 3 and 4 are closed with plugs using the forceps (Fig. 2a).
4. A drop of 6 μl of cell suspension is added on top of filling port 1. Do not inject the cell suspension into the channel (Fig. 2b) (see Note 13).
5. Cell suspension is sucked through filling port 2 into and all the way through the channel until the liquid front is touching the pipet tip (Fig. 2c) (see Note 14).
6. Carefully both plugs are removed and caps are put on all four filling ports (Fig. 2d). The humidifying chamber is closed.
7. By (phase contrast) microscopy the number of cells inside the observation area is controlled. There should be ca. 200–500 cells in the entire observation field (Fig. 3a). The reservoirs R1 and R2 need to be free of cells.

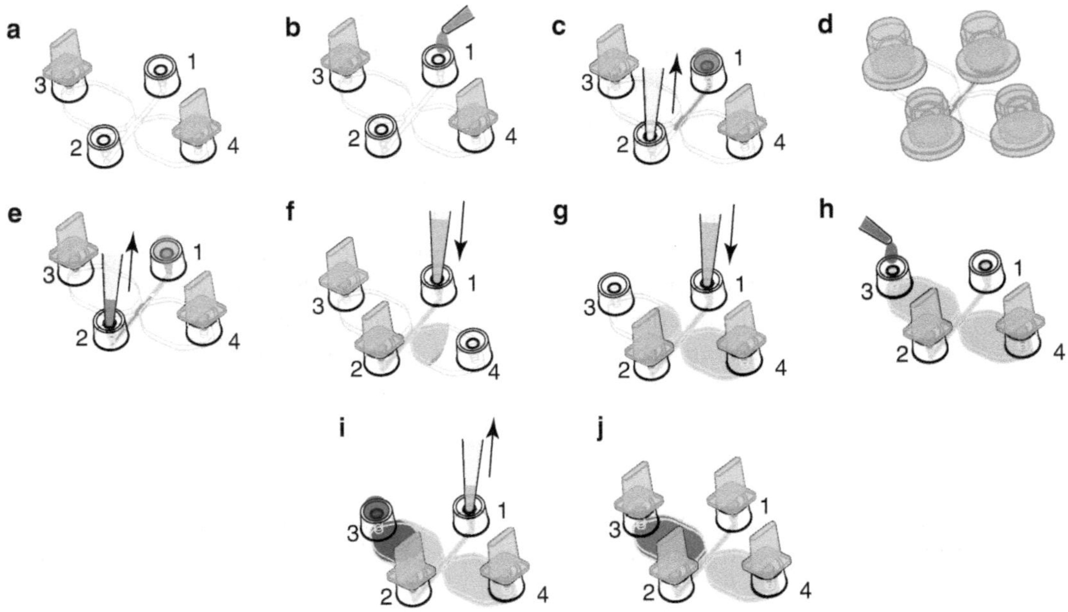

Fig. 2. Step by step procedure for preparing one chamber of μ-Slide Chemotaxis.

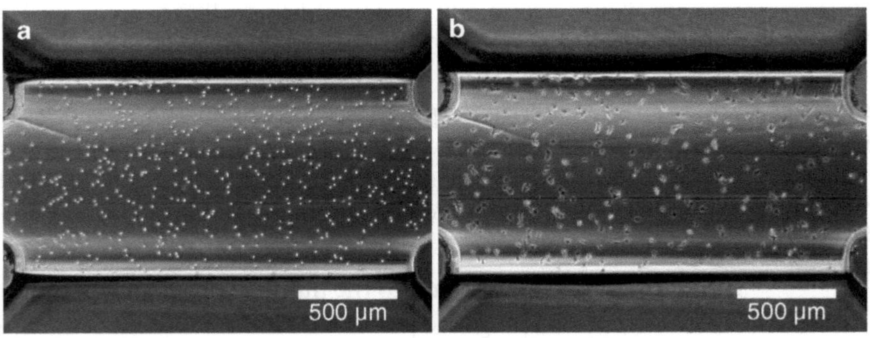

Fig. 3. Microscopic images of the observation area with recommended cell density right after cell seeding (**a**) and after 4 h in the incubator (**b**). The *bright lines* are the boarders of the observation area of one experimental chamber. In the picture one reservoir is on top and the other one below the observation area.

8. The humidifying chamber with the slide is placed in the cell culture incubator for 4 h (see Note 15).
9. After incubation, cell attachment and cell density are controlled by microscopy (Fig. 3b).
10. The caps are taken off and filling ports 3 and 4 are closed with plugs. For removing non-attached cells and *full medium*, the channel is rinsed twice with 10 μl of *starving medium*. For rinsing a 10-μl drop of *starving medium* is added on top of filling port 1 and sucked through the channel using filling port 2 (Fig. 2e) (see Note 16).
11. The plug from filling port 4 is moved to filling port 2. 40 μl of *starving medium* is slowly injected into filling port 1.

This way reservoir R1 is filled until the upper edge of filling port 4 (Fig. 2f) (see Note 17).

12. The plug from filling port 3 is moved to filling port 4. The reservoir R2 is filled with 40 μl *starving medium* to the upper edge of filling port 3 (Fig. 2g).
13. Filling port 3 is controlled to be completely filled but without any bulky drop of liquid on top of the flat surface in order to avoid unwanted dilution of the *chemoattractant solution*. An 18 μl drop of *chemoattractant solution* is placed on top of filling port 3 (Fig. 2h) and 18 μl of liquid are sucked from filling port 1 (Fig. 2i) (see Note 18).
14. All filling ports are closed using the plugs and the chemotaxis sample is ready for microscopy (Fig. 2j).

3.2.2. Preparation of the Negative Control Experiment (−/−)

1. Repeat steps 1–12 of Subheading 3.2.1.
2. All filling ports are closed using the plugs and the negative control is ready for microscopy.

3.2.3. Preparation of the Positive Control Experiment (+/+)

1. Repeat steps 1–9 of Subheading 3.2.1.
2. Repeat steps 10–12 in Subheading 3.2.1 but using *balanced chemoattractant solution* instead of *starving medium* and *chemoattractant solution*.
3. All filling ports are closed using the plugs and the positive control is ready for microscopy.

3.3. Video Microscopy

1. The slide is placed into the stage top incubator (5% CO_2, 37°C). Mechanical fixation of the slide helps avoiding focus drift.
2. The image sections are chosen to be in the centre of each of the three observation areas displaying the entire observation fields. In case the observation area appears tilted on the monitor the camera needs to be adjusted. The direction of the gradient is noted. In Figs. 3 and 4 the gradient is pointing upwards, as the upper reservoir is filled with chemoattractant.
3. The three different *x*–*y*–*z*-positions are taught into the microscope software and the time lapse parameters are controlled to take one picture every 10 min (see Note 19) for 18 h. This results in 109 images. The samples are only illuminated during image acquisition. Ca. 30 min after starting time lapse imaging thermal equilibration of the microscopy stage is reached and the focus of all three imaging positions is controlled.
4. After finishing the experiment the images of all three experiments are exported to single image files in tif-format with 8 bit resolution into different folders.

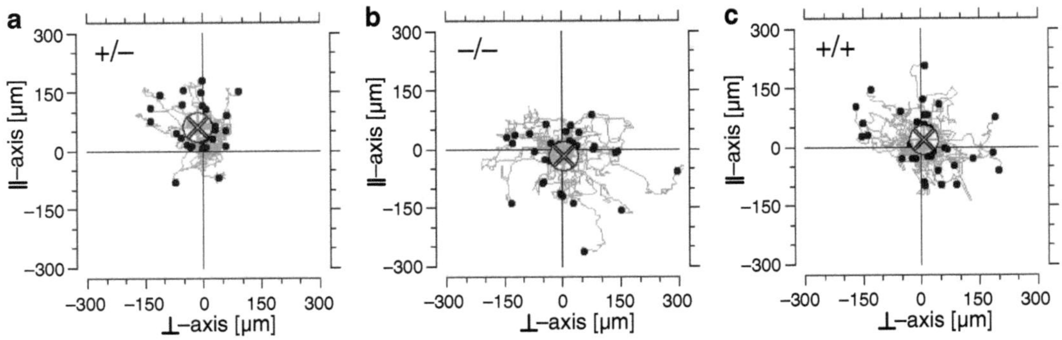

Fig. 4. Each of the shown plots contains the trajectories of 40 HUVEC cells, with the position of the centre of mass after 18 h represented by a cross. Through transformation all trajectories start at the origin in order to visualise the movements of all cells relative to each other. Data are shown for representative single experiments. (**a**) In the chemotaxis experiment (+/–) the centre of mass is clearly shifted in direction of the gradient pointing to the *top* of the graph. In the negative control (–/–) (**b**) and positive control experiments (+/+) (**c**) the entire volumes are either filled with *starving medium* or with *balanced chemoattractant solution.*

3.4. Data Analysis

1. Before the movie is watched for the first time the first image of each image stack is printed. At least 40 cells which are uniformly distributed over the entire cell population are marked in order (1) to ensure homogeneity and (2) to avoid tracking the same cell twice (see Note 20).
2. Images number 1 to number 109 representing 18 h of experiment are imported to ImageJ using the File\Import>Image Sequence\Image Sequence dialog. Following import parameters are used: Number of Images 109; Starting Image 1; Increment 1.
3. The "Manual Tracking" plugin is used for tracking all marked cells on the printout. The settings of the "Manual Tracking" plugin have no effect on data analysed at a later timepoint with the "Chemotaxis and Migration Tool."
4. The resulting data table file is saved as .txt file.
5. The .txt file is imported into the program "Chemotaxis and Migration Tool."
6. For the marked dataset the calibration parameters are typed into the displayed input fields ["*x*/*y* calibration" for pixel size (see Note 21), "time interval" 10 min for time between images, and "number of slices" 109 which is the number of images of the tracked image stack].
7. The data is plotted ("Plot data") as shown in Fig. 4 and all resulting parameters ("Measured values") are saved to a text file.
8. The chemotactic potential is derived from the values Forward Migration Index FMI (14) and Rayleigh test (15) that are

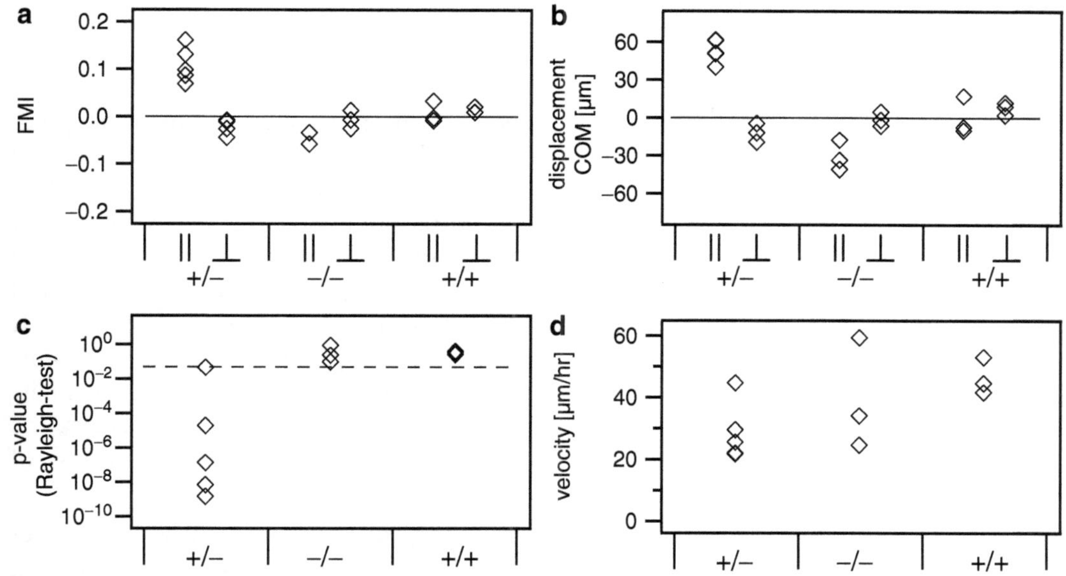

Fig. 5. Analysed migration parameters of one set of HUVEC experiments. Each *square* represents a data point (or its ||- or ⊥-component) of a single measurement. (**a**) Forward migration indices in parallel ($FMI^{||}$) and perpendicular ($FMI^{\perp}$) directions are compared to the FMI-values of negative (–/–) and positive controls (+/+). The *horizontal line* indicates zero. All FMI except $FMI^{||}$ of the chemotaxis experiment (+/–) are close to 0, indicating arbitrary movements. The $FMI^{||}$ (+/–) varies between 0.07 and 0.16 representing directed migration in case of HUVEC experiments. (**b**) The displacement of the centre of mass (COM) is shown from the same experimental set as A. Like in the case of the FMI, all COM are close to 0. Only the displacement in parallel direction ($COM^{||}$) of the chemotaxis experiment differs clearly from 0. (**c**) Corresponding *p*-values of the Rayleigh test in logarithmic scale. The *dashed line* indicates the $p = 0.05$ threshold. Only in the (+/–) case the *p*-value is smaller than 0.05, showing an inhomogeneous cell distribution. (**d**) Average cell velocities of HUVEC for the chemotaxis (+/–) and control experiments (–/–, +/+). The control measurements show the influence of the chemoattractant to the motility. In this data set there is no significant change of motility due to differing chemoattractant concentrations.

plotted as shown in Fig. 5. Displacement of centre of mass, velocity, and directness (Euclidean distance between starting and end point divided by the total length of the distance moved) as well as other parameters can be used to further analyse migration in more detail. The software calculates various values automatically.

9. For data interpretation the FMIs of the chemotaxis experiments and the control experiments are compared (see caption of Fig. 5). In case of chemotaxis all FMI components of the control experiments (–/–, +/+) and the perpendicular $FMI^{||}$ of the chemotaxis experiment (+/–) should be around 0. Values of the $FMI^{||}$ (+/–) significantly different from 0 represent a chemotaxis effect (Fig. 5a).

10. The *p* values of the Rayleigh test of all experiments are compared to judge for homogeneous distributions (Fig. 5c).

4. Notes

1. For other cell types following parameters might be adapted. Time interval: The cell velocity can be derived from positive control experiments (+/+) and delta *t* is calculated according to Note 19. Cell concentration: For different cells it might be advantageous to change the number of cells seeded into the observation area. Surface coating: The surface coating should mimic the in vivo migration situation of the cell type used. It can also be used for investigating binding effects to specific adhesion motives. Chemoattractant: Starting with chemoattractant concentrations used in other chemotaxis assays, the best suited concentration is derived by a series of experiments.
2. The observation area of the slide can be coated also with other coatings. Follow steps 3–6 of Subheading 3.2.1 using coating solution instead of cell suspension. Then follow step 10 of Subheading 3.2.1 using appropriate rinsing solution instead of *starving medium*. All rinsing solution is sucked out of the channel by an aspiration device. Afterwards the slide is air dried by storing it for several hours under sterile conditions. Slides containing remaining liquid can hardly be filled without producing air bubbles.
3. The *balanced chemoattractant solution* should have the average concentration of chemoattractant that is expected to be inside the observation area in case of the chemotaxis experiment. The given concentration of 11.25 ng/ml is calculated as followed: the volume of each of the reservoirs is 40 µl. For setting up the gradient 18 µl of the *chemoattractant solution* containing 50 ng/ml is exchanged in reservoir R2. The resulting concentration after diffusive equilibration in R2 is therefore (18 µl × 50 ng/ml)/40 µl. In reservoir R1 the concentration of chemoattractant is 0 ng/ml. In the centre of the observation area the concentration is the arithmetic mean of the concentrations in R1 and R2 because the system is symmetric. That is 1/2(18 µl × 50 ng/ml)/40 µl = 11.25 ng/ml.
4. It is absolutely necessary to use only recommended pipet tips because others are shaped slightly different. This leads in most cases to problems through air bubble formation during filling the slide. Other recommended pipet tips are from Greiner Bio-One #s 739261, 739280, 739290, 772288, Axygen #s TR-222-C, TR-222-Y, and Starlab TipOne RPT S1161-1800. The particular properties of these pipet tips are described in the following. First, when pressed onto the filling ports the tips need to make a sealed connection. The connection must be tight enough to apply positive or negative

pressure without losing liquid or sucking in air bubbles. Second, the tip needs to have a shape such that it does not reach more than 1 mm into the filling port. In case it does reach further in the formation of air bubbles is hard to avoid. Typically, the suitable pipet tips are bevelled. But not every kind of bevelled tips is suited for the assay.

5. Any inverted microscope with phase contrast and automated image acquisition can be used. Light should be switched on only during image acquisition in order to minimise phototoxic effects. A motorised stage is very convenient for increased throughput by parallelisation of video microscopy.
6. Other low magnifying phase contrast objectives can also be used. It is recommended to observe a maximum of the observation field as shown in Fig. 3.
7. Also other incubation units can be used that provide temperature control of 37°C and 5% CO_2 concentration. No humidity control is needed because the chambers are water and air tight sealed by the plugs.
8. Using Windows Vista or Windows 7 ImageJ should be installed in a public folder like C:\Users\Public\ImageJ\ImageJ.exe with access rights to all users also without administrator's rights. The path of the Manual Tracking plugin then would be C:\Users\Public\ImageJ\plugins\Manual_Tracking.class.
9. In case no Windows PC is available a MAC compatible Java version of the ibidi "Chemotaxis and Migration Tool" is available as an ImageJ plugin. It can be downloaded from http://www.ibidi.de/applications/chemotaxis/chemotaxis_tool.jar and must be placed into the plugin folder of the ImageJ installation folder.
10. For getting familiar with the preparation protocol it is recommended performing all steps using medium stained with food colouring before starting with cells in the sterile bench. Cell medium with the common amount of phenol red is barely visible in such small geometries.
11. For one planned experiment we recommend preparing a bit more material than needed. Therefore, we suggest equilibrating three slides and ca. three times the required amount of media inside the incubator (5 ml *full medium*, 5 ml *starving medium*, 50 μl *chemoattractant solution*, and 200 μl *balanced chemoattractant solution*).
12. The bottom of the μ-Slide can be scratched when handled without care.
13. It is very important to fill only the observation channel with cell suspension. Injection of liquid into the filling port at this

point leads to spilling of cell suspension into the reservoirs. In this case, a high cell density on the reservoirs' bottom is obtained so that inhomogeneous cell distribution leads to migration overlaying potential chemotactic effects.

14. Make sure that the pipet tip has a sufficient sealing with the conical edge of the filling port. It is recommended to use a 20-μl pipet for this step as it is fitting with the correct pipet tips and is made for handling small liquid volumes. The quality of the pipet is important.
15. During incubation of the slide the humidity inside the incubator should be maintained at high level. Frequent opening of the door might lead to evaporation of the slide filling disturbing the cell environment.
16. It is very important that there is no air bubble trapped within the filling port. For avoiding bubbles the filling ports need to be filled completely before a plug or further liquid is applied. In case an air bubble is trapped it can be removed stoking with a thin 10 μl pipet tip. The air bubble can also be sucked in using a 10-μl pipet tip.
17. Each time liquid is added or removed through filling port 1 liquid flows over the observation area thus over the cells. The resulting shear stress in this narrow channel can stress the cells or even detach them. In order to minimise such effects all liquids are filled in slowly. In case of strong adherent cells one reservoir might be filled within 10 s. In case of slightly attaching cells filling time could be up to 1 min.
18. Due to hydrostatic pressure the *chemoattractant solution* might start to flow passively into the reservoir R2. In this case less than 18 μl need to be removed actively by pipetting from filling port 1 in order to avoid air bubbles in reservoir R2.
19. The time interval between two images should be chosen to allow a cell movement not exceeding 40% of its average diameter. We estimate that the maximal speed of a cell is twice the average speed. Delta $t = 0.4 \times 0.5 \times$ cell diameter (25 μm)/ average velocity (0.5 μm/min) = 10 min. It is recommended to use the velocity values derived from the positive control measurement for optimising time interval between images.
20. For statistical reasons at least 30 cells need to be tracked that are remaining in the observation field during the entire experiment (18 h). Some cells will be lost through cell death, cell division, and cells leaving the observation field. Such cells are not used for data analysis. Therefore, we recommend starting with at least 40 cells.
21. An easy way for pixel size calibration is taking a picture of an object of known size like squares in a Neubauer chamber. The effective pixel size is calculated by dividing the object size [μm] by the number of pixels.

Acknowledgments

This work was supported by the BMBF grants "Schwerpunkt-programm Mikrosystemtechnik (BioMST)" and "Optische Technologien (Biophotonik)."

References

1. Cazes, A., Galaup, A., Chomel, C., Bignon, M., Brechot, N., Le Jan, S., Weber, H., Corvol, P., Muller, L., Germain, S., and Monnot, C. (2006) Extracellular Matrix-Bound Angiopoietin-Like 4 Inhibits Endothelial Cell Adhesion, Migration, and Sprouting and Alters Actin Cytoskeleton, *Circ Res* **99**, 1207–1215.
2. Harris, M. P., Kim, E., Weidow, B., Wikswo, J. P., and Quaranta, V. (2008) Migration of isogenic cell lines quantified by dynamic multivariate analysis of single-cell motility, *Cell Adh Migr* **2**, 127–136.
3. Lin, F., and Butcher, E. C. (2006) T cell chemotaxis in a simple microfluidic device, *Lab Chip* **6**, 1462–1469.
4. Fisher, P. R., Merkl, R., and Gerisch, G. (1989) Quantitative analysis of cell motility and chemotaxis in Dictyostelium discoideum by using an image processing system and a novel chemotaxis chamber providing stationary chemical gradients, *J Cell Biol* **108**, 973–984.
5. Schneider, L., Cammer, M., Lehman, J., Nielsen, S. K., Guerra, C. F., Veland, I. R., Stock, C., Hoffmann, E. K., Yoder, B. K., Schwab, A., Satir, P., and Christensen, S. T. (2010) Directional cell migration and chemotaxis in wound healing response to PDGF-AA are coordinated by the primary cilium in fibroblasts, *Cell Physiol Biochem* **25**, 279–292.
6. Lamalice, L., Le Boeuf, F., and Huot, J. (2007) Endothelial cell migration during angiogenesis, *Circ Res* **100**, 782–794.
7. Cooper, C. R., and Pienta, K. J. (2000) Cell adhesion and chemotaxis in prostate cancer metastasis to bone: a minireview, *Prostate Cancer Prostatic Dis* **3**, 6–12.
8. Zigmond, S. H. (1988) Orientation chamber in chemotaxis, *Methods Enzymol* **162**, 65–72.
9. Boyden, S. (1962) The chemotactic effect of mixtures of antibody and antigen on polymorphonuclear leucocytes, *J Exp Med* **115**, 453–466.
10. Eiseler, T., Doppler, H., Yan, I. K., Kitatani, K., Mizuno, K., and Storz, P. (2009) Protein kinase D1 regulates cofilin-mediated F-actin reorganization and cell motility through slingshot, *Nat Cell Biol* **11**, 545–556.
11. Molina-Ortiz, I., Bartolome, R. A., Hernandez-Varas, P., Colo, G. P., and Teixido, J. (2009) Overexpression of E-cadherin on melanoma cells inhibits chemokine-promoted invasion involving p190RhoGAP/p120ctn-dependent inactivation of RhoA, *J Biol Chem* **284**, 15147–15157.
12. Rothmeier, A. S., Ischenko, I., Joore, J., Garczarczyk, D., Furst, R., Bruns, C. J., Vollmar, A. M., and Zahler, S. (2009) Investigation of the marine compound spongistatin 1 links the inhibition of PKCalpha translocation to nonmitotic effects of tubulin antagonism in angiogenesis, *FASEB J* **23**, 1127–1137.
13. Zigmond, S. H., Foxman, E. F., and Segall, J. E. (2001) Chemotaxis assays for eukaryotic cells, *Curr Protoc Cell Biol* **Chapter 12**, Unit 12 11.
14. Foxman, E. F., Kunkel, E. J., and Butcher, E. C. (1999) Integrating conflicting chemotactic signals. The role of memory in leukocyte navigation, *J Cell Biol* **147**, 577–588.
15. Fisher, N. I. (1993) *Statistical Analysis of Circular Data*, Cambridge University Press, New York.

Chapter 14

Live Cell Fluorescence Microscopy Techniques

Shawn A. Galdeen and Alison J. North

Abstract

The use of fluorescent tags for in vivo tracking of proteins has provided an array of new data on cell function. Correspondingly, a variety of new methods utilizing these fluorescent tags have been developed. These methods must take into account all of the concerns of keeping live samples in conditions as close to physiological norms as possible, including temperature, CO_2 levels, media composition, and reduction of phototoxic effects. The microscope itself should also be designed to maximize the benefits and minimize the risks inherent in these methods. We provide an overview of these concerns.

Key words: Microscopy, Microscope, Live cell, Microinjection, Imaging, Confocal, Fluorescence, Phototoxicity

1. Introduction

The discovery of fluorescent proteins that can be endogenously produced as a fusion with a protein of interest has revolutionized the study of cell function. In particular, live cell fluorescence imaging studies are central to research efforts investigating almost all aspects of cell biology. The primary workhorse fluorophore for these studies has been enhanced green fluorescent protein (eGFP). More recently, other proteins have been developed such as engineered red fluorescent proteins (RFPs), the wide range of fluorophores produced by Roger Tsien's lab (most notably the red emitter mCherry), and far-red emitters such as Katushka (1) and mNeptune (2). The advent of these proteins has allowed for multiple fluorophore studies as well as fluorescence resonance energy transfer (FRET), multichannel fluorescence recovery after photobleaching (FRAP), and photoactivation applications (for more information on these techniques, see *Chapters 26 and 27*, this volume).

Claire M. Wells and Maddy Parsons (eds.), *Cell Migration: Developmental Methods and Protocols*,
Methods in Molecular Biology, vol. 769, DOI 10.1007/978-1-61779-207-6_14, © Springer Science+Business Media, LLC 2011

Researchers have also begun to perform longer-term timelapse studies, particularly to investigate processes that take place over one or more cell cycles. Examples of this include examination of centrosome duplication, golgi inheritance, and aneuploidy.

A serious problem encountered by all researchers using live cell methods is photodamage due to the high intensity light used to excite fluorescent proteins. In regard to prolonged observation of fluorescent proteins, the issue has not been whether excitation is damaging to cells (it is) (3–5), but rather how long the cell will tolerate observation. Advances in live cell fluorescent imaging are measured by how much longer a cell will tolerate observation over what was previously possible. The use of neither spinning disk confocals nor back-thinned, electron multiplication CCD cameras have completely removed this limitation.

In this chapter, we provide the reader with a basic set of tools and tips for maximizing observation time while minimizing photodamage. A familiarity with fluorescence as well as a basic understanding of the widefield and confocal microscope is assumed. For review on this information, *see Methods in Cell Biology: Volume 81* and the *Handbook of Confocal Microscopy, Third Edition.*

1.1. Light Sources and Phototoxicity

While it is clear that excitation of fluorophores leads to phototoxic effects, a more insidious cause is the excitation light itself (3–5). Cells expressing endogenous levels of fluorescently tagged protein will generally have very little fluorophore present, such that a low amount of excitation light is sufficient to saturate it. Since light cannot be directed only to the tagged protein, bathing the sample with a large excess of light at the excitation wavelength is unavoidable. However, careful choice of illumination sources can limit the resulting phototoxicity.

In the widefield microscope, several options for illumination are possible. Mercury, xenon, and metal halide lamps all emit pan-chromatic light that must be filtered to allow only light of the desired wavelengths to reach the sample. No standard filter set is able to completely eliminate light of unwanted wavelengths, especially in the damaging near-UV and IR regions. As a result, photodamage can be reduced by including additional UV and IR-blocking filters in the fluorescence light pathway. While this reduces phototoxic effects, one must then avoid fluorophores that require excitation in the blocked portions of the spectrum (such as some photoactivatable fluorophores). Alternatively, the blocking filters can be removed during photoactivation, then replaced for the remainder of the imaging period.

LED illumination sources are a more recent addition to the imaging market. These sources provide an advantage in the widefield microscope since their small emission bandwidth means that unwanted wavelengths do not exist. This advantage is tempered by the need for separate LEDs for each fluorophore of interest.

One solution is to include LEDs of multiple wavelengths in the same illuminator, but this reduces the available power at each wavelength and re-introduces the need for rigorous filtration. If sufficient wavelengths are available for the experiment, LEDs are an excellent choice for illumination.

Confocal systems use laser illumination for sample excitation. These sources produce the narrowest band of wavelengths, and thus produce no phototoxic effects from leaked light. However, the intensity of light per unit area is far higher than widefield illumination, even when the laser is operated at very low power. As a result, the danger in live cell confocal microscopy is from photodamage with the excitation wavelengths themselves producing phototoxic effects. This damage can be limited by illuminating the cell as sparingly as possible. Alternatively, confocal systems that use different means to attenuate the laser strength (e.g., spinning-disk systems) or reduce dwell time at a given location (such as resonance scanning systems) will correspondingly reduce phototoxic effects in the specimen.

Because all fluorescent illumination sources produce phototoxic effects, it is crucial to collect a transmitted light image of the sample in parallel with the fluorescence channels during any time-lapse session. Although the halogen lamp used for transmitted light collection in most widefield systems is pan-chromatic, the intensity is far lower than fluorescent illumination. When operated at low power, these lamps have no discernable effect on cell health (6). This set of control images will provide information about changes in morphology that could indicate phototoxic effects on the sample. It may be long after such morphologic effects are evident that changes in the fluorescence signal occur; if they are missed, the researcher could be potentially misled into imaging dead or dying cells.

1.2. Filter Sets

Fluorescence imaging typically requires a set of three filters. The excitation filter selectively passes wavelengths from the illumination source required to excite the fluorophore(s) of interest. Emitted fluorescence is then selectively passed through the dichromatic (dichroic) mirror to the emission pathway, where any stray excitation light is subsequently blocked by the emission (or barrier) filter.

It is critical to carefully select the excitation and emission filters to match the fluorophore(s) of interest in the sample. If the allowed wavelength band is too narrow, the researcher needlessly prevents light from reaching the sample or detector; if it is too broad, phototoxicity (in the case of the excitation filter) or excessive background and crossover (in the case of the emission filter) are possible. Since most detectors of sufficient sensitivity for fluorescence imaging cannot discriminate color, any inappropriate light that reaches the detector will appear as "signal" in the resulting image. To reduce this possibility, it is useful in both widefield and confocal systems to first observe the fluorescence channels through the oculars, taking quick and careful note of any background or channel crossover that is present. The same information should also be collected by imaging singly labeled samples with the filters to be used for the additional fluorophore(s). With this information, images of multiply labeled samples can be interpreted appropriately.

It is useful to consider the utility of separate filter wheels for excitation and emission filters, as well as dichroic beamsplitters, when constructing a system. A larger set of combinations is possible with this approach, allowing the researcher to easily accommodate a wider range of fluorophores without purchasing additional full filter sets. While faster than full filter cubes, separate filter wheels still require multiple operations to change from one configuration to another, which may restrict some rapid acquisition protocols.

When rapid multicolor imaging is desired, one approach is the use of multipass filters. These filter sets allow light appropriate for multiple fluorophores to pass, minimizing filter changes between fluorescent channels. Crosstalk may occur, however, and each bandwidth passed by a multipass filter is restricted in increasing proportion to the number of bandwidths accommodated by the filter. Because of these drawbacks, multipass filters should be avoided unless required. Fast switching can also be achieved using, for example, lasers coupled through AOTFs for rapid excitation switching or a DualView emission device, so that no filter changes are necessary at all.

When crosstalk among multiple fluorophores is unavoidable, a spectral imaging system can be used. These systems use different combinations of prisms and filters to segment the image into smaller "bins" of discrete wavelengths. Many then use mathematical algorithms to determine what proportion of signal in each image pixel is derived from each of the fluorophores in question (as well as any background fluorescence). For this mathematical reconstruction to proceed correctly, samples containing no expressed fluorophore as well as ones with only single fluorophores should first be imaged. It is also good practice to generate these singly labeled strains to ensure that the constructs are not toxic.

1.3. Environmental Controls

When imaging live cells, a major challenge is providing a cellular environment favorable to the sample. Careful selection of temperature, gas mixture, and media composition can mean the difference between a failed experiment and a successful one. Fortunately, modern imaging systems allow control of all of these critical elements.

1.3.1. Temperature

Temperature control is particularly problematic in live cell imaging. Many samples require temperatures well above ambient, with requirements ranging from 25 to 40°C. In contrast, computer-controlled microscopes require cool temperatures to operate effectively. To resolve this dichotomy, many live cell microscopes are fitted with a chamber that covers the stage and only some of the stand. This partial coverage produces a temperature gradient from inside to outside the chamber, resulting in flex of the stand and loss of focus over time.

Two options exist for resolving this problem. The first is to allow the stand to come to equilibrium before imaging. This is time-consuming, as equilibrium is not reached for 4–6 h after the chamber is turned on. Heat-shock experiments are also exceedingly difficult using this method, since any change in the set temperature induces further flex and loss of focus.

A second option is the focal-plane stability systems which can now be integrated on many microscopes designed for live cell imaging. These systems (known as "Perfect Focus," "Ultimate Focus," "Definite Focus," or "ZDC" depending on the microscope company) monitor the absolute position of the coverslip relative to the objective in real time; any subsequent change in position is corrected, bringing the sample back in focus. This obviates the need for a long warm up period prior to imaging.

The removal of microscope flex as a concern also allows the researcher to consider newer stage-top incubators that have a much smaller volume. These systems often control temperature and humidity, may have optional gas mixing apparatus, and work well with focal stability systems. Alternatively, some companies (e.g., Bioptechs and Warner Instruments) produce small stagetop inserts that heat electrically or by water circulation, but only heat the sample itself and not any area of the stage around it. Both the insert-type and stage-top incubators must be used with a focal-plane stability system for timelapse imaging beyond an hour or two. These systems never reach an effective equilibrium with the cooler microscope stand; as a result, flex-induced changes in focus occur throughout the imaging period.

1.3.2. Gas Mixtures

The prevalence of tissue culture in live cell research requires maintenance of a certain percentage of CO_2 (usually 5%) in the imaging media. The researcher may also wish to change gas percentages to determine the effects of anoxia on a given sample. Several systems exist for external gas mixing that is then provided to the sample. Many of these systems first bubble the gas through water, providing humidity that prevents media evaporation during long imaging sessions. These systems can be incorporated into a stagetop incubator that also provides temperature and/or media composition control, allowing a complete solution for environmental control that can be adapted to any microscope. Because these systems are often expensive, it is useful to investigate whether the sample can survive in commercially available CO_2-independent media.

1.3.3. Media Composition

In addition to CO_2-independence, other factors are important in media composition for live cell imaging. Phenol red is often added to provide an indication of pH. This additive is autofluorescent, and should be avoided in imaging media. If pH monitoring is desired, a small sample of imaging medium can be removed and measured with indicator paper or a pH meter.

Recent research indicates that removal of vitamins from imaging media may prevent photobleaching of GFP and certain fluorophores derived from it, presumably due to a reduction in available acceptors of free electrons from the fluorophore (7). A medium of this composition is now commercially available, and should be considered if significant photobleaching is observed.

The researcher may also wish to add drugs or change media during the imaging session. Adding large volumes of media should be avoided, since the temperature difference between the dish and the new medium will lead to sample flex and loss of focus. Focal stability systems will prevent this problem. Another option is use of a perfusion system to cycle a larger volume of medium through the sample. Such systems can also be used to wash drugs in and out of samples, allowing pulse-chase experiments. While this smaller cycling volume prevents the temperature changes that lead to coverslip flex, the pumping action of such systems can cause periodic deformation of the imaging dish. Again, focal stability systems should address this problem.

1.4. Dishes for Imaging

Once environmental controls have been addressed, the chamber or dish that will hold the cells must also be chosen. Many possibilities exist, in both custom and commercially available designs:

Glass-bottomed dishes. These commercially available dishes, usually 35 mm in diameter, have a hole of varying size removed from the bottom plastic and replaced with a coverslip. It is critical to choose the #1.5 coverslip thickness, unless the objective has a correction collar (see Note 9).

Glass-bottomed multiwell chambers. The same companies that make glass-bottomed culture dishes often make multiwell plates and chambers with coverslip bottoms. These can increase efficiency by allowing observation of multiple experimental conditions in a single round of imaging. A variety of well sizes and numbers are available. In some stage configurations, access to the outside wells may be obscured by the stage apparatus hitting the objective, particularly when using immersion optics. Moreover, points with distant locations may have very different focal positions due to miscasting of the plate or mishandling. These differences in *z*-position may exceed the ability of the stand to accommodate them, particularly when z-drives of limited travel (such as piezo-driven inserts) are employed. Testing chamber fit and construction is therefore recommended prior to use. Chambers with glass slides as bases instead of coverslips are inappropriate for live cell imaging.

Custom sealed chambers. The researcher may use a sealed preparation for imaging, particularly if CO_2 control is not available. Most machine shops can make an aluminum or stainless-steel chamber that is the same size as a standard microscope slide but slightly thicker (1 in. by 3 in., 1/8 in. thickness), ensuring proper fit on nearly any microscope stage. A hole slightly smaller than a standard coverslip (18–20 mm^2) allows coverslips to be adhered to the top and bottom of the chamber with vacuum grease or VALAP (see Note 12). If vacuum grease is used, a UV lamp should be used to sterilize the constructed chamber. After one coverslip is adhered, grease is added to the remaining side, the chamber is sterilized, imaging medium is added, and the final coverslip (on which the cells of interest have been grown) is adhered to complete the chamber. Cells prepared in this way at roughly 25% confluency will live 48–72 h before a medium change is required.

Plastic dishes. If long-working distance objectives are available, signal brightness is not an issue, and only low-magnification imaging is necessary, then using tissue culture plastic dishes is an inexpensive solution for imaging. Because plastic randomizes polarized light, however, transmitted light images can only be collected by phase-contrast. Since the phase objective is partially masked in the back focal plane, use of this method will reduce detected fluorescence. If plastic is the only substrate on which a sample will grow, another option is thin plastic substrates such as those produced by Ibidi GmbH (Martinsried, Germany). The bottom of these dishes is a plastic substrate of coverslip thickness with optical properties equivalent to borosilicate glass. The extra expense of using these dishes is often justified by the increased signal and resolution that can be obtained with high numerical aperture, shorter working distance immersion lenses.

1.5. Fluorophore Choices in Live Cells

An important consideration in live cell imaging is the choice of fluorophore and expression system. A wide range of in vivo fluorophores are available, and sorting through them all may seem daunting. A few basic considerations can reduce the field significantly:

Fluorophore class. Two major classes of fluorophore dominate the available options for in vivo studies. The first are those derived from green fluorescent protein (GFP; originally isolated from jellyfish). These include eGFP, plus Emerald, Venus, Ypet, YFP, Cerulean, and several others (8). Folding of the GFP variants is comparatively slow, producing a lag between protein expression and initial fluorescence. Their photostability and brightness are excellent, however, making them a good choice for long-term studies.

The second major class of fluorophores is those derived from the monomeric form of dsRed (isolated from reef coral) (9). These fluorophores include most of those developed by Roger Tsien's lab, such as mCherry, mOrange, and mPlum. mCherry, noted for its very rapid folding time ($t_{1/2} = 15$ min at 37°C) (8), is a good choice for detecting expression quickly. However, many of the dsRed variants suffer from decreased brightness and increased rates of photobleaching compared to eGFP, presenting detection problems for low-copy number proteins or long-term studies with frequent exposures.

Finally, several other fluorophores are used for fluorescence studies. Dronpa (from coral) and KFP1 (from sea anemone) are both photoconvertible fluorophores that can be "switched on and off" by illumination with specific wavelengths for studies of new protein production at specific timepoints. Kaede (from coral) and KikGR (artificially produced) are examples of photoconvertible fluorophores that change emission spectra from one color to another when illuminated with specific wavelengths. These fluorophores are useful when tracking two temporally discrete populations of protein at the same time.

When combining expression of multiple fluorophores in the same sample, the researcher should take care that sufficient separation of emission spectra is present to prevent crosstalk. Common pairs of proteins that fulfill this requirement are eGFP and mCherry as well as Cerulean and Venus; for three expressed fluorophores, consider Cerulean, Venus, and mCherry.

Colocalization studies. Assessing colocalization of two fluorescently tagged proteins is a common question in imaging studies. It is important to perform such analysis by quantitative, not merely qualitative, methods. If a red-emitting fluorochrome is approximately co-located with a green-emitting one, an overlay of their pseudocolored images may appear "yellow." However, if the green emitter is much brighter than the red, the overlay would appear green to our comparatively insensitive eyes. A line profile, where a graph displays the intensity of both signals at every pixel along a line across the image, is far more convincing. Better still, the percentage of colocalization across the whole image, or within a region of interest, can be calculated using commercially available software. Before performing such analyses, one must check the imaging system for chromatic aberrations that could cause two colocalized proteins to appear in different focal planes in the images (10).

Expression vector. Regardless of the fluorophore chosen, the expression vector used can have a large effect on both the signal strength of a given fluorescent tag and its biological relevance.

Careful mediation of these two opposing interests is a key part of effective fluorescence imaging. Excessive amounts of fluorophore are toxic to many cells, and overexpression of the protein of interest may have detrimental effects, yet the signal must be detectable at acceptable exposures.

The most straightforward choice is use of the native promoter for expression. In models amenable to knock-in strategies, this produces a tagged protein at the same level as a wild-type cell, obviating problems of excessive protein production. Failing that possibility, the researcher must carefully consider the experimental question. If localization only is of interest, then slight overexpression via a higher-expressing promoter will benefit signal detection. However, excessive overexpression may lead to mislocalization caused by inappropriate binding of the protein to other substrates. If tracking protein kinetics and/or interactions is the data of interest, a lower-expression system provides more physiologically relevant data.

Introduction method. Various methods can be used to introduce fluorescent protein constructs into cells. The two most common approaches are transfection and microinjection. Transfection can be performed using a variety of commercial kits, though the critical part is assessing what is an "appropriate" level of expression. It is important to decide whether to invest the extra time in making stably transfected cell lines, which will provide more uniform levels of expression than transiently expressing cells. The advantages and disadvantages of each approach, as well as protocols for successful transfection, are too great a topic for the purposes of this chapter and have been covered elsewhere.

Microinjection is less commonly used, partly because it is more technically challenging, and partly because it is a more time-consuming approach if large numbers of labeled cells are required. For high-resolution live cell imaging, where a limited number of cells will be followed by time-lapse microscopy, it can provide a number of advantages over transfection. First, one can target cells of a certain type, shape, size, or developmental stage. Second, if multiple constructs are co-injected, one can be confident that all injected cells contain all constructs in equal ratios. Third, one can inject a mixture of labeled antibodies and cDNA constructs. Fourth, one can closely control the amount of construct introduced to each cell. Fifth, if labeled antibodies are injected, one can proceed with imaging almost immediately after injection; if cDNA constructs are used, expression is typically detectable within a few hours. This is particularly useful in situations where expression of a protein interferes with normal cellular processes. While a protocol for microinjection is included in this chapter, the specifics of the injection process itself (type of injection needle, angle of injection, pressure and time of injection, etc.) will depend upon the available equipment and recommendations within the corresponding equipment manual.

2. Materials

2.1. Bio-cleaned and Acid-Etched Coverslip Preparation

One box #1.5 coverslips (commonly 22 × 22 mm; larger coverslips are easier to break during the protocol).

1 L or larger glass beaker.

Plastic spatula.

Bath sonicator.

Liquinox or other mild detergent suitable for sonication.

~400 mL 100% ethanol.

(Optional) ~300 mL 1 M HCl.

(Optional) Heated water bath for acid etching.

2.2. Preparing Live Cells for Microscopy

Bio-cleaned coverslips.

Imaging chamber (see Subheading 1.4).

Phenol Red-free imaging media.

(Optional) VALAP (see Note 12) or sterile vacuum grease for chamber sealing.

2.3. Microinjection

Microinjection buffer 1: 75 mM KCl, 10 mM K phosphate buffer, pH 7.5.

Microinjection buffer 2: 100 mM Glutamic acid, 140 mM KOH, 1 mM DTT, 1 mM Mg_2SO_4, pH to 7.2 using citric acid.

3. Methods

3.1. Bio-cleaned and Acid-Etched Coverslip Preparation

1. In a large glass beaker, mix warm water and 2–3 drops of a mild detergent suitable for sonication until uniformly distributed. Add one box of coverslips to this solution, one coverslip at a time (see Note 1).
2. Gently mix the coverslips with a plastic spatula or by swirling, then bath sonicate for 5 min (see Note 2).
3. Rinse the coverslips five times with distilled water, agitating and flipping them during each rinse (see Note 3).
4. Refill the beaker with distilled water and bath sonicate for 5 min.
5. Again, rinse the coverslips five times with distilled water, agitating and flipping them during each rinse. Drain off the distilled water.
6. (Optional) Incubate bioclean coverslips in 1 M HCl at 50–60°C for 4–16 h with occasional gentle mixing (see Notes 4–6). Rinse the coverslips five times with distilled water, agitating

and flipping them during each rinse. Drain off the distilled water.

7. Add a small amount of 100% ethanol (100 ml max) to the rinsed and drained coverslips to soak up excess water. Agitate and rotate thoroughly. Drain.
8. Gently move coverslips to a storage jar and cover with 100% ethanol. Replenish ethanol as required to ensure complete coverage.
9. Remove and flame individual coverslips using sterile forceps (see Notes 7 and 8).

3.2. Preparing Cells for Live Imaging

1. Coverslips are often shipped with residues from manufacturing still present on their surfaces. Take care to wash coverslips carefully and (if needed) acid-etch them to increase surface area for better cell adhesion (see Protocol 3.1, Note 9).
2. Choose a mounting medium and chamber appropriate to your sample type (see Subheading 1.4) and prepare them as required.
3. Grow cells directly on the coverslip surface, or mount your preparation in contact with the coverslip if possible. If possible, use imaging media that does not contain Phenol Red (see Note 10).
4. If sealant is required for the preparation, use sterile vacuum grease or VALAP (see Notes 11–13).
5. Mount your preparation on the microscope. If multiday imaging is required, change the culture media every 48–72 h, carefully using a razor blade or scalpel to open the chamber if it is sealed (see Note 14).

3.3. Widefield Fluorescence Imaging of Live Cells

1. Mount the preparation on the microscope stage, choosing your preferred magnification objective. Focus on the cells *in the transmitted light channel* (see Notes 15 and 16).
2. As briefly as is practical, view the individual fluorescence channels through the oculars to confirm the presence of your fluorophore(s) as well as any crossover or autofluorescence present in each channel (see Note 17).
3. If using a motorized stage for multipoint visiting, find and mark locations in the transmitted light channel (if possible) using the oculars or the camera. If using the oculars, return to each location later using the camera and ensure parfocality, again using the transmitted light channel. Mark 2–4 additional fields for observation at the end of your multipoint series (see Note 18).
4. Using 1–3 of your additional marked fields (more if the sample is more variable), determine an appropriate exposure time for each fluorophore in the sample. Decide also the location of

the starting focal plane for the sample, as well as what z-stack size and number of steps will be used (if any) (see Note 19).

5. Moving back to your location(s) of interest, adjust the exposure times for each channel to those you determined in step 4, and quickly adjust the focus to the location you determined. Remove the test locations from step 4 from your multipoint visitation queue.
6. Set up your timelapse experiment, including a paired transmitted light image at each location. If possible, collect the final stage location in transmitted light *only* as a positive control for the health of the preparation (see Note 20).
7. After imaging, determine the division time of cells in your control field and compare it to those in your experimental fields to ensure a lack of phototoxic effects. If cell division is not present or not observable, use overall morphology and cell density as an indicator (see Note 21).

3.4. Confocal Fluorescence Imaging of Live Cells

1. Mount the preparation on the microscope stage, choosing your preferred magnification objective. Focus on the cells *in the transmitted light channel* (see Notes 15 and 16).
2. As briefly as is practical, view the individual fluorescence channels through the oculars to confirm the presence of your fluorophore(s) as well as any crossover or autofluorescence present in each channel (see Note 17).
3. Using 1–3 fields of cells that you will not image later, determine the approximate focal plane of interest, then set the gain, offset, and laser power for each of your fluorophore channels, as well as the total z-stack thickness. Set the pinhole to 1 Airy unit for a good compromise between *z*-axis resolution and sufficient light detection (see Note 22) (for review, see ref. 10).
4. If using a motorized stage for multipoint visiting, find and mark locations in the transmitted light channel using the oculars or the detector(s). If using the oculars, return to each location later using the detector(s) and ensure parfocality, again using the transmitted light channel.
5. Set up your timelapse experiment, including a paired transmitted light image at each location as a control for cell health (see Note 23).
6. After imaging, determine the division time of cells in your control field and compare it to those in your experimental fields to ensure a lack of phototoxic effects. If cell division is not present or not observable, use overall morphology and cell density as an indicator (see Note 21).

3.5. Microinjection

1. Cells are grown on coverslips in plastic dishes or in glass-bottomed dishes until they are at the required confluency. Replace their regular growth medium with CO_2-independent medium (see Subheading 1.3) and leave them in a CO_2-free incubator for several hours or overnight to equilibrate.
2. Prepare the directly conjugated antibody (see Note 24), the required cDNA (s) (see Note 25), or both together, diluted to an appropriate concentration in microinjection buffer (see Note 26 and Subheading 2.3). In our hands, an injection stock at concentrations around 0.1 mg/ml of cDNA or 0.7 mg/ml of labeled antibody or Fab fragments will give suitable levels of expression or labeling, but appropriate concentrations will depend upon a number of factors such as the brightness of the fluorophore or the efficiency of antibody binding.
3. Inject the cells with a volume of injection solution equivalent to approximately 10% of the cellular volume. Unless it is necessary to commence imaging immediately, first leave the cells for at least 4 h to recover after injection. Expression of injected cDNA constructs usually commences within 3–6 h.

3.6. Mathematical Quantification of Colocalization

1. Collect your data images. During the same imaging session, use subresolution beads multiply labeled with fluorophores of spectra similar to those used in the experiment. Use these bead images to determine any spectral shift present in your system (measured in pixels in X, Y, and Z) (see Note 27).
2. After opening your data using a software package that can calculate Pearson's and/or Manders' correlation coefficients (see Note 28), overlap the channels and correct for any spectral shift determined in step 1.
3. Carefully consider the amount (if any) of background to subtract from the image. A reasonable guess may start with measuring background values in a portion of the field of view that should not contain any signal; however, this does not account for autofluorescence of the sample or other noise sources that specifically overlap the data area. After determining the values for each channel, subtract it from the images (see Note 29).
4. Threshold on the remaining fluorescence in each channel (or select the regions of interest) and calculate Pearson's or Manders' correlation coefficient (see Note 30). Verify any biological conclusions using a separate second methodology (see Note 31).
5. Finally, calculate M1 and M2, the coefficients representing the percentage of pixels from each channel that form the colocalized area. These two values may provide valuable information concerning the biology of the two proteins, but they can only be calculated if some colocalization is present.

4. Notes

1. Since it is critical that both surfaces of all coverslips are cleaned, one must take care not to allow them to adhere to one another in the wash solution. The easiest way to prevent this outcome is to ensure that they are added one at a time.
2. Remember that the coverslips are quite thin and fragile. Use of a stir bar will result in both inefficient mixing and shattering of an unacceptably large number of coverslips, and should be avoided. Plastic spatulas are preferred because of their beveled edge and less rigid composition, thus leading to less coverslip chipping during the protocol.
3. Again, be wary of the fragility of the coverslips. Do not give them a mighty shake, or you will have only shards to work with!
4. This step is intended to allow one to create etched coverslips for better cell adhesion or later coating if required. Agitation need only be performed two to four times to ensure access to all coverslip surfaces.
5. Longer incubations in the acid solution will necessarily result in thinner coverslips and thus greater difficulty in handling later. Beware of incubations longer than 14–16 h, as the unacceptably thin coverslips produced will be very difficult to handle and may begin to cause aberrations at high magnification unless a corrected objective is used.
6. When incubating the coverslips in acid, *do not* cover the beaker with parafilm, as the acid fumes will near-permanently affix the parafilm to your beaker. Instead, incubate uncovered in a fume hood.
7. Forceps can be flamed prior to use to provide sufficient sterilization. EM forceps are a good choice for this task.
8. If improper coverslip removal technique has been followed or the coverslip jar has been left uncovered for too long, a layer of shiny oil may develop on the surface of the ethanol. If this occurs, subsequent coverslip removal will coat the clean coverslips with this oil, effectively returning them to their state prior to cleaning. In this case, discard the ethanol and re-clean the coverslips and their container thoroughly.
9. Because nearly all objectives in use are corrected for them, it is *critical* to use #1.5 (0.17 mm thickness) coverslips for microscopy of any kind. While other "numbers" can be used, the differing thicknesses of these alternative coverslips will inevitably lead to greater spherical aberration of the sample and thus lower resolution. This problem can be partially resolved if your objective has a correction collar designed for this purpose.

10. While Phenol Red is a convenient indicator of media pH (and thus the need for changing media), it does produce autofluorescence in some channels and may promote phototoxic effects in some preparations. Since it is not required for cell growth, avoiding it removes the possibility of these detrimental outcomes.
11. If a sterile tube of vacuum grease is not available, apply vacuum grease to the surface to be sealed prior to addition of cells and place under a UV lamp for 10 min to sterilize. A 5-mL syringe works well as an applicator for small amounts of vacuum grease – hold it directly against the surface to be sealed and smear on a tiny amount as you press down on the syringe plunger.
12. VALAP is a 1:1:1 combination of vaseline, lanolin, and paraffin wax. To apply VALAP, heat up the surface of a small spatula and use it to melt a tiny well in your container of solid. Remove a bit of the resulting liquid, and drag the edge of the spatula along the edge of the coverslip to be sealed. Wait for this side to cool to a solid, and repeat with the remaining sides.
13. *Do not, under any circumstances, keep an entire liquid solution of heated VALAP near your microscope.* This solution can heat up sufficiently to then be deposited on all the optical surfaces of the microscope, necessitating extensive (and possibly expensive) cleaning and/or replacement. Heating the end of a spatula and depressing it into the solid VALAP will melt enough for sealing 1–2 sides of a coverslip.
14. If appropriate humidity is provided in the imaging chamber, the preparation may be unsealed. If using this approach, be wary of bacterial contamination, and check the pH of the media once per day by removing a few microliters and placing on a strip of indicator paper as an early sign of contamination (or look for visible signs in the collected transmitted light images).
15. Focusing: For upright microscopes, focus as close as possible to the preparation and then move *away* from it while looking through the oculars to find the focal plane. Using this method helps to ensure that you do not break the thin coverslips.
16. Focusing: For inverted microscopes, use the above method for air objectives. For immersion objectives, add a small amount of the appropriate immersion media to the surface of the objective, mount the sample, and focus up until the flat top surface of the objective (usually slightly larger than the glass of the objective front face) is barely immersed. This should be very close to the focal plane, and final focusing can be accomplished by looking through the oculars.
17. If long observations are necessary (e.g., if it is the first time viewing a new construct or using a new microscope), either

prepare multiple samples, using one for initial observations, or choose fields for initial observations that are different from those to be used for long-term imaging. This strategy will help to reduce long-term phototoxic effects.

18. If using a nonmotorized stage, proceed in a similar fashion, using fields other than the one to be used for the final imaging for the setup steps that follow.
19. Be very cautious when setting up a z-stack for live imaging. Steps should be reduced to the barest minimum required to provide the data of interest, keeping in mind that every additional step increases the probability of irreversible photodamage to the cells.
20. If using a nonmotorized stage, collect an image of a defined location in transmitted light before and after imaging as a control for preparation health. Mark this position with an objective scribe or a permanent marker dot adjacent to the location so that you can find it again after imaging. Use cell density and morphology as your indication instead of division time.
21. Remember, the most insidious phototoxic effects do not result in cell death, but merely in damage to a cell that may perturb obtained results. Cell arrest or cell cycle delay work well as indicators of possible photodamage, so pay close attention to the results from this control experiment!
22. If you have an older system with multiple pinholes, take care to ensure that they are all set to give the same optical slice thickness instead of the same number of Airy units.
23. Keep in mind that transmitted light images are often collected on confocal microscopes using one of the laser lines. Phototoxic effects are possible as a result, and the researcher may wish to check additional fields after the experiment is completed to ensure that global toxic effects did not occur.
24. Commercial kits are available from several companies to directly conjugate antibodies to fluorescent labels. Be aware that the chemistry of these different kits varies and therefore one may prove more suitable for your antibody than another. The antibody must be purified and concentrated prior to conjugation and no traces of azide can be present! The use of Fab fragments, rather than whole antibody, will address any concerns that the injected antibody might cause cross-linking of the target proteins in injected cells.
25. Achieving a “suitable” expression level of cDNAs in your cells is critical. We recommend searching for cells in which only a very faint level of signal is detected. In addition, injecting a single cDNA could disrupt the balance of a cell if its level of expression is critical to one binding partner. When using

microinjection of cDNAs to study GFP-labeled cytokeratins in epithelial cells, we found it critical to inject an equal concentration of cDNA encoding its partner cytokeratin in an unlabeled form (e.g., labeled K8 and unlabeled K18). Only then did the expressed GFP-tagged protein form a "normal"-looking array of intermediate filaments in the cells.

26. Two suggested recipes for microinjection buffers have been provided. We typically use the first, but for cells that do not tolerate high levels of chloride ions, the second recipe may be more suitable.

27. Spectral shift is a lack of registration between fluorophores of differing wavelengths emitted from the same location. The two most frequent causes are spherical aberration in the sample (leading to lensing effects and causing spectral separation in *X*, *Y*, or *Z*), and defects in the optical path itself (e.g., strained objectives, misaligned excitation laser delivery, objectives incompletely corrected for chromatic aberrations, and poorly registered filter cubes). While optical path defects can be corrected for to some extent (by, e.g., having a technician register the filter cubes to be used for colocalization analysis), spherical and some chromatic aberrations may always be present. Thus, it is critical to routinely check for spectral shift in the sample prior to analyzing colocalization.

28. ImageJ (the free image analysis software available from the National Institutes of Health) has a variety of plug-ins available to calculate both Pearson's and Manders' correlation coefficients. One excellent plug-in is available from the Wright Cell Imaging Facility at Toronto Western Research Institute (on the web at http://www.uhnresearch.ca/facilities/wcif/imagej/colour_analysis.htm).

29. While this step may seem vague, there is very little consistency in how background subtraction amounts are determined for a given sample, since so many variables are present. It pays to be cautious, however; too much subtraction may artificially limit the amount of colocalization, while too little will almost certainly skew the percentage in the positive direction.

30. Care should be taken when choosing the coefficient to calculate for overall colocalization. Pearson's coefficient is widely accepted. Values range from −1 to 1, and any positive value indicates some level of colocalization. Negative values are less informative, as they simply indicate an inverse relationship of some form, not necessarily only from exclusion. For example, one common effect producing a negative Pearson's coefficient is a large difference in intensities between the channels. If such a problem is present, Manders' correlation coefficient (sometimes called the "overlap coefficient", or *r*) is less sensitive to intensity differences and could be used instead.

31. As with any research method, colocalization analysis provides data that is only as good as the sample and analysis itself. Verification of any correlations measured by another method (e.g., deleting a suspected binding protein and observing a loss of any overlap, or using immunoprecipitation or sedimentation approaches to confirm members of a protein complex) is an important additional step that will help ensure that the researcher is not being misled.

References

1. Scherbo D. *et al.* (2007) Bright far-red fluorescent protein for whole-body imaging. *Nat. Methods* 4: 741–6.
2. Shu X. *et al.* (2009) Mammalian expression of infrared fluorescent proteins engineered from a bacterial phytochrome. *Science* 324: 804–7.
3. Tyrrell RM, Werfelli P, Moraes EC (1984) Lethal action of ultraviolet and visible (blue-violet) radiations at defined wavelengths on human lymphoblastoid cells: action spectra and interaction sites. *Photochem. Photobiol.* 39: 183–9.
4. Sparrow JR, Zhou J, Cai B (2003) DNA is a target of the photodynamic effects elicited in A2E-laden RPE by blue-light illumination. *Invest. Ophthalmol. Vis. Sci.* 44: 2245–51.
5. Chu R, Zheng X, Chen D, Hu DN (2006) Blue light irradiation inhibits the production of HGF by human retinal pigment epithelium cells in vitro. *Photochem. Photobiol.* 82: 1247–50.
6. Uetake Y and Sluder G (2004) Cell cycle progression after cleavage failure: mammalian cells do not possess a "tetraploidy checkpoint". *J. Cell Biol.* 165: 607–8.
7. Bogdanov A *et al.* (2009) Cell culture medium affects GFP photostability: a solution. *Nat. Methods* 6: 859–60.
8. Shaner NC, Steinbach PA, and Tsien RY (2005) A guide to choosing fluorescent proteins. *Nat. Methods* 2: 905–9.
9. Baird GS, Zacharias DA, and Tsien RY (2000) Biochemistry, mutagenesis, and oligomerization of DsRed, a red fluorescent protein from coral. *PNAS* 97: 11984–9.
10. North AJ (2006) Seeing is believing? A beginners' guide to practical pitfalls in image acquisition. *J. Cell Biol.* 172: 9–18.

Chapter 15

Measuring Invasion in an Organotypic Model

Veronika Jenei, Maria L. Nystrom, and Gareth J. Thomas

Abstract

Organotypic cultures are in vitro models that can be used to study the interactions between tumour and stromal cells. Collective tumour cell invasion in organotypic assays resembles that seen in human tissues in vivo, suggesting physiological relevance. A qualitative, pathological description of such invasion may be inadequate, and there is therefore a need to accurately quantify the degree of invasion. Although the simplest method to quantify invasion is to measure maximum invasive depth, this ignores the importance of the pattern of tumour invasion, which often reflects tumour aggressiveness. We use image analysis software to analyse organotypic invasion objectively, taking into account the average depth of tumour invasion, and the number and area of invading tumour islands. The product of these parameters is termed the "invasion index," which maximises differences in invasion and also reflects the invasive pattern of the gel in a way that none of the individual parameters does alone.

Key words: Invasion, Stroma, Fibroblast, Organotypic culture

1. Introduction

Historically, most tumour invasion assays have been monocellular, where carcinoma cells have been studied in isolation. Generally, these assays have "traded" physiological relevance for ease of repetition, and their inherent disadvantage is that they grossly oversimplify the complex process of invasion. Stromal cells are of fundamental importance in cancer cell invasion (1, 2), and a more relevant assay therefore should incorporate stromal components such as fibroblasts, and also reproduce the 3D characteristics of the relevant organ.

Currently, the "organotypic" culture model is the most physiologically relevant in vitro quantitative assay used to study tumour cell invasion (3). In organotypic culture, tumour cells are grown

Claire M. Wells and Maddy Parsons (eds.), *Cell Migration: Developmental Methods and Protocols*,
Methods in Molecular Biology, vol. 769, DOI 10.1007/978-1-61779-207-6_15, © Springer Science+Business Media, LLC 2011

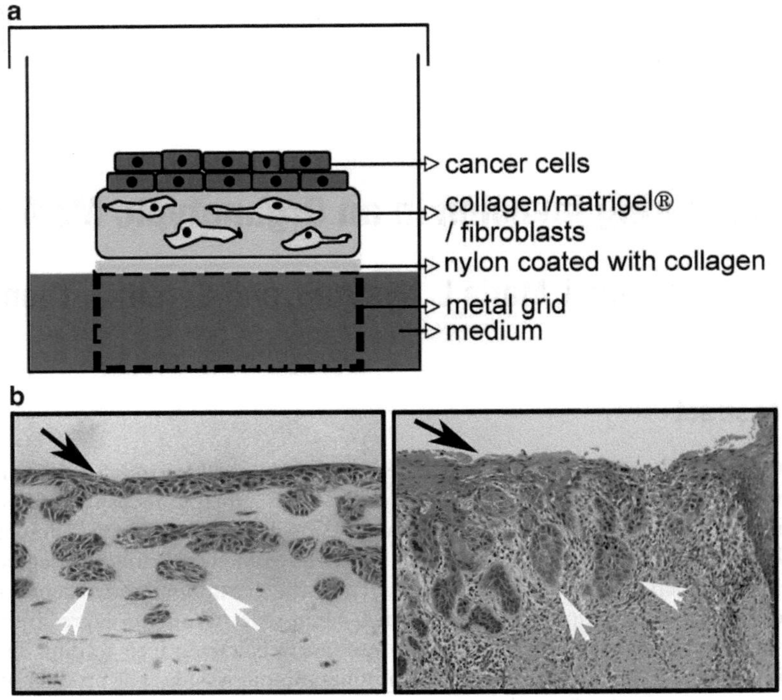

Fig. 1. Organotypic gel. (**a**) Organotypic cultures are grown at an air/liquid interface using a metal grid support covered with a collagen-coated nylon disc. The cells of interests are seeded on top of a collagen/Matrigel® gel containing fibroblasts (or other stromal cells). (**b**) *Left panel*: Oral squamous carcinoma cells invading in an organotypic culture. After 14 days in culture gels are fixed in formal saline, embedded into paraffin, sectioned, and stained with H&E. After staining the middle one fifth of the gel is photographed using a microscope equipped with a digital camera at ×100 magnification. *Right panel*: H&E-stained section of a human squamous cell carcinoma. The pattern of invasion is similar to that seen in organotypic culture in vitro. *Black arrows* show surface epithelium; *white arrows* show invading tumour islands.

at an air/liquid interface on collagen/Matrigel® matrices populated with stromal cells (see Fig. 1a). Such models allow the study of interactions between different cell types using chemical inhibitors or gene silencing within tumour and/or stromal cells.

Fibroblasts are most commonly used as the stromal component in organotypic invasion assays. It is now appreciated that fibroblasts exhibit phenotypic heterogeneity, differing not only between organ systems, but also within a given anatomical site (4). When investigating carcinoma invasion, it is therefore advisable to use anatomically matched fibroblasts. In many of our investigations, we have found tumour cell invasion to be fibroblast-dependent; omitting fibroblasts from the assay leads to loss of the invasive phenotype, and different fibroblast cell lines support carcinoma invasion to variable degrees (3).

The pattern of invasion produced in organotypic cultures is similar to invasion patterns in human tissues in vivo (see Fig. 1b).

A qualitative, pathological description of such invasion may be inadequate and there is therefore a need to accurately quantify the degree of invasion. This may be especially important when comparing gels in which there are subtle differences in invasion, or when assessing the efficacy of potential inhibitors of invasion, which are unlikely to give an "all or nothing" effect (5). The simplest method for quantifying invasion is simply to measure maximum depth, or take an average of several depth measurements. However, measuring invasion in this way ignores the importance of the pattern of tumour invasion and may not be fully objective. In many tumours, the pattern of invasion is highly prognostic (e.g. oral cancer); tumours that invade widely as small, discohesive nests, or strands behave significantly more aggressively than tumours composed of large, cohesive islands of cells, with a well-demarcated invasive front (6). A simple depth measurement will not distinguish between these different patterns in organotypic culture. We therefore used image analysis software to develop an objective method for analysing organotypic invasion, taking into account the average depth of tumour invasion, and the number and area of invading tumour islands. The product of these parameters is termed the "invasion index," which maximises differences in invasion and reflects the invasive pattern of the gel in a way that none of the individual parameters does alone (3).

2. Materials

2.1. Cell Culture

1. Standard keratinocyte growth medium (KGM): Minimum Essential Medium of Eagle (α-MEM) supplemented with 10% heat-irradiated foetal calf serum (FCS; Biosera, Ringmer, UK), 1.8×10^{-4} M adenine, 1×10^{-10} M cholera toxin, 5 μg/ml insulin, 0.5 μg/ml hydrocortisone, 10 ng/ml epidermal growth factor (Sigma-Aldrich®, Gillingham, UK), and 2 mM L-glutamine.
2. Fibroblast growth medium (FGM): of Dulbecco's Modified Eagle's Medium (DMEM) supplemented with 10% heat-irradiated FCS and 2 mM L-glutamine.
3. Phosphate-buffered saline (PBS) to rinse monolayers of cells before detaching them.
4. Solution of trypsin/EDTA (0.05% trypsin (w/v)/5 mM EDTA) is used to detach cells.

2.2. Organotypic Culture

1. Cells: fibroblasts re-suspended in FGM at a concentration of 5×10^6/ml and keratinocytes re-suspended in α-MEM containing 10% FCS and 2 mM L-glutamine at a concentration of 5×10^6/ml.
2. Matrigel® (BD Biosciences, Bedford, MA, USA).

3. Type I collagen (concentration 4 mg/ml), prepared from rat tails (Upstate Cell Signalling Solutions, Milton Keynes, UK).
4. 10× concentrated DMEM.
5. Heat-irradiated FCS.
6. 0.1 M sodium hydroxide to neutralise the gel mixture.
7. Sterile 6- and 24-well plates.
8. Sterile (autoclaved) nylon discs (100 μm pore size; 2.5 cm^2, Tetko Inc, New York, USA).
9. 1% glutaraldehyde (Sigma-Aldrich®, Gillingham, UK) diluted in PBS.
10. Autoclaved steel grids made from 2.5 cm^2 squares of stainless steel mesh with the edges bent down to form 4–5 mm high "legs."
11. Sterilised forceps and sterile spatula.
12. Cell counter: haemocytometer or a CASY automated cell counter (Shärfe System, Reutlingen, Germany) can be used.

2.3. Processing and Quantifying Organotypic Cultures

1. Formal Saline: 9 g sodium chloride, 100 ml formaldehyde, 900 ml water.
2. Scalpels.
3. 70% Ethanol.
4. Haematoxylin/eosin (H&E).
5. Antibodies: mouse monoclonal antibody against human pan-cytokeratin AE1/AE3 (Dako Cytomation, High Wycombe, UK).
6. Microscope equipped with a digital camera.
7. Optilab Pro (Graftek Imaging Inc, Texas, USA).

3. Methods

The preparation of organotypic cultures is carried out in three steps over 3 days.

3.1. Preparation of Organotypic Basement Gels

1. On day 1, gels are prepared using 3.5 volumes of Type I collagen and 3.5 volumes of Matrigel®, which should be mixed on ice with 1 volume of 10× DMEM, 1 volume of FCS, and 1 volume of FGM in which fibroblasts have been suspended (see Notes 1 and 2; 5×10^5 fibroblasts/gel, final concentration of collagen is 1.8 mg/ml). The solution is neutralised as necessary (by drop-wise addition of 0.1 M sodium hydroxide until the solution turns from yellow to red), and mixed by gentle pipetting (taking care to prevent the introduction of bubbles).

2. 1 ml of this solution is pipetted into each well of a 24-well plate and allowed to polymerise for 30 min at 37°C.
3. Following polymerisation, 1 ml of FGM is added to each well and the gels are left for 18 h at 37°C to equilibrate.

3.2. Seeding Keratinocytes on Gels and Preparation of Nylon Sheets

1. After 18 h, the medium is aspirated from the wells and 5×10^5 keratinocytes (re-suspended in 1 ml α-MEM supplemented with 10% FCS and L-glutamine) are added to each well. Other epithelial cell types can also be used (see Note 3). Chemical inhibitors can be introduced at this stage. Alternatively, cells may have been pre-treated with siRNA (see Notes 4 and 5).
2. Additionally on day 2, a collagen-coated nylon sheet is prepared which is used to support the gel on the metal grid, and prevent lateral contraction of the gel. For this, a fibroblast-free collagen gel mix is prepared as follows: 7 volumes of Type I collagen, is mixed on ice with 1 volume of 10× DMEM, 1 volume of FCS, and 1 volume of FGM (made as above for the gel but omitting the fibroblasts). 0.1 M NaOH is again used to neutralise the mixture, added drop-wise. This mixture is then carefully spread onto the sterile (autoclaved) nylon discs (250 μl collagen gel mixture/2.5 cm^2 nylon sheet of 100 μm pore size) and allowed to polymerise at 37°C for 15–30 min. The gel-coated nylon sheets are then fixed in 1% glutaraldehyde diluted in PBS for 1 h at 4°C, washed four times in PBS and twice in DMEM supplemented with 10% FCS and L-glutamine. Nylon sheets can be left in the final DMEM wash at 4°C until required.

3.3. Raising Organotypic Cultures onto Steel Grids and Processing

1. On day 3, sterile steel grids (made from 2.5 cm^2 squares of stainless steel mesh with edges bent to form 4–5 mm high "legs") covered with the previously prepared collagen-coated nylon sheets are placed in six-well plates.
2. The medium covering the gels is carefully aspirated without disturbing the surface of the gels, thus allowing the gels to be removed from the 24-well plate using a sterile spatula.
3. These are placed onto individual collagen-coated nylon discs resting on the steel grids ensuring that the keratinocyte-covered surface of the gel is facing upwards (see Fig. 1).
4. After the gels are placed on the grids, sufficient growth medium is added to reach the undersurface of the grid (i.e. the gel is not covered), allowing the epithelial layer to grow at an air–liquid interface. If keratinocytes are used for the organotypic culture, the growth medium used is KGM from which cholera toxin has been omitted otherwise, the normal growth medium of the studied cell line. This initial time point when the gels are raised on the metal grids is defined as day 1 of organotypic culture.
5. The medium should be changed every 2–3 days.

3.4. Processing of Organotypic Gels

Organotypic cultures are usually harvested and then histopathologically processed between 10 and 14 days. If RNA interference is used, the time point may be earlier (typically day 6; see Note 5).

1. Using forceps, samples are carefully picked up (touching only the still-adherent nylon membrane) and fixed in formal saline. 24 h later, the formal saline should be changed to 70% ethanol.
2. Samples are bisected and processed to paraffin blocks.
3. Sections can be cut and stained using standard haematoxylin and eosin (H&E), or used for immunochemistry (cytokeratin staining is useful for highlighting carcinoma cells).

3.5. Quantifying Invasion in Organotypic Culture: Definition of the Invasion Index

After sections have been stained, the invasion in the organotypic culture can be quantified using image-analysis software as described in Fig. 2. It is important to stress that due to changes in physical constraints at the edge of the gel, only the central area of the gel should be used for analysis (4 mm radius). In general, serial or stepped sections within this 4 mm radius give highly reproducible results and we have shown that one section within this radius is representative of the entire gel (3).

In order to accurately quantify invasion, it is essential to discriminate between the two different cellular components of the assay. This is particularly important if tumour cells invade singly, rather than in cohesive groups. In such cases, it can be difficult to distinguish between tumour cells and fibroblasts in H&E-stained sections. Such difficulties can be overcome by immunostaining sections with pan-cytokeratin antibody AE1/AE3 (see Note 6 and Fig. 2a) which labels the carcinoma cells (cytokeratin-positive) but not the fibroblasts (cytokeratin-negative). Comparison of immunostained sections with the corresponding H&E stained sections confirms that staining for cytokeratins does not underestimate tumour volume.

Fig. 2. Quantifying invasion using the invasion index. (**a**) Section of a CA1 oral squamous cell carcinoma organotypic gel (day 9) stained with cytokeratin antibody. The colour photograph was recorded and standardised to micrometre readings. (**b**) Using Optilab Pro (Graftek), the image was changed to greyscale and saturated areas rendered *red* (**shown as grey in the figure**). All non-saturated areas were rendered *black* to give a binary (two-colour) image. The software "sees" all *red* areas as particles and measures number, perimeter, and area of individual particles. (**c**) As a measure of mean depth of invasion, the software also creates a convex image and calculates the Mean Cord Y, i.e. the mean vertical thickness, of the convex image.

Sections are photographed using a microscope equipped with a digital camera at ×100 magnification (×10 objective, ×10 eyepiece). Where tumour invasion exceeds the photographic field several images can be taken and a composite image generated using Adobe Photoshop or other digital image editing software. A colour image is recorded of the cytokeratin-stained section, standardised to a 100 micrometre reading (Fig. 2a) and converted to greyscale (Optilab Pro software, Graftek; Fig. 2b). Using the "threshold" function, the immunostained areas are converted to red, particles and the unsaturated areas rendered black to create a binary (two-colour) image (Fig. 2b). This is then subjected to two "clean-up" procedures to remove background "noise" – the first using the "low pass 3" function to remove all particles less than 3 pixels in size. Then, comparing the binary image to the immunostained image, any further artefacts should be removed manually. From this binary image, the function "convex" can generate a polygon figure that encompasses all the red particles and delineates the invasive front of the epithelium (Fig. 2c). Then the depth of the polygon is calculated by measuring vertical "cords" every 70 μm going from left to right across the whole image. The average length of the cords gives a value for mean invasion depth and is referred to as Mean Cord Y (MCY). The software registers all red areas as tumour and generates a "particle report" for each image. After excluding the largest particle, which is the overlying surface tumour epithelium, the software also counts the number of particles (N) and calculates the sum of areas (A) of the particles. The "Invasion Index" is calculated as $\text{MCY} \times N \times A$ and provides a quantitative value for tumour invasion.

Figure 3 shows a validation of this model in a simple time-course experiment. A time course for the oral cancer cell line CA1 co-cultured with human foreskin fibroblasts was performed, in which identical gels were harvested at time points between days 3 and 14. Sections were immunostained for cytokeratin and binary images generated, which were then subjected to quantitative analysis as described above. It shows that there was a step-wise incremental increase in invasion over time.

In all analyses, the largest particle (representing the surface epithelium) should be eliminated as it does not represent invading cells. Although each of the individual parameters (MCY, N, A) gives an incremental increase over time reflecting the increased invasion (see Fig. 3), these differences are maximised by combining them, and the maximal difference between consecutive time points is better reflected by "*Invasion Index*." The "Invasion Index" also clearly reflects the invasive pattern of the gel in a way that none of the individual parameters do alone (Fig. 4).

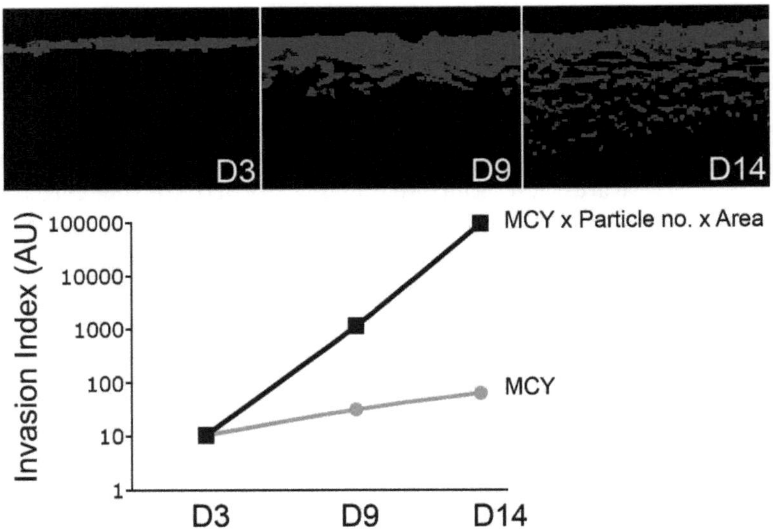

Fig. 3. Time course of invasion quantified using the invasion index. Organotypic cultures generated with CA1 oral squamous carcinoma cells and human foreskin fibroblasts (HFF) were fixed after 3, 9, and 14 days, processed to paraffin blocks, immunostained with pan-cytokeratin antibody and photographed. For each time point the field of view from left to right was 2.4 mm. Binary images were then generated and the invasion index was calculated. Although the depth of invasion (MCY), number of invading tumour islands (particles), and area of invading tumour all increase with time, maximal difference is shown by using the equation MCY × particle No. × area of invading particles shown in the histogram. Data are normalised to D3 values, which were arbitrarily set at 10.

4. Notes

1. Matrigel® and collagen should be kept on ice at all times and the components of the gels should also be mixed on ice in order to avoid early polymerization.
2. Other type of stromal cells can also be added to the organotypic gel, however, it should be stated that the origin of fibroblasts/stromal cells can and will most probably affect the invasion. As has been shown by Nystrom et al., different fibroblast cell lines support invasion to varying degrees and also the number of fibroblast affects cancer cell invasion. The quality of the fibroblasts is very important for the successful outcome of the organotypic culture. Fibroblast proliferation (particularly that of primary fibroblasts) slows down after multiple passages, and the cells finally senesce. Late passage fibroblasts often support invasion poorly, and it is important that healthy, proliferating fibroblasts are used.
3. If other cell types (rather than squamous carcinoma cells) are used, the cells should be re-suspended in their respective growth medium (e.g. DMEM) supplemented with 10% FCS and L-glutamine.

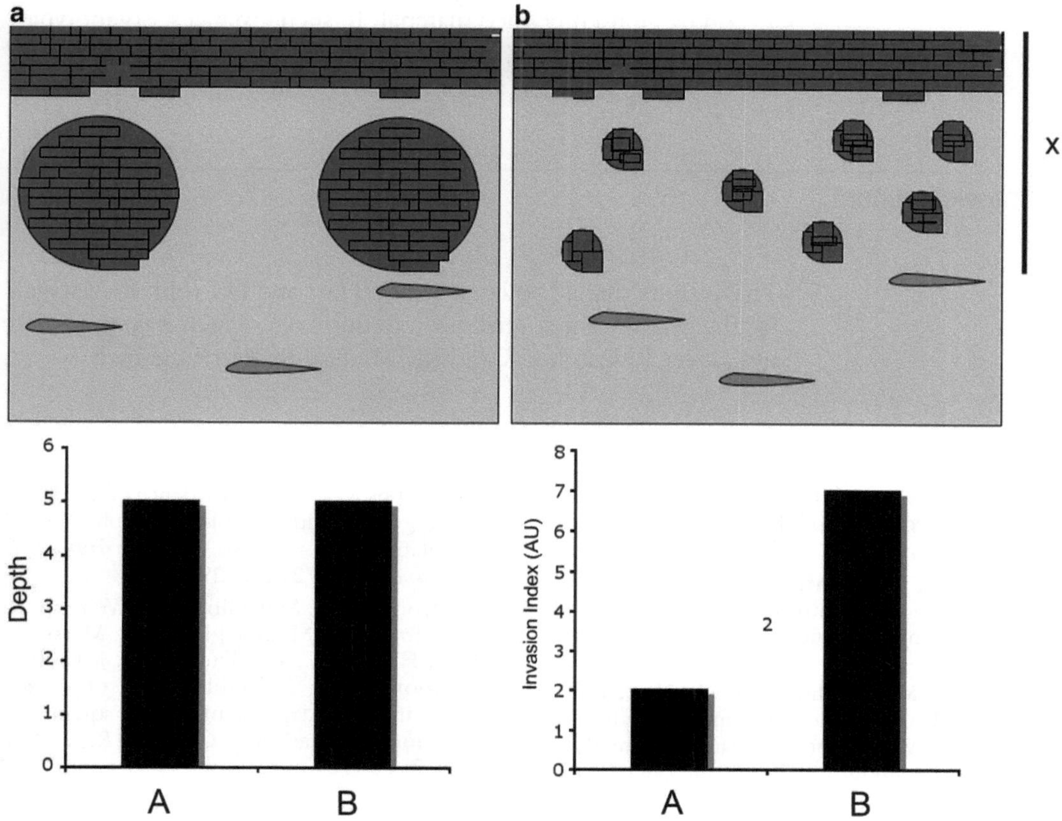

Fig. 4. Invasion index reflects the invasive pattern of the gel. Schematic representation of organotypic cultures showing similar invasive depth and tumour area, but differing patterns of invasion. Simple depth or area measurements would not distinguish between these cultures, however, the invasive pattern in (**b**) is significantly more aggressive. Using the invasion index, which takes into account the number of invading tumour islands (particles), highlights the difference between the cultures.

4. If blocking antibodies or chemical inhibitors are used, tumour cells should be placed in medium containing either the blocking antibody or the chemical inhibitor at the required concentration for 30 min prior to plating. Inhibitors should also be added to the growth medium whenever the cultures are fed such that the supplements are present throughout the entire experiment.
5. To assess the effect of gene silencing on invasion, cells transfected with siRNA are used 24 h post-transfection. It is important to determine duration of siRNA effect prior to performing the organotypic experiment. This will indicate the optimum time point to fix the gel. For such experiments, gels are usually harvested after 6 days of organotypic culture (7 days post-transfection).
6. Immunostaining is a useful adjunct to this type of assay, and is limited only by the availability of suitable antibodies that work

on paraffin-embedded material. In such instances, organotypics can be flash frozen, rather than fixed, and staining carried out on frozen sections (although this is more technically difficult).

Acknowledgement

The authors thank Professor Ian R. Hart and Dr. John F. Marshall for their significant contribution to this work, which was supported by Cancer Research UK and the Medical Research Council.

References

1. Liotta, L. A., and Kohn, E. C. (2001) The microenvironment of the tumour-host interface. *Nature* **411**, 375–379.
2. De Wever, O., Demetter, P., Mareel, M., and Brack, M. (2008) Stromal myofibroblasts are drivers of invasive cancer growth. *Int. J. Cancer* **123,** 2229–2238.
3. Nystrom, M. L., Thomas, G. J., Hart, I. R., Stone, M., McKenzie, I. C., and Marshall, J. F. (2005) Development of a quantitative method to analyse tumour cell invasion in organotypic culture. *J. Pathol.* **205**, 468–475.
4. Fries, K. M., Blieden, T., Looney, R. J., Sempowski, G. D., Silvera, M. R., Willis, R. A., and Phipps R. P. (1994) Evidence of fibroblast heterogeneity and the role of fibroblast subpopulations in fibrosis. *Clin. Immunol. Immunopathol.* **72**, 283–292.
5. Nystrom, M. L., McCulloch, D., Weinreb, P. H., Violette, S. M., Speight, P. M., Marshall, J. F., Hart, I. R., and Thomas, G. J. (2006) Cyclooxygenase-2 inhibition suppresses $\alpha_v\beta_6$ integrin–dependent oral squamous carcinoma invasion. *Cancer Res.* **66**, 10833–10842.
6. Woolgar J. (2006) Histopathological prognosticators in oral and oropharyngeal squamous cell carcinoma. *Oral. Oncol.* **42,** 229–239.

Chapter 16

Analysis of Cell Migration Using *Caenorhabditis elegans* as a Model System

Ming-Ching Wong, Maria Martynovsky, and Jean E. Schwarzbauer

Abstract

The nematode *Caenorhabditis elegans* is an excellent model system in which to study long-distance cell migration in vivo. This chapter describes methods used to study a subset of migratory cells in the hermaphrodite nematode, the distal tip cells. These methods take advantage of the organism's transparent body and the expression of green fluorescent protein to observe cell migration and behavior. Additionally, the availability of nematode mutants and gene knockdown techniques that affect cell migration allow the analysis and comparison of wild-type and aberrant migratory paths. Methods for nematode growth and maintenance, strain acquisition, observation and live imaging, gene knockdown, and analysis of cell migration defects are covered.

Key words: *C. elegans*, Cell migration, Mutants, RNAi, Distal tip cells, Live imaging, Green fluorescent protein

1. Introduction

Caenorhabditis elegans is a relatively simple animal that has a finite number of cells, a transparent body, and that executes larval development in only 36 h. In addition to these advantages, a subset of cells undergo long-distance migrations that are easily observed during larval stages (Table 1), making *C. elegans* an excellent model organism for studying cell migration. Here, we describe methods to study cell migration in *C. elegans*, focusing on the migration of hermaphroditic distal tip cells (DTCs). Beginning at the L2 larval stage, two DTCs migrate longitudinally away from the mid-body along the ventral basement membrane, then turn to migrate dorsally, and turn again to migrate longitudinally back toward the mid-body along the dorsal basement membrane.

Claire M. Wells and Maddy Parsons (eds.), *Cell Migration: Developmental Methods and Protocols*, Methods in Molecular Biology, vol. 769, DOI 10.1007/978-1-61779-207-6_16, © Springer Science+Business Media, LLC 2011

Table 1
Cells that undergo long-distance migration in the *Caenorhabditis elegans* hermaphrodite and the gene promoters that are active in these cells

Migratory cell(s)	Active promoter
ALM neurons	*mec-4* (16)
CAN neurons	*ceh-23* (17)
QL	*unc-73* (18)
PQR neuron	*osm-6* (19)
HSN neurons	*unc-86* (20)
QR	*unc-73* (18)
AQR neuron	*osm-6* (19)
AVM neuron	*mec-4* (16)
SDQR neuron	*unc-119* (21)
Embryonic coelomocytes	*hlh-8* (22)
M cell	*hlh-8* (22)
Sex myoblasts	*hlh-8* (22)
Gonadal distal tip cells	*lag-2* (23)

The migratory paths of the DTCs are reflected in the shape of the gonad arms. Wild-type DTC migration yields two U-shaped gonad arms in the hermaphrodite (Fig. 1a, b).

The transparency of *C. elegans* combined with the use of the green fluorescent protein (GFP) and other inherently fluorescent protein variants has greatly aided the visualization of migrating cells. A typical GFP transgene consists of a promoter that is active in the migrating cell of interest driving the expression of GFP (Fig. 1c), which then enables the analysis of cell migration in live animals and bypasses fixation procedures. Expression of a cellular protein fused to GFP can also be used to follow protein subcellular localization or to provide a more detailed view of cell shape changes during migration (Table 1, see Note 1).

The ease of forward and reverse genetic approaches in *C. elegans* allows one to dissect the functions of specific genes during cell migration. Certain genetic mutant strains exhibit cell migration defects, which, when characterized, have revealed the roles of key molecules that are required for cell movement or pathfinding (1–4). In addition, RNA interference (RNAi) is a straightforward approach to reduce the specific expression of a particular gene of interest. The availability of RNAi libraries covering ~94% of the 19,000 predicted ORFs in the *C. elegans* genome makes it feasible to phenocopy the absence of genes for which mutants

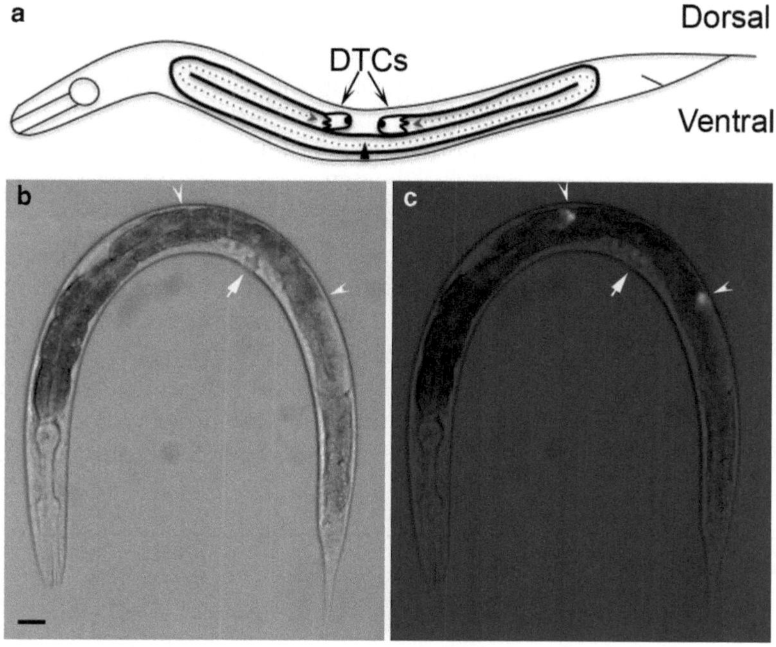

Fig. 1. Distal tip cell migration and gonad morphology in the *Caenorhabditis elegans* hermaphrodite. (**a**) Schematic diagram of a late L4 animal. At the L2 larval stage, DTCs initiate migration on the ventral side (at the *triangle*, which represents the vulva) and follow the paths indicated by the dotted arrows. Migration terminates on the dorsal side opposite the vulva, yielding two U-shaped gonad arms. DIC image (**b**) and fluorescence superimposed on DIC image (**c**) of an L4 nematode carrying the *lag-2p::GFP* transgene. The *lag-2* promoter is active in DTCs and drives the expression of GFP in these cells. Arrowheads indicate the DTCs and arrows indicate the location of the vulva. Anterior is to the left and ventral is down in all images. Scale bar represents 20 μm.

have not yet been generated (5–8). Because RNAi can be initiated at different times during development, one can control the knockdown of a gene in a temporal fashion and, therefore, bypass earlier requirements for a gene. For example, there are many genes that are essential for embryonic development, but determining the roles for these genes at later stages of development is complicated by the lethal nature of the mutants. One can allow wild-type animals to develop past the embryonic stage and then reduce gene expression by RNAi during postembryonic development to analyze its role in cell migration. More recently, tissue-specific RNAi strains have been developed based on mutant strains that are unable to respond to RNAi treatment. In such a genetic background, the introduction of a transgene that confers RNAi sensitivity in a tissue of interest allows RNAi to affect a single type of cell or tissue (9).

Phenotypes induced by RNAi depend on the level of knockdown of gene expression and one potential disadvantage of the RNAi technique is incomplete knockdown. Genetic strains with increased sensitivity to RNAi treatment have been characterized, such as the *rrf-3(pk1426)* mutant strain (10) and the *eri-1(mg366)* mutant strain, which shows increased effects in most tissues including the nervous system (11).

In general, the study of cell migration in *C. elegans* relies on imaging (bright-field, differential interference contrast (DIC), fluorescence, and time-lapse video microscopies), cell-labeling (transgenic GFP expression), and genetic manipulation (mutants and RNAi). These topics will be addressed here, as the following sections will describe how to acquire wild-type, mutant, and transgenic *C. elegans* strains (Subheading 3.1), how to grow and maintain *C. elegans* in culture (Subheading 3.2), how to observe cell migration in *C. elegans* using a microscope (Subheading 3.3), how to immobilize nematodes to make cell migration movies (Subheading 3.4), and how to knock down the expression of specific genes using RNAi (Subheading 3.5).

2. Materials

2.1. Nematode Growth and Maintenance

1. Wild-type, mutant, and transgenic *C. elegans* strains.
2. Nematode growth medium (NGM) plates: 2.5 g peptone, 17 g agar, and 3 g NaCl. Add water to a total volume of 1 L and autoclave to dissolve the agar and sterilize the medium. Once the medium has cooled to 55°C, add the following sterile solutions: 1 mL of 1 M CaCl, 1 mL of 1 M $MgSO_4$, 25 mL of 1 M KH_2PO_4, pH 6.0, 1 mL of 5 mg/mL cholesterol in ethanol, and 5 mL of 10, 000 units/mL nystatin suspension. Stir to mix well. Aliquot agar medium into 60-mm plastic Petri plates (see Note 2).
3. OP50 *Escherichia coli* bacteria strain.
4. Luria broth (LB) bacterial growth medium: 10 g tryptone, 5 g yeast extract, and 10 g NaCl. Add water to a total volume of 1 L, stir to mix, and autoclave to sterilize.
5. 10-mL borosilicate glass pipet and pipet bulb or pipet controller.
6. A wire worm pick: a 1 in. length of 28-gauge platinum wire melted to the end of a borosilicate glass Pasteur pipet (see Note 3).
7. A metal spatula.
8. Alcohol burner.
9. 95% Ethanol.
10. 20°C incubator.
11. A dissecting microscope.

2.2. Observation of Worms Using Microscopy

1. Small 2.5-mL pipet bulb.
2. Borosilicate glass Pasteur pipet.

3. Glass coverslips (22 × 60 mm).
4. M9 buffer: 3 g KH_2PO_4, 6 g Na_2HPO_4, 5 g NaCl, and 1 mL of 1 M $MgSO_4$. Add water to a total volume of 1 L. Sterilize by autoclaving.
5. 2% Agarose in M9 buffer.
6. M9 buffer with NaN_3: add NaN_3 to M9 for a 0.08 M final concentration.
7. M9 buffer with 0.1% tricaine and 0.01% levamisole: Prepare 10% solutions of tricaine (made fresh every time) and levamisole: 100 mg in 1 mL of M9. Add 10 μL of 10% tricaine and 1 μL of 10% levamisole to 1 mL of M9 (see Note 4).
8. A microscope with fluorescence and DIC optics and at least 10× and 40× objective lenses. In addition to this equipment, video microscopy requires a 60× objective lens, a computer-controlled motorized focus-drive unit, and camera (see Note 5).

2.3. RNAi Treatment of Worms by Feeding

1. N2 or RNAi-sensitized worm strains, such as *rrf-3 (pk1426)* (10), and tissue-specific RNAi strains, such as the hypodermis-specific strain, *rde-1(ne219); lin-26p::rde-1* (9).
2. HT115 (DE3) *E. coli* strain carrying the pL4440 vector. One clone should carry a control pL4440 vector, which can either be the empty vector or a vector with 500–1,000-bp fragment of GFP, depending on the nature of the experiment. The other clone should contain the pL4440 vector with a 500–1,000-bp fragment of your gene of interest.
3. LB bacterial growth medium with 100 μg/mL ampicillin.
4. 60 mm NGM plates with 1 mM IPTG.
5. Bleach solution: 4.5 mL water, 2 mL household bleach (5% solution of NaOCl), and 0.5 mL of 10 M NaOH.
6. 1.5-mL microcentrifuge tubes.
7. M9 buffer.
8. Bench-top microfuge.
9. Vortex.
10. Bunsen burner.
11. 23°C incubator.

3. Methods

3.1. Acquiring Worm Strains

Wild-type, mutant, and transgenic strains can be requested from the *C. elegans* Genetics Center (CGC; http://www.cbs.umn.edu/CGC/) or from individual laboratories. To order strains from the CGC, there is a required annual fee and a small fee per strain requested.

3.2. Nematode Growth and Maintenance

C. elegans are grown on OP50 bacterial lawns on NGM plates. OP50 bacteria can also be obtained from the CGC.

3.2.1. Seeding NGM Plates with OP50 Bacteria

1. Using sterile technique, inoculate LB medium without antibiotics with OP50.
2. Grow the OP50 culture for 2 days at room temperature or overnight at 37°C with shaking.
3. Using a sterile technique, place ~100 μL of the liquid OP50 culture in the center of an NGM plate using a sterile glass pipet.
4. Spread this volume of OP50 culture using the tip of the pipet (see Note 6).
5. Allow the bacteria to grow overnight at least.

3.2.2. Transferring Individual Worms Between NGM Plates ("Picking")

1. Briefly flame the wire pick to sterilize it and wait for several seconds to let the pick cool.
2. Using a dissecting microscope, identify the worm you wish to transfer.
3. Gently slide the flat end of the pick under a single worm and lift it off the agar.
4. Still using the dissecting microscope, quickly transfer the animal to a new, seeded NGM plate to avoid desiccation of the animal. Gently lower the pick to the agar without disturbing the surface, and let the worm crawl onto the new plate (see Note 7).
5. Be sure to sterilize the pick each time to avoid bacterial/fungal contamination and cross-contamination of strains.
6. Incubate the worms at 20°C.

3.2.3. Transferring Large Numbers of Worms Between NGM Plates ("Chunking")

1. Dip a clean metal spatula into 95% ethanol and flame to sterilize.
2. Use the spatula to cut a chunk of the agar from the NGM plate that contains worms.
3. Remove the chunk of agar and quickly transfer the chunk of agar to a fresh, seeded NGM plate.
4. Sterilize the spatula between each transfer to avoid bacterial/fungal contamination and cross-contamination of strains.
5. Incubate the worms at 20°C.

3.3. Observation and Analysis of Cell Migration Using Microscopy

3.3.1. Preparation of an Agarose Pad and Mounting Animals onto the Pad

1. Heat 2% agarose solution in a microwave oven to melt.
2. Using a small pipet bulb and a glass Pasteur pipet, drop 2 small drops of the molten agar onto a 22 × 60 mm coverslip.
3. Quickly place another 22 × 60 mm coverslip over the drop of agarose, but perpendicular to the first cover slip (Fig. 2).
4. Allow the agarose to solidify, then gently slide the bottom slide out from under the agarose, leaving the agarose pad

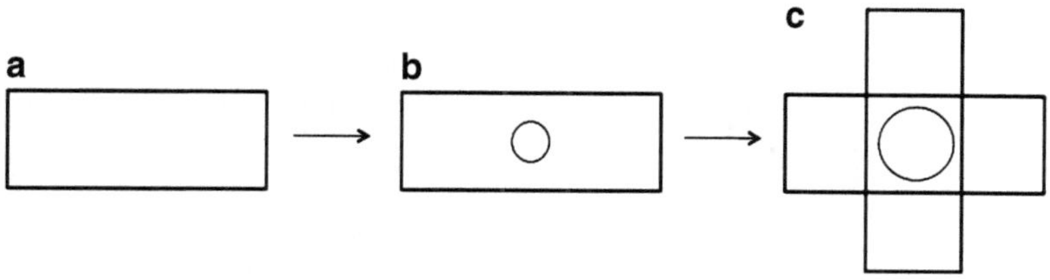

Fig. 2. Diagram of the preparation of the agarose pad. (**a**) A single 22 × 60 mm coverslip. (**b**) The same 22 × 60 mm coverslip with two drops of molten 2% agarose on it (*circle*). (**c**) A second 22 × 60 mm coverslip placed on top of the molten 2% agarose, perpendicular to the first coverslip. The drop of agarose will spread out upon placement of the second coverslip.

stuck to the top coverslip. Write on the coverslip with the agarose pad on it to indicate which side the agarose pad is on.

5. Place 10 μL of M9 with NaN_3 on the agarose pad (see Note 8).
6. Using a wire worm pick and a dissecting microscope, transfer animals at the desired developmental stage to the drop of M9 (see Note 9).
7. Transfer animals before the drop of M9 evaporates (see Note 10).
8. Place a coverslip slowly onto the drop of M9 (see Note 11).

3.3.2. Observing Animals Under the Compound Microscope

1. Place the coverslip with agarose pad and nematodes onto the stage of the microscope. Ensure that the slide has been placed such that animals are being visualized through a coverslip and not the side that has both the coverslip and agarose pad.
2. Under low (10×) magnification and bright field, locate and focus on an animal. To observe GFP expression in a transgenic animal, switch to the fluorescence source and switch off bright field (see Note 12).
3. Switch to 40× magnification to observe the cells of interest. This magnification will be needed to observe gonad shape.
4. Repeat steps 2 and 3 to observe other animals.

3.3.3. Analysis and Scoring of Cell Migration Defects

1. When scoring migration defects in either mutant or RNAi-treated animals, it is important to have observed wild-type migration of the same cell type. In the case of DTC migration, the shape of the hermaphroditic gonad reflects the DTC migratory path (Figs. 1a and 3).
2. In wild-type animals, the U-shaped gonad morphology reflects migration of the DTC on the ventral side, the turn from ventral to dorsal, and then migration back toward the vulva along the dorsal side (Fig. 1). Perturbations of DTC migration can result in a variety of phenotypes, including little to no DTC migration (Fig. 4c), excessive DTC migration and turning (Fig. 4d), or pathfinding defects, in which the DTC stops short on the dorsal side (Fig. 4e) or exhibits a meandering migration path, resulting in a gonad arm with a

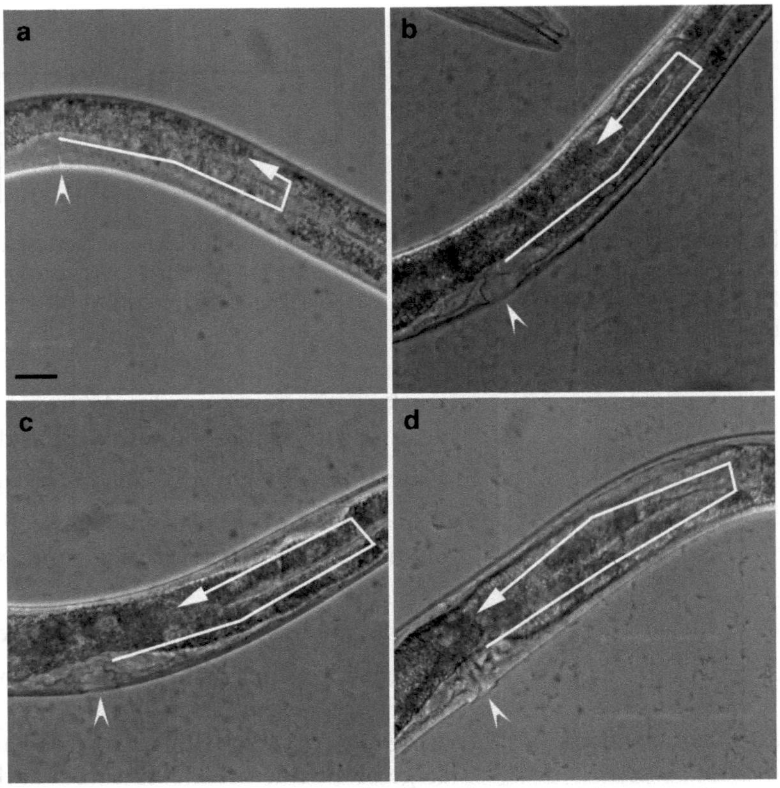

e

Temperature (°C)	Hours from hatching to larval stage transition. L2-L3	L3-L4	L4-adult
20	24	34	46
25	18.5	26	35.5

Fig. 3. Staging and developmental timing of the *Caenorhabditis elegans* hermaphrodite. (**a-d**) DIC images showing vulval morphology and the posterior gonadal arm at different stages of development. *Arrowheads* indicate the location of the vulva and *arrows* indicate the progression of DTC migration. Late L3 (**a**), early L4 (**b**), mid L4 (**c**), and late L4 (**d**) larval stages of N2 hermaphrodites are shown. By late L3 (**a**, *arrow*), the DTC has completed the ventral-to-dorsal turn and is starting to migrate on the dorsal side. The DTC continues to migrate in this direction throughout the L4 stage (**b–d**, *arrows*). The morphology of the vulva also changes as it develops. The vulval precursor appears as a relatively simple structure during late L3 (**a**, *arrowhead*) and progressively becomes more complex during the L4 stage (**b–d**, *arrowheads*). Anterior is to the left and ventral is down in all images. Scale bars represent 20 μm. (**e**) *Table* showing the temperature dependence of *C. elegans* developmental timing. Development occurs more rapidly at warmer temperatures (24).

wave-like morphology (Fig. 4f). Observation and analysis of these phenotypes can give important information about potential roles of the gene of interest. For example, *gon-1* is required for DTC migration, and worms lacking *gon-1* expression exhibit no DTC migration (2).

3. It is important to score both the penetrance and the qualitative aspects of the phenotypes. Penetrance refers to the more

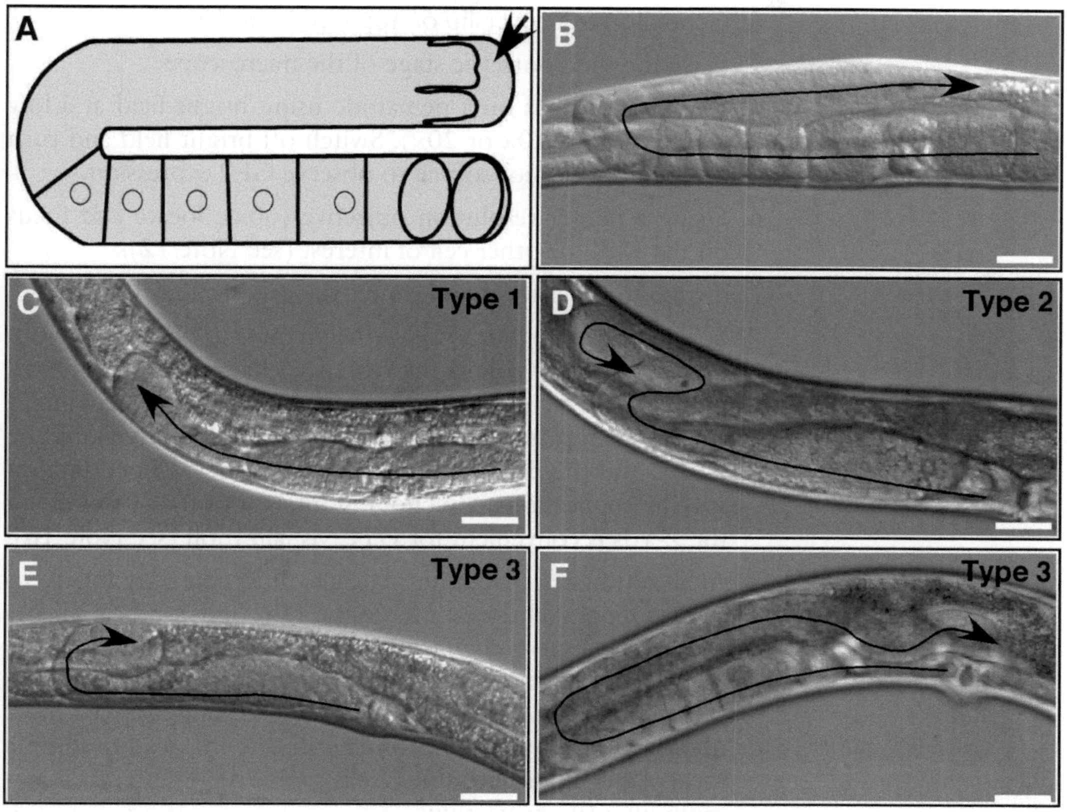

Fig. 4. Examples of wild-type and defective DTC migration paths and gonad morphologies. (**a**) *Diagram* showing one arm of the hermaphrodite gonad. An *arrow* indicates the DTC. Cuboidal cells are maturing oocytes. Ovals are fertilized embryos. (**b-f**) DIC images showing wild-type morphology (**b**) and three types of defects (**c-f**). Treated *rrf-3(pk1426)* hermaphrodites were grown on *Escherichia coli* HT115(DE3) carrying an empty vector (**b**) or RNAi targeting vectors for *act-1* (**c**), *ced-5* (**d**), *pat-3* (**e**), or *dyn-1* (**f**). *Arrows* indicate DTC migration paths. Anterior is to the left, ventral is down in all images except (**c**) and (**e**) in which anterior is to the right. *Bars*, 20 μm. Adapted with permission from Journal of Cell Science (http://www.jcs.biologists.org) and originally published in Cram et al. (2006) (doi: 10.1242/jcs.03274) (25).

quantitative aspect of the analysis, which is the proportion of cells that exhibit a migratory defect. This information is important to determine the severity of the migration defect and gives information about the role of the gene of interest during cell migration (see Note 13).

4. When scoring cell migration, it is useful to make drawings (and/or take pictures) of the cell migration pathway, and record the anatomical landmarks in relation to the migrating cells. Keeping a record of the experiment is extremely useful for future analysis.

3.4. Video Microscopy of Cell Migration in Nematodes

1. Place 15 μL of M9 with 0.1% tricaine and 0.01% levamisole on a 2% agarose pad on a coverslip (see Note 14).
2. Using a worm pick and a dissecting microscope, transfer animals to the drop of M9 and let sit for 5–10 min for the anesthetic to take effect.

3. Gently lower a coverslip on top (see Note 15).
4. Place the slide onto the stage of the microscope.
5. Locate and focus on a nematode using bright field at a low magnification (10× or 20×). Switch off bright field and turn on the fluorescence source to observe GFP expression.
6. Using a higher resolution objective (60×), locate and focus on the DTC or other cell of interest (see Note 12).
7. Switch to the camera and center the cell in the field of view.
8. Specify the timing interval for the pictures to be taken in both bright-field (DIC) and fluorescence channels and the total number of time points. If multiple focal planes are required, specify the top and the bottom limits for sectioning along the *Z*-axis, and the distance between focal planes. Depending on specific applications, a sectioning interval of 0.2–1 μm along the *Z*-axis is convenient for DTC visualization (see Note 16).
9. Analyze images pertaining to each time point separately to track temporal changes. Merge the DIC and confocal micrographs to observe changes in position of the cell within the context of the nematode body, or analyze images obtained in the confocal channel separately to track changes in cell morphology. For visualization of data in 3D, images in multiple focal planes along the same *Z*-axis can be represented as maximum intensity projections.
10. Convert the sequences of image files into movies using available software. We have used Quicktime software with success; however, the choice of software should be based on what is appropriate and available for each specific operating system.

3.5. RNAi Treatment of Worms

This RNAi feeding protocol spans 5 days, from spreading bacteria, inducing expression of double-stranded RNA, nematode egg preparation, hatching and growth on RNAi bacteria, and analysis. The HT115 (DE3) *E. coli* strain carrying fragments of the *C. elegans* genome in the pL4440 vector can be ordered from Geneservice™ (http://www.geneservice.co.uk/products/rnai/).

3.5.1. Spreading HT115 Bacteria onto IPTG NGM Plates

1. Using a sterile technique, inoculate LB containing ampicillin with the appropriate HT115 clone (see Note 17).
2. Grow the culture overnight and no longer than 18 h at 37°C while shaking (see Note 18).
3. Using a sterile technique, place 150 μL of the culture in the center of an IPTG containing NGM plate.
4. Gently move the plate in a brief circular motion to spread the culture, but not so much to have it touch the edge of the plate.
5. Let the liquid dry and incubate at room temperature overnight to induce expression of double-stranded RNA.

3.5.2. Egg Preparation and RNAi

1. Rinse a NGM plate of gravid adult hermaphrodites with water. Pipet the water across the plate a few times to loosen all worms from the plate. Then transfer the liquid to a microfuge tube.
2. Spin the tube at 13793×*g* for 10 s in a microfuge. Aspirate the supernatant, preserving the loose worm pellet at the bottom.
3. Add 1 mL of bleach solution to the tube (bleach solution should be made fresh).
4. Vortex the tube for 2 s at the maximum setting and incubate at room temperature for 3 min. Then vortex the tube again and incubate at room temperature for 2 more min.
5. Spin the tube at 13793×*g* for 10 s. Aspirate the supernatant, again preserving the pellet at the bottom.
6. Add 1 mL M9 buffer to rinse the pellet and spin the tube at 13,000 rpm for 15 s. Aspirate the supernatant.
7. Repeat step 6.
8. Resuspend the pellet in approximately 200 μL M9 buffer.
9. Aliquot between 3 and 5 μL onto a coverslip. Using a dissecting microscope, count the number of eggs on the coverslip to determine the concentration of eggs in M9 buffer.
10. Depending on the concentration of eggs in the buffer, aliquot the appropriate volume to transfer approximately 200 eggs per plate. Use the IPTG NGM plates that were spread with the RNAi bacteria the day before.
11. Place in 23°C incubator and incubate for 48–50 h.
12. Analyze the worms using the methods described in Subheading 3.3.

4. Notes

1. Transgenes exist in two forms: extrachromosomal arrays and integrated arrays. Extrachromosomal arrays (*Ex* in *C. elegans* nomenclature) consist of multiple repeats of the transgene and are not integrated into the genome. The transmission of extrachromosomal arrays is variable, and therefore selecting individuals carrying the GFP transgene to propagate the strain is necessary for maintenance. Integrated arrays (*Is* in *C. elegans* nomenclature) are transgenes that have been integrated into the genome. Once the integrated array is homozygous, all animals should express GFP.
2. By aliquoting 12 mL of medium in each 60-mm Petri plate, you can expect to make approximately 100 NGM plates.

Also, by aliquoting an equal amount of NGM media into each plate, the depth of agar in all plates is similar, which eliminates large adjustments in focus when switching between plates to look at worms under the dissecting microscope.

3. When fashioning a worm pick, flatten the free end of the platinum wire. This flattened end can be used as a scoop to pick up animals and allows them to easily crawl off the pick onto the agar.
4. Sodium azide treatment stops nematode movement completely, but it is toxic to the animal. Tricaine and levamisole are less effective than azide at inducing paralysis, but nematodes can be rescued after prolonged anesthesia (also see Note 14). To recover anesthetized animals, apply plenty of M9 by placing the tip of the pipet next to the edge of the top coverslip, gently slide the coverslip off, and add another drop of M9 on top of the nematode of interest. Wait until the worm starts moving before placing it on a plate with food.
5. Spinning disk confocal microscopy reduces the danger of phototoxicity or photobleaching and is advisable for live nematode imaging experiments lasting for several hours.
6. When spreading the OP50 liquid culture on NGM plates using the tip of a glass pipet, be sure that the tip of the pipet is smooth and does not break the surface of the agar. Nematodes will burrow into the agar at a broken surface, making observation and recovery of these worms difficult. Also, avoid spreading the bacteria to the edge of the plate to make it less likely that worms will crawl up the side of the plate, dry out, and die.
7. Alternately, the pick can be used to pick up a small amount of bacteria. Worms can stick to bacteria on a pick and can be transferred to a new plate by touching the bacteria and worm to the agar, allowing the worm to crawl away.
8. Agarose pads help immobilize the specimen for observation.
9. The staging of the animal is crucial when analyzing the end point of cell migration. It is important that animals with aberrant phenotypes are compared to wild-type phenotypes at the same stage since gonad size and the position of the ventral-dorsal turn are related to overall body size. DTC migration is best scored at the late L4 stage, which is the stage in which DTCs have completed their migration (Fig. 3). Depending on which cell types are being scored, *C. elegans* postembryonic staging can be done by observing vulval development (Fig. 3).
10. If possible, move more than one animal at a time. Also, shake or quickly drag the wire pick in the M9 drop to release worms rapidly into the drop, as opposed to letting the worm swim

off the pick, which is a slower process. If necessary, add another drop of M9 if you notice evaporation; however, try to keep this to a minimum to avoid concentrating the M9 buffer too much. It is important to not lose too much of the M9 volume to avoid bubbles forming between the coverslips. Bubbles can impede proper visualization of the animals.

11. Try to drop one side of the coverslip over the animals first and then slowly lower the coverslip from one end to the next. This can prevent bubble formation.
12. DTC observation can be facilitated by using the *lag-2p::GFP* strain. If using this strain, switch off the bright field and turn on the fluorescence source to locate the DTCs.
13. In some cases, the phenotype will be subtle and a more careful quantification is necessary for scoring cell migration defects. For example, a subset of defects exhibit a U-shaped path back to the dorsal mid-body, but the path is shorter than that of wild type. This means that the ventral to dorsal turn occurs at a location that is closer to the mid-body than in a wild-type animal. This difference in size can be quantified by determining the ratio of the length of the ventral migration path (from the vulva to the turn) to the length of the worm (from the vulva to either the tip of the tail or the tip of the head). The ratio between these two values can be compared between wild-type animals and genetically manipulated animals to determine whether the migratory paths of DTCs are shortened.
14. Tricaine plus levamisole anesthetic treatment provides an amenable and straightforward way for preventing nematode movement; however, it also slows down nematode developmental processes. For example, the DTC ventral to dorsal turn is completed in 1 h in untreated L3 larvae, but requires ~6 h in anesthetized hermaphrodites. A similar effect has been observed in neuron migration (12). The recommended concentration is optimal for keeping animals alive, but paralyzed for several hours. Nematodes can survive anesthesia for 9 h or more and be recovered for subsequent analysis. Also, nematodes can be re-anesthetized after at least 2 h of recovery. In addition to anesthesia, nematodes can be immobilized by other methods that require special equipment (cooling to 4°C or microfluidic chambers) or do not allow recovery of animals after analysis (gluing) (12–15).
15. In order to prevent desiccation, seal the coverslip with Vaseline around the periphery of the agarose pad, leaving one corner open for aeration.
16. Nematodes treated with levamisole and tricaine will exhibit slight movements that may change the focal plane of the DTC. Readjust the settings every hour if necessary. The specimen

should not be illuminated in between photographic recordings to avoid heat shock and/or photobleaching.

17. When planning for an RNAi experiment, it is useful to have both a negative and a positive control. HT115 clones carrying an empty L4440 plasmid or with a fragment encoding GFP can be used as a negative control. When scoring DTC migration, a useful positive control is a clone carrying a fragment of the gene *gon-1* in the L4440 plasmid. *C. elegans* treated with *gon-1* RNAi exhibit DTCs that are unable to migrate, resulting in dramatically truncated gonad arms. This phenotype is easily scored using a dissecting microscope.
18. Overgrowth of RNAi bacterial strains for longer than 18 h results in the reduction of RNAi effect.

Acknowledgments

This work was supported by grants from the NIH (R01 GM059383 and NIGMS Cell Migration Consortium U54 GM064346). M.C.W. is supported by a postdoctoral fellowship from the New Jersey Commission on Cancer Research (10-2409-CCR-EO). M.M. was supported by a Predoctoral Training Grant in Genetics and Molecular Biology (T32 GM007388).

References

1. Nishiwaki, K. (1999) Mutations affecting symmetrical migration of distal tip cells in *Caenorhabditis elegans. Genetics.* **152**, 985–997.
2. Blelloch, R., Anna-Arriola, S. S., Gao, D., Li, Y., Hodgkin, J., and Kimble, J. (1999) The gon-1 gene is required for gonadal morphogenesis in *Caenorhabditis elegans. Dev. Biol.* **216**, 382–393.
3. Meighan, C. M., and Schwarzbauer, J. E. (2007) Control of C. elegans hermaphrodite gonad size and shape by vab-3/Pax6-mediated regulation of integrin receptors. *Genes Dev.* **21**, 1615–1620.
4. Reddien, P. W., and Horvitz, H. R. (2000) CED-2/CrkII and CED-10/Rac control phagocytosis and cell migration in *Caenorhabditis elegans. Nat. Cell. Biol.* **2**, 131–136.
5. Fraser, A. G., Kamath, R. S., Zipperlen, P., Martinez-Campos, M., Sohrmann, M., and Ahringer, J. (2000) Functional genomic analysis of C. elegans chromosome I by systematic RNA interference. *Nature.* **408**, 325–330.
6. Kamath, R. S., Fraser, A. G., Dong, Y., Poulin, G., Durbin, R., Gotta, M., Kanapin, A., Le Bot, N., Moreno, S., Sohrmann, M., Welchman, D. P., Zipperlen, P., and Ahringer, J. (2003) Systematic functional analysis of the *Caenorhabditis elegans* genome using RNAi. *Nature.* **421**, 231–237.
7. Rual, J. F., Ceron, J., Koreth, J., Hao, T., Nicot, A. S., Hirozane-Kishikawa, T., Vandenhaute, J., Orkin, S. H., Hill, D. E., van den Heuvel, S., and Vidal, M. (2004) Toward improving *Caenorhabditis elegans* phenome mapping with an ORFeome-based RNAi library. *Genome Res.* **14**, 2162–2168.
8. Ahringer J. (June 6, 2006) Reverse genetics, In *WormBook*, ed. The C. elegans Reseach Community, WormBook doi/10.1895/wormbook.1.47.1, http://wormbook.org.
9. Qadota, H., Inoue, M., Hikita, T., Koppen, M., Hardin, J. D., Amano, M., Moerman, D. G., and Kaibuchi, K. (2007) Establishment of a tissue-specific RNAi system in *C. elegans. Gene.* **400**, 166–173.

10. Simmer, F., Tijsterman, M., Parrish, S., Koushika, S. P., Nonet, M. L., Fire, A., Ahringer, J., and Plasterk, R. H. (2002) Loss of the putative RNA-directed RNA polymerase RRF-3 makes *C. elegans* hypersensitive to RNAi. *Curr. Biol.* **12**, 1317–1319.
11. Kennedy, S., Wang, D., and Ruvkun, G. (2004) A conserved siRNA-degrading RNase negatively regulates RNA interference in *C. elegans. Nature.* **427**, 645–649.
12. Knobel, K. M., Jorgensen, E. M., and Bastiani, M. J. (1999) Growth cones stall and collapse during axon outgrowth in *Caenorhabditis elegans. Development.* **126**, 4489–4498.
13. Suzuki, H., Kerr, R., Bianchi, L., Frokjaer-Jensen, C., Slone, D., Xue, J., Gerstbrein, B., Driscoll, M., and Schafer, W. R. (2003) In vivo imaging of *C. elegans* mechanosensory neurons demonstrates a specific role for the MEC-4 channel in the process of gentle touch sensation. *Neuron.* **39**, 1005–1017.
14. Rohde, C. B., Zeng, F., Gonzalez-Rubio, R., Angel, M., and Yanik, M. F. (2007) Microfluidic system for on-chip high-throughput whole-animal sorting and screening at subcellular resolution. *Proc. Natl. Acad. Sci. USA.* **104**, 13891–13895.
15. Podbilewicz, B., and Gruenbaum, Y. (2006) Live Imaging of *Caenorhabditis elegans*: Preparation of Samples. In *Cold Spring Harb. Protoc.* doi:10.1101/pdb.ip19.
16. Driscoll, M., and Chalfie, M. (1991) The mec-4 gene is a member of a family of *Caenorhabditis elegans* genes that can mutate to induce neuronal degeneration. *Nature.* **349**, 588–593.
17. Forrester, W. C., Perens, E., Zallen, J. A., and Garriga, G. (1998) Identification of *Caenorhabditis elegans* genes required for neuronal differentiation and migration. *Genetics.* **148**, 151–165.
18. Honigberg, L., and Kenyon, C. (2000) Establishment of left/right asymmetry in neuroblast migration by UNC-40/DCC, UNC-73/Trio and DPY-19 proteins in *C. elegans. Development.* **127**, 4655–4668.
19. Collet, J., Spike, C. A., Lundquist, E. A., Shaw, J. E., and Herman, R. K. (1998) Analysis of osm-6, a gene that affects sensory cilium structure and sensory neuron function in *Caenorhabditis elegans. Genetics.* **148**, 187–200.
20. Finney, M., and Ruvkun, G. (1990) The unc-86 gene product couples cell lineage and cell identity in *C. elegans. Cell.* **63**, 895–905.
21. Maduro, M., and Pilgrim, D. (1995) Identification and cloning of unc-119, a gene expressed in the *Caenorhabditis elegans* nervous system. *Genetics.* **141**, 977–988.
22. Harfe, B. D., Vaz Gomes, A., Kenyon, C., Liu, J., Krause, M., and Fire, A. (1998) Analysis of a *Caenorhabditis elegans* Twist homolog identifies conserved and divergent aspects of mesodermal patterning. *Genes Dev.* **12**, 2623–2635.
23. Henderson, S. T., Gao, D., Lambie, E. J., and Kimble, J. (1994) lag-2 may encode a signaling ligand for the GLP-1 and LIN-12 receptors of *C. elegans. Development.* **120**, 2913–2924.
24. Epstein, H. F., and Shakes, D. C., (Eds.) (1995) *Caenorhabditis elegans: Moderm Biological Analysis of an Organism.* Vol. **48**, Academic Press, Inc., San Diego.
25. Cram, E. J., Shang, H., and Schwarzbauer, J. E. (2006) A systematic RNA interference screen reveals a cell migration gene network in *C. elegans. J. Cell. Sci.* **119**, 4811–4818.

Chapter 17

Drosophila Hemocyte Migration: An In Vivo Assay for Directional Cell Migration

Carolina G.A. Moreira, Jennifer C. Regan, Anna Zaidman-Rémy, Antonio Jacinto, and Soren Prag

Abstract

This protocol describes an in vivo assay for random and directed hemocyte migration in *Drosophila*. *Drosophila* is becoming an increasingly powerful model system for in vivo cell migration analysis, combining unique genetic tools with translucency of the embryo and pupa, which allows direct imaging and traceability of different cell types. In the assay we present here, we make use of the hemocyte response to epithelium wounding to experimentally induce a transition from random to directed migration. Time-lapse confocal microscopy of hemocyte migration in untreated conditions provides a *random cell migration* assay that allows identification of molecular mechanisms involved in this complex process. Upon laser-induced wounding of the thorax epithelium, a rapid chemotactic response changes hemocyte migratory behavior into a *directed migration* toward the wound site. This protocol provides a direct comparison of cells during both types of migration in vivo, and combined with recently developed resources such as transgenic RNAi, is ideal for forward genetic screens.

Key words: *Drosophila*, Hemocytes, In vivo Cell Migration, Chemotaxis, Microscopy

1. Introduction

Directed cell migration is a key mechanism that occurs throughout development and in homeostatic processes, such as wound healing and immune surveillance. Also, it is fundamental for the progression of clinically relevant pathologies such as vascular and chronic inflammatory diseases, and cancer metastasis (1, 2). Once cells sense a chemotactic signal, responding alterations in the actin cytoskeleton change the cellular behavior from *random* to *directed* migration. Because of their biological importance, the mechanisms

Claire M. Wells and Maddy Parsons (eds.), *Cell Migration: Developmental Methods and Protocols*, Methods in Molecular Biology, vol. 769, DOI 10.1007/978-1-61779-207-6_17, © Springer Science+Business Media, LLC 2011

and properties underlying cell migration have been intensively studied in vitro for the past few decades. This allowed the characterization of important molecular mechanisms controlling this complex process. However, the molecular mechanisms of cell migration in vivo remain poorly understood, mainly due to technical challenges.

By definition, *cell migration* refers to the movement of a cell or group of cells and this displacement can be classified either as *random* or *directional migration*. In the former situation, the migratory behavior shows no sign of preference in directionality. For instance during *chemokinesis*, cells migrate in a nondirectional pattern, which results from stimulation of motility mechanisms by a uniform motogenic signal. Conversely, *directional migration* occurs when the direction of migration of a cell is dictated by an asymmetric environmental factor. The type of directional migration depends on the nature of the asymmetric factor. For example, *chemotaxis* occurs when a cell migrates toward (or away from) a gradient (3, 4); *haptotaxis* is a substrate-dependent migration where the cell either responds to gradients of cellular adhesion sites or to other guidance signals anchored in the extracellular matrix (ECM) (4). Both types of directional migration have been intensively studied using in vitro techniques such as the wound-healing assay, and the Boyden and Dunn chamber assays (5, 6). These in vitro techniques have provided important insights regarding cell migration mechanisms. Although major improvements have been made such as the use of 3D matrices (7, 8), in vitro assays remain artificial approximations and are, therefore, potentially unable to identify important signaling and molecular mechanisms. Therefore, it is necessary to extend our knowledge of cell migration by employing in vivo studies.

In this protocol, we describe a method to control spatially and temporally the transition from random to directed cell migration in vivo. *Drosophila* is our model organism of choice mainly due to translucency of the pupa that allows for live confocal imaging, and the availability of sophisticated genetic tools. Among these is the GAL4/UAS system, which allows the expression of fluorescently tagged proteins in specific cells and defined tissues (9). Like the embryo, the pupa has numerous advantages for studying cell migration in vivo: (1) it is easy to work with since it is stationary and translucent, allowing straightforward microscopy techniques; (2) development proceeds normally after mounting on a glass slide; (3) the GAL4/UAS system is fully functional with high expression levels of GAL4 protein and subsequently high expression levels of fluorescently tagged proteins increasing the signal-to-noise ratio (10). Also, the expression levels can be controlled through changes in temperature as GAL4 folding is temperature sensitive (10). Moreover, the pupa presents some advantages over the embryo which has been used for a previously described hemocyte migration assay (11): it allows the generation of mosaic flies (12), and the efficient knockdown of genes of interest by cell type-specific expression of RNAi.

To analyze single cell migration, we image *Drosophila* hemocytes, which are immune circulatory cells analogous to vertebrate leukocytes in terms of features and functions (13). We use flies in which the hemocyte driver *srp*HemoGAL4 induces the expression of a nuclear-localized form of mCherry protein in the hemocytes 18–24 h after puparium formation (APF) (14). The majority of the hemocytes migrate on the collagen type IV-containing basement membrane connected to the epithelium. With a laser, we induce a wound in the dorsal part of the thorax epithelium (distinguished by expression of ubiquitous DEcadherin-GFP), whereby most hemocytes cease random migration and initiate directed migration toward the site of damage. This process is clearly visible and quantifiable within the first 20 min after wounding (Fig. 1 and Table 1). Dependent on the distance from the wound, hemocytes reach the wound in approximately 20 min.

Here, we describe an assay for the analysis of single cell migration in two different contexts: before and after epithelial wounding in *Drosophila* pupa. Although here we have used a specific nuclear marker for hemocytes, a vast number of other fluorescently tagged proteins are available. The combination of these genetic tools with powerful confocal imaging techniques can shed light on important mechanisms underlying random and directed cell migration behavior in vivo.

2. Materials

2.1. Collecting and Mounting Drosophila Pupae

1. Fine paintbrush.
2. Standard *Drosophila* vials.
3. Yeast paste: 6.6 g yeast powder, 6 ml distilled water, and 360 μl 10% Nipagin M (dissolved in 100% ethanol).
4. Scalpel – 5-mm depth, 15° cutting edge.
5. Forceps – Dumont #5, Student.
6. Double-sided sticky tape (Tesa).
7. Microscope cover glass # 1.5 24 × 24 mm.
8. Microscope slides.
9. Quick-dry nail polish.
10. Halocarbon oil 700.
11. Dissecting scope.
12. Fluorescence stereomicroscope with GFP filter.

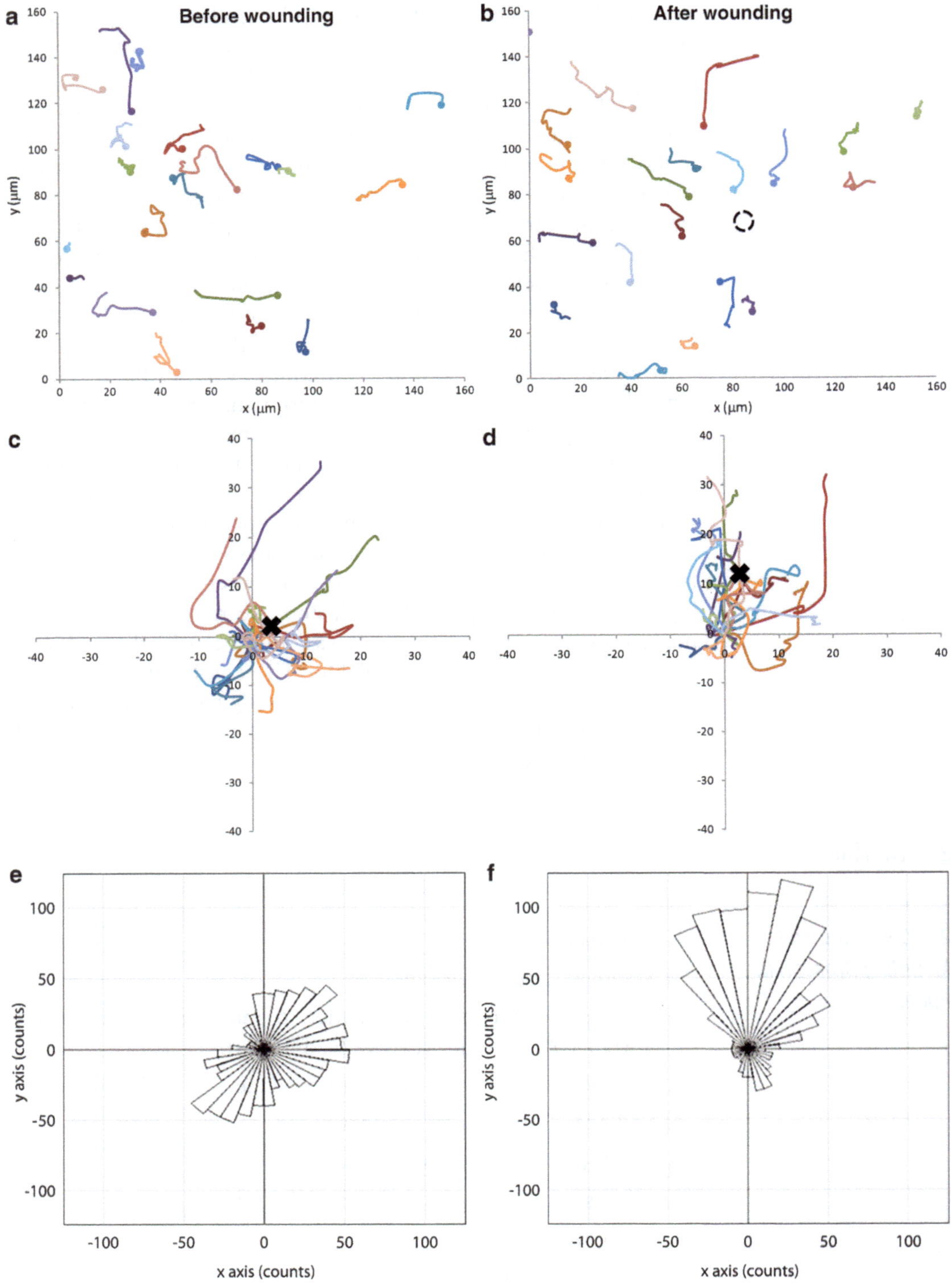

Fig. 1. Migratory behavior of *Drosophila* hemocytes before and after wounding. Cells were imaged and tracked for 15 min with 1-min interval, before and after wounding. (**a**, **b**) Trajectories of 20 hemocytes from a single field of view. End points of cell trajectories are indicated by *closed circles*. (**a**) Before wounding, hemocytes show a random migratory behavior. (**b**) After wounding, there is a change in their migratory behavior: cells display a more directed migratory behavior toward the wound site. The *dashed circle* indicates the center of the wound. (**c**, **d**) Tracks plotted in **a** and **b** were superimposed at (0,0). The wound site was rotated to the positive *y*-axis. All trajectories were rotated to maintain their relative position

Table 1
Quantification of *Drosophila* hemocyte behavior during random and directed migration

		Before wounding	After wounding
Mean speed (μm/min)		2.48	2.03
Directionality		0.46	0.56
Rayleigh test		0.628	0.002
Center of mass coordinates (μm)	*x*	3.37	2.86
	y	2.15	12.01
Forward migration index	*x*	0.09	0.07
	y	−0.04	0.35

20 wild-type cells were tracked for 15 min with a 1-min interval before and after wounding. The wound site was assigned in the positive *y*-axis (see Note 15)

2.2. Drosophila Lines

1. *w*; srpHemoGAL4, UAS-nlsmCherry* (expresses nuclear-targeted mCherry in the hemocytes).
2. *yw*; Ubi-DEcadherin-GFP* (expresses GFP in the epithelia).

2.3. Imaging and Ablation

1. Inverted or upright confocal microscope.
2. UV laser ablation system (DPSL-355/14; Rapp Opto-Electronic).

2.4. Post-acquisition Image Processing and Analysis

1. ImageJ software (http://www.rsbweb.nih.gov/ij/, see Note 1).
2. ImageJ plugins "Manual Tracking" and "Grouped ZProjector" (http://www.rsbweb.nih.gov/ij/) and "Chemotaxis and Migration Tool" (http://www.ibidi.de).
3. Microsoft Excel® and Notepad®.

3. Methods

3.1. Pupa Collection

1. Perform crosses of the two fly lines: ♂ *w*; srpHemoGAL4, UAS-nlsmCherry* x ♀*yw*; Ubi-DEcadherin-GFP* (see Note 2) in adequate food vials containing a small portion of yeast (see Note 3).
2. Incubate the vials at 18–29°C (see Notes 4 and 5).

Fig. 1. (continued) toward the wound center (see Note 15). Center of mass of all end points are marked with a cross. Comparison between CM positions indicates a change in the migratory behavior of the hemocytes after wounding, showing the transition from *random* to *directed migration* toward the wound site. (**e**, **f**) Rose plots were used to visualize the preferential direction of migration of the cells. Rose plot are calculated using default settings (interior angle = 66°, range interval = 10, see Note 14).

3. After 6–7 days, pupae will develop dependent on the incubation temperature.
4. Select a pupa at the desired stage (see Note 6 and Fig. 2e) using a fine paintbrush (see Note 7).

3.2. Pupal Dissection and Mounting

1. Add a 3–4-cm long strip of double-sided sticky tape to a microscope slide.
2. Place a pupa ventral side down on the tape using a forceps. Make sure that the pupa is well glued to the tape by applying pressure on the posterior part of the puparium (Fig. 2a).
3. With the scalpel, gently remove the operculum (the anterior lid) of the pupal cuticle (Fig. 2b). This allows you to insert the tip of the scalpel gently inside the cuticle case, open it from the inside out, and cut approximately halfway down (Fig. 2c, see Note 8). Remove the pupal cuticle from the dorsal side of the pupa (Fig. 2d). The debris of pupal cuticle can be glued to the tape so they will not interfere with the imaging.
4. Check for the Malpighian tubules position on a fluorescence stereomicroscope using an appropriate GFP filter. The Malpighian tubules migrate from the thorax to the abdomen during pupal development. At 20-h APF, the Malpighian tubules have just moved from the thorax, and are completely inserted in the abdomen (see Note 6 and Fig. 2e).

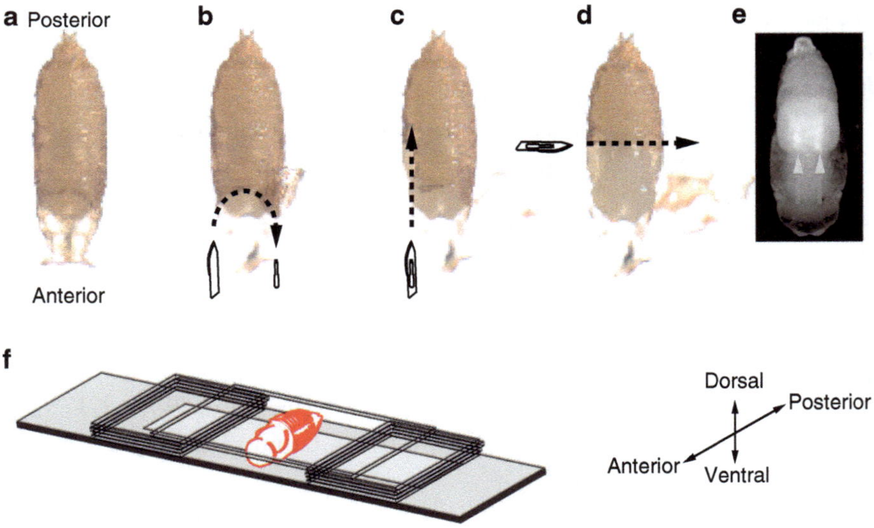

Fig. 2. Pupal dissection. *Dashed arrows* indicate direction of dissection of the pupal cuticle using a scalpel. (**a**) Pupa glued on a double-sided sticky tape, placed on a microscope slide, ventral side down. (**b**) Removal of the anterior part of the pupal cuticle. Pupal cuticle debris is glued to the tape. (**c**) Lateral cut of the pupal cuticle. The cut just reaches the beginning of the abdomen. (**d**) Removal of the pupal cuticle from the top of the thorax. (**e**) Malpighian tubules positioned at the beginning of the abdomen, posterior to the thorax. *Full arrows* indicate the Malpighian tubules. (**f**) Schematic drawing of pupa mounted on a microscope slide.

5. Build a coverslip bridge by placing five coverslips glued together in a stack using nail polish on either side of the pupa.
6. Place a very small drop of Halocarbon oil on a coverslip and gently place it to cover the pupal thorax (see Note 9). Fix the coverslip bridge using nail polish. The capillary motion of the nail polish will tighten the coverslip and apply a gentle pressure on the thorax (for schematic drawing, see Fig. 2f).

3.3. Hemocyte Chemotaxis Assay

3.3.1. Random Migration Assay

1. Place the slide with the mounted pupa as described in Subheading 3.2 on a confocal microscope (see Note 10) and focus using a 40× objective.
2. Use excitation/emission acquisition settings appropriate for mCherry imaging (561- or 594-nm excitation wavelength).
3. Find an appropriate region for imaging hemocytes. In our experiments, we consistently select an area on the pupal thorax away from the longitudinal anteroposterior axis as this area tends to be overcrowded with hemocytes. We consistently choose regions where the hemocytes are dispersed and subsequently are allowed independent, single-cell movements (for an example, see Fig. 3).
4. Set up time-lapse image acquisition series using a 10-μm Z-stack for 15 min with 1-min interval (see Note 11).

3.3.2. Wounding

1. Acquire a single image (approximately 5-μm Z-stack) of unwounded epithelium using excitation/emission acquisition settings appropriate for EGFP (488-nm excitation wavelength) (see Note 12).
2. Ablate a small patch of epithelium on the pupal thorax using a UV laser ablation system. The wound should be approximately 20 μm in diameter.
3. Acquire a single image of the epithelium after wounding using the same settings as in step 1.

3.3.3. Directed Migration Assay

1. Continue imaging the same region as in step 4 in Subheading 3.3.1 using the settings for mCherry imaging.

3.4. Post-acquisition Analysis

1. Open time-lapse series and the epithelia stacks in ImageJ.
2. Use the "Grouped ZProjector" plugin to project each Z-stack series into a single frame.
3. Track the hemocyte nuclei using Manual Tracking (Plugin > Tracking > Manual Tracking; see Note 13). Save the results as a *.xls file.
4. In order to determine the center of the wound, outline the wound area in the projected image of the epithelium using the Elliptical Selection Tool.

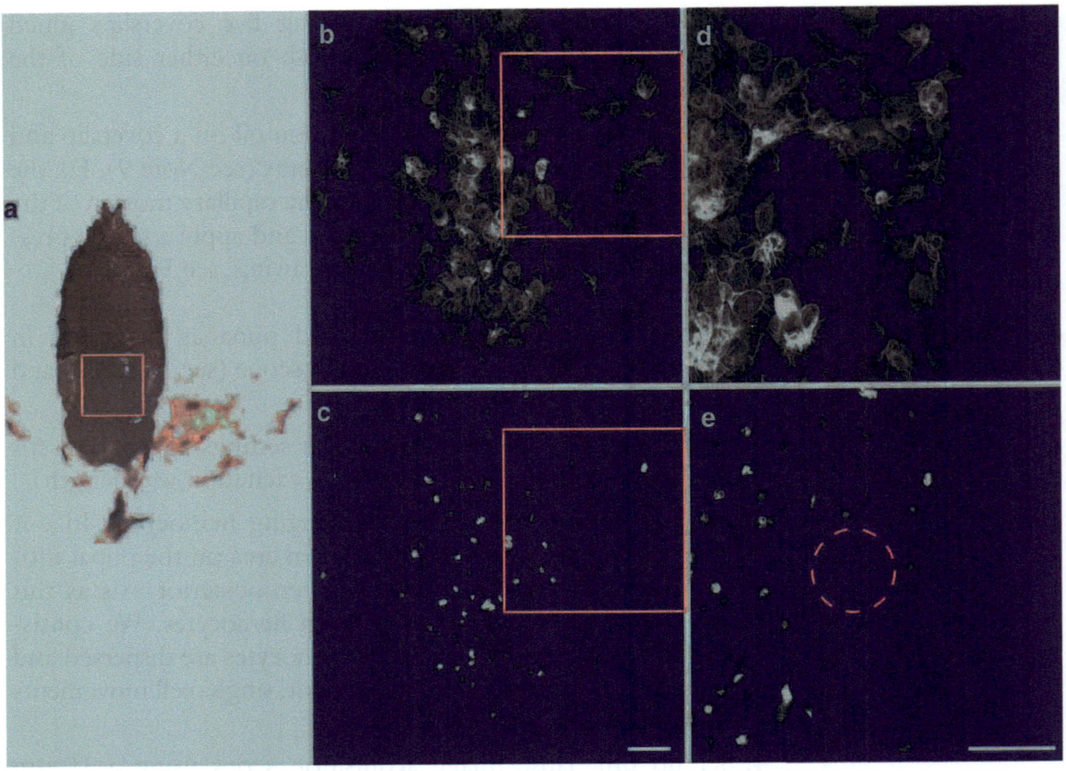

Fig. 3. Hemocyte imaging. Region of interest is consistently selected in areas where the hemocytes appear dispersed. Here are shown hemocytes expressing cytoplasmic GFP and nuclear mCherry under the control of *srp*HemoGAL4 to present both the hemocyte morphology and the nuclear localization for tracking. For regular measurements, we only follow hemocyte nuclear localization using *srp*HemoGAL4, UAS-nlsmCherry pupa. (**a**) Pupa glued on a double-sided sticky tape, placed on a microscope slide. (**b**, **c**) Confocal images of GFP channel (**b**) and mCherry channel (**c**) using 20× magnification of thorax area indicated by the *box* in **a**. (**d**, **e**) Confocal images of GFP channel (**d**) and mCherry channel (**e**) using 40× magnification of the area indicated in **c**. *Circle indicates* site for wounding. Bars indicates 20 μm.

5. Tick the "Centroid" parameter in the "Set Measurement" (Analyze > Set Measurements), "Measure" (Analyze > Measure), and save the coordinates corresponding to the Centroid of the wound.
6. For comparing hemocyte migration behavior before and after wounding, employ the Chemotaxis and Migration Tool plugin (Plugins > Chemotaxis tool; see Note 14), which provides basic statistical tools to analyze cell migratory behaviors. However, caution is needed when using this software as it is developed for particular experimental settings. In the case of the experiments presented in this protocol, all data had to be transformed in Excel. Nuclear trajectories are superimposed to (0,0) and rotated in accordance to their initial directionality toward the wound center (see Note 15). Data are saved as a *.txt file before importing it for analysis using the Chemotaxis and Migration Tool plugin. For consistency, the wound center is located in the positive *y*-axis.

4. Notes

1. A bundled ImageJ software package, Fiji, is available from http://www.pacific.mpi-cbg.de/wiki/index.php/Main_Page. This package includes a comprehensive selection of commonly used plugins, allowing for direct import of the main image formats.
2. We consistently express GAL4/UAS-nlsmCherry paternally, and Ubi-DEcadherin, or in the case of knockdown the UAS-RNAi, maternally. This will ensure that there is no contamination from non-virgins in the cross, as only progeny containing the intended paternal contribution will express nlsmCherry.
3. Yeast is used as a supplement for flies and increases their fertility.
4. The temperature for incubation can vary according to the experiment. We incubate UAS-RNAi lines at 29°C to increase the efficiency of the GAL4 induction. If the RNAi is lethal, we decrease the incubation temperature to 18–25°C for 6 days, and change them to 29°C only 24 h prior to the experiment.
5. We change fly crosses to new vials daily to facilitate accurate staging of the pupa. Also, the staging of pupal development improves in non-crowded conditions.
6. We consider that the puparium starts forming when the post-feeding larva everts the anterior spiracles, stops moving completely, and sticks to the sides of the culture vial. We routinely use pupa at 18–24 h APF when placed at 25–29°C. During this period, *srp*HemoGAL4-positive hemocytes start to appear as individual cells and are highly motile. After 24 h APF, hemocytes start to aggregate and are, therefore, not appropriate to analyze in a directed migration assay. Because pupal developmental progress depends on parameters such as temperature, we use the Malpighian tubules position for a more accurate staging (15). The tubules can be detected after dissection due to high autofluorescence using a fluorescence stereomicroscope and filter settings appropriate for GFP. During the time window of our experiments, the tubules appear at the beginning of the abdomen, posterior to the thorax.
7. Sometimes it may be useful to wet the paintbrush slightly to remove the pupa from the vial sidewalls.
8. The cuticle does not need to be completely removed. Consistently, we only remove the anterior part of the cuticle, which is sufficient to expose the thorax.

9. The amount of halocarbon oil used to cover the pupal thorax is of extreme importance. The pupal thorax should be covered but not soaked in oil as this impairs hemocyte migration.
10. Both upright and inverted microscopes can be used in this protocol, as the sample is fixed to the slide.
11. The stack size is a compromise between spatial information and image acquisition time. To minimize exposure to laser during image acquisition that can cause photobleaching and production of reactive oxygen species, we keep the dwell time between acquisitions longer than the acquisition time. Therefore, we routinely acquire images at 512 × 512 pixels, 11 planes with 1 μm Z-interval totaling a Z-stack of 10 μm. As the stacks will be projected, we undersample according to the Nyquist sampling theorem.
12. Comparing images of unwounded and wounded epithelia improves the recognition of laser-induced wounds.
13. Automatic Tracking plugin for ImageJ does not allow editing of erroneous tracking. With the acquisition settings we suggest here, the Manual Tracking plugin is practicable. For advanced use, please refer to the Manual Tracking user guide.
14. For further details, please consult the "Chemotaxis and Migration Tool" user guide (download from http://www.ibidi.de/applications/ap_chemo.html).
15. To rotate the tracks and keep their relative directionality toward the wound site, the following steps should be followed for each individual track:
    (a) Translocate trajectories to common origin (0,0): for individual tracks, calculate for every time point $(x_{tT}, y_{tT}) = (x_t, y_t) - (x_{t=0}, y_{t=0})$
    (b) Translocate wound site: calculate $(x_{wT}, y_{wT}) = (x_w, y_w) - (x_{t=0}, y_{t=0})$
    (c) Calculate the norm of the wound site relatively to the origin (0,0): $\text{norm} = \sqrt{(x_{wT}^2 + y_{wT}^2)}$
    (d) Calculate y_{wt}/norm
    (e) Calculate $\theta = a\cos(y_{wt} / \text{norm})$
    (f) Calculate the rotated (x_{tTR}, y_{tTR}) and (x_{wTR}, y_{wTR}) according to the following:
    - If $x_{wt} \leq 0$

$$\begin{cases} x_{tTR} = \cos(\theta) \times x_{tT} + \sin(\theta) \times y_{tT} \\ y_{tTR} = -\sin(\theta) \times x_{tT} + \cos(\theta) \times y_{tT} \end{cases}$$

$$\begin{cases} x_{wTR} = \cos(\theta) \times x_{wT} + \sin(\theta) \times y_{wT} \\ y_{wTR} = -\sin(\theta) \times x_{wT} + \cos(\theta) \times y_{wT} \end{cases}$$

- If $x_{wt} > 0$

$$\begin{cases} x_{tTR} = \cos(\theta) \times x_{tT} - \sin(\theta) \times y_{tT} \\ y_{tTR} = \sin(\theta) \times x_{tT} + \cos(\theta) \times y_{tT} \end{cases}$$

$$\begin{cases} x_{wTR} = \cos(\theta) \times x_{wT} + \sin(\theta) \times y_{wT} \\ y_{wTR} = -\sin(\theta) \times x_{wT} + \cos(\theta) \times y_{wT} \end{cases}$$

Note: these calculations presume that norm≠0, i.e., center of wound is different from the origin of the cell track.

(x, y)	*x* and *y* coordinates
t	Time point
w	Wound coordinates
T	Translocated coordinates
R	Rotated coordinates

Acknowledgments

The authors would like to acknowledge Marco Antunes for suggestions on hemocytes, Jan Stühmer for suggestions on cell tracking, and Luis Almeida for input on analysis. This work was supported by FCT PTDC/SAU-OBD/101259/2008 (S.P.) and SFRH/BD/62345/2009 (C.G.A.M.), ARC and FCT SFRH/BPD/44613/2008 fellowships (A.Z.R.), and EMBO ALTF 178-2009 (J.R.).

References

1. Luster, A. D., Alon, R., and von Andrian, U. H. (2005) Immune cell migration in inflammation: present and future therapeutic targets, *Nat Immunol* **6**, 1182–1190.
2. Yamaguchi, H., Wyckoff, J., and Condeelis, J. (2005) Cell migration in tumors, *Current Opinion in Cell Biology* **17**, 559–564.
3. Rørth, P. (2009) Collective Cell Migration, *Annual Review of Cell and Developmental Biology* **25**, 407–429.
4. Petrie, R. J., Doyle, A. D., and Yamada, K. M. (2009) Random versus directionally persistent cell migration, *Nat Rev Mol Cell Biol* **10**, 538–549.
5. Chen, H. C. (2005) Boyden chamber assay, *Methods Mol Biol* **294**, 15–22.
6. Zicha, D., Dunn, G., and Jones, G. (1997) Analyzing chemotaxis using the Dunn direct-viewing chamber, *Methods Mol Biol* **75**, 449–457.
7. Fraley, S. I., Feng, Y., Krishnamurthy, R., Kim, D.-H., Celedon, A., Longmore, G. D., and Wirtz, D. (2010) A distinctive role for focal adhesion proteins in three-dimensional cell motility, *Nat Cell Biol* **12**, 598–604.
8. Gaggioli, C., Hooper, S., Hidalgo-Carcedo, C., Grosse, R., Marshall, J. F., Harrington, K., and Sahai, E. (2007) Fibroblast-led collective

invasion of carcinoma cells with differing roles for RhoGTPases in leading and following cells, *Nat Cell Biol* **9**, 1392–1400.

9. Brand, A. H., and Perrimon, N. (1993) Targeted gene expression as a means of altering cell fates and generating dominant phenotypes, *Development* **118**, 401–415.
10. Duffy, J. B. (2002) GAL4 system in *drosophila*: A fly geneticist's swiss army knife, *Genesis* **34**, 1–15.
11. Stramer, B., and Wood, W. (2009) Inflammation and Wound Healing in *Drosophila*, *Methods Mol Biol*, pp 137–149.
12. Xu, T., and Rubin, G. M. (1993) Analysis of genetic mosaics in developing and adult *Drosophila* tissues, *Development* **117**, 1223–1237.
13. Wood, W., and Jacinto, A. (2007) *Drosophila* melanogaster embryonic haemocytes: masters of multitasking, *Nat Rev Mol Cell Biol* **8**, 542–551.
14. Bruckner, K., Kockel, L., Duchek, P., Luque, C. M., Rorth, P., and Perrimon, N. (2004) The PDGF/VEGF Receptor Controls Blood Cell Survival in *Drosophila*, *Developmental Cell* 7, 73–84.
15. Bainbridge, S., and Bownes, M. (1981) Staging the metamorphosis of *Drosophila* melanogaster, *J Embryol Exp Morph* **66**, 57–80

Chapter 18

Measuring Inflammatory Cell Migration in the Zebrafish

Philip M. Elks, Catherine A. Loynes, and Stephen A. Renshaw

Abstract

A key feature of inflammatory cells is the ability to migrate to a site of injury or infection quickly and efficiently. Infectious agents can then be taken up by these inflammatory cells, preventing established infection. Inflammatory cell migration is driven by a complex interaction between inflammatory cells and their environment. In order to maintain health, inflammation needs to resolve, allowing the surrounding tissues to recover and heal. These processes are not fully understood and have been difficult to study in cell culture due to the complex interactions between cell types. We therefore use a range of techniques in near-transparent zebrafish (*Danio rerio*) larvae to study these migration events in a whole-organism, *in vivo* model. Using a transgenic zebrafish line that specifically marks neutrophils with green fluorescent protein, *Tg(mpx:GFP)i114*, we are able to follow neutrophil behaviour at a single cell level. Using these methods, the cellular processes involved in all phases of inflammation can be studied and better understood.

Key words: Neutrophil, Migration, Inflammation, Zebrafish, In vivo

1. Introduction

Zebrafish (*Danio rerio*) have a unique place in the study of immune cell migration. In their larval stages, they are approximately 3 mm long, but have a well-developed immune system, with cell types with close homology to human neutrophils, macrophages, eosinophils and mast cells (1, 2). Over recent years, the zebrafish has emerged as a key model organism for the study of developmental processes (3, 4), and the tools of the developmental biologists have been seized by those of us interested in human disease, and turned to problems beyond development. Three features of the zebrafish make it a particularly suitable model system for the study of *in vivo* cell behaviour: 1. The larvae are transparent, possible, not visible visualisation of a wide range of physiological events possible.

Claire M. Wells and Maddy Parsons (eds.), *Cell Migration: Developmental Methods and Protocols*,
Methods in Molecular Biology, vol. 769, DOI 10.1007/978-1-61779-207-6_18, © Springer Science+Business Media, LLC 2011

In particular, cells of the immune system are readily visualised by DIC microscopy. 2. Zebrafish transgenesis has meant that individual cell populations can be labelled with fluorescent markers, making them visible during processes such as inflammation (5). 3. Gene knockdown using morpholino-modified antisense constructs ("morpholinos") has enabled genetic manipulations possible (6). When combined together, these three features make for a powerful model system.

Neutrophils and macrophages are key effectors of the immune response, and we depend on their activity to defend us against environmental pathogens from day to day. They have a number of unique features, but perhaps the most pertinent here is the rapidity with which they migrate. Tissue macrophages may control initial infectious or injurious challenges, but if these efforts cannot control the stimulus, neutrophils are recruited, which take up and kill bacteria. Inflammatory macrophages arrive shortly after, clear cell debris, and orchestrate the resolution of the immune response (7).

Much can be learned about these interesting cells by their study *in vivo*, and the zebrafish provides a model where all phases of the inflammatory response can be identified and studied at a single cell resolution (8). However, zebrafish inflammatory cells also provide a model for the study of more generic aspects of cell migration. The ability of the zebrafish to allow *in vivo* imaging, combined with the facility to manipulate the migrating cells genetically, makes this model a powerful resource for the study of cell migration in general.

Leukocytes can be drawn to sites of induced inflammation by sterile tail transection of zebrafish larvae (5). This not only allows the study of the kinetics of inflammation initiation and resolution, but also of individual cell behaviours during inflammation, including multiple features of cell migration, speed, and cell-to-cell contact. The size of injury and, therefore, inflammatory response can be varied with either a small "nick" in the fin or a more extensive injury with complete tailfin transection (Fig. 1). Zebrafish larvae have complete regrowth of the tail fin within 3 days of transection without affecting survival, allowing the process of inflammation to be studied over several days after injury. The role of environmental

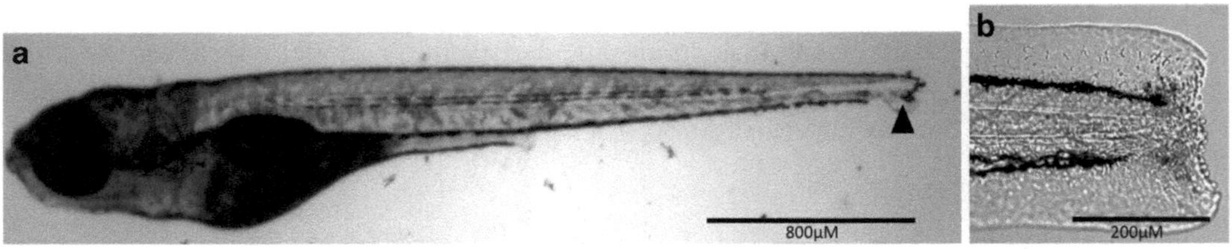

Fig. 1. (**a**) Lateral view of a 3-day post-fertilisation zebrafish larva with caudal tail transection. Gap in pigment along the ventral side is indicated by arrowhead. (**b**) Magnified image of caudal tail transection.

pathogens, LPS, etc., in inducing inflammation is controversial. Some groups have found that addition of bacterial contaminants leads to improved recruitment of inflammatory cells (9). Performing the injury under aseptic conditions, in our hands, does not change the pattern or degree of neutrophil recruitment. For ease, therefore, we do not perform these experiments under strict sterile conditions, but seek only to maintain a clean and contaminant-free environment. Additional stimulation of recruitment can be achieved by the use of chemoattractant molecules such as f-Met-Leu-Phe (fMLP, see below).

The migration of neutrophils toward the site of tail transection can be modulated by application of a chemoattractant during the tail transection protocol. These molecules can be administered by two main methods. They can be administered locally to the site of injury by coating the blade used to transect the tail in the required concentration of chemical. It should be noted that this method may lead to diffusion of the applied compound into the surrounding medium, altering the concentration at the site of injury. It is, however, a useful method to observe the effects of a gradient emanating from a source, i.e. the site of injury. Chemoattractants can be used to increase the number of neutrophils accumulating at the injury site. This can then produce a larger resolution effect by 24-h post-injury, allowing smaller affects to be detected in treatment groups. The second method is to administer the compound directly to the embryo medium. This has the advantage of maintaining a known concentration; however, the compound may not permeate the site of injury depending on the size of the compound.

Zebrafish larvae are near transparent, allowing for easy visualisation of individual cells. Leukocytes can be visualised throughout the process of inflammation in larvae by using differential interference contrast, DIC. Using this method to study induced inflammation negates the need for fluorescent reporter lines, and therefore genetic manipulation, providing rapid experimental data (10). However, for a more detailed study of different cell types, postmortem stains, real-time *in vivo* vital dyes, and fluorescent reporters are required. Both postmortem stains described below are neutrophil specific as they react with endogenous peroxidases within, not with the neutophils. The vital dye (neutral red) specifically stains macrophages, being pinocytosed exclusively by this cell type (11).

In order to visualise neutrophils within the zebrafish larvae, a fluorescent reporter of a suitable transgenic line (e.g. the GFP of *Tg(mpx:GFP)i114* line), or a postmortem fluorescent stain (e.g. TSA stain) can be used. In order to count neutrophils over multiple timepoints, it is recommended that a low-power dissecting microscope is used. However, to image for automated counting or publication, a higher power inverted microscope connected to a camera will give higher quality images.

The *Tg(mpx:GFP)i114* transgenic line combined with the transparency of zebrafish larvae gives the opportunity to investigate the behaviour of neutrophils in a living organism. Advances in microscopy and computational technology have allowed a number of neutrophil behaviours to be studied within living zebrafish larvae. To obtain high quality and high resolution images, we recommend the use of confocal microscopy. Confocal microscopy focuses a laserbeam onto a single point in a tissue using a pinhole. Scanning across and through the tissue allows a sharp three-dimensional picture to be built up. High quality images are obtained because out-of-focus light is not detected by this method. There is, however, a speed of acquisition issue, as scanning through the tissue of interest can take a few minutes. For many applications, this time delay is not a problem. However, for imaging of rapidly moving cells such as neutrophils, the speed of acquisition of traditional confocal microscopy can be too slow and the neutrophil migrates during the scan. Spinning disc confocal microscopy or similar technologies can circumvent this problem. Instead of having a single variable pinhole, spinning disc confocal microscopy employs two discs with an array of specifically arranged pinholes. When the discs are spun, the pinhole created can move very rapidly across a volume of tissue, allowing a confocal image to be built up extremely rapidly. Spinning disc microscopy acquisition has allowed high quality time-lapse imaging of neutrophil movement in the zebrafish tail transection assay (12, 13).

A variety of software packages for the acquisition and analysis of neutrophil time-lapse data are available. Our experience is predominantly with the Volocity™ software package (Version 5.3, Improvision) and it is this method that will be described below (see Subheading 3.8). Volocity™ allows detailed analysis of captured images and different modules are available for different applications. Volocity Quantitation™ allows cells labelled with GFP to be individually tracked over a time-lapse series. Volocity™ will automatically track cells (saving tedious manual tracking of individual cells) dependent on a number of parameters manually set by the user (including GFP intensity, volume, distance moved, and timespan). From this information, a number of cell migration parameters can be extracted, such as neutrophil speed and meandering index. Since the time between each image in the time lapse is known, the speed of the cells can be calculated. We have shown that using chemoattractants, such as fMLP, at the tail transection significantly increases the speed of neutrophil migration towards the tail injury site (Fig. 2). Meandering index is a measure of how direct a cell moves during its migration, and is calculated by dividing the displacement (straight line distance from start to finish) by total path length. Therefore, if a neutrophil travelled in a perfectly straight line to the final destination, the meandering index would be 1. Any deviation from a straight line will decrease the meandering index.

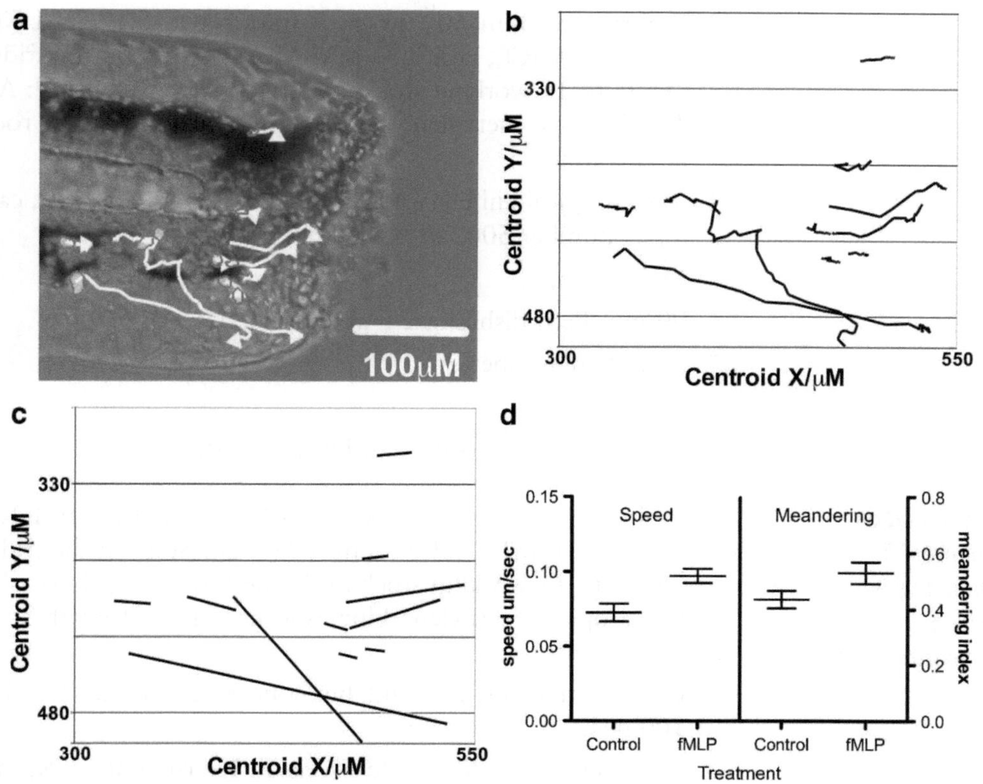

Fig. 2. Tracking of neutrophils over the first hour post-injury using spinning disc microscopy and Volocity™ software package. (**a**) A photomicrograph showing the movement of neutrophils in a *Tg(mpx:GFP)i114* larva over the first hour post-injury. (**b**) A graph of neutrophil movement in the same larva as in A, generated by the Volocity™ software. (**c**) A graph of the overall vector of neutrophil movement in the same larva generated by the Volocity™ software. (**d**) A graph showing the neutrophil speed and meandering index over the first hour post-injury as an average of a number of different larvae. Treatment with the chemoattractant, fMLP, increases both the speed and meandering index of neutrophils.

After injury, cell numbers can be assessed at the site of inflammation very quickly and accurately at any number of time points during the inflammatory process. When assessing numbers in a small region of interest or in thin tissues, for example in the fins, cells can be counted by eye using a dissecting microscope with speed and accuracy at a low magnification.

2. Materials

2.1. Tail Transection

1. Zebrafish larvae 2–5 days post-fertilisation (dpf) in embryo medium maintained according to standard protocols (14).
2. Tricaine, MS-222: 400 mg powder, make up to 100 ml with water, buffer to pH 7 using 1 M Tris pH 10. Use at a final concentration of 0.017%. Store in the dark.

3. Embryo medium 60× stock: 5 mM NaCl, 0.17 mM KCl, 0.33 mM $CaCl_2$, and 0.33 mM $MgSO_4$. Store in the fridge. Dilute to 1× working stock in distilled water. Optional: Add few drops of methylene blue as a fungicide. Store at room temperature.
4. Microsurgical knife (World Precision Instruments, UK, catalogue number 500249) (see Note 1).
5. Masking tape.
6. 90-mm Petri dishes or 24-well plates.
7. 3 ml Pasteur pipettes.
8. Pair of forceps.
9. Dissecting microscope: e.g. Leica MZ10F.

2.2. Stimulation of Neutrophil Chemotaxis by Addition of Chemoattractant

1. Embryo medium 60× stock: 5 mM NaCl, 0.17 mM KCl, 0.33 mM $CaCl_2$, and 0.33 mM $MgSO_4$. Store in the fridge. Dilute to 1× working stock in distilled water. Optional: Add few drops of methylene blue as a fungicide. Store at Room temperature.
2. Chemoattractant, at a suitable concentration, e.g. fMLP at 100 nM.
3. Microsurgical knife (World Precision Instruments, UK, catalogue number 500249).

2.3. Neutrophil Identification Using Diaminobenzidine: Postmortem Stain

1. 4% (w/v) paraformaldehyde (PFA) in 1×PBS. Dissolve at 70°C. Store at 4°C. Keep for 1 week only.
2. Diaminobenzadine tetrahydrochloride (DAB). Make up stock to 40 mg/ml in water and store in fridge.
3. Hydrogen peroxide 30% stock. Keep at 4°C.
4. 1×Phosphate-buffered saline (PBS).
5. Triton X-100.
6. 1.5 ml microcentrifuge tubes.
7. Bright-field dissecting microscope with 100× magnification.

2.4. Neutrophil Identification Using Tyramide Signal Amplification: Postmortem Stain

1. 4% (w/v) PFA in 1×PBS. Dissolve at 70°C. Store at 4°C. Keep for 1 week only.
2. TSA Plus cyanine 3 system (PerkinElmer).
3. 1.5 ml microcentrifuge tubes.
4. 1× PBS with 0.1% Tween20.
5. Epiflourescence microscope with 100× magnification and a red filter suitable for visualising cyanine-3.

2.5. Macrophage Identification Using Neutral Red: Vital Dye

1. Embryo medium 60× stock: 5 mM NaCl, 0.17 mM KCl, 0.33 mM $CaCl_2$, and 0.33 mM $MgSO_4$. Store in the fridge. Dilute to 1× working stock in distilled water. Store at room

temperature. Do not add any methylene blue as described previously in Subheading 2.2.1.

2. Neutral red solution. Dilute to working solution of 2.5 μg/ml in embryo medium without methylene blue.
3. Zebrafish larvae from 24-h post-fertilisation.
4. Petri dishes or multiwell plates.
5. Bright-field dissecting microscope with 100× magnification.
6. 3 ml Pasteur pipettes.

2.6. Fluorescence Microscopy

1. Leica MZ10F dissecting microscope or similar, with a suitable UV light source and a GFP-PLUS filter set.
2. Tricaine, MS-222: 400 mg powder, make up to 100 ml with water, buffer to pH 7 using 1 M Tris pH 10. Use at a final concentration of 0.017%. Store in the dark.
3. Embryo medium 60× stock: 5 mM NaCl, 0.17 mM KCl, 0.33 mM $CaCl_2$, and 0.33 mM $MgSO_4$. Store in the fridge. Dilute to 1× working stock in distilled water. Add few drops of methylene blue as a fungicide. Store at room temperature.
4. 90-mm Petri dishes.
5. Zebrafish larvae carrying a fluorescent reporter transgene (see Note 2).

2.7. Spinning disc Confocal Microscopy Mounting

1. PerkinElmer UltraVIEW VoX spinning disc confocal system, and a suitable inverted microscope.
2. Tricaine, MS-222: 400 mg powder, make up to 100 ml with water, buffer to pH 7 using 1 M Tris pH 10. Use at a final concentration of 0.017%. Store in the dark.
3. Embryo medium 60× stock: 5 mM NaCl, 0.17 mM KCl, 0.33 mM $CaCl_2$, and 0.33 mM $MgSO_4$. Store in the fridge. Dilute to 1× working stock in distilled water. Add few drops of methylene blue as a fungicide. Store at room temperature.
4. 0.5–1% low melting point agarose.
5. 40-mm Petri dishes.
6. High vacuum grease
7. 25-mm circular number 0 coverslips.
8. An eye-brow hair attached to a 1 ml pipette tip (see Note 3).

2.8. Cell Tracking

1. High quality data acquired from a spinning disc microscope.
2. Volocity™ software package (Version 5.3, Improvision).
3. A processing computer, preferably with Windows 64-bit operating system and at least 4Gb of RAM.

2.9. Timecourse Counting

1. Leica dissecting fluorescent microscope with 80× magnification and not with a suitable filter set.
2. Tricaine, MS-222: 400 mg powder, make up to 100 ml with water, buffer to pH 7 using 1 M Tris pH 10. Use at a final concentration of 0.017%. Store in the dark.
3. Embryo medium 60× stock: 5 mM NaCl, 0.17 mM KCl, 0.33 mM $CaCl_2$, and 0.33 mM $MgSO_4$. Store in the fridge. Dilute to 1× working stock in distilled water. Add few drops of methylene blue as a fungicide. Store at room temperature.
4. 90-mm Petri dishes.
5. Plastic Pasteur pipettes.
6. Zebrafish larvae carrying a fluorescent reporter transgene.

3. Methods

3.1. Tail Transection

1. Zebrafish larvae are anaesthetised with Tricaine by adding it directly to the embryo medium. One 90-mm Petri dish of 50 larvae should be anaesthetised with 1 ml of stock Tricaine in 25 ml of embryo medium at any time to ensure that larvae are not over-anaesthetised. Wait until the larvae are no longer swimming to ensure that they are fully anaesthetised. This will normally take 1–2 min.
2. Using a 3 ml Pasteur pipette, remove 20 anaesthetised larvae from the embryo medium and place on a piece of masking tape adhered to a clean Petri dish lid (see Note 4).
3. Remove excess liquid from the tape using the Pasteur pipette (other groups use a thin glass pipette) so that the larvae are gently resting on their side on the tape. This prevents them from being moved when the microsurgical knife comes into contact with the remaining liquid, allowing for more accurate transection. There will still be a small amount of liquid remaining keeping the larvae anaesthetised. Spread the larvae along the tape using the pipette so that they do not overlap.
4. Place the lid with the prepared larvae under bright field on a dissecting microscope and focus in on the caudal tail. Identify the circulatory loop, where the dorsal aorta turns back and becomes the axial vein. At 3dpf, in some strains of zebrafish, there is a gap in the pigment on the ventral side, beyond the end of the circulatory loop, that can be used to indicate the position of transection (Fig. 1). This pigment gap can be absent at different ages or in different strains; therefore, identification of the circulatory loop is more reliable as a marker.

5. Using a microsurgical knife, transect the tail in a single cut from dorsal to ventral side. Remove the section of fin immediately posterior to the circulation loop for consistent injury between fish. For a more minor injury, use a pair of forceps to gently pinch the caudal fin and release without tearing any tissue away. This will recruit less leukocytes to the site of injury, which may be more suitable for studying individual cells. Transect tails of the remaining larvae on the tape.
6. Tilt the Petri dish lid with the now tailfin-transected larvae over a Petri dish containing fresh embryo medium, without anaesthetic, and gently wash fresh medium over the tape to dislodge the larvae. Allow larvae to recover at 28°C. Continue with the remaining anaesthetised larvae from step 2.

3.2. Stimulation of Neutrophil Chemotaxis by Addition of Chemoattractant

3.2.1. Local Administration

1. Immerse the scalpel blade into the working concentration of chemoattractant (see Note 5).
2. Transect the zebrafish tail as described above (see Subheading 3.1) with the chemoattractant on the surgical blade.
3. Repeat the immersion of the scalpel blade between each tail transection (see Note 6).

3.2.2. Immersion

1. Transect the zebrafish tail as described above (see Subheading 3.1).
2. Immediately transfer the larvae into embryo medium with a suitable concentration of chemoattractant in solution. Be sure to ensure that the chemoattractant is both fully in solution and well mixed.

3.3. Neutrophil Identification Using DAB: Postmortem Stain

1. Place up to 50 anaesthetised larvae in a 1.5 ml microcentrifuge tube and remove excess liquid. Add 1 ml of 4% PFA in PBS and leave the 1.5 ml microcentrifuge tube horizontally overnight at 4°C. This ensures that all larvae fix in a straight position rather than curled from being in the bottom of the microcentrifuge tube. The minimum fixation time is 30 min at room temperature.
2. Remove the fixative and wash 1×5 min at room temperature (22°C) in PBS/0.1% Triton X-100 to remove any remaining fixative and place on a rocking table.
3. Add 1 ml of PBS/0.1% Triton X-100 with 0.5 mg/ml diaminobenzadine and 0.0003% hydrogen peroxide to the fixed samples. Stain at room temperature for 10 min.
4. Remove stain and wash for 3×5 min in PBS/Triton at room temperature (22°C) on a rocking table.
5. Visualise the stain on a bright-field microscope (see Note 7).

3.4. Neutrophil Identification Using Tyramide Signal Amplification: Postmortem Stain

1. Place up to 50 anaesthetised larvae in a 1.5 ml microcentrifuge tube and remove excess liquid. Add 1 ml of 4% PFA in PBS and leave the tube on its side overnight at 4°C. This ensures that all larvae fix in a straight position. The minimum fixation time is 30 min at room temperature.
2. Remove the fixative and wash thoroughly for 3 × 10 min in 1× PBS at room temperature (22°C) on a rocking table to ensure that no residual PFA remains.
3. Briefly wash the sample by adding 50 μl of amplification diluent provided in the TSA Plus kit to each 1.5 ml tube.
4. Add fresh amplification diluent to the sample containing 1:50 dilution of cyanine-3 amplification reagent (see Note 8). Place at 28°C for 10 min covered in foil.
5. Remove stain and wash in 1 × PBS/0.1% Tween20 for 3 × 10 min at room temperature (22°C) on a rocking table. Amplification reagents can cause larvae to stick together, so adding Tween to PBS helps separate them without damage. Keep samples in foil throughout washes to prevent bleaching of the fluorophore.
6. Visualise on a suitable fluorescence microscope.

3.5. Macrophage Identification Using Neutral Red: Vital Dye

1. Add neutral red at a final concentration of 2.5 μg/ml directly to zebrafish larvae medium in a Petri dish or multiwell plate.
2. Keep at 28°C in the dark for 4–8 h. This allows macrophages to pinocytose enough neutral red to visualise under a dissecting microscope.
3. Once macrophages are clearly red, remove excess neutral red solution using a Pasteur pipette, and wash 3 × 5 min with fresh embryo medium at room temperature (22°C) on the bench.
4. Using a dissecting microscope, visualise red cells with brightfield light.

3.6. Fluorescence Microscopy

1. Anaesthetise the zebrafish larvae using 0.017% Tricaine in embryo medium.
2. The larvae should lie on their side on the bottom of the dish to enable the site of injury to be accessed clearly. If the larvae are not lying on their side, then nudge them gently using a blunt pointer until they are.
3. Turn on the UV light source of the dissecting microscope, ensuring that the larvae are viewed through a suitable filter for the fluorophore used, in this case GFP. When zebrafish larvae are not viewed, close the shutter of the UV source in order that the fluorescent protein does not get bleached.
4. Neutrophils can be visualised by their bright green fluorescence.

5. For imaging at higher power on an inverted microscope, the larvae either must be kept very still, or for higher stability can be mounted in low melting point agarose (see Subheading 3.7).

3.7. Spinning disc Confocal Microscopy Mounting

1. Anaesthetise the zebrafish larvae using 0.017% Tricaine in embryo medium.
2. Place a 40-mm diameter Petri dish with a 12-mm diameter drilled hole in the centre upside down on the bench (see Note 9). A 25-mm diameter circular coverslip is then fixed over the hole using high vacuum grease. Excess vacuum grease should be carefully wiped off using tissue to stop contamination of microscope equipment.
3. Flip the Petri dish upright again, transfer the larvae onto the attached coverslip, and remove as much liquid as possible.
4. Cover the larvae with approximately 1 ml of preheated (and cooled to 37°C by keeping in a water bath until ready to mount larvae) 1% low melting point agarose containing 0.017% Tricaine to keep the larvae anaesthetised (see Note 10).
5. Manipulate the larvae so that the tails are flat to the coverslip using the eyebrow hair. This must be done immediately and very quickly, before the agarose sets (see Note 11). Depending on the temperature of the room and the starting temperature of the agarose, the setting time may differ. We find that a maximum of approximately 20 larvae can be successfully manipulated by an experienced operator in the normal setting time of 1% low melting point agarose.
6. Once the agarose has set, put some embryo medium without methylene blue in the 40-mm Petri dish, containing 0.017% Tricaine to ensure that the agarose does not dry out. The grease prevents any leaking.
7. Carefully load the 40-mm Petri dish onto the stage of the spinning disc microscope.
8. Image the GFP of the *Tg(mpx:GFP)i114* neutrophils using the 488-nm laser line and a 10–40× lens. Capture a *z*-stack through the depth of the tail to build up a high quality extended focus image.
9. To make a time-lapse video, image the tail every 2–5 min to obtain images for automated tracking (see Subheading 3.8).

3.8. Cell Tracking

1. Perform tail transection as described above (see Subheading 3.1).
2. Mount larvae in agarose as described above (see Subheading 3.7).
3. Capture a confocal *z*-series obtained every 2–5 min to produce a time-lapse series (see Subheading 3.7).

4. Load the time-lapse images acquired from the confocal microscope onto the Volocity Quantitation™.
5. Use the cell tracking function of software such as Volocity™ to track the neutrophils.
6. The following parameters successfully track neutrophils in the *Tg(mpx:GFP)i114* line at 10 or 20 times magnification in Volocity™:
   - "Find objects by intensity" low – 15%, high – 100%.
   - "Exclude objects by size" >35,000 μM^3.
   - "Exclude objects by size" <1,000 μM^3.
   - "Fill holes in objects."
   - "Track objects", use the shortest path with a maximum distance of 50 μM.
   - "Filter with time" <10 frames.
7. Using these settings, create a "measurement item" using all time points.
8. In the analysis tab of the measurement item, migration behaviours such as neutrophil speed and meandering index can be extracted.

3.9. Timecourse Counting

3.9.1. Pre-swim Bladder Inflation (<4dpf)

1. Remove the zebrafish larva from the Petri dish or well using a 3 ml Pasteur pipette and place in a new Petri dish containing embryo medium and Tricaine at 0.017%. The larva should lie on its side on the bottom of the dish to enable the site of injury to be seen clearly.
2. To count cells at the site of injury, use a fluorescent microscope with 80× magnification. Firstly, focus in on the tail injury using bright-field illumination. Identify the location of the circulatory loop, where the dorsal aorta turns back on itself to become the axial vein. It is convenient and reproducible if only leucocytes posterior to this loop are considered to be involved in the inflammatory response.
3. Select the appropriate filter for fluorescent imaging and allow fluorescent light to be transmitted at the same time as the bright-field light. Gradually turn the bright-field light off, enabling the fluorescent cells to be seen at the same time. Pinpoint a fluorescent cell immediately after the circulatory loop. A mental line can be drawn dorsal to ventral, parallel to the transection injury. Count all fluorescent cells between the circulatory loop and the point of injury.
4. The larva should be returned to fresh embryo medium, without anaesthetic, to recover for future time points. Maintain the larvae at 28°C until the next time point.

3.9.2. Post-swim Bladder Inflation (>4dpf)

1. Remove the zebrafish larva from the Petri dish or well using a 3 ml Pasteur pipette and place in a new Petri dish containing embryo medium and Tricaine at 0.017%. Once the swim bladder is inflated, the larva will float, making visualisation of the site of injury more difficult.
2. Once anaesthetised, remove larva in a drop of medium to a pre-wet (using embryo medium), but empty Petri dish.
3. Remove the excess liquid just until the larva lies on its side and the fluorescent cells at the site of injury are clearly identifiable.
4. To count cells at the site of injury, use a fluorescent microscope with 80× magnification. Firstly, focus in on the tail injury using bright-field illumination. Identify the location of the circulatory loop, where the dorsal aorta turns back on itself to become the axial vein. It is convenient and reproducible if only leucocytes posterior to this loop are considered to be involved in the inflammatory response.
5. Select the appropriate filter for fluorescent imaging and allow fluorescent light to be transmitted at the same time as the bright-field light. Gradually turn the bright-field light off, enabling the fluorescent cells to be seen at the same time. Pinpoint a fluorescent cell immediately after the circulatory loop. A mental line can be drawn dorsal to ventral, parallel to the transection injury. Count all fluorescent cells between the circulatory loop and the point of injury.
6. Add a drop of fresh embryo medium to the larva once the count is complete and return to the dish or well with fresh embryo medium, without anaesthetic. Maintain at 28°C until the next time point. This allows data for a complete timecourse of inflammation to be obtained (Fig. 3).

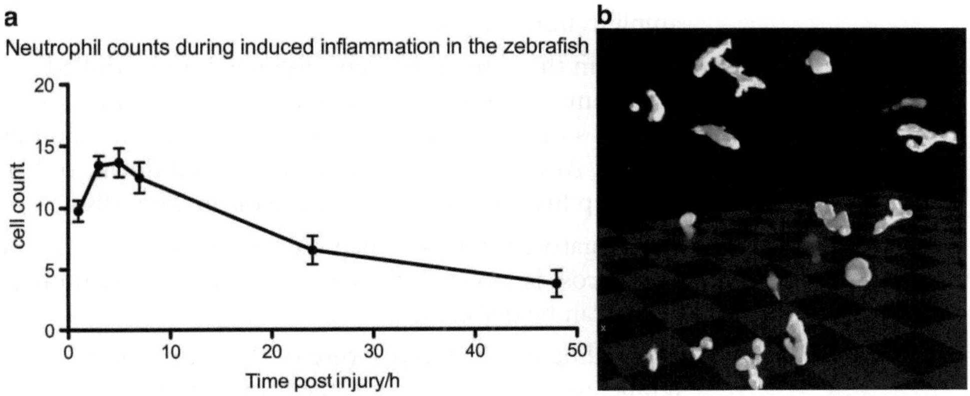

Fig. 3. (**a**) Neutrophil numbers at the site of injury following complete tail transection throughout a 48-h inflammatory timecourse. (**b**) High-power image of neutrophils in the *Tg(mpx:GFP)i114* transgenic zebrafish after migration to a site of inflammation.

4. Notes

1. This particular microsurgical knife is very small and allows for precise injury without further damage to the larva. Larger blades can cause a larger and more variable injury to the larva.
2. For neutrophils, *Tg(mpx:GFP)i114* or similar is recommended, as this line has high specificity for neutrophils, and labels a high proportion of neutrophils. Less specific lines exist. For macrophages, unpublished lines exist which label this lineage (15, 16).
3. Attach the hair to the pipette tip using a small piece of tape as this is more delicate than a metal seeker.
4. Autoclave tape will repel the medium, so try to use masking tape or a similar non-repellant tape. Masking tape is ideal as it allows light to be transmitted, enabling visualisation of the zebrafish larva.
5. Some chemoattractants are more viscous than other. Make sure to remove excess from the blade by gently scraping against the side of the tube after immersion.
6. It is difficult to avoid contamination of working solution with cells from previous transections; therefore, use a different aliquot of chemoattractant per experiment.
7. DAB is converted to an orange-brown substrate by peroxidases; therefore, neutrophils are clearly visible at sites of inflammation. This stain is mostly suited for thin tissue staining and ideal for looking in zebrafish fins where there is little pigment.
8. A total of 50 larvae in a 1.5 ml tube can be sufficiently covered with 49 μl of amplification diluent and 1 μl cyanine amplification reagent.
9. A hole in the base of the Petri dish needs to be drilled so that a 25-mm coverslip fits over it with approximately 2–5 mm overlap so that it can be sealed. We use a 12-mm drill bit to do this. Any excess plastic needs to be filed down so that the coverslip fits snug to the underside of the Petri dish.
10. If the agarose is too hot, then the tails of the larvae curl up. If the agarose is too cold, then it sets before manipulation of the larvae can be performed.
11. Depending on the temperature of the room and the starting temperature of the agarose, the setting time may differ. We find that a maximum of approximately 20 larvae can be successfully manipulated in the normal setting time of 1% low melting point agarose.

Acknowledgements

SAR and CAL are funded by an MRC Senior Clinical Fellowship to SAR (reference number: G0701932). PME is funded by a project grant from the Wellcome Trust (reference number: WT082909MA). Microscopy studies in our laboratory are supported by a Wellcome Trust grant to the MBB/BMS Light microscopy facility (GR077544AIA), and the laboratory is supported by an MRC Centre grant (G0700091).

References

1. Lieschke, G.J., Oates, A.C., Crowhurst, M.O., Ward, A.C., Layton, J.E. (2001) Morphologic and functional characterization of granulocytes and macrophages in embryonic and adult zebrafish. *Blood* **98**, 3087–96.
2. Lieschke, G.J., Currie, P.D. (2007) Animal models of human disease: zebrafish swim into view. *Nat Rev Genet* **8**, 353–67.
3. Trede, N.S., Langenau, D.M., Traver, D., Look, A.T., Zon, L.I. (2004) The use of zebrafish to understand immunity. *Immunity* **20**, 367–79.
4. Carradice, D., Lieschke, G.J. (2008) Zebrafish in hematology: sushi or science? *Blood* **111**, 3331–42.
5. Renshaw, S.A., Loynes, C.A., Trushell, D.M., Elworthy, S., Ingham, P.W., Whyte, M.K. (2006) A transgenic zebrafish model of neutrophilic inflammation. *Blood* **108**, 3976–8.
6. Nasevicius, A., Ekker, S.C. (2000) Effective targeted gene 'knockdown' in zebrafish. *Nat Genet* **26**, 216–20.
7. Savill, J.S., Wyllie, A.H., Henson, J.E., Walport, M.J., Henson, P.M., Haslett, C. (1989) Macrophage phagocytosis of aging neutrophils in inflammation. Programmed cell death in the neutrophil leads to its recognition by macrophages. *J Clin Invest* **83**, 865–75.
8. Martin, R.T., Bartman, T. (2009) Analysis of heart valve development in larval zebrafish. *Dev Dyn* **238**, 1796–802.
9. Brown, S.B., Tucker, C.S., Ford, C., Lee, Y., Dunbar, D.R., Mullins, J.J. (2007) Class III antiarrhythmic methanesulfonanilides inhibit leukocyte recruitment in zebrafish. *J Leukoc Biol* **82**, 79–84.
10. Redd, M.J., Cooper, L., Wood, W., Stramer, B., Martin, P. (2004) Wound healing and inflammation: embryos reveal the way to perfect repair. *Philos Trans R Soc Lond B Biol Sci* **359**, 777–84.
11. Herbomel, P., Thisse, B., Thisse, C. (2001) Zebrafish early macrophages colonize cephalic mesenchyme and developing brain, retina, and epidermis through a M-CSF receptor-dependent invasive process. *Dev Biol* **238**, 274–88.
12. Loynes, C.A., Martin, J.S., Robertson, A., Trushell, D.M., Ingham, P.W., Whyte, M.K., Renshaw, S.A. Pivotal Advance: Pharmacological manipulation of inflammation resolution during spontaneously resolving tissue neutrophilia in the zebrafish. *J Leukoc Biol* **87**, 203–12.
13. Elks, P.M., van Eeden, F.J., Dixon, G., Wang, X., Reyes-Aldasoro, C.C., Ingham, P.W., Whyte, M.K.B., Walmsley, S.R., and Renshaw, S.A. (2011) Activation of Hif-1alpha delays inflammation resolution by reducing neutrophil apoptosis and reverse migration in a zebrafish inflammation model. *Blood.* In press.
14. Nusslein-Volhard C, D.R. (2002) *Zebrafish: A Practical Approach.* Oxford University Press, Oxford.
15. Ellett, F., Pase, L., Hayman, J.W., Andrianopoulos, A., and Lieschke, G.J. (2010) mpegl promoter transgenes direct macrophage-lineage expression in zebrafish. *Blood.* In press.
16. Gray, C., Loynes, C.A., Whyte, M.K.B., Crossman, D.C., Renshaw, S.A., and Chico, T.J.A. (2011) Simultaneous intravital imaging of macrophage and neutrophil behaviour during inflammation using a novel transgenic zebrafish. Thrombosis and Haemostasis 105.

Chapter 19

Border Cell Migration: A Model System for Live Imaging and Genetic Analysis of Collective Cell Movement

Mohit Prasad, Xiaobo Wang, Li He, and Denise J. Montell

This volume includes electronic supplemental material that will appear on the website http://extras.springer.com/wells.

Abstract

Border cell migration in the *Drosophila* ovary has emerged as a genetically tractable model for studying collective cell movement. Over many years border cell migration was exclusively studied in fixed samples due to the inability to culture stage 9 egg chambers in vitro. Although culturing late stage egg chambers was long feasible, stage 9 egg chambers survived only briefly outside the female body. We identified culture conditions that support stage 9 egg chamber development and sustain complete migration of border cells ex vivo. This protocol enables one to compare the dynamics of egg chamber development in wild type and mutant egg chambers using time-lapse microscopy and taking advantage of a multiposition microscope with a motorized imaging stage. In addition, this protocol has been successfully used in combination with fluorescence resonance energy transfer biosensors, photo-activatable proteins, and pharmacological agents and can be used with widefield or confocal microscopes in either an upright or inverted configuration.

Key words: Border cell migration, *Drosophila* stage 9 egg chambers, Organ culture, Collective cell migration, Time-lapse live imaging

1. Introduction

Collective cell migration refers to the concerted movement of groups of cells. Unlike single moving cells such as fibroblasts or fish keratinocytes, collectively migrating cells maintain some level of adhesion among themselves during movement (1, 2). Though this kind of cellular movement is characteristic of several physiological processes during embryonic development (3), wound healing and tumor metastasis (1), it has been studied less extensively than the movements of single cells. Recently, a number of model systems have emerged for the study of collective movement

Claire M. Wells and Maddy Parsons (eds.), *Cell Migration: Developmental Methods and Protocols*,
Methods in Molecular Biology, vol. 769, DOI 10.1007/978-1-61779-207-6_19, © Springer Science+Business Media, LLC 2011

using the powerful combination of genetic manipulations and live imaging (4, 5). One of these, border cell migration in the *Drosophila* ovary, is the focus of this chapter.

Drosophila females bear a pair of ovaries within the abdomen (Fig. 1). Each ovary consists of 15–20 strings of egg chambers of increasing stages of maturity, called the ovarioles. At the tip of each ovariole resides the germarium, which contains germline and somatic stem cells and their immediate progeny. Egg chambers assemble in the germarium when somatic follicle cells surround a cyst of 16 interconnected germline cells, one of which develops into the oocyte while the other 15 differentiate as support

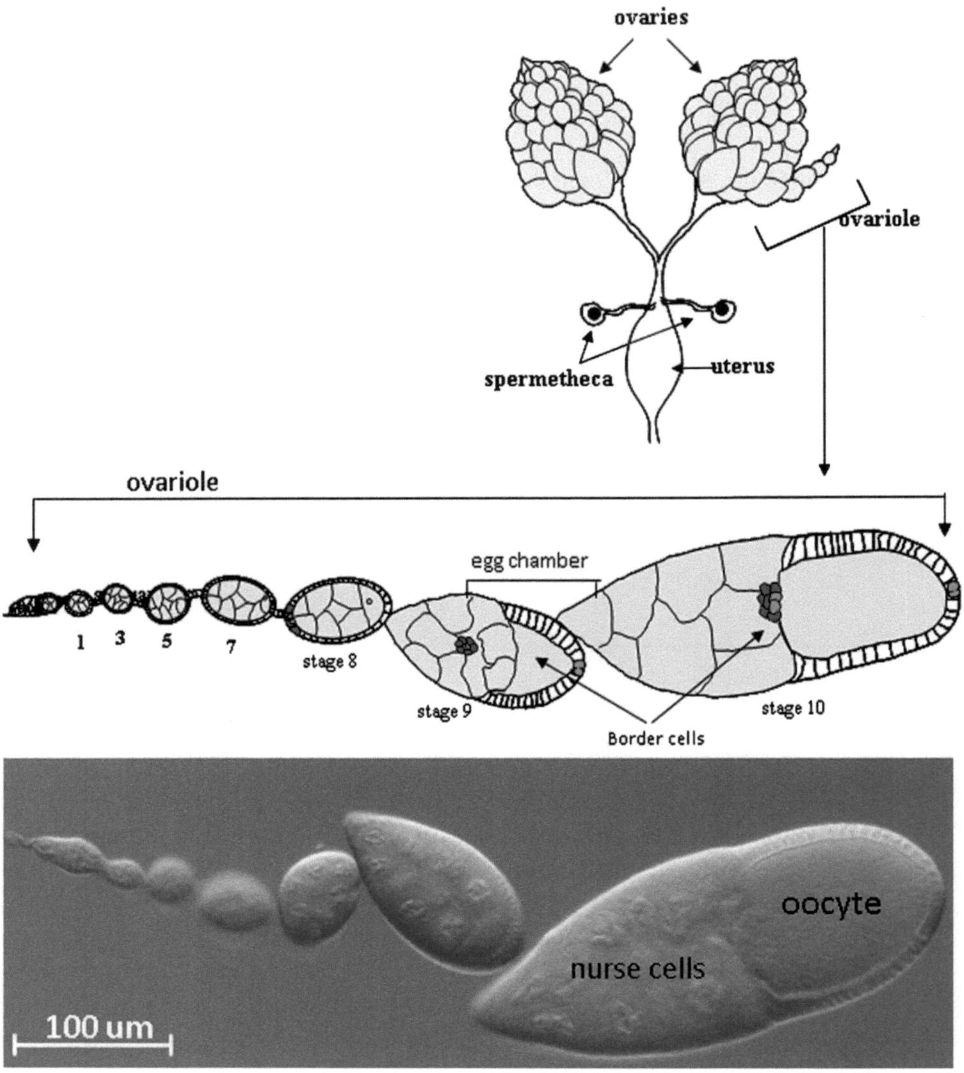

Fig. 1. Anatomy of the Drosophila ovary. *Top* – Schematic drawing of a pair of ovaries dissected from female fruit fly. A schematic drawing of an enlarged single ovariole containing egg chambers of the indicate stages of development. The *bottom panel* shows a DIC image of an ovariole with similar stages of egg chamber development.

cells called nurse cells (6). Egg chambers bud off of the germarium and then grow and progress through 14 developmental stages (7). Whereas germline cells do not divide further, follicle cells continue to undergo mitotic divisions until stage 6 when they switch to endoreplication without cytokinesis (8). During early oogenesis at each end of each egg chamber a pair of specialized follicle cells differentiates into the polar cells (9). The polar cells secrete a cytokine, Unpaired, which activates JAK-STAT signaling in nearby follicle cells (10). In late stage 8 and early stage 9, anterior follicle cells (4–6 in number) that perceive the highest level of JAK-STAT signal round up (11, 12). These cells are the border cells.

One or two of the cells extend protrusions in between the nurse cells (Movie 1). Some of these protrusions retract right away but sooner or later a protrusion attaches stably to the nurse cells and the border cell cluster detaches from the other follicle cells and from the basement membrane that surrounds the egg chamber (13, 14). The border cells migrate directly down the center of the egg chamber toward the oocyte, in response to secreted signals. One such signal is the PDGF- and VEGF-related factor 1 (PVF1), which binds to a receptor tyrosine kinase, PVR, expressed on the border cells (15, 16). PVR functions redundantly with epidermal growth factor receptor (EGFR) (15). Three ligands for the EGFR are expressed in the oocyte (16, 17), Spitz, Keren, and Gurken, all of which are TGFα homologs. Spitz and Keren mRNAs are distributed throughout the oocyte at stage 9 and these two ligands can redirect border cells when either one is misexpressed (16). Thus, these ligands seem to promote migration of the border cells to the oocyte. When the border cells get very near to the oocyte, they turn and move toward the dorsal side (17) (Movie 1). Grk mRNA and protein are restricted to the dorsal/anterior corner of the oocyte and promote the dorsal turn (17). It is unlikely that border cells sense Grk until they get near the oocyte because there is no dorsal bias to the migration before that point (16). Moreover, when Grk is expressed ectopically it is not sufficient to redirect border cells during the posterior migration (16).

The border cells cover a distance of approximately 150–200-μm in 4–6-h (13). Their migration speed is variable but may be a little faster in the beginning and a little slower near the end (14, 18). In the migrating cluster, individual border cells can change relative position within the group, while the polar cells remain in the center (13, 14). Until 2007, border cell migration was studied exclusively in fixed tissue due to the lack of suitable culture conditions for stage 9 egg chambers. Recently, we identified the culture conditions and subsequently optimized the imaging conditions for capturing the complete migration while minimizing phototoxicity (13) (Movie 1). This protocol has enabled more detailed phenotypic analysis and use of pharmacological agents, fluorescence resonance energy transfer (FRET) probes, and photoactivatable proteins (12, 19). In addition, this protocol can be used for studying other aspects of oogenesis including epithelial morphogenesis of follicle cells (20), RNA localization in the oocyte (21), and actin dynamics in nurse cells (22). Key features of the protocol are optimization of pH and addition of insulin, which may generally enhance cultures of *Drosophila* tissues including imaginal discs. Longer-term cultures (>6-h) may require perfusion systems to allow media exchange.

2. Materials

2.1. Culture Reagents and Equipment

1. *Drosophila* Schneider's medium (Invitrogen, cat. no. 11720-034) supplemented with 15% fetal bovine serum (FBS).
2. Insulin, a 10-mg/ml (Sigma, cat. no. I5550) stock solution was prepared in acidified water (see Note 1) and used to supplement the culture media to a final concentration of 0.2-mg/ml.
3. (Optional) FM4-64 (Molecular Probes, cat. no. T-3166) lipophilic dye dissolved in water or DMSO (see Note 2) used for labeling plasma membranes of all cells.
4. A fluorescent reporter that marks the border cells, e.g., *slbo*-GAL4;UAS-mCD8GFP (23, 24) or *slbo*-GAL4 UAS-MoesinGFP (13).
5. Streptomycin/penicillin (Invitrogen, cat. no. 15140-122) 10,000-U/ml of penicillin G–sodium, 10,000-mg/ml streptomycin sulfate in 0.85% saline.
6. Greiner Lumox culture dish hydrophilic (50-mm) (Sigma, cat. no. Z376744).
7. Fisher brand no. 1 coverslips (Fisher, cat. no. 12-542-B).
8. Two pairs of very fine surgical forceps (FST-5 biologie) and a dissecting microscope.
9. Depression glass slides (Fisher Scientific, cat. no. 12 565B).
10. Halocarbon oil 27 (Sigma, H 8773).
11. Poly-D-lysine (Sigma, P1024), 10-mg/ml in PBS.
12. Dry baker's Yeast dissolved in MilliQ water to form a thick paste.
13. Heat-filter KG1 (Chroma technologies).
14. BG38 IR suppression filter (Chroma technologies).
15. Neutral-density filters (ND 0.3) (Chroma technologies).
16. Widefield microscope. We use a Zeiss Axio Imager upright epifluorescence microscope and Plan-Apochromat 20×/N.A.

0.8 air objective. Illumination source is X-Cite 120 metal halide lamp. Filter sets: BP470/40, FT495, BP525/50 for Alexa 488/GFP; BP550/25, FT570, BP605/70 for Alexa 568/DsRed.

17. Confocal Microscope. We use a Zeiss 510-Meta inverted confocal microscope and Plan-Apochromat 63×/N.A. 1.4 oil objective with argon (488-nm) and HeNe (542-nm) lasers. Filter sets are FITC and TRITC.
18. Two pairs of forceps (Dumostar forceps #5).

2.2. Preparation of Culture Media

1. *Drosophila* S2 media was supplemented with 15% FBS and 0.6× streptomycin/penicillin. It was sterilized by passing through 0.2-μm filter.
2. The final pH of the media was adjusted to 6.95–7.00. Verify the pH of the media before every use (see Note 3).
3. Supplement with insulin to a final concentration of 0.2-mg/ml. Henceforth, the media with all the supplements would be referred as complete media.
4. (Optional) FM4-64 dye can be used at a final concentration of 9-μM to stain plasma membranes of all cells.

3. Method

3.1. Preparing Poly-D-Lysine Coated Coverslip

Five to ten Fisher brand No. 1 coverslips (22×22-mm), washed with milliQ water and dried. Layer the coverslip evenly with approximately 200-μl of 2-mg/ml of poly-D-lysine in PBS and dry it completely in a 37°C oven. Wash the coverslip couple of times in flowing tap water and subsequently rinse with MilliQ water and then dry.

3.2. Fattening Females Flies

Five to ten females (2–4 days old) of the desired genotype along with three to four males are transferred into a new fly vial with fresh fly food, a small amount of yeast paste and incubated at 25°C for 14–18-h (see Note 4).

3.3. Dissection of Ovaries to Egg Chambers

1. Immobilize 2–3 females flies either by exposure to CO_2 or anesthetic ether.
2. Fill the cavity of the depression slide with complete culture media and place the slide under the dissecting microscope.
3. Transfer one immobile fly to the cavity slide.
4. Hold the thorax or upper abdomen of the immobile fly using a pair of forceps, submerge it in the medium, while with another pair of forceps grasp and peel off a bit of abdominal cuticle at the middle of the abdomen (Fig. 2a).

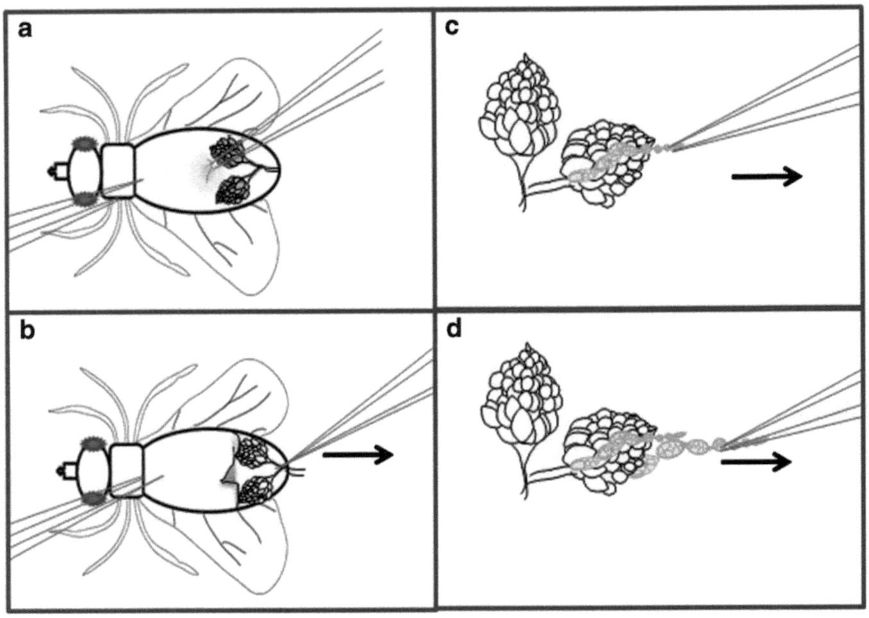

Fig. 2. Ovary dissection technique. Schematic representation of egg chamber dissection. (**a**) The female fly should be immobilized on its back using the left forceps. The right forceps are used to pinch the soft cuticle of the ventral side of the abdomen. (**b**) Pull the cuticle toward the right (*arrow*), revealing the ovaries. Grasp the common oviduct with the right forceps and pull to the right (*arrow*) to free the ovaries from the carcass. (**c**) Immobilize the ovary pair by pinching the oviduct using the left forceps. Use the right forceps to grasp the tip of an ovariole and slowly pull to the right (*arrow*). (**d**) Repeat until multiple ovarioles have been freed from the sheath.

5. This reveals a large pair of ovaries, which are white and opaque. Grasp the base of the ovaries (the common oviduct) and pull them away from the rest of the fly and into the medium (Fig. 2b). Remove the carcass using the forceps and discard it on a Kimwipe. Keep the ovaries covered in medium at all times.
6. Repeat this process for all female flies, accumulating the ovaries in the complete medium in the depression slide itself. Be careful to handle the ovaries very gently (see Note 5).
7. With one pair of forceps, hold the posterior part of a single ovary. Use another pair of forceps to gently grasp the anterior tip of the ovary (Fig. 2c). Older egg chambers reside at the posterior while the germarium and early stage egg chambers occupy the anterior tip.
8. Gently pull the ovarioles out of the muscular sheath. As you pull, you will observe strings of egg chambers pop out of the muscular sheath and sometimes you will also get individual egg chambers (Fig. 2c, d). Repeat this action until you see five to ten stage 9 egg chambers. Never touch the stage 9 egg chambers directly with the forceps! You can use early or very late stage egg chambers to manipulate the position of stage 9 chambers (see Note 5).

9. Use a plastic transfer pipette to transfer small number of egg chambers to four separate locations on a poly-D-lysine-coated glass coverslip. The distance between the locations should be large enough so that during multiposition imaging only the egg chamber being imaged is in the field of view (see Note 6).
10. As the egg chambers are settling down, use the forceps/transfer pipette to gently arrange them so that they are evenly spread on the coverslip and not too close together (see Note 7). Manipulate the stage 9 egg chambers by holding/touching late stage or early stage egg chambers. Allow a minute for the egg chambers to adhere to the coverslip.
11. In the meantime, break a 22×22-mm coverslip into two halves. Align them approximately 1-cm apart on a lumox dish with help of paint brush. Transfer approximately 60–80-μl of the complete media underneath and between the broken coverslip. Remove excess media from underneath the broken coverslip if they start to float.
12. With the help of the forceps very slowly flip over the coverslip so that the egg chambers face downward. Then gently lower this coverslip to form a bridge between the two coverslip pieces such that the egg chambers face the lumox dish membrane (see Note 8).
13. Gently remove excess media from around the coverslips, taking care not to compress the egg chambers.
14. Seal the edges of the coverslips with a thin layer of halocarbon oil 27 to minimize evaporation of the media during imaging (see Note 9).
15. For the upright microscope, the glass spacers and coverslips are mounted on the top of lumox dish; for the inverted microscope, the sample is mounted on the bottom of the dish.

3.4. Time-Lapse Imaging of the Egg Chambers

1. Transfer the lumox dish to a microscope stage equipped with a petri-dish holder.
2. Set up the center of the lumox dish as a reference point. Move to one of the locations and identify an egg chamber of the desired stage and mark its coordinates. Repeat this step for all the other locations avoiding egg chambers that have abnormal morphology. In addition avoid choosing egg chambers that are located near a germarium. The inherent pulsating movement of germarium might move the egg chamber during imaging.
3. Lower the light intensity using 25% Neutral Density filter. Include KG1 and BG38 filters to suppress heating of the sample (see Note 10). When imaging more than one egg chamber at a time, mark the position of the center of each border cell cluster and identify the correct exposure for each channel

individually for each egg chamber. Try not to exceed 150-ms for each channel. With this exposure setting one can collect 16-z sections, 1.25-μm apart at 20× magnification for each channel. Egg chambers generally tolerate a time interval of 2.25-min between successive frames when imaging four egg chambers with 16-z sections for each of two channels.

4. Start the time-lapse image capture. Use modest speed of the mobile stage. We have successfully used Axiocam MRm camera mounted on a Zeiss Axioimager (upright, widefield) microscope with 20× magnification for 5–6-h. Higher magnification like 40× can be used for shorter total imaging times.
5. Between the acquisitions, it is worthwhile to refocus on the border cells as they move. In addition, look for signs of normal development such as oocyte growth, outer follicle cell rearrangement, dynamic changes in gene expression (e.g., expression of the *slow border cells* enhancer in centripetal cells), and of course border cell movement.
6. The lumox dish can be reused. After the experiment, carefully remove the coverslips and wash away the halocarbon oil by rinsing several times with ethyl alcohol.
7. After imaging, one can process the images in different ways depending upon the available software. Movie 1 shows a maximal intensity projection of the 16-z slices over the entire time interval of acquisition.

4. Notes

1. Insulin powder dissolves in slightly acidic condition. For preparing acidic water, dilute 1-μl of concentrated hydrochloric acid to 1-ml with milliQ water.
2. If using DMSO for dissolving FM4-64, the final dilution of DMSO in complete medium should be 1:1,000. Higher concentrations of DMSO, impedes border cell migration.
3. The pH of the medium is critical for the experiment. Low pH impedes border cell migration and high pH leads to early degeneration of the egg chamber.
4. Age of the female fly is very important. Newly hatched flies do not fatten very well. While the ovaries of older fly have large number of mature egg chambers.
5. Gentle handling of the ovaries is critical as even inconspicuous damage to egg chambers inhibits border cell migration.
6. Optimum distance between the egg chambers of interest is desired so as to reduce phototoxicity.

7. Evenly spread out egg chambers with empty space between is preferable to accommodate the increase in size of the egg chambers during imaging.
8. While lowering the coverslip with attached egg chambers, one needs to be very careful that ends of coverslip should align on the bridge to avoid any damage to the egg chambers.
9. Do not use a Nail Polish.
10. Addition of heat-filter KG, infra red suppression filter (BG38), and 25% neutral density suppresses phototoxicity of the sample during long-term imaging.

References

1. Ilina, O., and Friedl, P. (2009) Mechanisms of collective cell migration at a glance. *J. Cell Sci.* **122**, 3203–8.
2. Aman, A., and Piotrowski, T. (2010) Cell migration during morphogenesis. *Dev. Biol.* **341**, 20–33.
3. Weijer, C.J. (2009) Collective cell migration in development. *J. Cell Sci.* **122**, 3215–23.
4. López-Schier, H. (2010) Fly fishing for collective cell migration. *Curr. Opin. Genet. Dev.* **20**, 428–32.
5. Rørth, P. (2009) Collective cell migration. *Annu. Rev. Cell Dev. Biol.* **25**, 407–29.
6. Spradling, A.C. (1993) Developmental Genetics of Oogenesis. In The Development of *Drosophila melanogaster*. M. Bate and A. Martinez-Arias, eds. (Cold Spring Harbor, N.Y.: Cold Spring Harbor Laboratory Press), pp. 1–70.
7. King, R.C. (1970) Ovarian Development in *Drosophila* melanogaster (Academic Press, New York).
8. Edgar, B.A., and Orr-Weaver, T.L. (2001) *Cell.* **105**, 297–306.
9. Ruohola, H., Bremer, K.A., Baker, D., Swedlow, J.R., Jan, L.Y., and Jan, Y.N. (1991) Role of neurogenic genes in establishment of follicle cell fate and oocyte polarity during oogenesis in *Drosophila*. *Cell.* **66**, 433–449.
10. Silver, D.L., and Montell, D.J. (2001) Paracrine signaling through the JAK/STAT pathway activates invasive behavior of ovarian epithelial cells in *Drosophila*. *Cell.* **107**, 831–41.
11. Montell, D.J., Rørth, P., and Spradling, A.C. (1992) Slow border cells, a locus required for a developmentally regulated cell migration during oogenesis, encodes *Drosophila* C/EBP. *Cell.* **71**, 51–62.
12. Starz-Gaiano, M., Melani, M., Wang, X., Meinhardt, H., and Montell, D.J. (2008) Feedback inhibition of Jak/STAT signaling by apontic is required to limit an invasive cell population. *Dev. Cell.* **14**, 726–38.
13. Prasad, M., and Montell, D. J. (2007) Cellular and Molecular Mechanisms of Border Cell Migration Analyzed Using Time-lapse Live-cell Imaging. *Dev. Cell.* **12**, 997–1005.
14. Bianco, A., Poukkula, M., Cliffe, A., Mathieu, J., Luque, C. M., Fulga, T. A., and Rorth, P. (2007) Two distinct modes of guidance signalling during collective migration of border cells. *Nature.* **448**, 362–65.
15. Duchek, P., Somogyi, K., Jekely, G., Beccari, S., and Rorth, P. (2001) Guidance of cell migration by the *drosophila* PDGF/VEGF receptor. *Cell.* **107**, 17–26.
16. McDonald, J. A., Pinheiro, E. M., Kadlec, L., Schupbach, T., and Montell, D. J. (2006) multiple EGFR ligands participate in guiding migrating border cells. *Dev. Biol.* **296**, 94–103.
17. Duchek, P., and Rorth, P. (2001) Guidance of cell migration by EGF receptor signaling during *Drosophila* oogenesis. *Science.* **291**, 131–33.
18. Prasad, M., Jang, A.C., Starz-Gaiano, M., Melani, M., and Montell, D.J. (2007) A protocol for culturing *Drosophila melanogaster* stage 9 egg chambers for live imaging. *Nat. Protoc.* **2**, 2467–73.
19. Wang, X., He, L., Wu, Y.I., Hahn, K.M., and Montell, D.J. (2010) Light-mediated activation reveals a key role for Rac in collective guidance of cell movement *in vivo*. *Nat. Cell Biol.* **12**, 591–7.
20. Horne-Badovinac, S., and Bilder, D. (2005) Mass transit: epithelial morphogenesis in the

Drosophila egg chamber. *Dev. Dyn.* **232**, 559–72.

21. Grunert, S. and St Johnston, D. (1996) RNA localization and the development of asymmetry during *Drosophila* oogenesis. *Curr. Opin. Genet. Dev.* **6**, 395–402.
22. Hudson, A.M., and Cooley, L. (2002) Understanding the function of actin-binding proteins through genetic analysis of *Drosophila* oogenesis. *Annu. Rev. Genet.* **36**, 455–88.
23. Rørth, P., Szabo, K., Bailey, A., Laverty, T., Rehm, J., Rubin, G.M., Weigmann, K., Milán, M, Benes, V., Ansorge, W., and Cohen, S.M. (1998) Systematic gain-of-function genetics in *Drosophila. Development.* **125**, 1049–57.
24. Wang, X., Bo, J., Bridges, T., Dugan, K. D., Pan, T. C., Chodosh, L. A., and Montell, D. J. (2006) Analysis of cell migration using whole-genome expression profiling of migratory cells in the *Drosophila* ovary. *Dev. Cell.* **10**, 483–95.

Chapter 20

Assessment of Development and Chemotaxis in *Dictyostelium discoideum* Mutants

Yulia Artemenko, Kristen F. Swaney, and Peter N. Devreotes

Abstract

Studies using the social amoeba *Dictyostelium discoideum* have greatly contributed to the current understanding of the signaling network that underlies chemotaxis. Since directed migration is essential for normal *D. discoideum* multicellular development, mutants with chemotactic impairments are likely to have abnormal developmental morphologies. We have used multicellular development as a readout in a screen of mutants to identify new potential regulators of chemotaxis. In this chapter, we describe how mutants generated by restriction enzyme-mediated integration (REMI) are analyzed, from assessment of development to detailed characterization of 3′,5′-cyclic adenosine monophosphate (cAMP)-induced responses. Two complementary approaches, plating cells either clonally on a bacterial lawn or as a population on non-nutrient agar, are used to evaluate multicellular development. Once mutants with aberrant developmental phenotypes are identified, their chemotaxis toward cAMP is assessed by both small population and micropipette assays. Furthermore, mutants are tested for defects in both general and specific signaling pathways by examining the recruitment of actin-binding $LimE_{\Delta coil}$ or PIP3-binding PH domains to the plasma membrane in response to cAMP stimulation.

Key words: Multicellular development, Differentiation, Under buffer assay, cAR1, Small population assay, Micropipette assay, CMFDA labeling, PIP3 dynamics, Actin

1. Introduction

Chemotaxis, or the directed migration of cells up a chemical gradient, is a conserved process that occurs in many organisms, ranging from bacteria to mammals. In humans, this fundamental process is essential for embryogenesis, wound healing, and the immune response. In addition, inappropriate migration contributes to the development and progression of a number of pathological conditions, such as tumor cell metastasis and chronic

Claire M. Wells and Maddy Parsons (eds.), *Cell Migration: Developmental Methods and Protocols*, Methods in Molecular Biology, vol. 769, DOI 10.1007/978-1-61779-207-6_20,

inflammation (reviewed in (1, 2)). A powerful model system for the study of chemotaxis is the social amoeba *Dictyostelium discoideum*. In this organism, chemotaxis is a vital process both at the vegetative growth stage and during a starvation-induced developmental program, during which single cells differentiate and aggregate to form a multicellular structure, the fruiting body, consisting of a stalk and a spore head.

Studies of *D. discoideum* have led to the identification of many chemotactic regulators. In differentiated cells, the predominant chemoattractant, 3′,5′-cyclic adenosine monophosphate (cAMP), binds to the G-protein-coupled cAMP receptor (cAR1) and activates several parallel pathways that trigger actin polymerization and pseudopod extension (reviewed in (3)). One key signaling molecule important for directed migration is phosphoinositide-3,4,5-trisphosphate (PIP3), produced from phosphoinositide-4,5-bisphosphate by phosphoinositide 3-kinase (PI3K) specifically at the leading edge (4, 5). PIP3 accumulation leads to the recruitment of downstream effectors, such as PKBA and cytosolic regulator of adenylyl cyclase (CRAC), that mediate actin polymerization and signal relay (6, 7). In addition, *D. discoideum* has separate PIP3-independent pathways, involving phospholipase A_2 and target of rapamycin (TOR) complex 2 (TORC2), that also contribute to actin polymerization and migration (8–10).

Despite recent insights into the network of chemotactic pathways, many signaling molecules remain unknown. One approach to identifying novel chemotactic regulators is to analyze a collection of mutant strains generated by restriction enzyme-mediated integration (REMI). In this method, a blasticidin resistance cassette is inserted at random into genomic restriction sites (e.g., DpnII) approximately once per cell, presumably disrupting the function of any gene(s) located near the insertion site (11, 12). Since directed migration is important for many stages in the *D. discoideum* developmental cycle, impaired chemotactic behavior often results in defective multicellular morphology, which provides a convenient readout that can readily be assessed in a high-throughput manner. Therefore, REMI mutants are first screened for developmental defects and are then assayed for chemotactic ability to determine whether the gene disrupted by blasticidin insertion is involved in directed migration.

In this chapter, we describe how a typical REMI mutant is analyzed, from the initial assessment of multicellular development to a detailed examination of chemotaxis. Many different approaches are available for monitoring chemotaxis, including the micropipette, small population, Zigmond chamber, under agar, and microfluidic assays. Here, we focus on the two techniques that we use most frequently in our laboratory, the small population and micropipette assays, which together assess many complementary aspects of chemotactic behavior. In addition, mutants are tested for their ability to generate PIP3 and polymerize actin in response to cAMP stimulation by monitoring the recruitment of PIP3-binding PH domains and filamentous actin-binding proteins to the plasma membrane.

2. Materials

2.1. Assessment of Developmental Phenotypes

2.1.1. Development on a Bacterial Lawn

1. HL5 medium: 10 g/L dextrose, 10 g/L proteose peptone, 5 g/L yeast extract, 0.965 g/L $Na_2HPO_4{\cdot}7H_2O$, and 0.485 g/L KH_2PO_4 in water (see Note 1). Unless otherwise indicated, HL5 medium also contains 0.03 g/L streptomycin. Autoclave the medium and store at room temperature.
2. 10 cm Petri dish.
3. SM/5 plates: 2 g/L dextrose, 2 g/L Bacto™ peptone, 0.2 g/L yeast extract, 2.31 g/L KH_2PO_4, 1 g/L K_2HPO_4, and 20 g/L agar in water. Autoclave the mixture, allow it to cool to approximately 60°C, and pour 30 ml of the agar solution per 10-cm Petri dish. Once the agar solidifies, store the plates at 4°C for up to 1 month.
4. SM plates: 10 g/L dextrose, 10 g/L Bacto™ peptone, 1 g/L yeast extract, 2.31 g/L KH_2PO_4, 1 g/L K_2HPO_4, and 20 g/L agar in water. Autoclave, cool, and pour the agar into plates as described in Item 3 above. Store the plates at 4°C for up to 1 month.
5. *Klebsiella aerogenes* on an SM plate: Prepare a suspension of *K. aerogenes* in HL5 medium without antibiotics, spread this onto an SM plate, and grow the bacteria at room temperature overnight for use the next day. These plates can also be stored at 4°C for up to 1 month after overnight growth.

2.1.2. Development on Non-nutrient Agar

1. 10× Phosphate buffer (PB): 13.4 g/L $Na_2HPO_4{\cdot}7H_2O$ and 6.8 g/L KH_2PO_4 in water. Store the 10× stock at 4°C and dilute it to 1× with water for use.
2. Development buffer (DB): Supplement 1× PB with 2 mM $MgSO_4$ and 0.2 mM $CaCl_2$ and store at 4°C.
3. DB agar: 1.5% (w/v) agar in DB. Autoclave the mixture and store at room temperature. Prior to use, reheat it in the microwave until the agar is dissolved.
4. 6-Well tissue culture plate or 35-mm tissue-culture dish.

2.2. Preparation of D. discoideum cells for Chemotactic Analysis: Cell Differentiation

1. DB (see Subheading 2.1.2).
2. cAMP: Prepare a 100 mM stock solution of cAMP sodium salt monohydrate in water. Use the 100 mM stock to prepare a 1 mM working stock in water. Store aliquots at –20°C.

These stock solutions can be freeze-thawed several times, although it is not recommended to reuse less concentrated solutions. To prevent cAMP degradation by environmental phosphodiesterases, surfaces that may come into contact with the solution should not be handled directly.

3. Peristaltic pump (Gilson Minipuls 3).
4. Manifold pump tubing (Fisherbrand; Accu-Rated PVC, ID 0.76 mm) connected on both ends by the tip of a 20-G 1.5 needle to 50 cm of 0.8-mm tubing (Saint Gobain Performance Plastics; Tygon R-3603).
5. Table-top timer.

2.3. Assessment of Differentiation

2.3.1. Under Buffer Assay

1. DB (see Subheading 2.1.2).
2. 6-Well tissue-culture plate or 35-mm tissue-culture dish.

2.3.2. Cell Lysis and Immunoblotting

1. 3× Sample buffer (such as the 3× sodium dodecyl sulfate (SDS) blue loading buffer): 187.5 mM Tris–HCl (pH 6.8), 6% (w/v) SDS, 30% glycerol, 125 mM DL-dithiothreitol, and 0.03% (w/v) bromophenol blue in water. Store aliquots at −20°C and dilute them to 1× in water for use (see Notes 2 and 3).
2. Criterion Tris–HCl 4–15% polyacrylamide gel.
3. Immun-Blot PVDF membrane (0.2 μm).
4. 20× TBST: 48 g/L Tris base, 160 g/L NaCl, and 2% Tween-20 in water. Adjust the pH to 7.6 and store the 20× stock at 4°C. Dilute the stock to 1× with water and store this at room temperature.
5. TBST/skim milk: 1× TBST with 5% (w/v) skim milk powder. Shake the mixture briefly until the milk powder is dissolved. Make a fresh solution prior to each use.
6. TBST/BSA: 1× TBST with 5% (w/v) bovine serum albumin (BSA) and 0.02% (w/v) NaN_3. Shake the mixture briefly until the BSA is dissolved and store at 4°C.
7. Anti-cAR1 primary antibody from rabbit.
8. Horseradish peroxidase (HRP)-conjugated anti-rabbit secondary antibody from donkey.
9. ECL Western blotting detection reagent.
10. Coomassie brilliant blue (CBB) staining solution: 40% (v/v) methanol, 10% (v/v) glacial acetic acid, and 0.1% (w/v) CBB in water. Store this at room temperature.
11. Destaining solution: 40% (v/v) methanol and 10% (v/v) glacial acetic acid in water. Store this at room temperature.

2.4. Analysis of Chemotaxis By Scoring or Video Acquisition of a Small Population Assay

2.4.1. Scoring Method for the Small Population Assay

1. 1× PB (see Subheading 2.1.2).
2. 10-cm Petri dish filled with noble agar: 1% (w/v) solution of Difco™ noble agar (BD Biosciences) in 1× PB. Boil the mixture until the agar is dissolved, allow it to cool slightly, and pour 10 ml of the agar solution per 10-cm Petri dish. Cover the plate and allow the agar to solidify for 30–60 min prior to use (see Note 4).
3. Drawn out Pasteur pipettes: Draw out 9″ glass Pasteur pipettes by holding them at each end so that the thin end of the pipette is over a flame. When the glass becomes pliable, pull the two ends apart to extend the thin portion of the pipette into a capillary. Cut off the very tip of the capillary using a razor blade to make a small opening.
4. 1×10^{-8} M, 1×10^{-7} M, and 1×10^{-6} M cAMP: Dilute the 1 mM cAMP stock solution (see Subheading 2.2) in 1× PB. Make fresh dilutions prior to each use.
5. Developed cells (see Subheading 3.2).

2.4.2. Video Acquisition of the Small Population Assay

1. 1× PB (see Subheading 2.1.2).
2. 1-well chambered coverglass filled with noble agar: 1% (w/v) solution of Difco™ noble agar (BD Biosciences) in 1× PB. Boil the mixture until the agar is dissolved, allow it to cool slightly, and pour 2 ml of the agar solution per 1-well Lab-Tek™ chambered coverglass (Thermo Scientific, Nunc). Allow the agar to solidify for 15–30 min prior to use (see Note 4).
3. 1×10^{-6} M cAMP: Dilute the 1 mM cAMP stock solution (see Subheading 2.2) in 1× PB.
4. Developed cells (see Subheading 3.2).

2.5. Analysis of Chemotaxis by a Micropipette Assay

2.5.1. Analysis of Unlabeled Cells

1. DB (see Subheading 2.1.2).
2. Femtotips.
3. Microloader tips.
4. FemtoJet microinjector (Eppendorf).
5. 1×10^{-5} M cAMP: Dilute the 1 mM cAMP stock solution (see Subheading 2.2) in DB that has been passed through a 0.22-μm filter.
6. 1-well Lab-Tek™ chambered cover glass (Thermo Scientific, Nunc).
7. Developed cells (see Subheading 3.2).

2.5.2. Simultaneous Analysis of Two Cell Lines Using CMFDA Labeling

1. 1× PB (see Subheading 2.1.2).
2. DMSO.

3. CellTracker™ Green 5-chloromethylfluorescein diacetate (CMFDA) (Molecular Probes): Prepare a 10 mM stock solution of CMFDA in DMSO. Store aliquots at −80°C.
4. Items 1–7 from Subheading 2.5.1.

2.6. Examination of PIP3 and Actin Dynamics in Chemotaxing Cells

2.6.1. Analysis of PIP3 and Actin Dynamics in Globally Stimulated Cells: Jumping Assay

1. DB (see Subheading 2.1.2).
2. 1×10^{-5} M cAMP: Dilute the 1 mM cAMP stock solution (see Subheading 2.2) in DB.
3. 8-well Lab-Tek™ chambered cover glass (Thermo Scientific, Nunc).
4. Developed cells (see Subheading 3.2) expressing GFP-tagged versions of either the PH-domain from CRAC (PH_{CRAC}) or a fragment of the filamentous actin-binding protein LimE ($LimE_{\Delta coil}$) for analysis of PIP3 or actin dynamics, respectively (5, 13).

2.6.2. Analysis of PIP3 and Actin Dynamics in a Gradient: Micropipette Assay

1. Developed cells (see Subheading 3.2) expressing GFP-tagged versions of either PH_{CRAC} or $LimE_{\Delta coil}$ (see Subheading 2.6.1).
2. Items 1–6 from Subheading 2.5.1.

3. Methods

3.1. Assessment of Developmental Phenotypes

In order to isolate strains with potential chemotactic defects from a population of REMI insertional mutants, cells are assessed for their ability to enter into the developmental program and form multicellular structures. One way to examine development is to clonally plate a population of REMI mutants on a bacterial lawn. Under these conditions, single vegetative cells proliferate as they feed on the bacteria, resulting in circular plaques that contain cells at every stage of the *D. discoideum* life cycle (reviewed in ref. 14). In the center of the plaque, where the bacterial food source has been completely depleted, the cells starve and initiate a developmental program in which they differentiate, aggregate, and ultimately form fruiting bodies (Fig. 1a). Further from the center of the plaque, cells undergo starvation later and, therefore, represent progressively earlier stages of the life cycle, including the mounding and streaming stages. At the very edge of the plaque, cells continue to feed on bacteria and remain in the vegetative state. Aberrant development leads to the enrichment of cells in particular stages of the life cycle and abnormal multicellular morphology, ranging from a complete inability to aggregate to an arrest at the streaming or mounding stages or to the formation of abnormal fruiting bodies. In addition, plaque morphology can be distinguished based on size, the shape of the border, and the extent to which the bacterial lawn is cleared by the cells before

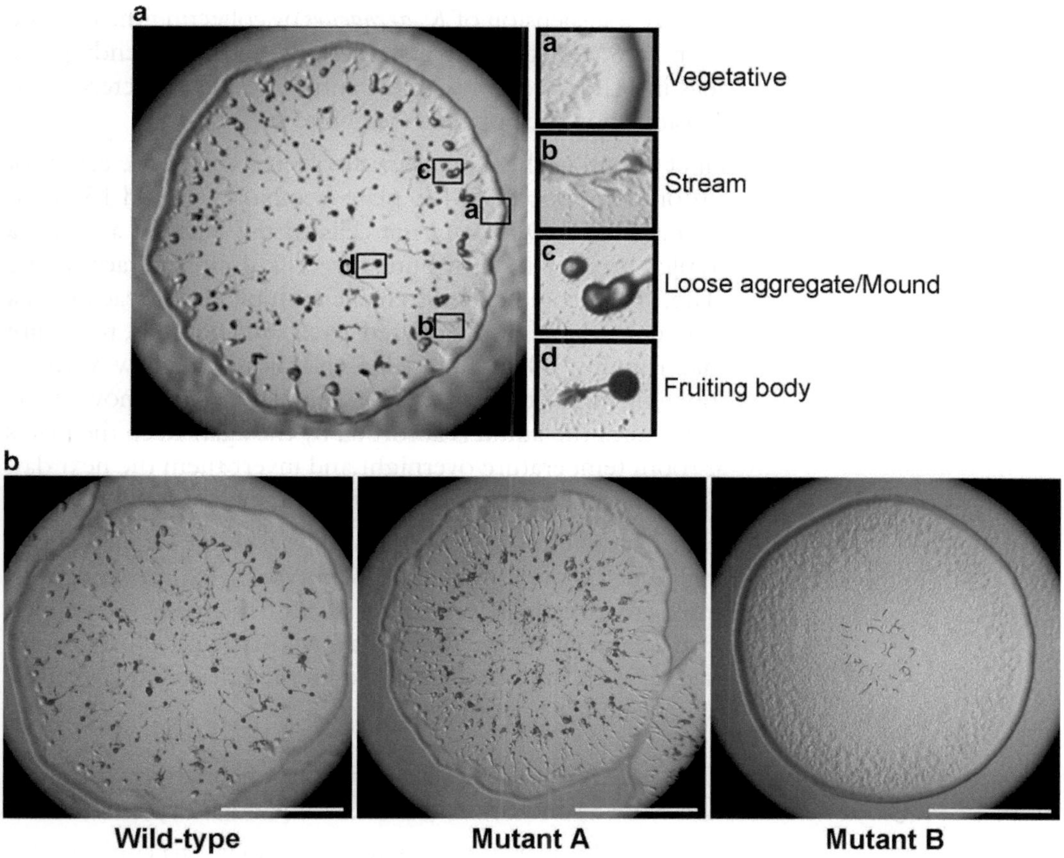

Fig. 1. *Dictyostelium discoideum* cells develop on a bacterial lawn. Single cells were plated clonally on a lawn of *K. aerogenes* and grown at room temperature for 4 days before the resulting plaques were photographed under a dissection microscope. (**a**) A representative wild-type plaque showing the various developmental stages of the *D. discoideum* life cycle. (**b**) Comparison of plaque morphology of wild-type and two mutant cell lines. Mutant A cells arrest at the streaming stage of development, whereas mutant B cells fail to aggregate completely. Scale bar, 5 mm.

they expand outward or undergo starvation. Figure 1b shows representative wild-type, stream-arrest, and aggregation-less plaques.

D. discoideum development can also be assessed by plating a population of cells on non-nutrient agar. In this assay, cells synchronously enter into the developmental program without proliferation, such that the entire population of cells represents the same stage of the life cycle at any given moment. This provides a useful tool by which mutant strains can be monitored over time as they progress through the developmental program.

3.1.1. Development on a Bacterial Lawn

1. Collect *D. discoideum* cells in the exponential growth phase (see Note 5) and count the cells using a hemocytometer. Dilute cell suspension to a final concentration of approximately 2 cells/μl in HL5 medium without antibiotics.

2. Prepare a suspension of *K. aerogenes* by collecting the bacteria from an SM plate using a sterile loop and resuspending the cells in HL5 medium without antibiotics. The bacterial suspension should be opaque (see Note 6).
3. Pipet 260 μl of the *K. aerogenes* suspension into the center of a room temperature SM/5 plate (see Note 7). Add 15–30 μl of the diluted *D. discoideum* cells to the bacteria and use a sterile spreader to distribute the cells uniformly across the plate. It is recommended to make multiple plates, each with a different volume of the diluted *D. discoideum* cells, to ensure that roughly 30–60 individual plaques will grow without touching each other on at least one plate. Avoid moving the plates until the liquid is absorbed by the agar. Keep the plates at room temperature overnight and invert them the next day. Continue to keep the plates at room temperature.
4. Four days after plating the cells, the resulting plaques can be analyzed for developmental phenotypes. Acquire images of representative plaques using a dissection microscope equipped with a camera. By the fourth day, wild-type cells form round plaques with large fruiting bodies in the center and vegetative cells around the periphery.

3.1.2. Development on Non-nutrient Agar

1. Collect *D. discoideum* cells in the exponential growth phase (see Note 5) and count the cells using a hemocytometer.
2. Melt the DB agar in the microwave, allow it to cool slightly, and pour 3 ml of the agar solution per well of a 6-well tissue-culture plate (or per 35-mm tissue-culture dish). Allow the agar to solidify for 1 h with the lid slightly off-set from the plate.
3. Centrifuge 2×10^7 cells in a 15-ml polypropylene tube at 360 rcf for 3–4 min. Discard the supernatant. To wash, resuspend the cell pellet in 15 ml of DB by gently vortexing and/or repeatedly pipetting the cells; repeat the centrifugation and discard the supernatant. Resuspend the resulting pellet in 1 ml of DB for a final density of 2×10^7 cells/ml.
4. Dilute the cells in 1 ml of DB, using at least two different densities for each cell line. Typically, densities of 1×10^6 cells/ml and 5×10^6 cells/ml are used; however, 1×10^7 cells/ml can also be used.
5. Plate 1 ml each of the diluted cells onto the solidified DB agar. Allow the cells to attach for approximately 10 min. Tilt the plate and aspirate the liquid. Keep the plate tilted and aspirate any remaining liquid after approximately 1 min. Leave the plate with an off-set lid until the cells acquire a "matte" finish, for approximately 20 min. Close the lid and leave the plate at room temperature until the cells reach the desired developmental stage.

6. Monitor the progression of the cells through the developmental program using a dissection microscope equipped with a camera to acquire images of the desired stage(s) of the life cycle. Wild-type cells typically begin streaming approximately 5 h after plating and form fruiting bodies within 24 h.

3.2. Preparation of D. discoideum Cells for Chemotactic Analysis: Cell Differentiation

Once mutants with aberrant developmental phenotypes have been identified, they are then assessed for defects in chemotaxis. Since vegetative cells are not responsive to cAMP, the cells must be differentiated prior to chemotactic analysis. This can be achieved by inducing the upregulation of a variety of developmental genes, including the cAMP receptor cAR1, through a combination of starvation and the addition of cAMP at regular intervals (15).

1. Collect *D. discoideum* cells in the exponential growth phase (see Note 5) and count the cells using a hemocytometer.
2. Centrifuge 8×10^7 cells in a 50-ml polypropylene tube at 360 rcf for 3–4 min. Discard the supernatant. To wash, resuspend the pellet in 50 ml of DB by gently vortexing and/or repeatedly pipetting the cells; repeat the centrifugation and discard the supernatant. Wash the pellet once more with 50 ml of DB. Resuspend the resulting pellet in 4 ml of DB for a final density of 2×10^7 cells/ml (see Note 8).
3. Transfer the cell suspension to a 50-ml Erlenmeyer flask (see Note 8). Initiate starvation by shaking the flask at 115 rpm on a rotary shaker for 1 h.
4. Connect one piece of tubing for each cell line to the peristaltic pump and wash the tubing with water or DB. Fill the tubing with a solution of 2 μM cAMP in DB, such that the final cAMP concentration is 50 nM when 100 μl of the cAMP solution is added to the cell suspension (see Note 9). Without turning on the pump, suspend one end of the tubing above each flask of cells and secure it with tape. Keep the other end of the tubing submersed in the cAMP solution.
5. Connect the peristaltic pump to a timer so that the pump will turn on for 5 s at 6-min intervals. The pump should be calibrated to release 100 μl of the cAMP solution with each pulse (every 6 min).
6. After 1 h of starvation, turn on the peristaltic pump to begin pulsing the cells with cAMP. Continue to shake the cells at 115 rpm on a rotary shaker. Wild-type cells are typically differentiated and ready for analysis after 4 h of cAMP pulsing (or 5 h of starvation).

3.3. Assessment of Differentiation

This section describes two assays for determining the extent of cell differentiation (see Subheading 3.2) prior to conducting assays of chemotaxis. Because cells are only competent to respond to cAMP

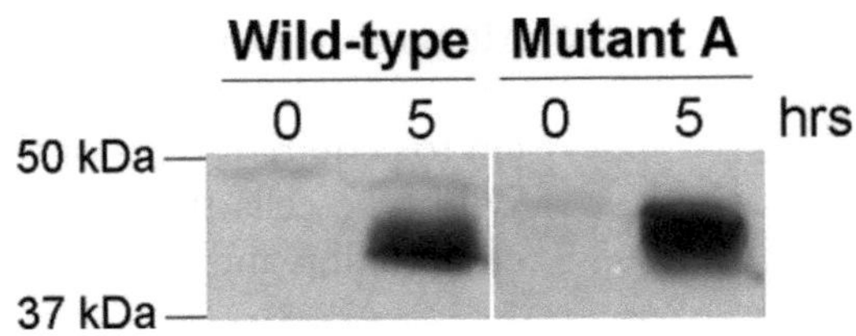

Fig. 2. cAMP receptor (cAR1) protein levels increase during the course of cell differentiation. An equal number of wild-type or mutant A cells were collected and lysed at 0 or 5 h after development was initiated. Proteins from approximately 1.5×10^6 cells were separated by SDS-PAGE and immunoblotted using an anti-cAR1 antibody. No cAR1 protein is observed for either cell line before development (0 h). However, a band corresponding to the endogenous cAR1 protein is observed at comparable levels for both wild-type and mutant A cells 5 h after development.

when they are differentiated, variability in the state of differentiation across cell lines may affect the results of chemotactic analyses. The "under buffer" assay is a method that allows for the rapid assessment of cell differentiation. Fully differentiated cells acquire an elongated and polarized morphology, form streams, and eventually aggregate when plated on a tissue-culture dish filled with buffer. Observing this behavior serves as a good indication that cells are differentiated and ready for chemotactic analysis.

Another method of assessing differentiation is to examine the expression of genes, such as cAR1, that are upregulated during development. To do this, cells are lysed and assessed for cAR1 expression, using protein levels as a readout, both before and after differentiation. Figure 2 shows a representative immunoblot of endogenous cAR1 protein levels for the wild-type and mutant A cell lines depicted in Fig. 1b. Wild-type cells show a dramatic increase in cAR1 protein levels after 5 h of starvation. Although mutant A has aberrant development on a bacterial lawn, these cells are capable of differentiating, as indicated by wild-type-like cAR1 levels, in response to pulses of exogenous cAMP. This suggests that any impairments observed for this mutant strain during chemotactic analysis will not be caused by defects in cell differentiation.

3.3.1. Under Buffer Assay

1. 5 h after initiating development (see Subheading 3.2) or prior to assaying chemotaxis, add 3.5×10^6 cells to 3 ml of DB in a well of a 6-well tissue culture plate. Pipet the cells up and down several times to dissociate any clumps.
2. Observe the behavior of the cells using an inverted microscope. Typically, differentiated wild-type cells will form streams within 15–30 min of plating and form tight aggregates after 2–3 h.

3.3.2. Cell Lysis and Immunoblotting

1. At the desired times after initiating development (see Subheading 3.2), collect 4×10^6 cells in a 1.5-ml Eppendorf tube (see Note 10). Centrifuge the cells at 300 rcf for 2–3 min.

2. Aspirate the supernatant and add 40 μl of 1× sample buffer to lyse the cells. Resuspend the pellet by vortexing. At this point, the lysates can either be analyzed directly by immunoblotting or stored at −20°C for analysis at a later date (see Note 11).
3. Load 10–15 μl of the cell lysates onto a 4–15% gradient polyacrylamide gel. Run the gel at 200 V for 60 min or at 150 V for 90 min.
4. Transfer the proteins onto a PVDF membrane at 75 V for 1 h 15 min.
5. Rinse the membrane with 1× TBST and incubate with a blocking solution of TBST/skim milk. Shake the membrane on a rotary shaker for at least 30 min at room temperature or overnight at 4°C. Rinse with 1× TBST after blocking.
6. Incubate the membrane with a 1:1,000 dilution of the primary antibody (anti-cAR1 from rabbit) in TBST/BSA. Shake the membrane on a rotary shaker for 1 h at room temperature (see Notes 12 and 13).
7. Wash the membrane with 1× TBST twice for 10 min each and incubate it with a 1:5,000 dilution of the secondary antibody (HRP-conjugated anti-rabbit) in TBST/skim milk. Shake the membrane on a rotary shaker for 45–60 min at room temperature.
8. Wash the membrane with 1× TBST once for 10 min and three times for 5 min each and detect by chemiluminescence using ECL Western blotting detection reagent. In wild-type cells, the cAR1 antibody specifically recognizes a band at approximately 40 kDa in developed (5 h) but not vegetative (0 h) cells. Occasionally, a size shift will be observed for this band, depending on the phosphorylation state of the cAR1 receptor at the particular moment that the cell samples were prepared.
9. To confirm equal sample loading, stain the membrane with Coomassie brilliant blue staining solution for approximately 10 min and then rinse it with water a few times. Wash the membrane with destaining solution until the desired contrast is achieved (see Note 14). Air-dry the membrane.

3.4. Analysis of Chemotaxis By Scoring or Video Acquisition of a Small Population Assay

One technique used to assess chemotaxis is the small population assay. In this assay, a drop of differentiated cells is placed next to a drop of cAMP on a hydrophobic agar surface. As the cAMP diffuses through the agar, it forms a gradient that causes the cells to migrate toward it. Since the cells cannot cross the boundary of their drop, they accumulate at the edge facing the cAMP. The extent of cell accumulation can be scored over time and compared between cell lines. Because of its simplicity, the small population assay is a convenient way to assess the chemotactic ability of multiple cell lines simultaneously at several different cAMP concentrations. Furthermore, the assay can be easily adapted for time-lapse microscopy, allowing the chemotaxis of individual cells to be tracked and quantified for several cell lines at once. For example, video acquisition of the small population assay was used to compare chemotaxis for wild-type and mutant A cells (Fig. 3). The cell behaviors that occurred during image

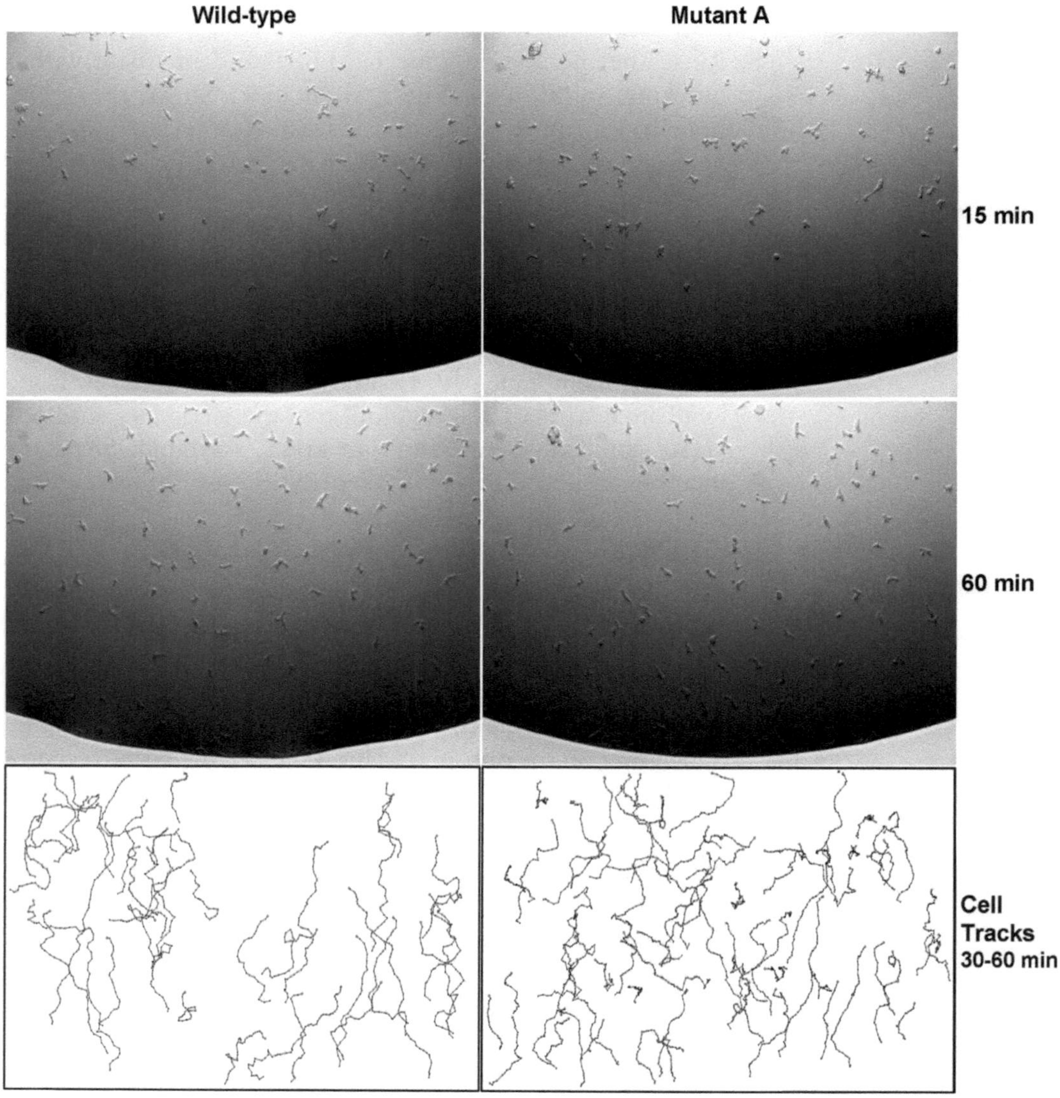

Fig. 3. The small population assay shows that mutant A cells have reduced speed during chemotaxis compared to wild-type cells. Drops of cells developed for 5 h were placed next to drops of 1 μM cAMP on a hydrophobic agar surface. Images of the cell drop borders facing the cAMP drops were acquired with a 10× objective and DIC illumination at 30-s intervals until 1 h after the drops were prepared. Frames taken 15 and 60 min after the drops were prepared are shown for both wild-type and mutant A cells. Tracks of the centroid of individual cells that migrated for at least 25 consecutive frames were obtained using Chemotaxis Quantification software (Version 1.2). Only frames acquired during the last 30 min of imaging were analyzed.

acquisition were quantified, demonstrating that mutant A has reduced chemotactic speed compared to wild-type cells (4.8 vs. 8.8 μm/min), whereas its chemotactic index and persistence are not significantly affected. One of the disadvantages of this assay is that cells are imaged at a low magnification, such that differences in cell morphology cannot easily be observed.

3.4.1. Scoring Method for the Small Population Assay

1. 5 h after initiating development (see Subheading 3.2), or when cells prepared for the under buffer assay appear differentiated (see Subheading 3.3.1), dilute the cells to a density of 1×10^6 cells/ml in 1× PB. Gently vortex the cells and pipet them up and down to dissociate any clumps.
2. Load the diluted cells into a drawn-out Pasteur pipette. Touch the pipette tip to the surface of a Petri dish filled with solidified noble agar to spot 10–15 small drops (approximately 1.5 μl each) of cells in a row, 3 mm apart from each other (see Note 15).
3. Use another drawn-out Pasteur pipette to spot a row of cAMP drops 3 mm below the row of cell drops. Typically, several different cAMP concentrations, including 1×10^{-8}, 1×10^{-7}, and 1×10^{-6} M, are used for each cell line. These can be tested on the same plate by spotting paired drops of cells and cAMP in separate rows for each cAMP concentration, such that the sets of pairs are spaced approximately 1 cm apart from each other. Alternatively, different cell lines at the same cAMP concentration can be tested on the same plate.
4. Use an inverted microscope with a 10× objective to examine the bottom border of each cell drop (facing the corresponding cAMP drop). Score the response of each cell drop as follows:

   No response (−): no cells have accumulated at the border of the drop

   Intermediate response (+/−): a small fraction of the cells has accumulated at the border of the drop

   Positive response (+): a large fraction of the cells has accumulated at the border of the drop, forming a continuous line of cells piled on top of each other

   Figure 4 shows a representative image for each of the three responses. For mutant analysis, scoring should be done at 15-min intervals for 1 h.
5. Calculate the relative chemotaxis score for each cell line as follows: [number of (+/−) drops × 0.5 + number of (+) drops]/total number of drops analyzed × 100%. The score for wild-type cells should be nearly 100% by the 1-h timepoint, since all or most of the drops should show a positive response marked by a large accumulation of cells at the border of the drop facing the cAMP gradient.

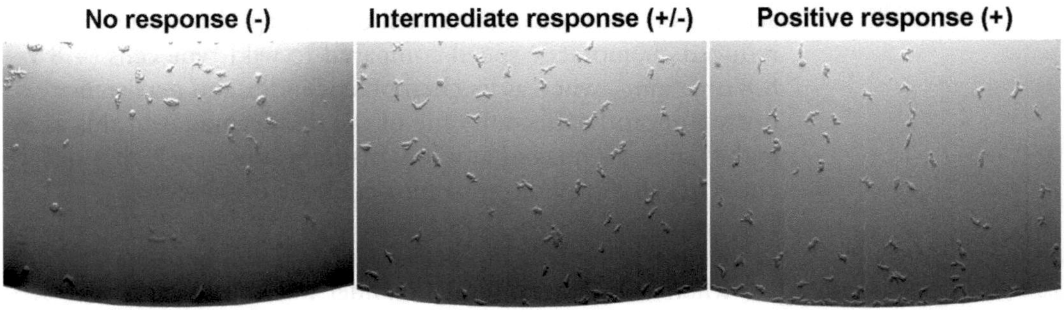

Fig. 4. Scoring for the small population assay can be used to assess chemotaxis. Drops of wild-type cells developed for 5 h were placed next to drops of 1 μM cAMP on a hydrophobic agar surface. Images of the cell drop borders facing the cAMP drops were acquired with a 10× objective and a 1.6× optovar using DIC illumination. Representative images are shown for drops with no response (–), an intermediate response (+/–), or a positive response (+).

3.4.2. Video Acquisition of the Small Population Assay

1. 5 h after initiating development (see Subheading 3.2), or when cells prepared for the under buffer assay appear differentiated (see Subheading 3.3.1), dilute the cells to a density of 6.7×10^5 cells/ml in 1× PB. Gently vortex the cells and pipet them up and down to dissociate any clumps.
2. Using a Pipetman for consistency, spot 1.5 μl drops of cells approximately 3 mm apart from each other on a 1-well chambered cover glass filled with solidified noble agar. Up to seven drops can fit onto a single 1-well chambered cover glass, allowing for the analysis of six different mutant strains and a wild-type control at the same time.
3. Spot 1.5 μl drops of cAMP in a row 3 mm below the cell drops. Typically, cAMP is used at a concentration of 1×10^{-6} M, although other concentrations can be tested as well.
4. Film the bottom edge of each cell drop (facing the corresponding cAMP drop) using an inverted microscope equipped with a moving stage and a camera (see Note 16). Images should be acquired using a 10× objective and DIC illumination at 30-s intervals for 1 h. During the course of the assay, wild-type cells migrate with directional persistence toward the border of the drop facing the cAMP gradient with a speed of approximately 10 μm/min.
5. The acquired images can be analyzed using Chemotaxis Quantification software (Version 1.2), which measures various parameters, including chemotactic speed, chemotactic index, and directional persistence (16). Typically, the last 30 min of the time-lapse acquisition (images acquired between 30 and 60 min after spotting the cells onto the agar) is analyzed.

3.5. Analysis of Chemotaxis By a Micropipette Assay

The micropipette assay is another way to analyze chemotaxis. In this assay, cells are imaged over time as they migrate toward a point source of cAMP, allowing for a detailed examination of the morphology and behavior of individual cells. A major disadvantage of this assay is that only one field of cells can be imaged at a time. To bypass this limitation, two cell populations can be differentially labeled, mixed together, and analyzed simultaneously. For example, one strain of cells can be labeled using the membrane-permeable fluorescent dye CMFDA, while a second strain can be treated with DMSO as a control (17). Figure 5 shows that

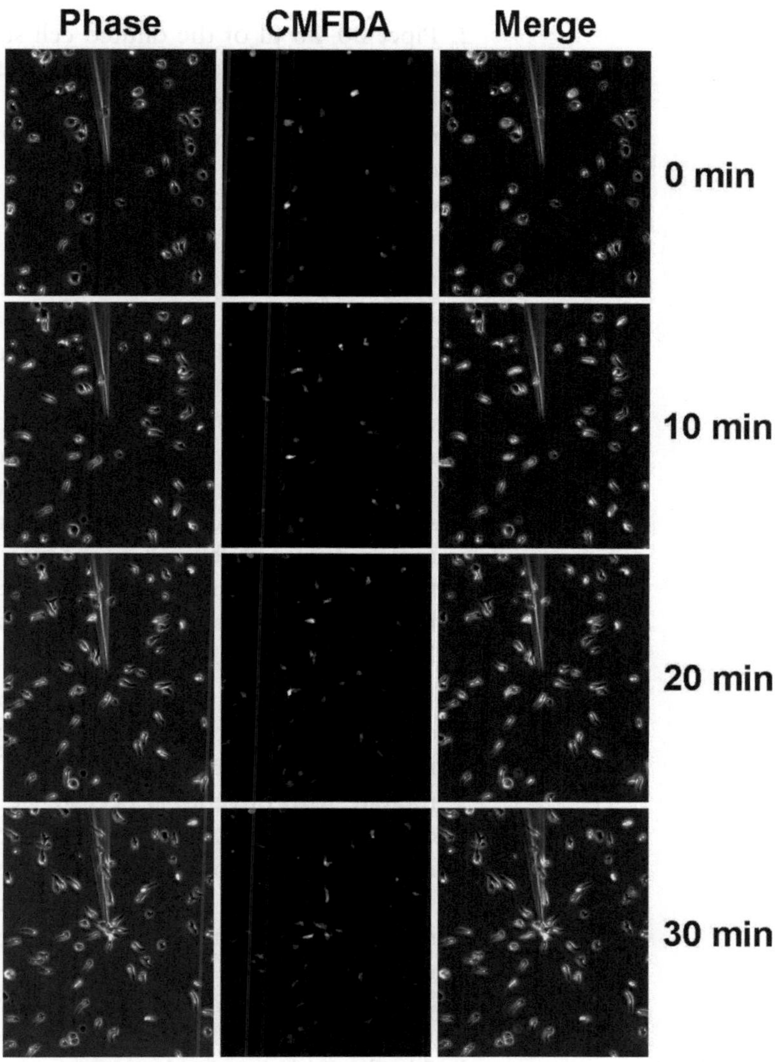

Fig. 5. Labeling with CMFDA does not affect the chemotaxis of wild-type cells. Cells developed for 5 h were labeled with CMFDA or treated with DMSO. An equal number of CMFDA-labeled and DMSO-treated cells were mixed, plated on a chambered cover glass, and exposed to a micropipette filled with 10 μM cAMP. Cells surrounding the micropipette were imaged at 30-s intervals with a 10× objective using both a fluorescein filter set and phase illumination. Sequential phase and epifluorescent images were overlaid at the indicated times, with labeled cells pseudo-colored green. After 30 min of exposure to the cAMP gradient, both labeled and unlabeled cells are able to reach the micropipette, suggesting that treatment with CMFDA does not affect chemotaxis.

wild-type cells labeled with CMFDA or treated with DMSO migrate equally well toward cAMP.

3.5.1. Analysis of Unlabeled Cells

1. 5 h after initiating development (see Subheading 3.2), or when cells prepared for the under buffer assay appear differentiated (see Subheading 3.3.1), dilute the cells to a few different densities (ranging from 1×10^5 to 4×10^5 cells/ml) in DB. Gently vortex the cells and/or pipet them up and down to dissociate any clumps (see Note 17).
2. Pipet 20–25 μl of the diluted cell suspensions onto the surface of a 1-well chambered cover glass. Several drops, of different cell densities and/or different cell lines, can be placed in the same chamber, spaced a few centimeter apart from each other. Allow the cells to adhere for 5–10 min before filling the chamber with 3–4 ml of DB.
3. Fill a femtotip (micropipette) with 1×10^{-5} M cAMP using a microloader tip and attach it to the microinjector system (see Note 18).
4. View the cells in the chambered cover glass with an inverted microscope to find a field appropriate for imaging. The field used for imaging should include enough cells for analysis, but not so many cells that they will come into contact with each other during the course of the assay. In addition, this field should include a central space for the micropipette that is mostly devoid of cells. Place the micropipette in the center of the field. Using the microinjector, allow the cAMP to flow from the micropipette tip with a compensation pressure of 30 hPa (see Note 19).
5. Image the cells using a 10× or 20× objective and either DIC or phase illumination. Acquire images every 20–30 s for a total of 30 min. During the course of the assay, wild-type cells migrate with directional persistence up the cAMP gradient, eventually reaching the tip of the micropipette. The resulting images can be analyzed using Chemotaxis Quantification software (Version 1.2), which measures various parameters, including chemotactic speed, chemotactic index, and directional persistence (16). Alternatively, cells can be imaged at a higher magnification to examine cell morphology in detail.

3.5.2. Simultaneous Analysis of Two Cell Lines Using CMFDA Labeling

1. 5 h after initiating development (see Subheading 3.2), or when cells prepared for the under buffer assay appear differentiated (see Subheading 3.3.1), centrifuge 2×10^6 cells at 300 rcf for 2 min and aspirate the supernatant. To wash, resuspend the cell pellet in 1× PB by gently vortexing and/or repeatedly pipetting the cells; repeat the centrifugation and discard the supernatant. Resuspend the pellet in 100 μl of 1× PB that contains either 100 μM CMFDA or DMSO. Incubate the cell

suspension at room temperature for 20–30 min, intermittently flicking the tube to mix the contents. Note that the cell suspension treated with CMFDA will gradually turn green.

2. Centrifuge the cell suspension and aspirate the supernatant. To wash, resuspend the pellet in 1× PB by gently vortexing and/or repeatedly pipetting the cells; repeat the centrifugation and discard the supernatant. Repeat the wash step once more. Resuspend the resulting pellet in 1× PB to obtain a final cell density of 1×10^6 cells/ml.

3. Dilute the cells to a few different final densities (ranging from 1×10^5 to 4×10^5 cells/ml) in DB, mixing equal volumes of CMFDA- and DMSO-treated cells. Proceed with steps 2–5 of the micropipette assay as described above (see Subheading 3.5.1); however, to visualize both cell populations, images should be acquired at each timepoint using both DIC or phase illumination and epifluorescence with a fluorescein filter set. Wild-type cells migrate toward the micropipette tip with similar directional persistence and speed when treated with either CMFDA or DMSO (see Fig. 5).

3.6. Examination of PIP3 and Actin Dynamics in Chemotaxing Cells

Chemoattractant stimulation leads to the rapid production of PIP3 at the membrane (5). In globally stimulated cells, this accumulation occurs transiently and uniformly around the cell membrane and is followed by a second, more prolonged peak in PIP3 production that corresponds to the generation of pseudopodia randomly around the cell periphery. In cells experiencing a chemotactic gradient, PIP3 production occurs specifically at the leading edge and is persistently maintained (4, 6, 18). In both cases, the accumulation of PIP3 results in the recruitment of PH domain-containing effector proteins, such as PKBA and CRAC, to the membrane (6, 7). These effectors, together with independent signaling pathways, lead to the polymerization of actin with temporal and spatial dynamics similar to those of the PIP3 response (18). Therefore, mutants can be assessed for defects in known chemotactic signaling pathways by monitoring the spatial and temporal changes in PIP3 production and actin polymerization that occur in response to cAMP stimulation. To do this, fluorescence time-lapse microscopy can be used to monitor cells that express GFP-tagged versions of either the PH domain from CRAC (PH_{CRAC}) or a fragment of the filamentous actin-binding protein LimE ($LimE_{\Delta coil}$), which serve as readouts for PIP3 production and actin polymerization, respectively (5, 13). Other proteins that bind PIP3 or actin with high affinity can also be fluorescently tagged and used for this assay. Figure 6 shows the PIP3 response of wild-type cells expressing PH_{CRAC}-GFP after global stimulation with cAMP.

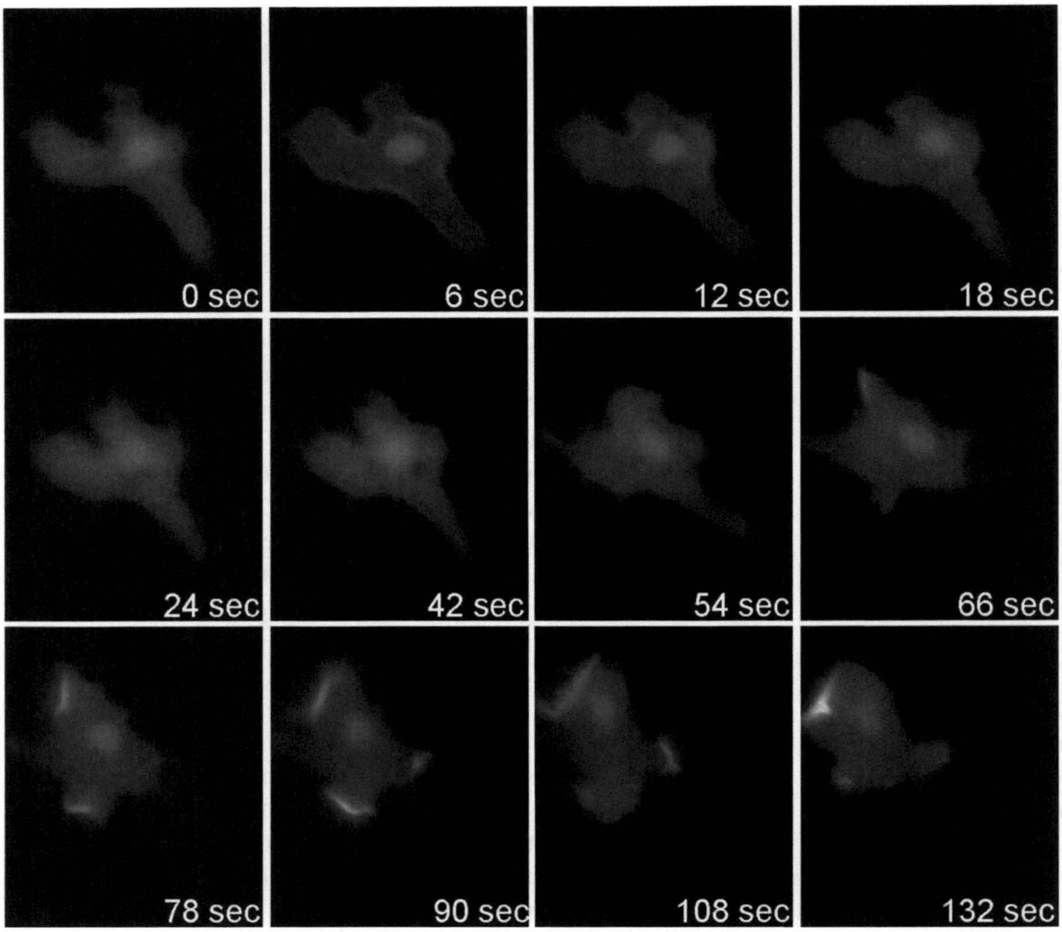

Fig. 6. PIP3 is produced in two phases upon cAMP stimulation. Wild-type cells expressing PH_{CRAC}-GFP as a readout for PIP3 levels were developed for 5 h, plated on a chambered cover glass, and imaged using an inverted microscope with a 40× oil objective. Epiflourescent images were acquired at 6-s intervals using a GFP filter set. Frames are shown for representative timepoints as indicated. After the second frame (t=0 s) was acquired, the cells were stimulated globally with 1 μM cAMP, resulting in an initial burst of PIP3 production and the rapid, transient recruitment of PH_{CRAC}-GFP to the cell membrane. This initial response is followed by a second, prolonged peak in PIP3 production with a corresponding enrichment of PH_{CRAC}-GFP on membrane protrusions.

3.6.1. Analysis of PIP3 and Actin Dynamics in Globally Stimulated Cells: Jumping Assay

1. Develop cells (see Subheading 3.2) that express GFP-tagged versions of either PH_{CRAC} (PIP3 dynamics) or $LimE_{\Delta coil}$ (actin dynamics). 5 h after initiating development, or when cells prepared for the under buffer assay appear differentiated (see Subheading 3.3.1), dilute cells to a final density of 1.5×10^5 cells/ml in 10 ml of DB. Gently vortex the cells and/or pipet them up and down to dissociate any clumps.
2. Seed 450 μl of the diluted cells into each well of an 8-well chambered cover glass (for several replicates). Let the cells adhere to the chamber for approximately 5–10 min.
3. Image the cells with an inverted microscope using either a 40× or a 60× oil objective. Acquire images at 5-s intervals for

30 frames (for a total of 2.5 min) using both DIC or phase illumination and epifluorescence with a GFP filter set at each timepoint. After imaging the basal state for two frames, stimulate the cells by adding 50 µl of 10 µM cAMP to the chamber for a final cAMP concentration of 1 µM (see Note 20). Upon stimulation, wild-type cells expressing fluorescently tagged PH_{CRAC} or $LimE_{\Delta coil}$ show a robust, transient recruitment of the GFP signal from the cytosol to the membrane, followed by a prolonged period during which the GFP signal is present at the tips of pseudopodia and membrane ruffles (see Fig. 6).

3.6.2. Analysis of PIP3 and Actin Dynamics in a Gradient: Micropipette Assay

1. Develop cells (see Subheading 3.2) that express GFP-tagged versions of either PH_{CRAC} (PIP3 dynamics) or $LimE_{\Delta coil}$ (actin dynamics). Proceed with steps 1–4 of the micropipette assay as described above (see Subheading 3.5.1).
2. Image the cells with an inverted microscope using a 40× oil objective. Acquire images every 20–30 s for a total of 20–30 min, using both DIC or phase illumination and epifluorescence with a GFP filter set at each timepoint. Wild-type cells expressing fluorescently tagged PH_{CRAC} or $LimE_{\Delta coil}$ show a pronounced and persistent enrichment of the GFP signal at the leading edge as they migrate toward the micropipette tip.

4. Notes

1. The water used in all cases is deionized and double distilled.
2. To ensure a uniform solution, place the sample buffer (3× or 1×) at 37°C for a few minutes and vortex it prior to use.
3. Other denaturing sample buffer recipes can also be used.
4. The agar plates should be made immediately prior to use since they can become dehydrated, causing the drops of cells and cAMP to dry out.
5. Grow *D. discoideum* axenically, without bacteria, at 22°C in HL5 medium. When preparing for experiments, grow cells in suspension culture at 200 rpm on a rotary shaker. Start a suspension culture so that the initial density is at least ~10^5 cells/ml. Experiments should be conducted using cells in the exponential growth phase, between 2×10^6 and 6×10^6 cells/ml. The length of time required to culture the cells to the necessary density can be estimated based on a doubling time of 8–12 h as observed for wild-type strains. For long-term maintenance, grow the cells on tissue-culture plates. Although this method does not provide as many cells for use in experiments, it provides better growth conditions compared to suspension

culture. For strains that cannot grow in suspension, for example, many mutants with cytokinesis defects, adherent cells from tissue-culture plates can also be used for experiments. In this case, aspirate the HL5 medium, add approximately 3 ml of fresh medium, and detach the cells from the surface of the plate by repeat pipetting. Collect the cell suspension from the plate and proceed as described for cells in a suspension culture. Unless otherwise indicated, all experiments are carried out at room temperature.

6. To prepare a source of bacteria, *K. aerogenes* cells can also be grown in suspension culture in HL5 medium without antibiotics overnight at room temperature for use the next day.
7. SM plates can be used instead of SM/5. In this case, plaques are analyzed 5 days after plating the cells. Additionally, the rate of plaque growth and the rate at which cells enter into the developmental program can be altered by plating more or less of the *K. aerogenes* suspension for the bacterial lawn.
8. The total number of cells used for differentiation can be altered as long as the volume used for resuspension is adjusted accordingly to maintain a final cell density of 2×10^7 cells/ml. For example, if 4×10^7 cells are used, they should be resuspended in 2 ml of DB. Cell volumes of less than 1 ml or greater than 10 ml should be avoided if possible. For volumes of 3 ml or less, 25-ml Erlenmeyer flasks should be used.
9. Final cAMP concentrations of 75–100 nM can also be used. The concentration of the cAMP solution must be adjusted if the volume of the cell suspension is altered (see Note 8). The stated concentration of 2 μM cAMP is based on a cell suspension volume of 4 ml.
10. For a full developmental timecourse, samples are typically prepared 0, 3, 4, 5, 6, and 7 h after starvation is induced. Alternatively, samples can be taken both at 0 h and immediately before conducting an experiment, usually at 5 h, to confirm that the cells being assessed have differentiated. When calculating the volume of the cell suspension to be removed at each timepoint, changes in cell density over time due to the addition of cAMP (1 ml every hour) should be taken into account. For a 4 ml cell suspension, the volume doubles to 8 ml after 4 h of pulsing (5 h of starvation), decreasing the cell density by half.
11. Do not boil the cell lysates prior to immunoblotting since this causes cAR1 receptors to aggregate.
12. The primary antibody dilution can be reused multiple times due to the addition of NaN_3. The antibody should be stored at 4°C between uses. With repeated use, the signal from the antibody becomes weaker, and higher exposure times may be necessary during detection.

13. If the primary antibody yields a weak cAR1 signal and strong background signals, cAR1 expression can alternatively be analyzed using CHAPS insoluble floating fractions (CHIFFs) rather than whole-cell lysates (19). Spin down 5×10^6 cells at 360 rcf for 3–4 min, discard the supernatant, and resuspend the cell pellet in TEB (40 mM Tris–HCl, pH 8.0, 2 mM EDTA, and 50 mM NaCl in water). Centrifuge the cell suspension, discard the supernatant, and resuspend the pellet in 600 μl of ice-cold TEB supplemented with a protease inhibitor cocktail (Roche). Add 300 μl of 60 mg/ml CHAPS in TEB with protease inhibitors and incubate the cells on ice for 5 min. Spin down the cells at maximum speed for 15 min. Remove the supernatant and resuspend the pellet in 80 μl of 1× sample buffer (without boiling). Load 5–10 μl of the sample on a gel and proceed with steps 3–9 as described (see Subheading 3.3.2).
14. A rolled up Kimwipe can be placed in the tray with the membrane to absorb the Coomassie dye and speed up the destaining process.
15. Drops can also be dispensed by pipetting 1.5 μl of liquid with a Pipetman; however, this requires a significantly longer amount of time since the pipette needs to be refilled with each drop.
16. The chambered cover glass should be handled carefully as it is being prepared for microscopy, since abrupt movement may cause the drops to spread and/or move.
17. It is not recommended to develop more than two or three different cell lines at a time for use with the micropipette assay. Once developed for 5 h, cells can be used for a maximum of roughly 2 h (the amount of time necessary to analyze two to three cell lines with this assay) before they form tight aggregates that cannot easily be dissociated. To test more than three cell lines on the same day, the initiation of development can be staggered. Alternatively, developed cells can be placed on ice and used for the assay at a later time, although the cells should not be left on ice for more than a few hours and require approximately 15 min to recover at room temperature in the chambered cover glass before imaging.
18. Handle the femtotip gently to prevent cracks or breaks. After filling it with the cAMP solution, the tip of the micropipette should be examined under a microscope, using a 10× objective, for the presence of air bubbles. Any observed bubbles can be dislodged by gently flicking the upper plastic portion of the micropipette while its tip points downward. Additionally, before placing it in the chambered cover glass, the femtotip should be examined under a microscope for the presence of any blockages that could prevent consistent

cAMP flow. This can be done using the cleaning function of the microinjector. When the micropipette is cleaned, drops of cAMP should be seen coming from the tip.

19. Alternatively, a 1×10^{-6} M cAMP solution can be used to fill the micropipette. In this case, the compensation pressure should be increased to 60 hPa.
20. The cAMP should be added very carefully to avoid disturbing the cells or microscopy chamber, which could shift the focus of the image. Gently add the cAMP by holding the pipette tip above the well containing the cells, rather than submerging the pipette tip in the liquid. This ensures that the cAMP will not hit the cells directly with enough force to cause them to detach from the surface of the well.

Acknowledgments

The authors would like to thank Huaqing Cai for helpful comments and suggestions. KFS is supported by the American Heart Association. This work was supported by National Institutes of Health Grants GM34933 and GM28007 to PND.

References

1. Kedrin, D., Van Rheenen, J., Hernandez, L., Condeelis, J., and Segall, J. E. (2007) Cell motility and cytoskeletal regulation in invasion and metastasis *J. Mammary Gland. Biol. Neoplasia* **12:** 143–152.
2. Luster, A. D. (1998) Chemokines - chemotactic cytokines that mediate inflammation *N. Engl. J. Med.* **338:** 436–445.
3. Swaney, K. F., Huang, C. H., and Devreotes, P. N. (2010) Eukaryotic chemotaxis: a network of signaling pathways controls motility, directional sensing, and polarity *Annu. Rev. Biophys.* **39:** 265–289.
4. Funamoto, S., Meili, R., Lee, S., Parry, L., and Firtel, R. A. (2002) Spatial and temporal regulation of 3-phosphoinositides by PI 3-kinase and PTEN mediates chemotaxis *Cell* **109:** 611–623.
5. Huang, Y. E., Iijima, M., Parent, C. A., Funamoto, S., Firtel, R. A., and Devreotes, P. N. (2003) Receptor-mediated regulation of PI3Ks confines PI(3,4,5)P3 to the leading edge of chemotaxing cells *Mol. Biol. Cell* **14:** 1913–1922.
6. Parent, C. A., Blacklock, B. J., Froehlich, W. M., Murphy, D. B., and Devreotes, P. N. (1998) G protein signaling events are activated at the leading edge of chemotactic cells *Cell* **95:** 81–91.
7. Meili, R., Ellsworth, C., Lee, S., Reddy, T. B., Ma, H., and Firtel, R. A. (1999) Chemoattractant-mediated transient activation and membrane localization of Akt/PKB is required for efficient chemotaxis to cAMP in Dictyostelium *EMBO J* **18:** 2092–2105.
8. Chen, L., Iijima, M., Tang, M., Landree, M. A., Huang, Y. E., Xiong, Y., Iglesias, P. A., and Devreotes, P. N. (2007) PLA2 and PI3K/PTEN pathways act in parallel to mediate chemotaxis *Dev. Cell* **12:** 603–614.
9. Kamimura, Y., Xiong, Y., Iglesias, P. A., Hoeller, O., Bolourani, P., and Devreotes, P. N. (2008) PIP3-independent activation of TorC2 and PKB at the cell's leading edge mediates chemotaxis *Curr. Biol.* **18:** 1034–1043.
10. van Haastert, P. J., Keizer-Gunink, I., and Kortholt, A. (2007) Essential role of PI3-kinase and phospholipase A2 in Dictyostelium discoideum chemotaxis *J. Cell Biol.* **177:** 809–816.
11. Kuspa, A., and Loomis, W. F. (1992) Tagging developmental genes in Dictyostelium by restriction enzyme-mediated integration of

plasmid DNA *Proc Natl Acad Sci USA* **89:** 8803–8807.

12. Kuspa, A. (2006) Restriction Enzyme-Mediated Integration (REMI) Mutagenesis *Methods Mol. Biol.* **346:** 201–209.
13. Bretschneider, T., Diez, S., Anderson, K., Heuser, J., Clarke, M., Müller-Taubenberger, A., Köhler, J., and Gerisch, G. (2004) Dynamic Actin Patterns and Arp2/3 Assembly at the Substrate-Attached Surface of Motile Cells *Curr. Biol.* **14:** 1–10.
14. Parent, C. A., and Devreotes, P. N. (1996) Molecular genetics of signal transduction in Dictyostelium *Annu. Rev. Biochem.* **65:** 411–440.
15. Klein, P. S., Sun, T. J., Saxe, C. L. r., Kimmel, A. R., Johnson, R. L., and Devreotes, P. N. (1988) A chemoattractant receptor controls development in Dictyostelium discoideum *Science* **241:** 1467–1472.
16. Kabacoff, C., Xiong, Y., Musib, R., Reichl, E., Kim, J., Iglesias, P. A., and Robinson, D. N. (2007) Dynacortin facilitates polarization of chemotaxing cells *BMC Biology* **5:** 53.
17. Knecht, D. A., and Shelden, E. (1995) Three-Dimensional Localization of Wild-Type and Myosin II Mutant Cells during Morphogenesis of Dictyostelium *Dev. Biol.* **170:** 434–444.
18. Chen, L., Janetopoulos, C., Huang, Y. E., Iijima, M., Borleis, J., and Devreotes, P. N. (2003) Two phases of actin polymerization display different dependencies on PI(3,4,5)P3 accumulation and have unique roles during chemotaxis *Mol. Biol. Cell* **14:** 5028–5037.
19. Xiao, Z., and Devreotes, P. N. (1997) Identification of detergent-resistant plasma membrane microdomains in dictyostelium: enrichment of signal transduction proteins *Mol. Biol. Cell* **8:** 855–869.

Chapter 21

Simple Experimental and Spontaneous Metastasis Assays in Mice

Gary M. Box and Suzanne A. Eccles

Abstract

Many steps of the metastatic cascade can be reproduced in simple in vitro assays such as tumour cell interactions with matrix proteins, proteolysis, chemotaxis, haptotaxis, and invasion into matrices or explanted tissues. Nevertheless, there are no fully adequate substitutes for the complexity of the in vivo process. Here, we describe two "experimental" metastasis assays to yield lung or liver colonies (mimicking established micrometastatic disease), and two spontaneous metastasis assays for breast and prostate carcinomas. Examples include either murine tumour cell lines in syngeneic immunocompetent mice or human tumour xenografts in immunodeprived mice.

Key words: Lung metastasis, Liver metastasis, Lymphatic metastasis, Mouse, Organ colonisation, Breast cancer, Prostate cancer, Orthotopic

1. Introduction

Metastasis is the major cause of treatment failure for most solid cancers and a greater understanding of the underlying molecular mechanisms is needed if we are to develop effective new treatments (1–3). Simple in vivo assays can mimic the late stages of metastasis (tumour cell circulation, lodgement, and organ colonisation) and may be appropriate for evaluation of therapies designed to target established disseminated disease.

These so-called experimental metastasis assays involve the introduction of tissue cultured tumour cells into the peripheral circulation. Most tumour cells (except those of lymphoreticular origin) arrest in the first capillary bed encountered, hence any superficial vein can be used to deliver cells to the lung with high

Claire M. Wells and Maddy Parsons (eds.), *Cell Migration: Developmental Methods and Protocols*, Methods in Molecular Biology, vol. 769, DOI 10.1007/978-1-61779-207-6_21, © Springer Science+Business Media, LLC 2011

efficiency based on the venous return to the lungs for oxygenation. For convenience, the lateral tail veins of the mouse are generally used. Similarly, liver metastases can selectively be generated by introducing appropriate tumour cell types (e.g. colon or pancreatic carcinomas) into the hepatic portal circulation. This can be achieved by injection into a mesenteric vein, but an easier method (described here) is injection into the spleen, from which blood flow is rapid and direct to the liver. This allows almost immediate resection of the spleen (with no adverse effects to the mouse) to prevent extensive local tumour growth.

However, the "gold standard" is the development of "spontaneous metastasis" from an orthotopic tumour (i.e. grown in the correct anatomical location), particularly for the evaluation of novel therapeutic agents destined for clinical use (4, 5). It has been shown that the correct tissue microenvironment (especially for human tumour xenografts) allows expression of enhanced metastatic potential compared with subcutaneous implants. Here, we give examples of breast carcinomas grown in the mammary fat pad (which, in the case of human tumour xenografts, can be excised to allow time for metastases to manifest) and carcinomas injected into the prostate gland (which are left in situ) and from which lymph node metastases develop.

Cells injected as a bolus into the circulation of a normal mouse may not experience the same microenvironment, or behave the same as those released gradually from an animal bearing an established primary tumour. For example it has recently been shown that primary tumours can selectively condition "pre-metastatic niches" in organs to dictate the sites of secondary disease (6, 7). All methods are, therefore, a balance of pragmatism and accuracy of reflection of the clinical reality, and each model should be selected according to the primary aims of the experimental study. In all cases, the highest standards of animal care and welfare should be adhered to, with the mildest possible procedure adopted. This is entirely consistent with high-quality preclinical cancer research (8).

2. Materials

2.1. Cell Culture

1. Dulbecco's modified Eagle's medium and RPMI-1640 medium, both supplemented with heat-inactivated 10% foetal bovine serum.
2. Versene and TrypLE™ (Gibco®, UK).
3. Hank's balanced salt solution (HBSS) without calcium chloride and magnesium chloride.

Table 1
Examples of cell lines from an authenticated source used as metastatic models

Name	Route	Metastatic site	Mouse strain	Selected reference	ATCC® number
Mouse tumour cell lines					
4T1 breast carcinoma	m.f.p.	l, lv, bn	BALB/c	(37)	CRL-2539
B16 F10 melanoma	i.v.	l	C57BL/6	(38)	CRL-6475
CT26 colon carcinoma	i.v.	l	C57BL/6	(39)	CRL-2638, CRL-2639
Lewis LL2 lung carcinoma	i.v.	l	C57BL/6J	(40)	CRL-1642
Human tumour cell lines					
MDA-MB-231 breast carcinoma	i.v., m.f.p, ic	l, lv, ln, bn	Athymic or SCID	(41)	HTB-26
MDA-MB-435[a]	i.v. and m.f.p	l, lv, ln	Athymic	(41)	HTB-129
HT-1080 sarcoma	i.v.	l	Athymic	(42)	CCL-121
PC-3 prostate carcinoma	Prostate	l, lv, ln, bn	Athymic	(28, 43)	CRL-1435
LS 174T colon carcinoma	isp	lv	Athymic	(44)	CL-188
HCT 116 colon carcinoma	isp	lv	Athymic	(45)	CCL-247

m.f.p mammary fat pad, *i.v.* intravenous, *isp* intra-splenic, *ic* intra-cardiac (beyond the scope of this chapter), *l* lung, *lv* liver, *ln* lymph node, *bn* bone

[a]Some controversy as to whether this cell line is a breast carcinoma or melanoma, but it is nevertheless a useful and reliable model (46–50)

4. Cell lines selected from the following (according to the model required, Table 1):
   (a) Mouse melanoma, B16 F10 (CRL-6475, American Type Culture Collection®).
   (b) Human fibrosarcoma, HT-1080 (CCL-121, American Type Culture Collection®).
   (c) Mouse mammary tumour, 4T1 (CRL-2539, American Type Culture Collection®).
   (d) Human colorectal adenocarcinoma, LS 174T (CL-188, American Type Culture Collection®).
   (e) Human prostate adenocarcinoma, PC-3 (CRL-1435, American Type Culture Collection®) (see Note 1).

2.2. Lung Colonisation

1. Mouse restrainer/platform designed for tail vein injections.
2. Sterile needle (gauge 25–27) and sterile 1-mL syringe.
3. 70% Isopropyl alcohol (swabbing disinfectant).
4. Sterile gauze swab.
5. Physical or chemical vasodilator (e.g. incandescent lamp or warm water (~43°C)); neat isopropyl alcohol or Vasolate (VetTec Solutions Ltd).

6. B16 F10 single cell suspension, at 0.1 million in 100 μL, to female C57BL/6 mice (syngeneic recipients) or HT-1080 single cell suspension, at 0.1 million in 100 μL, to female CrTac:NCr-*Foxn1*^{nu} athymic mice (human tumour xenograft).

2.3. Liver Colonisation

1. Sterile needle (gauge 30) and sterile 1-mL syringe.
2. 70% Isopropyl alcohol (swabbing disinfectant).
3. Sterile gauze swab.
4. Sterile cotton buds.
5. Gas anaesthetic (e.g. isofluorane) and vapouriser (both VetTech Solutions Ltd).
6. Sterile 4/0 mersilk suture (Ethicon, UK).
7. Sterile surgical instruments (e.g. scissors, forceps, wound clips, and applicator).
8. 10% Iodine solution.
9. LS 174T human colon carcinoma cell suspension at one million per100 μL.
10. Post-operative analgesic (e.g. buprenorphine).
11. Female CrTac:NCr-*Foxn1*^{nu} athymic nude mice.

2.4. Mammary Fat Pad Tumour-Derived Metastasis

1. Sterile needle (gauge 25) and sterile 1-mL syringe.
2. 70% Isopropyl alcohol (swabbing disinfectant).
3. Sterile gauze swab.
4. Hair removal clippers or depilatory cream.
5. Gas anaesthetic (e.g. isofluorane) and vapouriser (both VetTech Solutions Ltd).
6. Sterile 4/0 mersilk suture.
7. Sterile surgical instruments (e.g. scissors, forceps, wound clips, and applicator).
8. 10% Iodine solution.
9. 4T1 cell suspension at 0.1 million/10 μL.
10. Post-operative analgesic (e.g. buprenorphine).
11. Female syngeneic Balb/c mice.

2.5. Prostate Tumour-Derived Metastasis

1. Sterile needle (gauge 30) and sterile 1-mL syringe.
2. 70% Isopropyl alcohol (swabbing disinfectant).
3. Sterile gauze swab.
4. Gas anaesthetic (e.g. isofluorane) and vapouriser (both VetTech Solutions Ltd).
5. Sterile 4/0 mersilk suture.
6. Sterile surgical instruments (e.g. scissors, forceps, wound clips, and applicator).

7. 10% Iodine solution.
8. PC-3 cell suspension at 0.5 million/20 μL.
9. Post-operative analgesic (e.g. buprenorphine).
10. Male CrTac:NCr-*Foxn1*^{nu} athymic mice.

2.6. Histology Reagents

1. Bouin's fixative

   Add 15 volumes of saturated picric acid to 5 volumes of formalin, and then add 1 volume of glacial acetic acid.

2. India ink stain

India ink	15 mL
Water	85 mL
Ammonium hydroxide	150 μL

3. Feket's fixative

70% Ethanol	60 mL
Formalin	6 mL
Glacial acetic acid	1 mL

4. 10% Formol saline fixative (neutral buffered)

Formalin	10 mL
Water	90 mL
Sodium chloride	0.90 g
Sodium dihydrogen orthophosphate	0.40 g
Disodium hydrogen orthophosphate	0.65 g

3. Methods

All tissue-culture procedures should be performed in a class II laminar flow cabinet, applying strict aseptic techniques. Given the frequency of misidentification and cross-contamination (9, 10), it is essential that all cell lines are rigorously checked for their origins and genetic identity (11, 12). It is also important that cell lines are free from contamination with infectious agents such as mycoplasma, which can influence their biological behaviour and present a risk to handlers and animals (13, 14). Cells must, therefore, be routinely screened for mycoplasma using a highly sensitive polymerase chain reaction (PCR) method (15, 16). It is not good practice to maintain cells in antibiotics to ensure sterility, as this can mask low-level bacterial contamination, lead to outgrowth of antibiotic-resistant pathogens, and may also affect the behaviour of cells. Cells are maintained in a humidified incubator set at 37°C in an atmosphere of 5% carbon dioxide in air. The preferred culture flasks have a screw-cap lid with a 0.2-μm filter. Growth medium is replaced every 2–4 days depending on the growth rate and density of the cells. Cells should be harvested whilst in an exponential log phase of growth (50–70% confluent).

It is important to ensure that cells for intravenous delivery are not clumped, otherwise potentially lethal pulmonary embolism is a possibility. Numerous washings of cells are required to rid them of sticky residual serum components or DNA released from damaged cells. Some cells such as colon carcinomas also secrete a mucinous glycocalyx which can cause clumping and must be removed by vigorous washing. The harvesting of sub-confluent populations and their final resuspension in a vehicle free of both calcium and magnesium ions will help to ensure single cell preparations. Furthermore, mice that survive delivery of clumped cells develop more pulmonary foci than cohorts receiving the same total in a single cell preparation (17), and the variability between mice can be greater.

Athymic nude mice CrTac:NCr-*Foxn1^{nu}* (Taconic, USA), and C57BL/6 and BALB/c mice (Charles River, UK) at 6–8 weeks old are housed in individually ventilated cages and given sterile water and fed ad libitum.

3.1. Cell Culture

1. Pre-warm tissue culture reagents.
2. Decant all medium from culture vessel.
3. Remove traces of serum by rinsing with Versene using gentle rocking for 20 s and decant.
4. Add TrypLE™ dissociation solution (1 mL/75-cm^2 flask) and gently rock so that the entire monolayer is bathed.
5. Incubate for 5 min.
6. Dislodge cells from vessel by gentle tapping.
7. Gather cells by adding growth medium (containing serum to inactivate remaining trypsin) and decant to a tube for centrifuging (500×*g* for 5 min)
8. Resuspend the pelleted cells in HBSS to attain a single cell suspension and determine cell concentration using a cell counting chamber.
9. Adjust the concentration for the appropriate inoculum volume by centrifugation and resuspension in a small volume.
10. Keep cells on ice to maintain viability and prevent aggregation.

3.2. Animal Procedure for Lung Colonisation

1. Place a mouse in the restrainer.
2. Disinfect the whole tail using an alcohol-dampened swab.

3. Mix the tumour cell suspension with a sterile syringe and load with an excess volume.
4. Attach a sterile needle and, importantly, expel any air bubbles, leaving the intended volume for injection.
5. Rotate the tail to locate one of the two lateral tail veins.
6. Vasodilate the tail vein (see Note 2).
7. With the needle bevel facing upward and on the same plane as the tail vein, slide the needle in 2 mm. A slight pull on the syringe plunger should reveal a flash of blood in the needle hub confirming correct entry (see Note 3).
8. Push the plunger in slowly and deliver the cell suspension. The vein will change from dark to light as the cell suspension temporarily replaces the blood (Fig. 1). Any resistance or blanching will indicate the needle is not in the vein.
9. Remove the needle from the vein and with a sterile gauze swab, apply slight pressure to the injection site until bleeding has stopped.
10. Remove the mouse from the restrainer and return to its cage.
11. Observe the mouse for 5–10 min to ensure no recurrent bleeding or other ill effects.
12. Repeat using a fresh needle and new cell suspension for each mouse (loading a syringe with sufficient cells for multiple mice can lead to blood clotting in the needle between injections, and the delivery of variable numbers of cells; this applies to all such procedures).

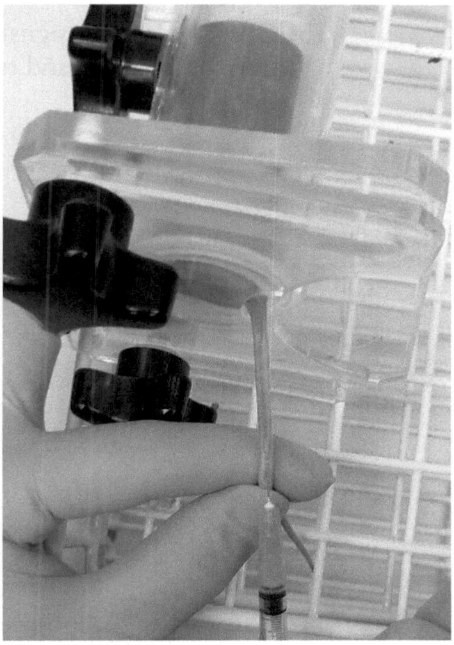

Fig. 1. Tail vein inoculation for lung metastases.

3.3. Animal Procedure for Liver Colonisation

1. Anaesthetize a mouse (see Note 4) and turn to the right lateral position.
2. If necessary, shave or depilate fur.
3. Disinfect the skin above the spleen with an alcohol-dampened swab.
4. Make an 8-mm cutaneous incision, perpendicular to the length of the spleen, and carry down through the abdominal wall.
5. Exteriorise the spleen onto a sterile swab saturated with sterile saline, blunt dissecting any pancreatic tissue.
6. Mix the tumour cell suspension with a sterile syringe and load with an excess volume.
7. Attach a sterile needle and, importantly, expel any air bubbles, leaving the intended volume for injection.
8. Slowly inject the single cell suspension under the spleen capsule, marked by a visible pale wheal (Fig. 2).
9. Remove the needle and stop injection site bleeding by applying gentle pressure with a sterile cotton bud for 30 s.
10. Ligate the splenic blood vessels with suture.
11. Remove the spleen by cutting distal to the ligatures.
12. Close the abdominal muscle with suture and swab the wound with iodine solution. Close the skin incision with surgical clips.
13. Administer post-operative analgesic according to the manufacturer's instructions.
14. Soak surgical instruments in a multipurpose disinfectant. Wash and rinse prior to autoclave sterilisation.
15. Observe daily for signs of post-operative ill health, inflammation, or infection and remove surgical clips after 7 days.

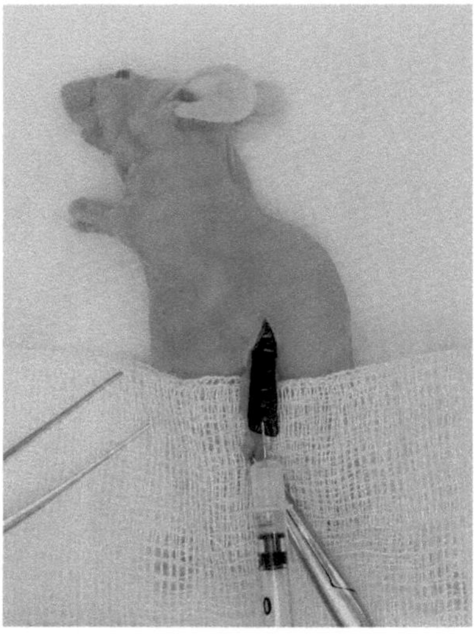

Fig. 2. Intra-splenic inoculation for liver metastases.

3.4. Animal Procedure for Mammary Fat Pad Tumour-Derived Metastasis

3.4.1. Injection of Tumour Cells

1. Anaesthetize a mouse and turn to the supine position.
2. If necessary, shave or depilate fur.
3. Disinfect the mammary glands with a dampened swab (see Note 5).
4. Make a 5-mm skin incision adjacent to a gland and identify the fatty tissue (Fig. 3).
5. Mix the tumour cell suspension with a sterile syringe and load with an excess volume. Attach a sterile needle and expel any air bubbles.
6. Direct the bevel of the needle upwards and inject slowly.
7. Close the incision with suture.
8. Soak surgical instruments in a multipurpose disinfectant. Wash and rinse prior to autoclave sterilisation.
9. As the tumour grows, monitor routinely for potential skin ulceration.

3.4.2. Excision (Mastectomy) of Mammary Tumour

1. Do not allow tumours to grow to excessive sizes (max 8–10 mm mean diameter).
2. Anaesthetise a mouse and turn to the supine position.
3. If necessary, shave or depilate fur.
4. Disinfect the area of excision with an alcohol-dampened swab.

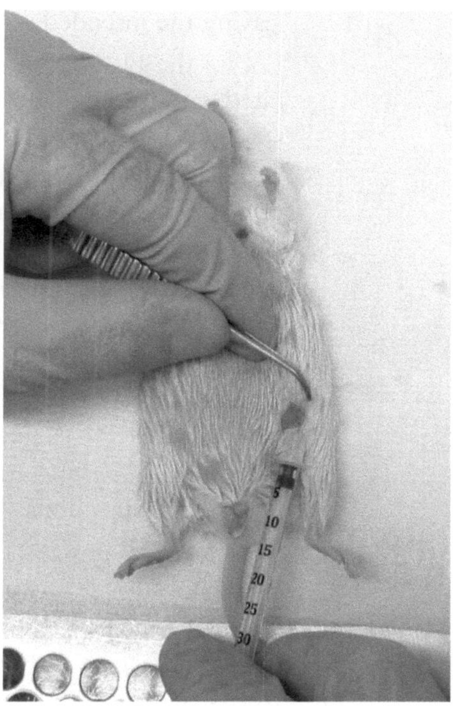

Fig. 3. Mammary fat pad inoculation.

5. With the aid of sterile surgical instruments, cut around the whole tumour mass, removing an adequate margin of skin tissue (see Note 6).
6. Swab the wound with iodine solution and repair with surgical clips.
7. Administer post-operative analgesic according to the manufacturer's instructions.
8. Soak surgical instruments in a multipurpose disinfectant. Wash and rinse prior to autoclave sterilisation.
9. Observe daily for signs of inflammation or infection and remove surgical clips after 7 days.
10. Continue observation for signs of ill health – lung metastases may manifest by shortness of breath or laboured breathing. Palpate regional (axillary) lymph nodes and abdomen for signs of metastasis.

3.5. Animal Procedure for Prostate-Derived Metastasis

1. Anaesthetise a male mouse and turn to the supine position.
2. If necessary, shave or depilate fur.
3. Disinfect the area of excision with an alcohol-dampened swab.
4. Make an 8-mm midline incision above the pubic symphysis.
5. Mix the tumour cell suspension with a sterile syringe and load with an excess volume.
6. Attach a sterile needle and, importantly, expel any air bubbles, leaving the intended volume for injection.
7. Locate the bladder and exteriorise onto a sterile swab saturated with sterile saline, cutting away connective tissue if necessary to aid release (Fig. 4).
8. Insert the needle into the ventral lobe of the prostate gland, exit into the dorsolateral lobe, and inject the cell suspension (see Note 7).
9. Check that there is no internal bleeding and then lift the abdominal muscle, allowing the bladder to retract.
10. Close the abdominal muscle with suture and swab the wound with iodine solution. Close the skin incision with surgical clips.
11. Administer post-operative analgesic according to the manufacturer's instructions.
12. Soak surgical instruments in a multipurpose disinfectant. Wash and rinse prior to autoclave sterilisation.
13. Observe daily for signs of post-operative ill health, inflammation, or infection and remove surgical clips after 7 days.
14. Palpate the abdomen for signs of lymph node metastases.

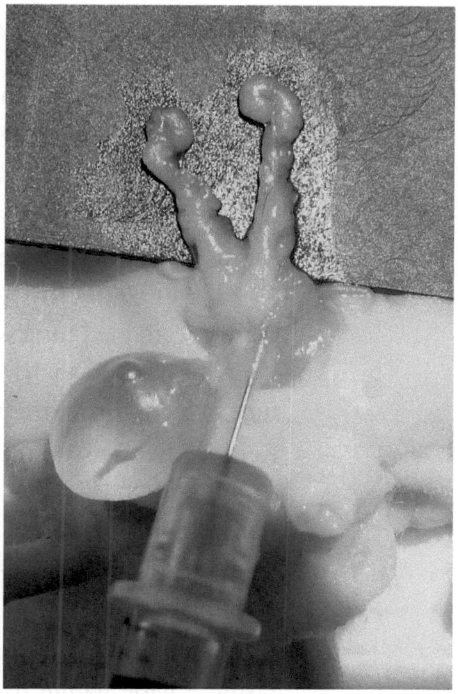

Fig. 4. Intra-prostatic injection.

3.6. Thoracotomy and Removal of Lung and Trachea En Bloc

1. A mouse is overdosed with anaesthetic and turned to the supine position.
2. Starting from the diaphragm, an incision is made upwards to the neck, exposing the thoracic cavity.
3. Hold the top of the trachea with forceps and dissect the lung, trachea, and heart en bloc and lay onto a saline-moistened sterile gauze swab.
4. Remove the heart and wash the lung in sterile water and return to the moistened swab.
5. Insert a large gauge needle (19 G) into the open end of the trachea and deliver approximately 1 mL of staining solution or fixative.

3.7. Staining of Lung Colonies for Quantitation

1. Bouin's method

   Immediately after sacrifice, remove the lungs and trachea en bloc.

   Wash the lungs in water.

   Using a wide gauge needle (19 G) and syringe, deliver fixative into the lung via the trachea.

   Fix overnight (<24 h) at room temperature by gently rotating in a 30-mL glass universal tube.

Remove fixed tissue to 70% ethanol for long-term storage at room temperature.

2. India ink and Feket's method

   Immediately after sacrifice, remove the lungs and trachea en bloc.

   Wash the lungs in water.

   Using a wide gauge needle (19 G) and syringe, deliver ink into the lungs via the trachea.

   Wash the stained lung in Feket's solution.

   Fix overnight (<24 h) in Feket's solution at room temperature by gently rotating in a 30-mL plastic universal tube.

   Remove fixed tissue to 70% ethanol for long-term storage at room temperature.

3. Formol saline fixation

   Immediately after sacrifice, remove the lung with the trachea intact.

   Wash the lung in water.

   Using a wide gauge needle (19 G) and syringe, fix the lung via the trachea.

   Fix for 4–24 h (optimise depending on colony size) at room temperature by gently rotating in a 30-mL plastic universal tube.

   Remove fixed tissue into sterile saline for short-term storage or to 70% ethanol for longer periods at room temperature.

3.8. Quantitation of Experimental and Spontaneous Metastases

The principal organ for development of "experimental metastases" following intravenous delivery is the lung. However, extrapulmonary disease is possible, so full autopsy examining the viscera, lymph nodes, and brain is advised. Mice showing abnormal balance or behaviour in particular merit sectioning and histology of the brain.

The time taken to develop metastases and the expected distribution of disease may differ from published information. The variability will come from different mouse strains; age and sex of individuals; their housing conditions; and health status. Variations in culture conditions and handling of cell lines can significantly alter their behaviour, and each operator should establish their own standard procedures and note the in vivo growth and metastatic rates for each cell line, so that any deviations can be carefully monitored. Changes in serum batches can lead to alterations in cell behaviour. Tumorigenicity and metastatic phenotype can change during long-term passage, or by altering harvesting procedures. Establish the "baseline" properties of the cell line and bank multiple early passage aliquots to avoid genetic drift or adaptation to tissue culture, which can reduce in vivo malignant potential.

A time course study to assess the rate and extent of metastasis development provides the observer the opportunity to standardise operating procedures and to define humane end points for termination, prior to animals developing severe symptoms. Death as an end point (survival study) is unacceptable (18).

For experimental studies, cohorts of 8–10 mice or fewer are generally sufficient to apply nonparametric analysis (e.g. Mann–Whitney rank sum test) for comparative analysis of groups, e.g. to test the effects of therapeutic agents (19).

3.8.1. Lung Metastases

Tumour lung colony quantitation is performed by careful gross examination. For animals with large numbers of lesions, one lobe can be examined and the total number approximated by multiplication. Surface lesions can be counted under a dissecting microscope or removed with a micro curve-ended forceps for counting, sizing, and weighing. Most tumours develop at the surface of the lung, and this can give a representative reflection of total tumour burden even if there are additional tumours deep within the lung parenchyma.

Discrimination of tumours is assisted by the use of fixation and staining (see Subheading 3.6): Bouin's fluid provides a contrast of white lesions against yellow lung tissue (20), whereas India ink stains the lung blue (21, 22). The B16 melanoma is a popular choice for experimental lung metastasis assays, as the melanin content of the tumour renders the black lesions easily visible against the normal lung without the need for staining (although non-pigmented lesions may also occur) (23). However, in all cases, microscopic lesions also need to be considered (see Note 8). An alternative is to section formalin-fixed lungs, take 5-μM sections, and stain with haematoxylin and eosin (H&E) for scoring by light microscopy (Fig. 5).

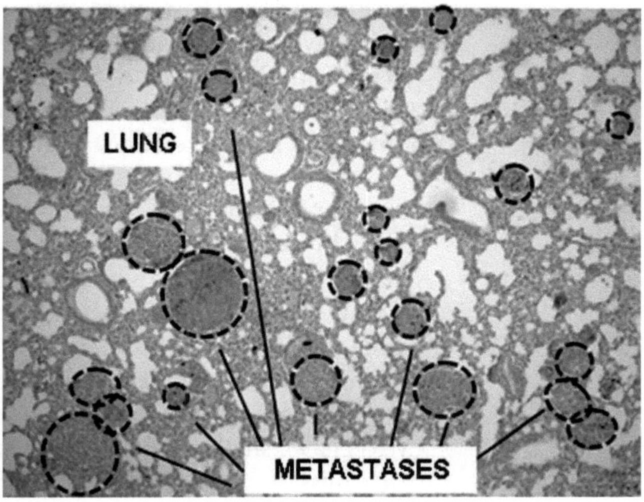

Fig. 5. Lung metastases.

A crude method, when tumour burden is extensive or colonies have coalesced and cannot be individually discriminated, is to remove and weigh the lungs of each mouse and express the results as average lung weight ± standard deviation. The lung weight of an average 25-g mouse is approximately 200 mg (although this should be determined from age-matched normal mice of the same age and strain) and hence some estimate of the total tumour burden can be adduced. However, this method is unreliable and not recommended since the presence of tumours can cause inflammation, oedema, and congestion, which can add to the mass of the tissue.

Quantitative PCR offers a precise method of detection and quantification of metastatic tumour burden in human xenograft models where microscopic lesions are common. Human and mouse primers can be designed to amplify a unique, species-specific, conserved, and non-transcribed sequence in the mouse and human genome (24). Alternatively, amplifying a specific mRNA transcript using real-time TaqMan PCR (25) offers another approach.

Computer-assisted video imaging is a technique that utilises mathematical models sensitive to subtle changes in size, shape, position, and tissue characteristics affected by disease (26). Increasingly, cells engineered to express fluorescent proteins or luciferase are being used to allow semi-quantitative, dynamic, non-invasive optical imaging of internal tumours, e.g. (22) and see Chapters 22 and 23, this issue.

3.8.2. Lymph Node Metastases (Breast and Prostate Models)

Serial sectioning and H&E staining are an option for microscopic analysis but a quantitative RT-PCR TaqMan approach offers a more sensitive and cost-effective approach (27). It is highly reproducible and quantitative, eliminating associated risks encountered with other PCR-based detection methods. Inokuchi et al. (25) utilised this approach and highlighted the sensitivity compared to conventional H&E staining and immunohistochemical staining by amplifying cytokeratin-19 mRNA transcripts from breast cancer patients. Havens et al. identified early microscopic lodgings and metastatic homing events through real-time PCR using human *Alu* sequence TaqMan probes (28).

3.8.3. Liver Metastases and Other Sites

Liver can be handled as for lung and surface or internal tumours scored by eye, dissection, or microscopic means (see Fig. 6). No staining is required as the pale tumour colonies are easily distinguishable from the reddish-brown liver parenchyma. Invasive end points, as described above, are an option but non-invasive (e.g. bioluminescent or fluorescent optical) imaging modalities should be considered. Developments in magnetic resonance imaging (MRI) (29), small-animal computed tomography (30), and positron emission tomography (31) offer a different end point, and multimodality imaging, combining both molecular and anatomical data, is the favoured approach (32, 33).

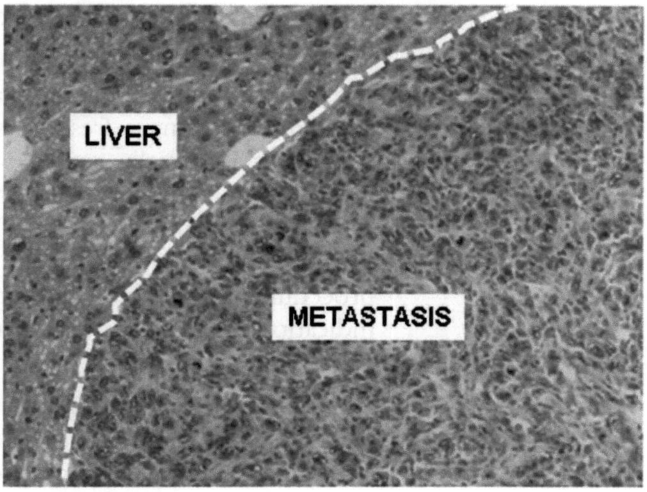

Fig. 6. Liver metastasis.

4. Notes

1. Recent genetic testing has revealed a high frequency of misidentification and cross-contamination of commonly used cell lines (9, 10). It is, therefore, essential that all human and mouse cell lines are sourced from a reputable establishment rigorously checked for their genetic identity (11, 12). Cell line authenticity is an increasing prerequisite by publishers who now often require short tandem repeat profiling as the measure for confirmation (34). It is also important that cell lines are free from contamination with infectious agents such as mycoplasma and pathogenic viruses which can influence their biological behaviour (35) and present a risk to handlers and animals (13, 14, 16, 36).
2. Incandescent lamp – place a well-ventilated mouse cage approximately 20 cm from a lamp, with a 150 W light bulb, for approximately 5 min.

   Warm water – take the tail of a restrained mouse and immerse it in a 15-mL tube containing water at approximately 43°C (<50°C) for 30–60 s.

   Vasolate or swabbing alcohol – dampen a gauze swab with the liquid and rub a lateral tail vein of a restrained mouse. Vasodilation is instant and remains so for several minutes.

   Mouse tail illuminators – specially designed restrainers sit on a light source that has an illuminated slot for the tail to be placed.
3. A prudent method is to first puncture the tail vein close to the tip, and if unsuccessful, continue towards the base of the tail

to a proximal point. If the chosen lateral vein collapses, move to the contralateral vein.

4. A mouse is restrained manually and gently, but firmly, held to a gas mask passing a fresh flow of oxygen (2 L/min). The vapouriser is initially set to 5% to induce (approximately 30–60 s) and lowered to 2% to maintain anaesthesia throughout the surgical procedure. Tail and footpad reflexes are checked before any incision is performed. Throughout the procedure, a constant vigil is paramount, watching for any indication related to under-sedation or laboured breathing if over-sedated.

   A mammary fat pad injection procedure will require 5–8 min of anaesthetic, with the mastectomy requiring between 5 and 15 min, the time depending on whether the tumour has invaded the muscle wall.

   An intra-prostatic procedure will require between 7 and 12 min, the time dictated by the amount of fat and connective tissue needed to be cut away.

   An intra-splenic procedure will require 10–12 min, the time lengthening with persistent bleeding.

5. Mice have ten mammary glands: four inguinal (posterior) and six thoracic (anterior). The best policy is to inject either the lower thoracic fat pads (closest to the inguinal set) or the highest inguinal fat pads since these are furthest from local lymph nodes. The more anterior site is reported to be more favourable for growth. This procedure ensures, at macroscopic examination, that lesions that develop post-excision of the primary tumour are genuine local lymph node metastases and not due to recurrence of primary tumour remnants.

6. Failure to remove the entire primary tumour will result in regrowth. Some animal project licences may not allow the opportunity to repeat the excision and force an unnecessary cull.

7. The dorsal and lateral (dorsolateral) lobes of the prostate gland are found between the ventral and anterior lobes, at the junction of the urinary bladder, urethra, and seminal vesicles. The inoculum site should "bleb", confirming correct delivery. No bleeding will occur for this injection.

8. The method of euthanasia should be considered. Carbon dioxide causes minute haemorrhaging (petechiae) in the lungs, potentially adding confusion when identifying small tumour foci during histological examination. Cervical dislocation causes clots in the lung. The cleanest method is to overdose with anaesthetic.

All animal procedures should be performed in accordance with UK Home Office regulations under the Animals (Scientific Procedures) Act 1986 or other national guidelines and the recently published NCRI guidelines for animal experimentation (18), which all major publishers have endorsed. This publication offers clear guidance on maximum volumes for inoculation at different sites, further examples of robust metastasis models, procedures, and humane end points. A further useful reference (regarding stringent reporting of animal experiments) is also recommended (www.nc3rs.org.uk/reportingguidelines).

Acknowledgements

This work was funded by Cancer Research UK Grants CA309/A2187, CA309/A8274. We acknowledge NHS funding to the NIHR Biomedical Research Centre.

References

1. Steeg, P. S., Anderson, R. L., Bar-Eli, M., Chambers, A. F., Eccles, S. A., Hunter, K., Itoh, K., Kang, Y., Matrisian, L. M., Sleeman, J. P., D., T., Thompson, E. W., and Welch, D. R. (2009) Preclinical drug development must consider the impact on metastasis, *Clin Cancer Res* **15**, 4529–4530.
2. Eccles, S. A., and Welch, D. R. (2007) Metastasis: recent discoveries and novel treatment strategies, *Lancet* **369**, 1742–1757.
3. Talmadge, J. E., and Fidler, I. J. (2010) AACR centennial series: the biology of cancer metastasis: historical perspective, *Cancer Res* **70**, 5649–5669.
4. Lu, J., Steeg, P. S., Price, J. E., Krishnamurthy, S., Mani, S. A., Reuben, J., Cristofanilli, M., Dontu, G., Bidaut, L., Valero, V., Hortobagyi, G. N., and Yu, D. (2009) Breast cancer metastasis: challenges and opportunities, *Cancer Res* **69**, 4951–4953.
5. Talmadge, J. E. (2010) Models of metastasis in drug discovery, *Methods Mol Biol* **602**, 215–233.
6. Kaplan, R. N., Rafii, S., and Lyden, D. (2006) Preparing the "soil": the premetastatic niche, *Cancer Res* **66**, 11089–11093.
7. Psaila, B., Kaplan, R. N., Port, E. R., and Lyden, D. (2006) Priming the "soil" for breast cancer metastasis: the pre-metastatic niche, *Breast Dis* **26**, 65–74.
8. Osborne, N. J., Payne, D., and Newman, M. L. (2009) Journal editorial policies, animal welfare, and the 3Rs, *Am J Bioeth* **9**, 55–59.
9. Lacroix, M. (2008) Persistent use of "false" cell lines, *Int J Cancer* **122**, 1–4.
10. Nardone, R. M. (2007) Eradication of cross-contaminated cell lines: a call for action, *Cell Biol Toxicol* **23**, 367–372.
11. Parodi, B., Aresu, O., Bini, D., Lorenzini, R., Schena, F., Visconti, P., Cesaro, M., Ferrera, D., Andreotti, V., and Ruzzon, T. (2002) Species identification and confirmation of human and animal cell lines: a PCR-based method, *Biotechniques* **32**, 432–434, 436, 438–440.
12. Yoshino, K., Iimura, E., Saijo, K., Iwase, S., Fukami, K., Ohno, T., Obata, Y., and Nakamura, Y. (2006) Essential role for gene profiling analysis in the authentication of human cell lines, *Hum Cell* **19**, 43–48.
13. Harlin, H., and Gajewski, T. F. (2008) Diagnosis and treatment of mycoplasma-contaminated cell cultures, *Curr Protoc Cytom Appendix* **3**, Appendix 3C.
14. Ishikawa, Y., Kozakai, T., Morita, H., Saida, K., Oka, S., and Masuo, Y. (2006) Rapid detection of mycoplasma contamination in cell cultures using SYBR Green-based real-time polymerase chain reaction, *In Vitro Cell Dev Biol Anim* **42**, 63–69.
15. Pruckler, J. M., and Ades, E. W. (1995) Detection by polymerase chain reaction of all common Mycoplasma in a cell culture facility, *Pathobiology* **63**, 9–11.

16. Sung, H., Kang, S. H., Bae, Y. J., Hong, J. T., Chung, Y. B., Lee, C. K., and Song, S. (2006) PCR-based detection of Mycoplasma species, *J Microbiol* **44**, 42–49.
17. Fidler, I. J. (1973) The relationship of embolic homogeneity, number, size and viability to the incidence of experimental metastasis, *Eur J Cancer* **9**, 223–227.
18. Workman, P., Aboagye, E. O., Balkwill, F., Balmain, A., Bruder, G., Chaplin, D. J., Double, J. A., Everitt, J., Farningham, D. A., Glennie, M. J., Kelland, L. R., Robinson, V., Stratford, I. J., Tozer, G. M., Watson, S., Wedge, S. R., and Eccles, S. A. (2010) Guidelines for the welfare and use of animals in cancer research, *Br J Cancer* **102**, 1555–1577.
19. Festing, M., Overend, P., Das, R., Borja, M., and Berdoy, M. (2002) The design of animal experiments: reducing the use of animals in research through better experimental design, *Laboratory Animal Handbooks* **14**.
20. Ran, S., Mohamedali, K. A., Luster, T. A., Thorpe, P. E., and Rosenblum, M. G. (2005) The vascular-ablative agent VEGF(121)/rGel inhibitspulmonarymetastasesofMDA-MB-231 breast tumors, *Neoplasia* 7, 486–496.
21. Fong, S., Itahana, Y., Sumida, T., Singh, J., Coppe, J. P., Liu, Y., Richards, P. C., Bennington, J. L., Lee, N. M., Debs, R. J., and Desprez, P. Y. (2003) Id-1 as a molecular target in therapy for breast cancer cell invasion and metastasis, *Proc Natl Acad Sci USA* **100**, 13543–13548.
22. Miretti, S., Roato, I., Taulli, R., Ponzetto, C., Cilli, M., Olivero, M., Di Renzo, M. F., Godio, L., Albini, A., Buracco, P., and Ferracini, R. (2008) A mouse model of pulmonary metastasis from spontaneous osteosarcoma monitored in vivo by Luciferase imaging, *PLoS One* **3**, e1828.
23. Price, J. E., Tarin, D., and Fidler, I. J. (1988) Influence of organ microenvironment on pigmentation of a metastatic murine melanoma, *Cancer Res* **48**, 2258–2264.
24. Malek, A., Catapano, C. V., Czubayko, F., and Aigner, A. (2010) A sensitive polymerase chain reaction-based method for detection and quantification of metastasis in human xenograft mouse models, *Clin Exp Metastasis* **27**, 261–271.
25. Inokuchi, M., Ninomiya, I., Tsugawa, K., Terada, I., and Miwa, K. (2003) Quantitative evaluation of metastases in axillary lymph nodes of breast cancer, *Br J Cancer* **89**, 1750–1756.
26. Nielsen, B. S., Lund, L. R., Christensen, I. J., Johnsen, M., Usher, P. A., Wulf-Andersen, L., Frandsen, T. L., Dano, K., and Gundersen, H. J. (2001) A precise and efficient stereological method for determining murine lung metastasis volumes, *Am J Pathol* **158**, 1997–2003.
27. Gibson, U. E., Heid, C. A., and Williams, P. M. (1996) A novel method for real time quantitative RT-PCR, *Genome Res* **6**, 995–1001.
28. Havens, A. M., Pedersen, E. A., Shiozawa, Y., Ying, C., Jung, Y., Sun, Y., Neeley, C., Wang, J., Mehra, R., Keller, E. T., McCauley, L. K., Loberg, R. D., Pienta, K. J., and Taichman, R. S. (2008) An in vivo mouse model for human prostate cancer metastasis, *Neoplasia* **10**, 371–380.
29. Townson, J. L., Ramadan, S. S., Simedrea, C., Rutt, B. K., MacDonald, I. C., Foster, P. J., and Chambers, A. F. (2009) Three-dimensional imaging and quantification of both solitary cells and metastases in whole mouse liver by magnetic resonance imaging, *Cancer Res* **69**, 8326–8331.
30. de Kemp, R. A., Epstein, F. H., Catana, C., Tsui, B. M., and Ritman, E. L. (2010) Small-animal molecular imaging methods, *J Nucl Med* **51** *Suppl 1*, 18S–32S.
31. Wang, H., Zhang, J., Tian, J., Qu, B., Li, T., Chen, Y., Liu, J., and Wang, S. (2009) Using dual-tracer PET to predict the biologic behavior of human colorectal cancer, *J Nucl Med* **50**, 1857–1864.
32. Deroose, C. M., De, A., Loening, A. M., Chow, P. L., Ray, P., Chatziioannou, A. F., and Gambhir, S. S. (2007) Multimodality imaging of tumor xenografts and metastases in mice with combined small-animal PET, small-animal CT, and bioluminescence imaging, *J Nucl Med* **48**, 295–303.
33. Koo, V., Hamilton, P. W., and Williamson, K. (2006) Non-invasive in vivo imaging in small animal research, *Cell Oncol* **28**, 127–139.
34. Barallon, R., Bauer, S. R., Butler, J., Capes-Davis, A., Dirks, W. G., Elmore, E., Furtado, M., Kline, M. C., Kohara, A., Los, G. V., MacLeod, R. A., Masters, J. R., Nardone, M., Nardone, R. M., Nims, R. W., Price, P. J., Reid, Y. A., Shewale, J., Sykes, G., Steuer, A. F., Storts, D. R., Tomson, J., Taporewala, Z., Alston-Roberts, C., and Kerrigan, L. (2010) Recommendation of short tandem repeat profiling for authenticating human cell lines, stem cells and tissues, *In Vireo Cell Dev Biol Anim* **46**, 727–732.
35. Rodriguez-Cuesta, J., Vidal-Vanaclocha, F., Mendoza, L., Valcarcel, M., Gallot, N., and Martinez de Tejada, G. (2005) Effect of asymptomatic natural infections due to common mouse pathogens on the metastatic progression of B16 murine melanoma in C57BL/6 mice, *Clin Exp Metastasis* **22**, 549–558.

36. Young, L., Sung, J., Stacey, G., and Masters, J. R. (2010) Detection of Mycoplasma in cell cultures, *Nat Protoc* **5**, 929–934.

37. Aslakson, C. J., and Miller, F. R. (1992) Selective events in the metastatic process defined by analysis of the sequential dissemination of subpopulations of a mouse mammary tumor, *Cancer Res* **52**, 1399–1405.

38. Fidler, I. J., and Nicolson, G. L. (1976) Organ selectivity for implantation survival and growth of B16 melanoma variant tumor lines, *J Natl Cancer Inst* **57**, 1199–1202.

39. Karpatkin, S., Ambrogio, C., and Pearlstein, E. (1984) Lack of effect of in vivo prostacyclin on the development of pulmonary metastases in mice following intravenous injection of CT26 colon carcinoma, Lewis lung carcinoma, or B16 amelanotic melanoma cells, *Cancer Res* **44**, 3880–3883.

40. Mayhew, E., and Glaves, D. (1984) Quantitation of tumorigenic disseminating and arrested cancer cells, *Br J Cancer* **50**, 159–166.

41. Price, J. E., Polyzos, A., Zhang, R. D., and Daniels, L. M. (1990) Tumorigenicity and metastasis of human breast carcinoma cell lines in nude mice, *Cancer Res* **50**, 717–721.

42. Zijlstra, A., Mellor, R., Panzarella, G., Aimes, R. T., Hooper, J. D., Marchenko, N. D., and Quigley, J. P. (2002) A quantitative analysis of rate-limiting steps in the metastatic cascade using human-specific real-time polymerase chain reaction, *Cancer Res* **62**, 7083–7092.

43. Patel, S., Turner, P. R., Stubberfield, C., Barry, E., Rohlff, C. R., Stamps, A., McKenzie, E., Young, K., Tyson, K., Terrett, J., Box, G., Eccles, S., and Page, M. J. (2002) Hyaluronidase gene profiling and role of hyal-1 overexpression in an orthotopic model of prostate cancer, *Int J Cancer* **97**, 416–424.

44. Shah, V., Kumar, S., and Zirvi, K. A. (1994) Metastasis of human colon tumor cells in vivo: correlation with the overexpression of plasminogen activators and 72 kDa gelatinase, *In Vivo* **8**, 321–326.

45. Ishizu, K., Sunose, N., Yamazaki, K., Tsuruo, T., Sadahiro, S., Makuuchi, H., and Yamori, T. (2007) Development and characterization of a model of liver metastasis using human colon cancer HCT-116 cells, *Biol Pharm Bull* **30**, 1779–1783.

46. Zhang, Q., Fan, H., Shen, J., Hoffman, R. M., and Xing, H. R. (2010) Human breast cancer cell lines co-express neuronal, epithelial, and melanocytic differentiation markers in vitro and in vivo, *PLoS One* **5**, e9712.

47. Chambers, A. F. (2009) MDA-MB-435 and M14 cell lines: identical but not M14 melanoma?, *Cancer Res* **69**, 5292–5293.

48. Lorenzi, P. L., Reinhold, W. C., Varma, S., Hutchinson, A. A., Pommier, Y., Chanock, S. J., and Weinstein, J. N. (2009) DNA fingerprinting of the NCI-60 cell line panel, *Mol Cancer Ther* **8**, 713–724.

49. Lacroix, M. (2009) MDA-MB-435 cells are from melanoma, not from breast cancer, *Cancer Chemother Pharmacol* **63**, 567.

50. Rae, J. M., Creighton, C. J., Meck, J. M., Haddad, B. R., and Johnson, M. D. (2007) MDA-MB-435 cells are derived from M14 melanoma cells – a loss for breast cancer, but a boon for melanoma research, *Breast Cancer Res Treat* **104**, 13–19.

Chapter 22

Two-Photon Intravital Multicolour Imaging to Study Metastatic Behaviour of Cancer Cells In Vivo

Sylvia E. Le Dévédec, Wies van Roosmalen, Chantal Pont, Reshma Lalai, Hans de Bont, and Bob van de Water

Abstract

In the last decade, intravital microscopy on breast tumours in mice at single-cell resolution has resulted in important new insight into mechanisms of metastatic behaviour such as migration, invasion, and intravasation of tumour cells; angiogenesis; and the response of immune cells. This chapter describes the methods that can be used for analysing tumour cell motility in a mouse model of breast cancer metastasis. It includes protocols for generation of a labelled primary tumour, its imaging with two-photon microscopy, and the processing of time-lapse image data. Furthermore, we present a methodology, recently developed in our laboratory that combines multicolour imaging with an inducible cell model to study the role of a specific gene of interest in tumour cell motility in vivo. This protocol can be used to image the metastatic behaviour of different individual tumour cells within the same tumour microenvironment and correlate it with metastasis formation. Additional protocols for labelling macrophages to visualise blood flow and image analysis are also included.

Key words: Breast cancer, Rag2$^{-/-}$ $\gamma c^{-/-}$ mouse model, MTLn3-ErbB1 cells, Inducible lentiviral system, Two-photon microscopy, Image processing and analysis

1. Introduction

Cell migration is essential in a number of processes, including wound healing, angiogenesis, and cancer metastasis. Metastasis is the principal cause of death of cancer patients, but the process of tumour cell dissemination is poorly understood (1). Invasion of cancer cells from the primary tumour into the surrounding tissue is a critical step that requires increased cell motility. Multiphoton microscopy has been developed for in vivo imaging and using this technique, we are beginning to understand how invasive

Claire M. Wells and Maddy Parsons (eds.), *Cell Migration: Developmental Methods and Protocols*,
Methods in Molecular Biology, vol. 769, DOI 10.1007/978-1-61779-207-6_22, © Springer Science+Business Media, LLC 2011

tumour cells move (2–6). Multiphoton microscopy uses longer wavelengths (up to 1,200 nm) that are able to penetrate deeper into tissues, allowing us to image to a depth of more than 100 μm into the primary tumour. In addition, multiphoton excitation causes less photo-damage, permits good optical sectioning and 3D resolution, and, using the second harmonic generation phenomenon, allows non-invasive visualisation of the extracellular matrix (7). In recent years, intravital imaging has been used mainly to follow fluorescently labelled cells within the primary tumour, enabling the interaction of moving cells with their microenvironment to be studied (8–10). In particular, the ability to image tumour cells as they are invading the surrounding tissue allows the direct evaluation of tumour cell motility and allows experimental evaluation of specific mechanisms for tumour cell invasion. To enable the analysis of the role of candidate metastasis-regulating genes, we recently developed a methodology that combines the use of an inducible lentiviral expression system with a mouse breast cancer model and two-photon intravital multicolour imaging (11, 12). Using multicolour intravital imaging together with an inducible tumour cell model allows the direct comparison of metastatic behaviour of genetically modified breast cancer cell populations within the intact primary tumour of a mouse. In conclusion, this chapter describes methods that can be used for analysing tumour cell motility in a breast cancer mouse model, including (1) generation of a fluorescent mammary tumours and (2) their imaging, (3) labelling macrophages for blood flow visualisation, (4) simultaneous CFP and GFP imaging combined with an inducible cell model, and (5) quantitative image analysis (see Fig. 1 for overview).

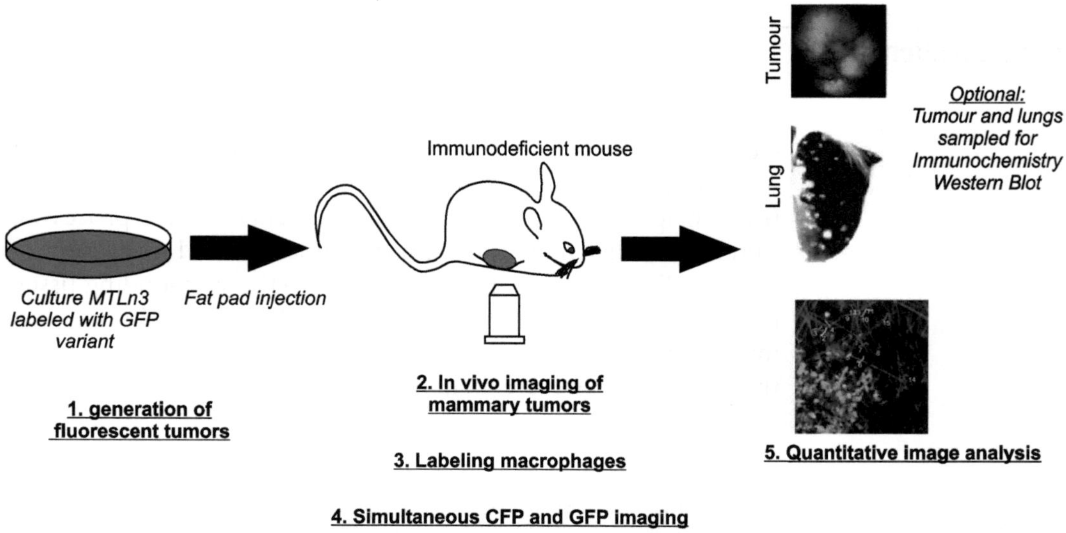

Fig. 1. Schematic view of the experimental setup of intravital imaging of cancer invasion using multiphoton microscopy.

2. Materials

2.1. Breast Cancer Metastasis Model

1. Six- to eight-week-old female Rag2$^{-/-}$ $\gamma c^{-/-}$ mice from in-house breeding. Animals are housed in individually ventilated cages under sterile conditions containing three mice per cage. Sterilised food and water are provided ad libitum (see Note 1) (11).
2. MTLn3-GFP-ErbB1 cells (available from J. Segall) (see Note 2) (13).

2.2. Inducible Cell System for Conditional Ectopic Expression of a Gene of Interest

1. The inducible lentiviral vector TREAutoR3 was kindly provided by J. Seppen (14). The lentiviral TREAutoR3-CFP vector was made as reported previously (12).
2. Virus production cell system with HEK cells (12, 15).
3. Doxycycline-HCl.
4. Flow cytometer (FACS) to sort the cells expressing the highest CFP fluorescence levels after 16 h of doxycycline induction.
5. During tumour development, at desired time points, CFP (or other gene) expression is induced by adding doxycycline (400 μg/ml in 2.5% (w/v) sucrose) in sterilised drinking water, while control animals received 1.5% (w/v) sucrose in their drinking water (12, 16).

2.3. Cell Culture and Injection

1. Minimum essential medium (MEM)-*Alpha* medium with 5% foetal bovine serum (FBS), with or without doxycycline (1,000 ng/ml).
2. Dulbecco's phosphate-buffered saline (without calcium and magnesium; DPBS).
3. 0.05% (w/v) Trypsin/EDTA.
4. Hemacytometer.
5. 1-ml syringes with 25-G needles.

2.4. Labelling Macrophages to Visualise Blood Flow

1. 20 mg/ml Texas Red-dextran (70 kDa) in DPBS.
2. 1-ml syringe with 25-G needle.
3. 70% (v/v) ethanol.
4. Heating lamp.
5. Mouse restrainer (homemade).

2.5. Tumour Preparation for Imaging

1. To measure tumour size over time, we use a Digital Caliper I-65.
2. 70% (v/v) ethanol.
3. Isoflurane.

4. Oxygen/anaesthesia apparatus (Burtons, unit 2 guardian Industrial Estate, Marden, Kent, UK, Vaporiser TEC3).
5. 100% Compressed oxygen tanks to be used in combination with vaporizer.
6. Surgical instruments.
7. Sterile compress.
8. Surgical tape.
9. Oculentum simplex ointment (PCH).
10. 1-ml syringes with 25-G needles for Nembutal (pentobarbital) injection.

2.6. In Vivo Imaging Microscope Setup

1. Coverslips Knittel 25-mm diameter (thickness nr. 1). New coverslips are used for each animal and are replaced where necessary to maintain clear imaging.
2. Grease for laboratories to seal the coverslips hermetically in the imaging box.
3. Anaesthesia/surgery box that ensures that the animal is sedated during surgery (Fig. 2a).

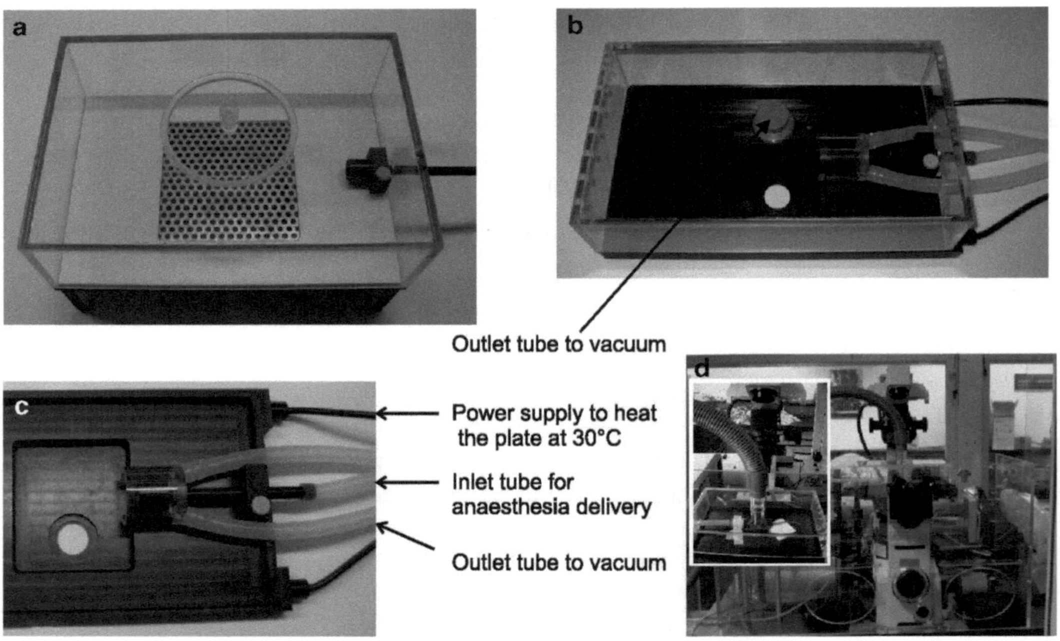

Fig. 2. In vivo imaging microscope setup: View from the *top* of the box that is used for anaesthesia and to prepare the tumour for imaging (**a**). View from the *top* of the imaging box that is used to immobilise the mouse for imaging (**b**). The *bottom* plate of the imaging box is heated via a power supply. A flow of isoflurane enters the imaging box through a nozzle and exits the box driven by a vacuum (**c**). The animal is placed in a 30°C temperature chamber on an inverted microscope and imaged with a 20× objective, and is maintained for the course of the imaging session (3 h) at 3.5–1.5% isoflurane to control breathing (**d**).

4. Imaging box that ensures that the primary tumour lies flat above the microscope objective. It also facilitates temperature and anaesthesia control through (1) an iron plate that can be warmed up at the desired temperature of 30°C and (2) a constant airflow of isoflurane. The base of the imaging box fits the Nikon microscope stage that we use (Fig. 2b, c). It is important to have the animal in a 30°C environment as tumour cell movement is reduced at low temperatures, while higher temperatures result in hyperventilation. The front of the box contains an inlet for the anaesthesia delivery and an outlet that is connected to a vacuum (Fig. 2d).
5. Two-photon confocal microscope: a 3-W Chameleon Vision II Ti-Sapphire femtosecond pulsed laser from Coherent is used in our setup. This laser is used to run a TiE microscope equipped with an A1R multiphoton scanhead system (Nikon). The system is equipped with non-descanned detectors. Previously, it has been found that a wavelength between 860 and 880 nm is optimal for imaging GFP, CFP, matrix fibres, and Texas Red at the same time (17). The inverted TiE Nikon microscope allows the placement of an animal on the automatic imaging platform and has the advantage of presenting a perfect focus system (PFS). The objectives used are as follows: 20× CFI Plan Apo 0.75NA (air); CFI Apo Lambda 40× 1.15 NA (water); and ApoTIRF 60× 1.49NA (oil). The transmission of these objectives is >60% for the wavelength range of 680–1,080 nm. The filters used are 450/80 (blue), 480/30 (CFP), 515/30 (green), and 620/60 (red).

3. Methods

Here, we describe the methods that we use for analysing MTLn3 tumour cell motility in our Rag2$^{-/-}$ γc$^{-/-}$ mouse model of breast cancer metastasis. Recently, we established an improved in vivo model using a Rag2$^{-/-}$ γc$^{-/-}$ mouse strain together with GFP-MTLn3 cells that have increased levels of the EGF receptor, which enables us to study tumour cell autonomous mechanisms underlying lung metastasis formation (11). Furthermore, we optimised a protocol that combines the use of multicolour intravital imaging together with an inducible cell model to compare metastatic behaviour of control and genetically modified breast cancer cell populations within the intact primary tumour of a mouse (12). In addition, to provide information about blood flow within and surrounding the tumour, we label the macrophages by injecting Texas red-dextran into the tail vein of a tumour-bearing animal. All protocols using live animals have been approved by the local

Institutional Animal Care and conform to governmental regulations regarding the care and use of laboratory animals.

3.1. Generation of Fluorescent Tumours by Injecting Tumour Cells into the Mammary Fat Pad

3.1.1. Preparation of Cells for Injection

1. Grow GFP-ErbB1-MTLn3 cells in α-MEM to 50–70% confluency in a TC75 culture flask. For all in vivo experiments, use the same passage number to maintain reproducibility. On day 1, thaw cells from vial containing approximately one million cells; on day 4, split the cells and take care that there are enough stock flasks (500,000 cells per animal); and on day 7, cells should be about 50–70% confluent for the injection.
2. On the day of injection, early in the morning rinse the flasks (TC75) three times, each time with 10 ml of PBS w/o Ca^{2+} or Mg^{2+} per flask.
3. Add 1 ml of trypsin/EDTA per flask and incubate at 37°C in incubator to detach cells. This should take no longer than 5 min. Check under microscope that all cells are floating after light tapping of the flask.
4. Neutralise the trypsin by resuspending the cells in 10 ml α-MEM/5% FBS per flask.
5. Collect the cells into one 50-ml sterile tube. Centrifuge cells for 5 min at 1,000 × *g*-force, room temperature.
6. Resuspend pellet in 10 ml PBS. Take an aliquot and count cells using a hemacytometer.
7. Centrifuge cells for 5 min at 1,000 × *g*-force, room temperature.
8. On the basis of the cell count, determine the appropriate volume for resuspending the pellet to obtain a cell density of 5×10^6 cells/ml. Resuspend cells in sterile PBS at 5×10^6 cells/ml, and recount to confirm actual density (inject 0.1 ml).
9. Transfer cell suspension to a sterile 1.5-ml microcentrifuge tube and keep on ice until injection (inject within 30 min). Always prepare at least double the amount of cells that are actually needed as you will lose a lot by filling the syringes.

3.1.2. Inject Cells into Xenograft

1. Prior to injection of the mice, make the nipple easily visible by shaving the nipple area. This should be done 24 h before injection to minimise the stress for the animals on the day of injection.
2. Work in a sterile hood and spray the area around the fourth (abdominal) nipple with 70% ethanol.
3. Fill syringe without needle with ~0.4 ml cell suspension. Remove air bubbles by gently tapping the syringe, while keeping the syringe upside down. Remove air, place needle (orange 25 G)

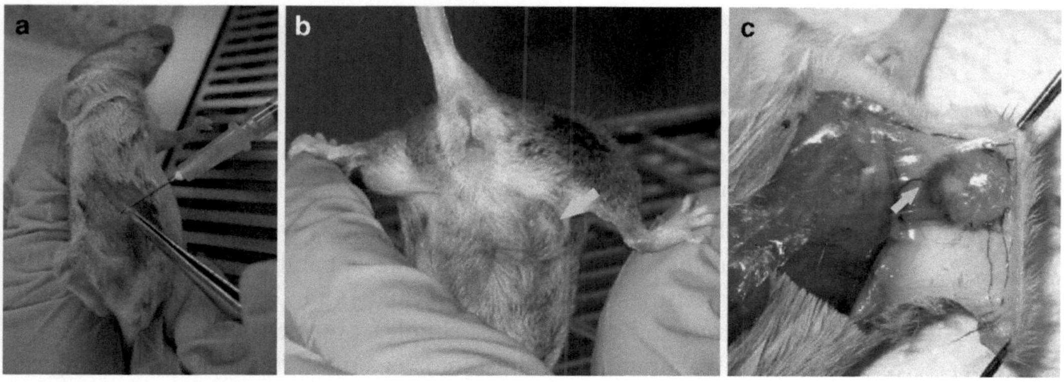

Fig. 3. Cell injection and tumour preparation for imaging: Injection of tumour cells into the mammary fat pad around the fourth nipple (**a**) GFP-ErbB1-MTLn3 tumour (*arrow*) in an immunodeficient mouse after 4 weeks of injection in the mammary fat pad. (**b**) The animal is placed under isoflurane anaesthesia and the tumour is exposed using skin flap surgery with little disruption of surrounding vasculature (*arrows*) (**c**).

on syringe, and push the plunger up till 0.1 ml to remove excess fluid.

4. With two people working together, inject 0.1 ml into the mammary fat pad. One person holds the animal and another inserts the needle subcutaneously at 20/30° angle in the mammary fat pad. If you are right handed, lift the skin with a pincette with the left hand and inject the cells with the right hand (the other way around for those left handed). Notice a bulge in the skin above the injected area. It is important that the needle only penetrates the skin and stays within the mammary fat pad, so no cells spread into the peritoneum (Fig. 3a). (If you are injecting by yourself, put the mouse under light anaesthesia using isoflurane).
5. Keep the animals in a sterile environment until the tumours are large enough to image (up to 10-mm diameter for mice, typically 3–4 weeks). Measure tumour size on a regular basis using a caliper to make sure it is growing. The tumour should be palpable after 1.5 weeks. After 3–4 weeks, the tumour should be about 10 mm in diameter and at that time point, the animals are ready to be imaged (Fig. 3b).
6. *Optional:* to induce the expression of a gene of interest using the autoR3 lentiviral vector, drinking water containing doxycycline + sucrose should be provided to the test group, while drinking water with sucrose only should be provided to the control group. Prepare a fresh stock for the whole experiment. Freeze at −20 C. Provide fresh water with or without doxycycline to the animal twice a week, starting at the chosen time (see Subheading 3.4).

7. Optional: another possibility to image early stage breast tumors intravitally for prolonged times is to use the mammary imaging window (see Note 3).

3.2. In Vivo Imaging of Mammary Tumours

3.2.1. Prepare Animal for Imaging

1. *Optional*: Inject Texas red-dextran to image blood flow (see next Subheading 3.3 and Note 4).
2. Place the animal under anaesthesia with 5% isoflurane in O_2 using an oxygen/anaesthesia apparatus combined with the anaesthesia/surgery box (N_2O=4–5 L/min, O_2=2–2.5 L/min) and apply oculentum simplex ointment (PCH) to the eyes. Isoflurane is administered via a mask to the animal during the time of surgery and imaging.
3. Place the animal in the surgery box and clean the tumour area with 70% ethanol. Perform a medial skin incision to expose the tumour (Fig. 3c).
4. Clear the area around the tumour from fat while keeping the vasculature intact as much as possible and minimise bleeding. Also, hairs should be completely removed as these will appear as black areas where present in the imaging field.
5. Transfer the animal in the pre-warmed imaging box (30°C) with the exposed tumour placed on the coverslip (objective) for imaging. If necessary, use surgical tape to hold the tumour in place. It also minimises drying out of the tumour. The tumour should be isolated away from the animal as much as possible to reduce vibrations due to breathing and secured to the bottom of the imaging box. This must be done carefully and without compressing the tumour too much (see Note 4).
6. Place the box on the microscope stage and adjust the stage so that the coverslip is above the objective. Focus on fluorescence of the GFP tumour and start imaging.
7. Maintenance of breathing and minimisation of drift and movement of the animal are crucial for successful time-lapse data collection. The best way to minimise breathing is by controlling the isoflurane level. If breathing is too rapid, vibrations can make interpretation of the acquired data difficult. Anaesthesia levels should be reduced or the position of the animal on the imaging platform should be adjusted to minimise drift and breathing artefacts. With time, less anaesthesia is required to maintain a low breathing rate; too much anaesthesia will kill the animal. Typical final isoflurane levels are 2–3.5%.
8. During imaging, the animal has to be continuously checked for regular breathing frequency. A too low breathing rate indicates that the animal is getting weak. It should not be too high as this disturbs the image acquisition and reduces image

quality due to numerous artefacts. The monitoring can be done visually, or by using the MouseOx pulse oximetry system (Starr life Sciences Corp) as done in other laboratories.

3.2.2. Perform Intravital Microscopy

The procedure is described for the Nikon A1R confocal microscope and should be used as a guideline to set up the experiment on your microscope. For one-colour GFP imaging, the animal is placed under anaesthesia as described above and imaged. On average, each imaging sequence lasts 30–60 min, and the same animal can be maintained on the microscope for up to 4 h.

1. First visualise the tumour through the eyepiece and 10× lens objective with the green fluorescence filter, focus in and locate major blood vessels which are visibly flowing and positioned flat in the same focal plane. Then switch to 20× lens objective and focus in again. Switch to PMT detection.
2. While imaging in the multiphoton mode, move around the tumour until a good imaging area is obtained. Good imaging fields for cell migration are areas around flowing vessels with well-defined cell boundaries and extracellular matrix. Flowing vessels can be visualised by labelling of macrophages with a fluorescent dextran or by viewing vessels through the oculars of the microscope. It is convenient for further analysis and processing to image at the outer tumour surface, where cells are invading the matrix, and to generate a *z*-series with steps of 50–100 μm into the tumour.
3. For the outermost plane of the *z*-series, focus in on the outside of the tumour by visualising the collagen fibre network in the second harmonic channel (Fig. 4a). Depending on the goal of the imaging session, the size of the *z*-series that is selected is defined by (1) the desired quality of each image, (2) the time-lapse interval desired between images at a particular plane, and (3) the *z*-axis resolution desired. The quality of each image is determined by the brightness of the fluorescence and the pixel dwell time. The time-lapse interval required is determined by the type of structures that are being imaged. Observing subcellular shape changes using a high-magnification (40× or 60×) objective may require 10- to 30-s time intervals. In general, reliable tumour cell tracking using lower magnification objectives (20×) can be done at 1- to 3-min intervals.
4. Time-lapse imaging is performed with NIS Elements using the ND acquisition protocol in the software. Adequate emission filters for GFP (*optional:* CFP) and second harmonic generation were selected and a simultaneous 3-channel time lapse was performed. Using an automatic stage in combination with the PFS, imaging 50 μm with 5-μm steps will require about 20 s and switching from one position to another

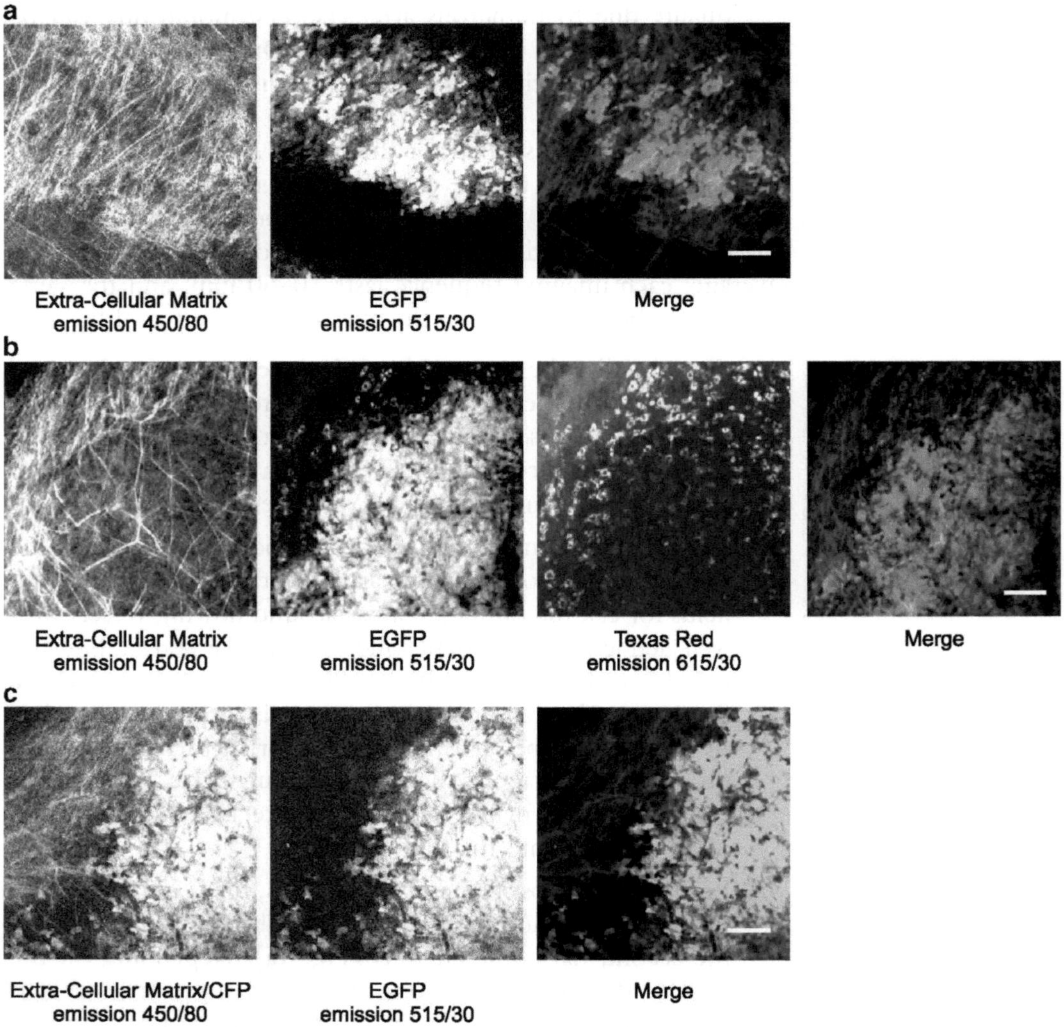

Fig. 4. Multiphoton imaging of carcinoma cells and extracellular matrix (ECM) in primary mammary tumour: Images of *z*-scan projection of tumour tissue imaged with multiphoton microscope. The GFP tumours were scanned every 10 μm for a total of 100 μm. Images shown are a projection of 40 μm (**a**). Simultaneous imaging of GFP, CFP, and collagen in vivo. A 860-nm laser light was used to capture GFP, CFP, and collagen second harmonic fluorescence in living tissue. For conditional expression of CFP-tagged focal adhesion-associated mutant proteins, we use the AutoR3 lentiviral construct that contains an integrated conditional rtTA (**b**). Simultaneous imaging of GFP, collagen, and macrophages in vivo (**c**). All scale bars are 100 μm.

will take another second, so we image about five positions for 30–60 min to increase our chance for visualising cell events and still be within less than 3 min per position frame. During each 3-min interval, a *z*-series of five images is taken at a spacing of 5 μm between images, beginning at the periphery of the tumour and moving into the tumour and performing this in five different areas of the tumour. For each position in the tumour, this image acquisition process is repeated for 30–60 min, resulting in several time-lapse three-dimensional *z*-series for analysis of tumour cell motility (see Note 5).

5. For time-lapse multiphoton imaging, as many *z*-planes should be taken as possible, given the above constraints on image quality and time-lapse interval. For detailed tracking of individual cells, 2.5- to 5-μm step sizes are best.
6. Euthanise the animal at the end of data collection.
7. *Optional:* collect blood from the right atrium of the heart to quantify intravasation events (see Note 6).
8. *Optional:* isolate first the tumour for measuring its weight and second the lungs for counting the metastases. Both tissues can be further processed for histology study (Fig. 6, see Note 7).
9. Perform image analysis (see Subheading 3.5).

3.3. Labelling Macrophages to Visualise Blood Flow

1. Warm up the tail vein of the tumour-bearing animal by immersing the tail for 1–2 min in warm (50–60°C) water. This causes the tail vein to dilate so that a 25-G needle can be inserted. The animals can also be placed under a heat lamp for 5 min.
2. Place mouse into restrainer by holding it by the tail, and close the restrainer very tight. The animal should not be able to move.
3. Clean the tail with 70% ethanol and look for the lateral vein for injection. Stabilise the vein by holding the tail between thumb and middle finger and securing the vein with the index finger.
4. Prepare a 1-ml syringe with a 25-G needle containing 100 μl of 20 mg/ml Texas Red-dextran. Inject the mouse by orienting the angle of the needle facing upward and forcing it gently through the skin, parallel to the vein, a few centimetres from the base of the tail. Successful insertions into the vein are indicated by easy sliding of the needle into the vein and easy injection of the solution without accumulation at the injection site.
5. Inject slowly so that 100 μl of the dextran solution enters the vein. If too much force is required for injecting the solution, it is an indication that the needle is not inserted into the vein. In this case, try again, moving slightly closer to the base of the tail with each attempt.
6. Place the mouse back into the cage for later imaging. Since the macrophages have to take up the dextran molecules by phagocytosis first, the animals can only be imaged 1–2 h after injection.
7. Image macrophages with a red filter while imaging the tumour with the green filter.

3.4. Simultaneous CFP and GFP Imaging

Recently, we presented a methodology that combines the use of inducible lentiviral expression systems and two-photon intravital multicolour imaging. We make use of multicolour intravital imaging together with an inducible cell model to compare metastatic behaviour of control and genetically modified breast cancer cell populations within the intact primary tumour of a mouse (12). It has previously been shown that it is possible to image two separate populations of cancer cells in vivo using the multiphoton microscope (17). Using the inducible cell model, we are able to image two populations that are distinct from each other for the expression of one gene and that can be optically discriminated through imaging of GFP and CFP fluorophore pair.

1. Sub-clone the gene of interest in the inducible lentiviral TREautoR3 construct by standard molecular biology techniques (12).
2. Produce lentiviral particles according to the standard protocol.
3. Transduce the cell line and expose the cells to doxycycline (10 ng/ml for 24 h should be enough) so that the successfully transduced cells start expressing the CFP-tagged protein. Sort with a flow cytometer (FACS) for the CFP-positive cells to ensure that your cell line is homogenous for the inducible expression of the protein. Next, culture the transduced cell line in a medium without doxycycline.
4. Inject the cells into the fourth mammary fat pad (see Subheading 3.1).
5. During tumour progression, add doxycycline in the drinking water of the animals at chosen time points to induce ectopic expression of the candidate gene.
6. The optimal wavelength for imaging CFP and GFP was determined to be around 860–880 nm (17). In our setup, we set the wavelength at 860 nm and use a 450/480-nm filter to image matrix and CFP, and a 515/530-nm filter to image CFP and GFP. Time-lapse images of CFP- and GFP-labelled MTLn3-ErbB1 generated tumours can be taken at 1–3-min intervals, as described above. The images are collected using NIS Elements software, as described above (Fig. 4b).
7. Perform image analysis (see Subheading 3.5).

3.5. Quantitative Image Analysis

The following steps describe the three typical procedures that can be performed to process and analyse the time-lapse *z*-series produced during the intravital imaging. For this, we make use of Image Pro Plus (version 7.0.1.658), a commercial image analysis & processing software developed by Media Cybernetics. The three procedures include the following:

1. Generating time-lapse movies from time-lapse *z*-series.
2. Eliminating *x*–*y* drift in time-lapse movies.
3. Manual cell tracking in time-lapse *z*-series.

3.5.1. Generating Time-Lapse Movies from Time-Lapse z-Series

As a result of the live cell imaging in the primary tumour of a mouse, we have a collection of time-lapse *z*-series in which one *z*-series is collected every 1–3 min. In order to visualise any cell event that is taking place in a particular image plane in the stack, it is necessary to generate time-lapse movies. Depth of field refers to the distance between the nearest and furthest object in an image in acceptably sharp focus, i.e. the zone of sharpest focus in front of, around, and behind the subject. With the extended depth of field (EDF) module in Image Pro, images that have been captured in a different *z*-axis are used to create an all-in focus image (Fig. 5a). We wrote a macro in Image Pro Plus that makes

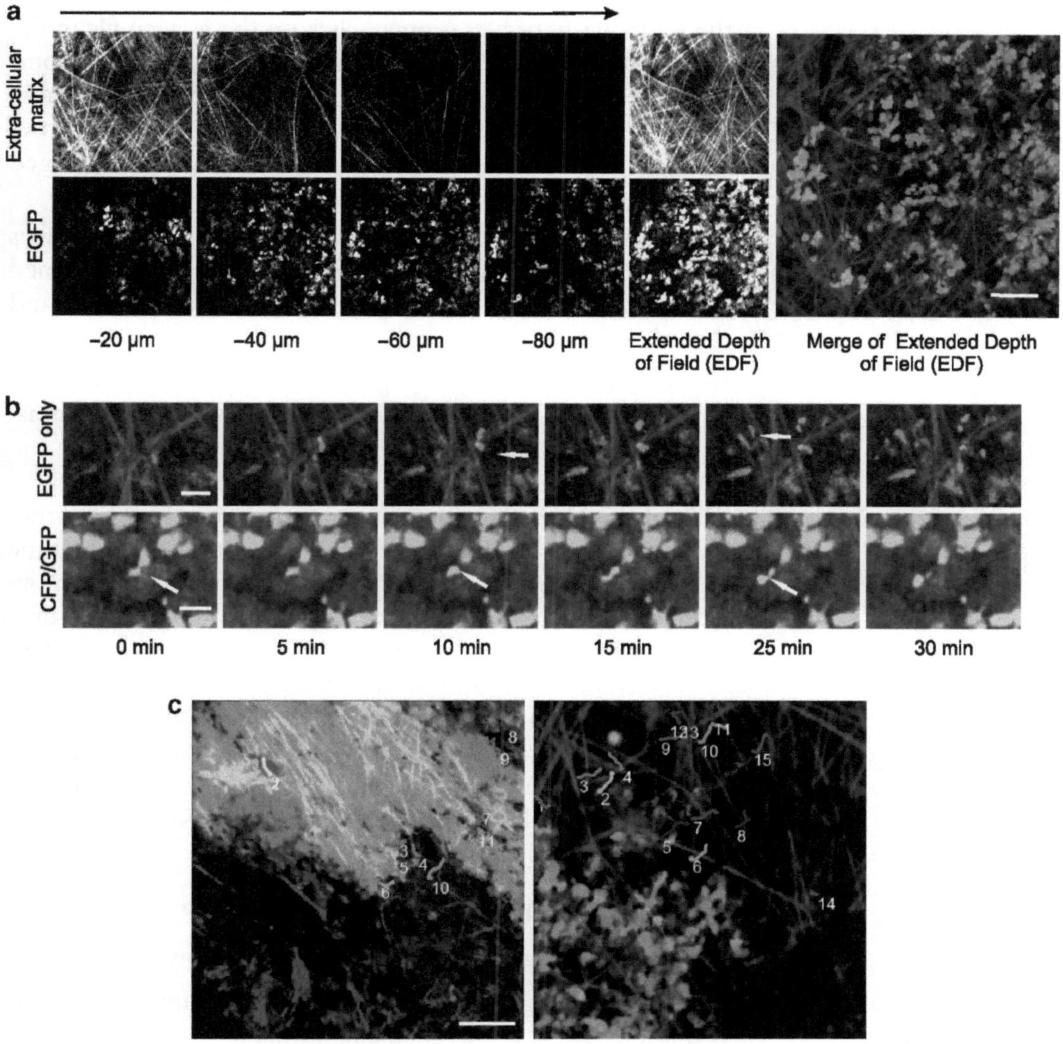

Fig. 5. Multiphoton time-lapse image analysis using Image Pro: With the extended depth of focus (EDF) module, images that have been captured in a different *z*-axis can be used to create an all-in focus image. Scale bar is 100 μm (**a**). Time-lapse images of carcinoma cells (GFP and CFP) as they extend protrusions along ECM fibres (*arrowheads*) or are crawling along counterparts. Images shown are at 5-min intervals. Scale bars are 20 μm (**b**). Manual cell tracking analysis. All individual tracks are represented with different colours. Scale bar is 100 μm (**c**).

use of this module to produce time-lapse movies. The macro is available for downloading on request.

1. In NIS Elements, extract .nd2 files into .tif for every channel, position, and time point. You will get one folder that contains all the .tif files of the experiment and that are automatically named (*namet01xy01z01c01.tif*, where *t* represents time, *xy* the position, *z* the focus plane, and c the channel).
2. Start ImagePro.
3. Run macro in Image Pro Plus "ExtendedDepth_TifFolder-Avi". A user dialogue window will pop up. In this window, a folder with .tif files can be selected and you can define how many *z*-stacks you want to use for applying EDF. When the folder with .tif files is open, click on the first .tif file in the folder. The macro will start automatically to generate a composite best focus image for each time point from the *z*-stacks and as a result, the time-lapse *z*-series are converted in time-lapse movies .avi files for each position that have been imaged during a session of 30–60 min.
4. Wait until all the .tif files are converted in .avi files. These files are saved in the same folder and are consequently named *namet01xy01z01c01 extendeddepth.avi* for each position and each channel.
5. In order to reduce memory space, it is then possible to delete all the tif files in the same folder, but always keep the original acquisition .nd2 file.

3.5.2. Eliminating x–y Drift in Time-Lapse Movies

The PFS of the Nikon microscope (or autofocus system on other types of microscope) avoids drift in the *z*-direction over the time course of an experiment. However, *x–y* drift in the time-lapse movies is still present. Eventual irregular animal breathing and/or drying of the tumour results often in a drift in the *XY* direction during image acquisition. To minimise this drift, we use the Alignment Correction feature in Image Pro that allows us to align images in a sequence. This feature is useful if you are working with images taken over time when the stage, well plate, or, in this particular case, tumour shifts a bit in the *XY* direction (see Note 8).

1. Start Image Pro.
2. Open movies to be aligned.
3. Select Align images from the Process drop-down menu.
4. Select the movies you want to align from the "input" window. It is also possible to select an anchor plane.
5. Wait until the aligned movies appear.

3.5.3. Image Analysis and Quantification

Because the data are collected as a time lapse in the *x,y* plane and a *z*-series, measurements of cell behaviour, such as polarity, motility, and intravasation, can be visualised (Fig. 5b).

1. Open Image Pro Plus.
2. Open the movie to be *analysed*.
3. Using the sequence toolbar, navigate through the movie with the slider bar or the play/stop control buttons.
4. First evaluate the movie for xy drift and image quality. If there is a xy drift, follow the procedure previously described. Depending on the image quality, it is possible to use different filters such as sharpen or high gauss filters to improve image quality. Apply these to the whole sequence.
5. Visually inspect the time lapses for moving cells. First, count the number of moving tumour cells and second, record their behaviour patterns (e.g. movement on or off matrix fibres, interaction with host cells, and movement towards blood vessels). Most MTLn3 cells will not move, but motile cells will move with speeds of up to 10 μm/min.
6. For further quantification, choose measure and track object. Open up the tracking tool.
7. Choose manual tracking for complete control. Find a moving cell and click on it frame by frame. Repeat for every moving cell in the time-lapse series. All track information are then displayed in a data table. With the tracking options, you can also select some specific measurements, e.g. distance, angle, and acceleration. The angle measurement can be useful to detect eventual specific cell orientation towards blood vessels (than can be also visualised with intravenous injection of Texas Red). Measurement statistics are also available.
8. Display tracks in tracking graphs. You can also save the tracks in the time sequence or make a snap measurement to have a recapitulative picture of all the tracks on the first frame of the movie (Fig. 5c).
9. Lastly, all or highlighted data can be easily exported to an excel sheet.

4. Notes

1. Rag2$^{-/-}$ $\gamma c^{-/-}$ mice and NOD/SCID/γcnull (NOG) mice share a similar immunodeficiency background (importantly absence of natural killer cells) and are also commercially available from Taconic Farms. Female 5- to 6-week-old BALB/c SCID/Ncr mice, Fisher 344 rats, or transgenic mouse tumour models (e.g. MMTV-pyMT) can be used as well for intravital imaging.
2. In this protocol, we make use of the well-established MTLn3 cell line, but it is possible to perform intravital imaging with

other xenografts and in particular with the human breast cancer cell line MDA-MB-231 (18).

3. Recently, a new technical development, the mammary imaging window (MIW), has been shown to be advantageous for studying cell movement and adhesion with high resolution. The use of photoconvertible fluorophores such as Kaede or Dendra2 allows a precise monitoring of cellular movement in vivo not just for hours, but for days (19, 20).

4. Although blood vessels can easily be identified through the eyepiece with the green fluorescence filter (black area with flooding hematocytes), it may be of interest to label them for direct imaging of intravasation events. Nonetheless, the labelling technique by injecting a dye in a tail vein is not easy to apply.

5. It is possible to excite the second harmonic signal with wavelengths of 760–960 nm, and since our interest is to study tumour cell migration, the excitation wavelength is selected based on the fluorescent pair CFP/GFP used to label the cells (860–880 nm for CFP and GFP), and the second harmonic image that is generated is then collected using a 450- to 480-nm-cutoff filter. Note that the second harmonic signal only reveals a very restricted subset of matrix molecules that may be present. Typically only large, type I collagen fibres are likely to be imaged. At the periphery of the tumour, elastic fibres may also be identified, but other structures including small fibres, basement membrane, fibronectin fibres, and other cell types are not imaged.

6. Before sacrificing the animal, blood from the right atrium of the heart can be sampled and plated in medium with antibiotics. If the studied cell line is highly metastatic, there should be growth of many tumour cell colonies after a week. The number of cell colonies per millilitre of blood is indicative of the intravasation events and can be correlated with the number of cell events quantified during intravital imaging.

7. The right lung was used to count the tumour burden. For approximate estimation, the right lungs were imaged with the fluorescent imaging unit IVIS (Fig. 6a). For detailed quantification, the flat side of the right lung was analysed with the immunofluorescence microscope. With a 10× objective lens (NA 0.25), we screened the flat surface of the lobe and counted the number of GFP- or CFP-positive metastases. It is possible to image the metastases in the multiphoton mode (Fig. 6b). Following this step, the right lung is cut into two pieces and fixated in 4% paraformaldehyde. The left lung is injected with ink solution and thereafter

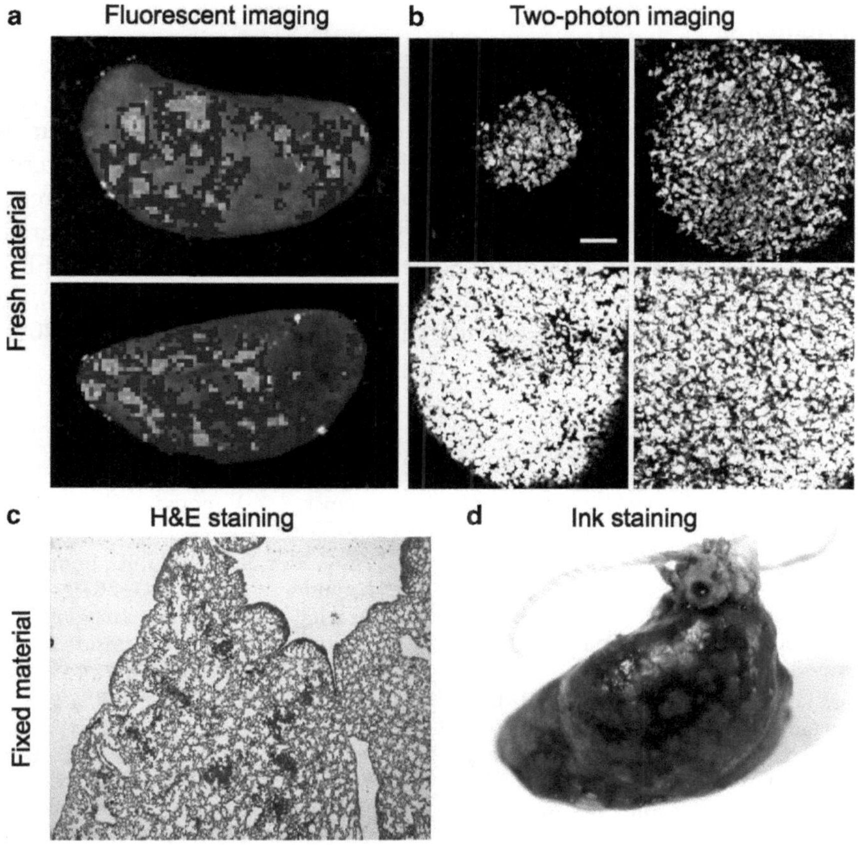

Fig. 6. Post-imaging analysis of the lung metastases: Lung metastases visualised with fluorescence lifetime imaging (FLI, GFP detection) after spontaneous metastasis assay with the MTLn3-GFP-ErbB1 cells (**a**). Lung metastases visualised with two-photon microscopy at 20× magnification (on the flat side of the left lung). Scale bar is 100 μm (**b**). H&E staining of lung tissue of MTLn3-GFP-ErbB1 cells (*right panel*) injected mice (**c**). Ink staining of lung tissue of mice bearing MTLn3-GFP-ErbB1 tumours (**c**).

destained in water and fixed in Fekete's solution (4.3% (v/v) acetic acid and 0.35% (v/v) formaldehyde in 70% ethanol) (Fig. 6d). Tumour metastases can be seen as pale nodules as they exclude the ink. The lung tumour burden is also determined by counting the number of surface metastases. The primary tumours and lungs from each mouse can be used for histological analysis. Samples are then fixed in formalin and embedded in paraffin, and 5-μm sections are stained with H&E (Fig. 6c).

8. Especially during the first 30 min of imaging, there is significant (typically >10 μm) drift. This *z*-series collection can compensate to some degree for this drift, but in the authors' experience, tracking a cell from one plane in a *z*-series to the neighbouring plane always generates an artefact for the time interval during which the shift occurs.

Acknowledgements

We thank J. Wyckoff for helpful suggestions for our intravital imaging work, M. Balzar for helpful advises concerning the multiphoton microscope, and L. Price for critically reading the manuscript. This work was financially supported by grants from the Dutch Cancer Society (UL 2007-3860), the EU FP7 Health Program Metafight (Grant agreement no.201862), and the Netherlands Organization for Scientific Research (902-21-229 and 911-02-022).

References

1. Chambers, A. F., Groom, A. C., and MacDonald, I. C. (2002) Dissemination and growth of cancer cells in metastatic sites. *Nat Rev Cancer* **2**:563–572.
2. Condeelis, J. and Segall, J. E. (2003) Intravital imaging of cell movement in tumours. *Nat. Rev. Cancer* **3**:921–930.
3. Condeelis, J. S., Wyckoff, J., and Segall, J. E. (2000) Imaging of cancer invasion and metastasis using green fluorescent protein. *Eur. J. Cancer* **36**:1671–1680.
4. Sahai, E. (2007) Illuminating the metastatic process. *Nat. Rev. Cancer* 7:737–749.
5. Wolf, K. and Friedl, P. (2005) Functional imaging of pericellular proteolysis in cancer cell invasion. *Biochimie* **87**:315–320.
6. Friedl, P. (2004) Dynamic imaging of cellular interactions with extracellular matrix. *Histochem. Cell Biol* **122**:183–190.
7. Kedrin, D., Wyckoff, J., Sahai, E., Condeelis, J., and Segall, J. E. (2007) Imaging tumor cell movement in vivo. *Curr. Protoc. Cell Biol.* **Chapter 19**: Unit 7.
8. Friedl, P. and Wolf, K. (2003) Tumour-cell invasion and migration: diversity and escape mechanisms. *Nat. Rev. Cancer* **3**:362–374.
9. Friedl, P. and Gilmour, D. (2009) Collective cell migration in morphogenesis, regeneration and cancer. *Nat Rev Mol Cell Biol* **10**:445–457.
10. Kedrin, D., van Rheenen, J., Hernandez, L., Condeelis, J., and Segall, J. E. (2007) Cell Motility and Cytoskeletal Regulation in Invasion and Metastasis. *J. Mammary. Gland. Biol. Neoplasia.* **12**:143–152.
11. Le Dévédec, S. E., van Roosmalen, W., Maria, N., Grimbergen, M., Pont, C., Lalai, R., and van de, W. B. (2009) An improved model to study tumor cell autonomous metastasis programs using MTLn3 cells and the Rag2(-/-) gammac (-/-) mouse. *Clin Exp. Metastasis* **26**:673–684.
12. Le Dévédec, S. E., Lalai, R., Pont, C., de Bont, H., and van de, W. B. (2010) Two-Photon Intravital Multicolour Imaging Combined with Inducible Gene Expression to Distinguish Metastatic Behaviour of Breast Cancer Cells *In vivo. Mol Imaging Biol.*
13. Xue, C., Wyckoff, J., Liang, F., Sidani, M., Violini, S., Tsai, K. L., Zhang, Z. Y., Sahai, E., Condeelis, J., and Segall, J. E. (2006) Epidermal growth factor receptor overexpression results in increased tumour cell motility *in vivo* coordinately with enhanced intravasation and metastasis. *Cancer Res.* **66**:192–197.
14. Markusic, D., Oude-Elferink, R., Das, A. T., Berkhout, B., and Seppen, J. (2005) Comparison of single regulated lentiviral vectors with rtTA expression driven by an autoregulatory loop or a constitutive promoter. *Nucleic Acids Res.* **33**:e63.
15. Seppen, J., Barry, S. C., Klinkspoor, J. H., Katen, L. J., Lee, S. P., Garcia, J. V., and Osborne, W. R. (2000) Apical gene transfer into quiescent human and canine polarized intestinal epithelial cells by lentivirus vectors. *J. Virol.* **74**:7642–7645.
16. van Nimwegen, M. J., Verkoeijen, S., van Buren, L., Burg, D., and van de, W. B. (2005) Requirement for focal adhesion kinase in the early phase of mammary adenocarcinoma lung metastasis formation. *Cancer Res.* **65**:4698–4706.
17. Sahai, E., Wyckoff, J., Philippar, U., Segall, J. E., Gertler, F., and Condeelis, J. (2005) Simultaneous imaging of GFP, CFP and collagen in tumours *in vivo* using multiphoton microscopy. *BMC. Biotechnol.* **5**:14.
18. Sahai, E., Garcia-Medina, R., Pouyssegur, J., and Vial, E. (2007) Smurf1 regulates tumour

cell plasticity and motility through degradation of RhoA leading to localized inhibition of contractility. *J Cell Biol* **176**:35–42.

19. Kedrin, D., Gligorijevic, B., Wyckoff, J., Verkhusha, V. V., Condeelis, J., Segall, J. E., and van Rheenen, J. (2008) Intravital imaging of metastatic behaviour through a mammary imaging window. *Nat. Methods* **5**:1019–1021.

20. Gligorijevic, B., Kedrin, D., Segall, J. E., Condeelis, J., and van Rheenen, J. (2009) Dendra2 photoswitching through the Mammary Imaging Window. *J Vis. Exp* pii: 1278. doi: 10.3791/1278.

Chapter 23

Measuring Angiogenesis in Mice

Bernardo Tavora, Silvia Batista, and Kairbaan Hodivala-Dilke

Abstract

Angiogenesis is the formation of new blood vessels from pre-existing vessels. This process is essential during embryonic development, wound healing, and regeneration as well as during several pathological conditions such as cancer. In this chapter, we describe an assay to measure angiogenesis in vivo in mice – the subcutaneous sponge assay.

Key words: Angiogenesis, Endothelial, Migration, Vessels

1. Introduction

Angiogenesis, the formation of new blood vessels from pre-existing vessels, contributes to several pathological conditions such as cancer (1, 2). During angiogenesis, endothelial cells migrate and proliferate in response to several proangiogenic growth factors (3–7).

The molecular and cellular basis of angiogenesis has been studied, not only to understand better the role of this process during development, morphogenesis, and organogenesis but also as a powerful tool to evaluate the therapeutic potential of pre-clinical drugs to treat human diseases that depend on a change in vascular blood supply. Pathological angiogenesis differs in several aspects from physiological angiogenesis. For example, tumour blood vessels are usually leaky, tortuous, and display an irregular shape, resulting in inefficient blood flow (8–10).

Several assays have been used to study angiogenesis using the mouse (*Mus musculus*) as a model organism. Ideally, angiogenic assays should be reproducible, robust, provide reliable readouts

Claire M. Wells and Maddy Parsons (eds.), *Cell Migration: Developmental Methods and Protocols*,
Methods in Molecular Biology, vol. 769, DOI 10.1007/978-1-61779-207-6_23, © Springer Science+Business Media, LLC 2011

and multi-parameter assessment as well as allow the inclusion of negative and positive controls. Angiogenic assays are usually subgrouped into two main groups: in vivo and in vitro assays. This chapter focusses on an in vivo angiogenic assay: the subcutaneous sponge assay. This assay has been adapted from the subcutaneous sponge assay developed in rats (*Rattus norvegicus*) (11). The sponge assay allows testing the angiogenic response on normal or transgenic mice to different compounds such as growth factors, inhibitors, siRNA, and other soluble compounds. This method has been shown to be an objective and reproducible assay to measure angiogenesis in vivo (12–16).

2. Materials

2.1. Animals

1. Mice (*Mus musculus*) (9–12 weeks old).

2.2. Anaesthesia

1. Isofluorane (100 % w/v; Animalcare; Dunnington, York, UK).
2. Carprofen-Rimadyl (Rimadyl, Pfizer Animal Health; Surrey, UK).

2.3. Subcutaneous Sponge Assay

1. Cylindrical polyether polyurethane sponges $1 \times 0.5 \times 0.8$ cm (Caligen Foam, Accrington, UK).
2. Phosphate-buffered saline (PBS).
3. Vascular endothelial growth factor 164 (VEGF) 10 ng/ml diluted in PBS.
4. Trocar.
5. GLUture (Abbott Laboratories, Chicago, IL, USA).
6. 27 G 1/2" (0.4×13 mm) needles (BD Microlance, Oxford, UK).
7. 1 ml syringes (BD Plastipak, Oxford, UK).
8. Ethanol: dilute ethanol absolute to 70% (v/v) in distilled water.

2.4. Immunostaining for Endothelial Marker Endomucin or Laminin

1. Formaldehyde: prepare 4% formaldehyde (v/v) diluted in PBS.
2. Ethanol: dilute ethanol absolute to 70% (v/v) in distilled water.
3. Mounting solution Prolong Gold Antifade reagent with DAPI.
4. Microscope slides.
5. Cover slips.
6. Anti-endomucin antibody (Clone V.7C7, Santa Cruz Biotechnology, Santa Cruz, CA, USA).

7. Anti-laminin antibody (Sigma-Aldrich, Dorset, UK).
8. Rat IgG_{2a}, K (clone R35-95 BD Biosciences, Oxford, UK).
9. Anti-Rat Alexa 456 (Molecular Probes, Eugene, OR, USA).
10. Anti-Rabbit Alexa 488 (Molecular Probes, Eugene, OR, USA).
11. Polyclonal Rabbit Anti-Rat Immunoglobulins/Biotinylated (DAKO, Ely, Cambridgeshire, UK).
12. Avidin/biotin Vector, blocking kit (Vector, Peterborough, UK).
13. ABC Kit Vectastain (Vector, Peterborough, UK).
14. Sigma Fast 3,3-Diaminobenzidine (DAB) tablets (Sigma-Aldrich, Dorset, UK).
15. Haematoxylin (Sigma-Aldrich, Dorset, UK).
16. DPX mountant for microscopy (BDH, VWR International, Poole, UK).
17. Acetone.
18. Xylene.

3. Methods

The subcutaneous sponge assay can be adapted to test the role of a specific protein in angiogenesis, for instance, by comparing wild type mice with mice that lack the expression of that protein or that express a mutated version of the same protein. Moreover, this assay can also be used to assess the effect of specific drugs on angiogenesis in vivo.

Injection of a growth factor is used to stimulate angiogenesis. Growth factors such as bFGF (basic fibroblast growth factor), HGF (hepatocyte growth factor), VEGF, and others can be used to stimulate the formation of new blood vessels into the subcutaneous sponge. The time course of treatment should be determined according to the properties of the compounds tested and the genetic background of the mice used.

Several characteristics of polyether sponges allow the use of this material in an in vivo assay: flexibility and hydrophilicity; capacity to be autoclaved without being physically or chemically modified; uniform pore size and pore intercommunication; resistant to formaldehyde treatment; and biocompatibility (17). Nevertheless, the greatest advantage of this assay when compared with other methods such as subcutaneous matrigel plugs and subcutaneous tumour growth relies on the specificity of the pro- and/or anti-angiogenic factors used. Moreover, the sponge material does not get reabsorbed, or dissolve like Matrigel or other such gel compounds can.

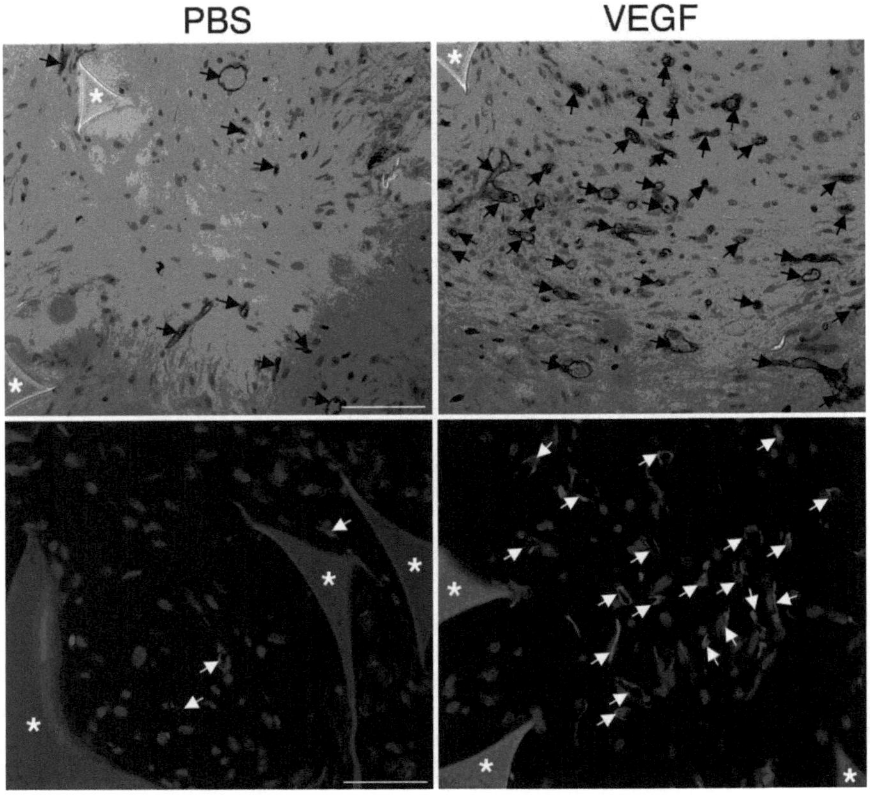

Fig. 1. Colourimetric and immunofluorescence detection of endomucin in subcutaneous sponges injected with either PBS or VEGF. Representative micrographs of endomucin-stained sponge sections show an increase in the blood vessel density within the subcutaneous sponges. *Arrows* indicate endomucin-positive vessels and *asterisks* indicate sponge spicules. Haematoxylin or DAPI nuclear staining. Scale bar = 50 μm.

In this chapter, we describe the basic method to detect the angiogenic response to VEGF in vivo (Fig. 1). VEGF was initially identified as a vascular permeability factor and is considered to be a major regulator of blood vessel formation (4–6, 18). VEGF is responsible for several responses in endothelial cells including proliferation, migration, survival, differentiation, tube formation, permeability, and vessel maintenance (6, 18–21).

3.1. Implantation of Subcutaneous Sponges

1. Polyether sponges (Caligen Foam, Accrington, UK) (1 × 0.5 × 0.8 cm) are cut to shape using a paper hole-punch.
2. Sponges are autoclaved to sterilise them.
3. On the day before the procedure, the flanks of the mice should be shaved using a clippers.
4. Before anaesthetizing the mice with isofluorane, the animals should be injected subcutaneously with 50 μl of the analgesic carprofen (5 mg/Kg). For a 25 mg mouse, dilute 0.4 ml of Rimadyl (Rimadyl, Pfizer Animal Health) in 3.6 ml of sterile water.

5. The mouse is then placed in the flow cabinet and anaesthetised with isofluorane. Meanwhile the skin over the flanks is wiped with 70% ethanol in aqueous solution before starting the surgical procedure.
6. Two sterile sponges are inserted subcutaneously in both flanks (one in each flank). The sponges are implanted, using a trocar, under sterile conditions. Briefly, the sponge is inserted longitudinally into the trocar and this is used to then insert it into a 2 cm skin flap at the posterior side of the flank. The wound is then sealed with GLUture (Abbott Laboratories, Chicago, IL, USA).
7. Using a 1 ml syringe with a 27 G ½" (0.4 × 13 mm) needle, the sponges are injected every other day with 100 μl of 10 ng/ml VEGF A_{164} or 100 μl of PBS as a negative control. $VEGFA_{164}$ is synthesised in house according to the method of Krilleke et al. (22).
8. After 7–40 days (depending on the genetic background of the mice), animals are culled by cervical dislocation, and the sponges are removed with the help of surgical forceps and scissors. After fixation in 4% formalin (diluted in PBS) overnight at 4°C, the sponges are transferred to 70% ethanol and paraffin embedded.

3.2. Immunofluorescent Staining for Detection of Blood Vessels

1. In order to evaluate the blood vessel infiltration within the sponges, 7 μm sponge sections are immunostained for an endothelial marker – endomucin or a basement membrane marker – laminin.
2. Formaldehyde-fixed sections are de-waxed by immersing in xylene for 4 and 3 min, followed by re-hydration in a gradient of ethanol diluted in distilled water (100%, 90%, 60%, and 50%) for 5 min in each solution.
3. After washing once in PBS, the samples are incubated for 1 h at room temperature in blocking solution (5% goat serum in PBS), followed by incubation (2 h at room temperature or overnight at 4°C) with 2 μg/ml anti-endomucin antibody (Santa Cruz Biotechnology, clone V.7C7, Santa Cruz, CA, USA) or 7 μg/ml anti-laminin (L9393, Sigma-Aldrich, St Louis, MO, USA) antibody diluted in 0.5% goat serum in PBS.
4. After washing three times in PBS for 5 min each, the samples are incubated for 1 h at room temperature with 2 μg/ml of secondary Alexa 488 anti-rabbit antibody (Molecular Probes, Eugene, OR, USA) or Alexa 546 anti-rat antibody (Molecular Probes, Eugene, OR, USA) diluted in 0.5% goat serum in PBS.
5. Finally after another three PBS washes for 5 min each, the sections are mounted in Prolong antifade reagent, already prepared DAPI.

6. An epifluorescence microscope equipped with a digital camera should be used to capture immunofluorescent images to quantify the number of blood vessels per unit area.

3.3. Colourimetric Immunostaining for Detection of Blood Vessels

1. Sponge sections are processed and de-waxed as described in Subheading 3.2. After washing in PBS, the slides should be incubated for 10 min in acetone at –20°C, followed by another wash in PBS and incubated with blocking solution (5% goat serum in PBS) for 30 min at room temperature.
2. The endogenous biotin or biotin-binding proteins should also be blocked by incubating the samples with Avidin D solution (Avidin/biotin Vector, blocking kit; Vector, Peterborough, UK) for 15 min at room temperature, followed by a brief rinse in PBS and another 15 min incubation with Biotin solution (Avidin/biotin Vector, blocking kit; Vector, Peterborough, UK).
3. After washing once in PBS, the samples are incubated for 2 h at room temperature or overnight at 4°C with 2 μg/ml of primary antibody (anti-endomucin antibody, Santa Cruz Biotechnology, clone V.7C7, Santa Cruz, CA, USA) diluted in 0.5% goat serum in PBS.
4. The sections are then washed three times in PBS for 5 min each and incubated with 2 μg/ml of secondary anti-rat biotinylated antibody (BA-9401, Vector Laboratories, Burlingame, CA, USA) diluted in 0.5% goat serum in PBS.
5. After washing twice in PBS, the sections are incubated for 30 min at room temperature with avidin:biotinylated enzyme complex – ABC Kit (Vector, Peterborough, UK).
6. After another two washes in PBS, the samples are incubated with the enzyme substrate – DAB prepared by dissolving one Sigma Fast DAB tablet (Sigma, St Louis, MO, USA) in 1 ml of distilled water, until the sections start to become brown (around 5 min).
7. The slides are then rinsed in running tap water for 2 min. Incubation with haematoxylin (Sigma, St Louis, MO, USA) for 1 min at room temperature is used to counterstain the nuclei.
8. After rinsing in running water for 1 min, the samples are dehydrated by immersing in gradient of ethanol diluted in distilled water (50, 60, 90, and 100%) for 5 min in each solution followed by a wash in xylene. Finally, the slides are mounted with cover glass using DPX mounting solution.
9. The total number of endomucin-positive blood vessels are counted under bright field across the entire sponge sections for multiple midline sections of multiple samples. This may

seem laborious, but we have found it to be the most accurate measure of blood vessel density. Using the same images, blood vessel density is calculated by dividing the number of blood vessels by the total sponge section area.

4. Notes

1. The background of mice should be taken into consideration when performing the subcutaneous sponge assay. It is established that different genetic backgrounds might result in differences in in vivo angiogenic assays (23). Usually, mice on a pure C57Black6 background will give a response within 10–15 days, 129 Sv mice within 20 days, and FVB mice within 20–35 days.
2. Even if the mice are from the same genetic background, there is always variability in between different animals. Therefore, when quantifying the number of blood vessels per unit of area it should at least be counted three to five sections per sponge and each experimental group should not be less than five animals.
3. It should also be noted that, occasionally, no vessels or other infiltrating cells are detectable in some areas of the sponge. We presume that this is due to a physical barrier of the compressed sponge. Therefore, it is important when measuring the sponge section area that this is done only in regions where infiltrated cells are visible by DAPI or haematoxylin staining.

References

1. Hicklin DJ, Ellis LM. Role of the vascular endothelial growth factor pathway in tumor growth and angiogenesis. J Clin Oncol 2005;23:1011–27.
2. Carmeliet P. Angiogenesis in health and disease. Nature medicine 2003;9:653–60.
3. Carmeliet P, Tessier-Lavigne M. Common mechanisms of nerve and blood vessel wiring. Nature 2005;436:193–200.
4. Keck PJ, Hauser SD, Krivi G, et al. Vascular permeability factor, an endothelial cell mitogen related to PDGF. Science (New York, NY 1989;246:1309–12.
5. Leung DW, Cachianes G, Kuang WJ, Goeddel DV, Ferrara N. Vascular endothelial growth factor is a secreted angiogenic mitogen. Science (New York, NY 1989;246:1306–9.
6. Yancopoulos GD, Davis S, Gale NW, Rudge JS, Wiegand SJ, Holash J. Vascular-specific growth factors and blood vessel formation. Nature 2000;407:242–8.
7. Gerhardt H, Golding M, Fruttiger M, et al. VEGF guides angiogenic sprouting utilizing endothelial tip cell filopodia. The Journal of cell biology 2003;161:1163–77.
8. Bergers G, Benjamin LE. Tumorigenesis and the angiogenic switch. Nat Rev Cancer 2003;3:401–10.
9. Bergers G, Song S. The role of pericytes in blood-vessel formation and maintenance. Neuro-oncology 2005;7:452–64.
10. Nunez G, del Peso L. Linking extracellular survival signals and the apoptotic machinery. Curr Opin Neurobiol 1998;8:613–8.
11. Mahadevan V, Hart IR, Lewis GP. Factors influencing blood supply in wound granuloma quantitated by a new in vivo technique. Cancer research 1989;49:415–9.

12. da Silva RG, Tavora B, Robinson SD, et al. Endothelial {alpha}3{beta}1-Integrin Represses Pathological Angiogenesis and Sustains Endothelial-VEGF. The American journal of pathology.
13. Robinson SD, Reynolds LE, Kostourou V, et al. Alphav beta3 integrin limits the contribution of neuropilin-1 to vascular endothelial growth factor-induced angiogenesis. The Journal of biological chemistry 2009; 284:33966–81.
14. D'Amico G, Robinson SD, Germain M, et al. Endothelial-Rac1 is not required for tumor angiogenesis unless alphavbeta3-integrin is absent. PLoS One;5:e9766.
15. Reynolds AR, Hart IR, Watson AR, et al. Stimulation of tumor growth and angiogenesis by low concentrations of RGD-mimetic integrin inhibitors. Nature medicine 2009;15:392–400.
16. Reynolds LE, Watson AR, Baker M, et al. Tumour angiogenesis is reduced in the Tc1 mouse model of Down's syndrome. Nature;465:813–7.
17. Lipsky MH, Lamberton P. Establishment of a neovascular bed in a collagen-impregnated polyurethane sponge. J Biomed Mater Res 1989;23:1441–52.
18. Carmeliet P. VEGF as a key mediator of angiogenesis in cancer. Oncology 2005;69 Suppl 3:4–10.
19. Connolly DT, Heuvelman DM, Nelson R, et al. Tumor vascular permeability factor stimulates endothelial cell growth and angiogenesis. The Journal of clinical investigation 1989;84:1470–8.
20. Gerber HP, McMurtrey A, Kowalski J, et al. Vascular endothelial growth factor regulates endothelial cell survival through the phosphatidylinositol 3'-kinase/Akt signal transduction pathway. Requirement for Flk-1/KDR activation. The Journal of biological chemistry 1998;273:30336–43.
21. Gupta K, Kshirsagar S, Li W, et al. VEGF prevents apoptosis of human microvascular endothelial cells via opposing effects on MAPK/ERK and SAPK/JNK signaling. Experimental cell research 1999;247:495–504.
22. Krilleke D, DeErkenez A, Schubert W, et al. Molecular mapping and functional characterization of the VEGF164 heparin-binding domain. The Journal of biological chemistry 2007;282:28045–56.
23. Rohan RM, Fernandez A, Udagawa T, Yuan J, D'Amato RJ. Genetic heterogeneity of angiogenesis in mice. FASEB J 2000;14:871–6.

Chapter 24

Time-Lapse Imaging of Chick Cardiac Precursor Cells

Junfang Song, Qiaoyun Yue, and Andrea Münsterberg

Abstract

The chick embryo is easily accessible and has therefore been widely used in developmental biology studies. In particular, the early embryo can be removed from the egg and cultured, which allows real-time observations and imaging. Here, we describe ex vivo electroporation followed by long-term time-lapse microscopy, image capture, and processing. We have applied this approach to characterise the migration route of cardiac progenitor cells (CPCs) in live embryos. The heart is the first organ to function during vertebrate development and it is essential for the continued growth and survival of the embryo. In the chick, cardiac progenitors have been mapped to the anterior and mid-primitive streak at Hamburger–Hamilton stage 3. However, until recently it was not possible to observe cell migration trajectories directly. Furthermore, we used grafting of beads or cell pellets or electroporation of expression plasmids to show that Wnt3a acts as a repulsive signal to guide the movement of cardiac progenitors.

Key words: Wnt3a, Cardiac precursors, Migration, Long-term time-lapse video, Chick embryo

1. Introduction

The chick embryo has been used as a leading developmental model for its easy adaptability to manipulation. It develops in the egg outside the mother, and thus allows a variety of manipulations, such as tissue transplantation and microdissection. Furthermore, it is amenable to over expression and knockdown studies through techniques including viral infection or the electroporation of morpholinos (MO) and expression vectors (1–3). In addition, the early embryo can be removed from the egg and cultured for about 2–3 days using early chick embryo (EC) culture (4, 5), combined with sophisticated imaging techniques (6, 7), and this makes the chick an attractive model to study cell migration behaviour.

Claire M. Wells and Maddy Parsons (eds.), *Cell Migration: Developmental Methods and Protocols*,
Methods in Molecular Biology, vol. 769, DOI 10.1007/978-1-61779-207-6_24, © Springer Science+Business Media, LLC 2011

The cells that will become the heart, cardiac progenitor cells (CPCs), are among the first cell lineages to be established as the mesoderm layer forms during gastrulation. In chick embryos, CPCs have been mapped in the anterior and mid-primitive streak at Hamburger–Hamilton (HH) stage 3 (8). After they have gastrulated through the primitive streak, cardiogenic cells occupy bilaterally symmetric regions in the lateral plate mesoderm. These future heart-forming regions are commonly named "bilateral heart fields." Subsequently, CPCs reach the anterior intestinal portal and contribute to this crescent-shaped structure. The primary heart tube develops by fusion at the ventral midline and starts beating by HH10 (9, 10).

Recent advances of in vivo electroporation have made possible to introduce specific genes into the developing embryo in a spatially and temporally restricted manner. Combined with EC culture (4, 5) and transplantation (11, 12), it is possible to label restricted cell populations with fluorescent proteins (e.g. green fluorescent protein (GFP)) and to carry out gain-of-function (overexpression or dominant-negative) experiments in the same region. If bicistronic vectors are used, the GFP-positive cells will also express the gene of interest. Alternatively, two expression plasmids can be co-injected, but this will result in mosaic expression as the plasmids may segregate. In addition, FITC-labelled antisense MO or siRNAs can be used to achieve knock-down of gene function (13, 14).

Here, we describe a method using electroporation, grafting, and long-term time-lapse video techniques to study the migration patterns of CPCs. We reported the live images and migration patterns of GFP-labelled CPCs from HH3 to HH10. In addition, we have demonstrated that their normal migration trajectories can be challenged by an ectopic source of Wnt3a (cell pellets) or by vector-mediated overexpression of Wnt3a (11). This approach can potentially be adapted to study the behaviour of other cell populations and tissues.

2. Materials

2.1. EC Culture

1. Bacterium-grade Agar.
2. Penicillin/streptomycin.
3. Thin albumen, collected from eggs.
4. Saline: 7.19 g NaCl dissolved in 1 L distilled water, autoclave and store at room temperature.
5. 30 mm Petri dish.
6. Fertile eggs obtained fresh from a commercial breeder/hatchery.
7. Incubator (38°C, e.g. Gallenkamp).

2.2. DNA Preparation

1. Fast green (1%; Sigma F7250).
2. High-speed midi-prep plasmid kit.

2.3. Micromanipulation of Chick Embryos

1. Glass capillaries (Harvard apparatus, 1.0 mm O.D. × 0.78 mm I.D.).
2. FLAMING/Brown micropipette puller (e.g. Intracel).
3. Injector (e.g. Eppendorf, FemtoJet).
4. Electroporator (e.g. Intracel, TSS20).
5. Upright brightfield microscopy (e.g. Zeiss, Stemi SV6/SV11, or Leica MZ75).
6. Upright fluorescent microscopy (as above with fluorescent attachment).
7. Tungsten needles.
8. Gas lighter for sharpening tungsten needles.

2.4. Cell Culture

1. Dulbecco's Modified Eagle's Medium (DMEM) supplemented with 10% foetal bovine serum (FBS).
2. Trypsin–EDTA.
3. PBS.

2.5. Long-Term Time-Lapse Video

1. Invert wide-field fluorescent microscope with temperature-controlled and mobile stage (Axiovert, Zeiss).
2. Six-well cell culture dishes.

2.6. Image Processing

1. Axiovison software (Zeiss).
2. Optimas VI software.
3. Image-Pro software (MediaCybernetics).
4. Image J (http://rsbweb.nih.gov/ij/).

3. Methods

CPCs ingress through the anterior and middle part of HH3 primitive streak and they arrive in the primary heart tube forming at the ventral midline by HH10. Our aim was to directly observe the migration route of prospective cardiac cells from HH3 to HH8-10. Image recording of these cells involved the following key steps: in vitro (EC) culture, gene transfer, microsurgery and grafting, image recording, and processing.

Our strategy was:

1. To take chick embryos into EC culture, we used a quick and efficient way for EC culture (4, 5). In this method, a filter paper carrier is used to hold the embryo under tension, and

the embryo is transferred onto a semi-solid substratum of agar–albumen in a 3 cm Petri dish, where it can grow and develop normally for up to 3 days.

2. To label CPCs in the primitive streak with GFP by targeted electroporation in EC culture using the bicistronic vector, pCAβ, which contains the chicken beta-actin promoter, a multiple cloning site followed by an internal ribosomal entry site (IRES), and the coding sequence for GFP. After detection of GFP (1–2 h), small regions of primitive streak are grafted to an untreated EC-cultured host. The host is selected to be stage matched with the donor and for healthy development in culture. Electroporation introduces transient but strong expression of the transgene within the targeted cells. It uses a brief electric pulse to create reversible pores in the plasma membrane, which allows the charged DNA to permeate into the cell (15). Injected DNA diffuses easily and this can result in a larger domain than intended being electroporated. To make up for this limitation, grafting is applied after GFP expression has become visible, thus a smaller region can be selected and microdissected.
3. To record movement patterns by time-lapse video microscopy. The system consists of a camera, wide-field inverted microscope, and imaging software. To support chick embryo development during imaging, an incubation chamber is also required. This provides embryos with a humidified and temperature-controlled environment. To obtain reliable and reproducible results, only HH stage 3b and 3c embryos were used for all micromanipulation unless specifically stated (16). In grafting experiment, the incision was made in the same position in both donor and host embryos, just posterior to the tip of the primitive streak. Time-lapse imaging for 24 h allows the recording of labelled embryos from HH3 until they develop into HH8 to HH10. Each time-lapse imaging experiment used four to six embryos simultaneously and was repeated at least three times.

3.1. In Vitro Culture for Time-Lapse Imaging

1. Fertile eggs are incubated at 38°C overnight (about 12–13 h) in a humidified incubator.
2. After the eggs reach stage 3b or 3c, EC cultures are prepared as described (see Note 1) (4, 5). Embryos grow on a semi-solid medium composed of saline, agar, and thin albumen (see Note 2).
3. Embryos on filter paper rings are cultured with the ventral side up at least until they reach stage 8 (see Note 3).

3.2. Preparation of DNA for Electroporation

1. pCAβ-IRES-GFP vector is used for transient transfection. Expression of any cDNA insert is driven by the chicken β-actin promoter, and GFP translation is initiated from an IRES (17).

2. Plasmid DNA is purified by an endotoxin-free plasmid midi-kit. We use a High-speed midi-prep plasmid kit available from Qiagen.
3. DNA solution is adjusted to a final concentration of approximately 200 ng/μl. Fast green (1%) is added in a ratio of 1:40 for visualisation.

3.3. Labelling Cardiogenic Cells by Electroporation

1. A glass micro-capillary pipette is pulled to generate two needles consisting of a fine outer diameter and a sharpened point. Twenty micolitres of DNA plasmid/fast green mixture is loaded into the needle by pipetting.
2. After mounting the needle into the injector, break the needle tip with fine forceps to the desired diameter (about 10 μm).
3. Set operating mode of the injector to "Manual." Press the foot control to trigger injection. Gravity and vacuum pressure draw the DNA solution to the pointed tip of the needle. Adjust injection pressure (300–500 hPa) to allow a slow, low-pressure delivery of volumes that range from 0.5 to 5 μl.
4. Position the glass needle over one side of the primitive streak using a micromanipulator. Move the needle downwards and stop once the tip enters the epiblast (Fig. 1a). Then inject the DNA solution into the space between epiblast and vitelline membrane (see Notes 4 and 5).
5. After injection, remove the glass needle. Gently apply a drop of saline on the surface of the embryo.
6. Arrange the electrodes and position the positive electrode (anode) right above the primitive steak omitting the tip (see Note 6). Slowly adjust the electrode down and stop once it contacts the saline solution (Fig. 1b).
7. Set the electroporator to deliver 5 pulses with 100 ms interval of 12 V and 50 ms duration (see Note 7).
8. Run the electroporation program twice. Conductance can be verified when bubbles form between cathode and anode.
9. Allow the recovery of electroporated embryos by incubation at 38°C for 1–2 h and then transfer them to 28°C overnight. The incubation at lower temperature permits expression of plasmid and GFP fluorescence develops, while the embryos remain arrested in their development at HH3.

3.4. Homochronic Grafting in Steak Region

1. Screen for GFP fluorescence by examining the electroporated embryos under a microscope. Healthy embryos with GFP-positive streak are chosen as donors for grafting (Fig. 1d).
2. Prepare fresh HH3 embryos in EC culture as host.
3. Use a sharp tungsten needle to dissect a small piece of fluorescent cells from the cardiogenic region of the primitive streak

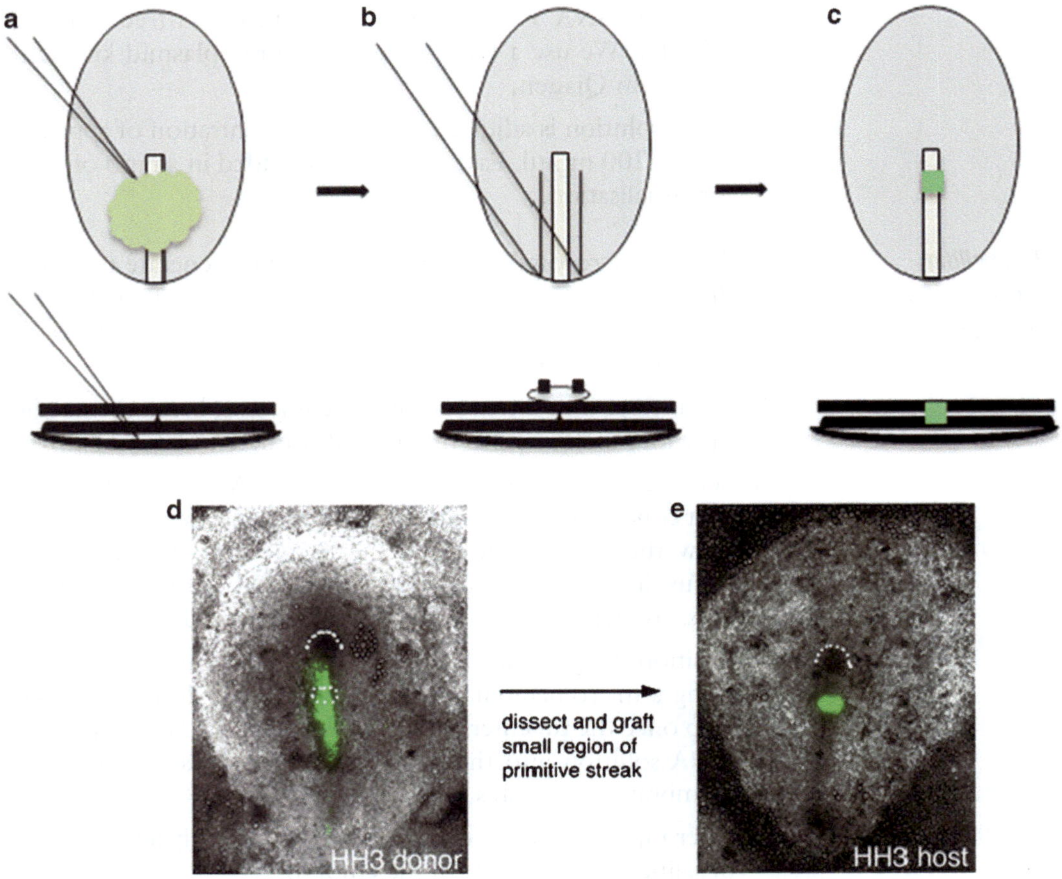

Fig. 1. Schematic illustration of the electroporation and grafting procedure. (**a**) HH3 embryos were placed in EC culture with their ventral side up. DNA/fast green mixture was injected into the space between epiblast and vitelline membrane as indicated in the transverse view. (**b**) The anode was positioned above the primitive steak for electroporation with GFP expression vector. (**d**) After 3–5 h incubation, HH3 donor embryos were expressing GFP fluorescence and used as donors. (**c**, **e**) Small grafts were dissected and transplanted into a stage-matched host. The white-dotted semi-circles show the anterior margin of the primitive streak.

(see Note 8). Transfer the tissue to a stage matched host embryo with a 20 μl micropipette.

4. An equivalent sized tissue is cut and removed from the host primitive streak at exactly the same position.
5. Gently use the needle to guide the fluorescent donor tissue into the excision, and position it into the hole carefully (Fig. 1c).
6. Culture grafted embryos at 38°C for imaging or other further operations. An example of the results produced is shown (Fig. 1e).

3.5. Implantation of Cell Pellets into the Future Heart-Forming Region

1. Grow RatB1 fibroblasts expressing Wnt3a (18) to 80–90% confluency in a 75 cm^2 tissue culture flask.
2. Remove culture medium and wash cells with PBS twice.
3. To dissociate them from the plastic surface, incubate cells with 0.5 ml Trypsin–EDTA for 2–3 min at 37°C.

4. Stop the digestion by adding 10 ml DMEM including 10% FBS. Gently wash cells off the flask surface and collect them with a 10 ml pipette.
5. One millilitre of cell suspension and 1 ml fresh medium are seeded into a 3 cm bacterial Petri dish. After 37°C overnight incubation, cells form aggregates of different sizes, which do not attach and can be dislodged easily from the bottom of dish.
6. Select and transfer a small cell aggregate to the surface of the host embryo by using a 20 μl micropipette (see Note 9).
7. Make an incision in the future heart-forming region (anterior-lateral) with a tungsten needle. Gently push the cell pellet into the incision.
8. Incubate implanted embryos first at 28°C for an hour. This allows the graft to heal and prevents rupture due to rapid growth whilst allowing the GFP plasmid to be expressed.
9. Transfer embryos to 38°C for long-term imaging. An example of the results produced is shown (Fig. 2).

3.6. Explant Culture and Cell Pellet Implantation in Area Opaca

1. Dissect a small piece of GFP-positive streak from donor embryos. Transfer the tissue to a HH4 host embryo.
2. With a tungsten needle cut an incision into the area opaca of the HH4 host and gently guide the GFP-labelled explant into the incision.
3. Prepare cell pellets as described above and transfer a small cell pellet to the same host embryo and implant it close to the explant. The distance between cell pellet and explant should be no more than twice the explant diameter.
4. The boundary between the area opaca and the area pellucida strongly attracts primitive streak cells. Thus, the cell pellet and explants should be positioned a sufficient distance away, approximately in the middle of the area opaca, and parallel to the boundary (see Note 10).
5. Culture host embryos at 28°C for an hour and then transfer them to 38°C for imaging (see Note 11). A schematic diagram for the experiment and an example of the results are shown (Fig. 3).

3.7. Live Embryo Imaging

1. Turn on temperature control module at least 30 min before setting up time-lapse movie.
2. Select six healthy embryos after micromanipulation for imaging. Transfer them from 3 cm Petri dishes into a six-well plate by picking up the filter paper with forceps.
3. To keep embryos humidified, fill the gaps in between the wells with the cotton soaked in water. Place six-well dish into the pre-warmed chamber (see Note 12).

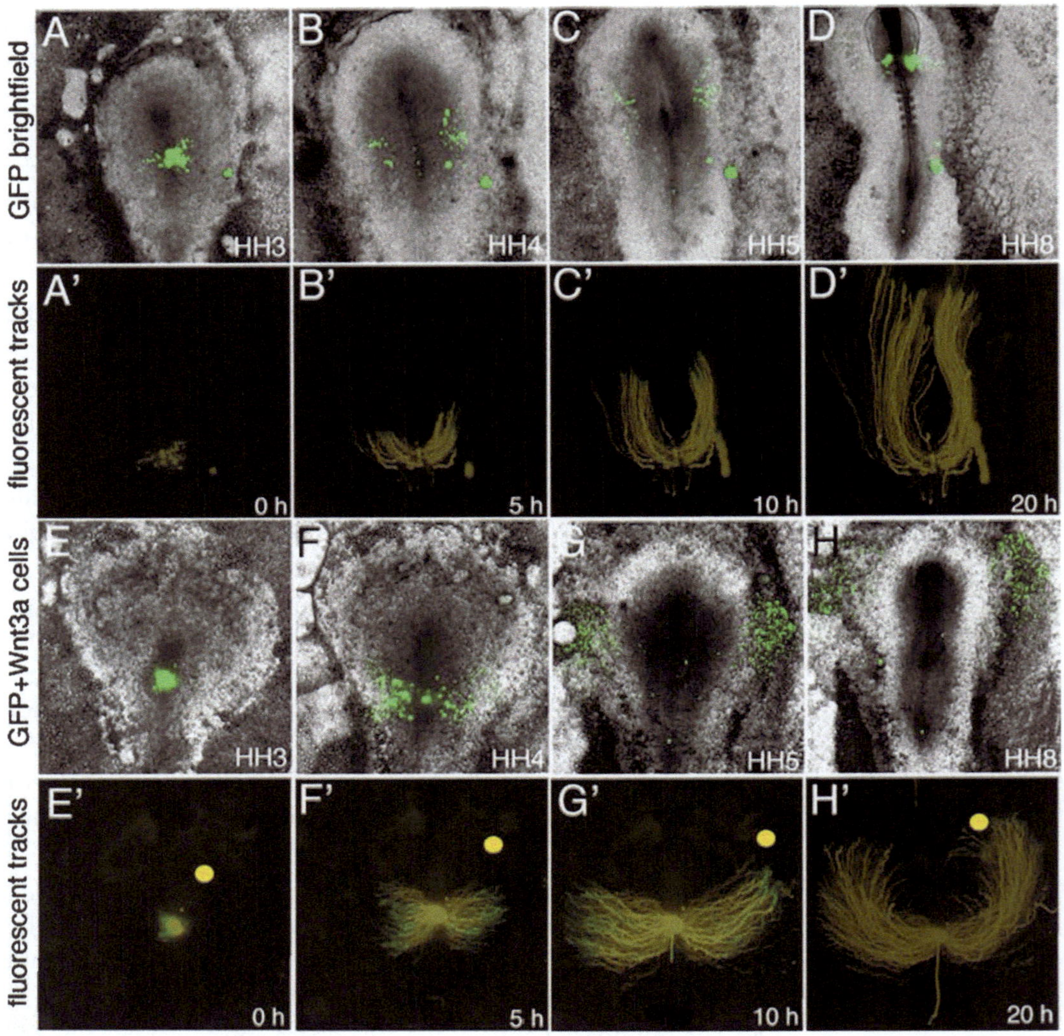

Fig. 2. Cardiac progenitors move on highly directed trajectories, which are controlled by Wnt3a, at least in part. (**A**–**D**) Bright-field images from long-term time-lapse recordings show the location of GFP-labelled CPCs at different stages of development, indicated on each panel. (**A'**–**D'**) Corresponding movement trajectories were generated and visualised using successive logical addition of images. Time elapsed after imaging began is indicated in hours. (**E**–**H**) Wnt3a expressing cells (*yellow dot*) were grafted into HH3 embryos to examine the response of GFP-labelled cardiac progenitors and the effects on their movement trajectories. Bright-field images showed that most cells failed to reach the midline and remained located in the area opaca and lateral mesoderm. (**E'**–**H'**) Tracked images indicated that the cells took a longer trajectory with a more pronounced lateral direction.

4. Turn on the appropriate lasers and set up the desired configuration on the microscope.
5. Open software and open a work area window. Set up the parameters for microscopy. The parameters have to be adjusted for different types of image acquisition software and different size of objects being imaged. For Axiovision (Zeiss) and ECs, choose 100% light port left (TV) on light path, 2.5× objective, and 1.0 on Optovar/Tabelens.

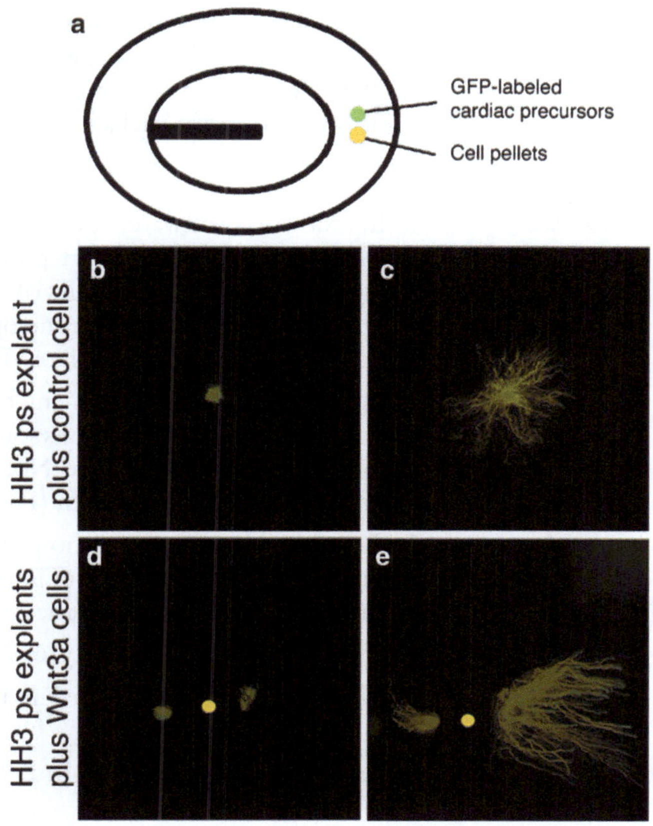

Fig. 3. Explants of CPCs migrate away from a Wnt3a-expressing cell pellet on the area opaca. (**a**) Schematic illustration of explant assay. GFP-positive primitive streak (ps) explants were cultured on the area opaca of an HH4 host embryo, either in the presence of control cell pellets (**b**, **c**) or adjacent to a cell pellet expressing Wnt3a (**d**, **e**). (**b**, **d**) Fluorescent image at the beginning of imaging and (**c**, **e**) movement trajectories. (**c**) HH3 primitive streak cells are highly motile and migrated in all directions away from a graft cultured on the area opaca and ignored the presence of a control cell pellet. (**e**) HH3 primitive streak cells migrated away from Wnt3a-expressing cell pellet (*yellow dot*).

6. For imaging configuration, choose "multi-dimensional acquisition" or equivalent in work area window. For image acquisition, choose time-lapse and position list at experiment section. Define two channels (bright field and eGFP) (see Note 13) and set autofocus. Choose 6 min for interval setting and 24 h duration (see Note 14).
7. To define the position of the first embryo, open live window from main menu and mark/find window from work area. Position the embryo to the centre of the live window and then manually focus. Click measure button in camera setting. Save the position in mark/find window.
8. Repeat step 7 to define the positions of the other five embryos in the six-well dish (see Note 15). This process may vary slightly between different software; however, the principle should be the same.

9. Click start button to start time-lapse imaging. During the imaging, the software creates a designated file (Zeiss format: "zvi") for each embryo in the desired location. The file format of other software packages will have a different designation (see Note 16).

3.8. Data Processing

1. In Axiovision software, open "zvi" file. Choose eGFP channel. Click File and then Export button. Export the one-channel movie (fluorescent) as a series of "jpg" files. Most image acquisition software should have the option to export image files in other formats, such as TIFF or JPEG.
2. Fluorescent object movement can be traced in two dimensions by several image analysis software packages (e.g. Volocity, Optimax, Image-Pro) (see Note 17). Some of the tracked images produced are shown (Figs. 2 and 3).
3. The following describes how to track migrating cell using successive logical addition of images with macros written with Image-Pro (MediaCybernetics):
    1. Open image-Pro Analyzer 7.0 software.
    2. Click Open button in File menu to open a sequence of images (see Note 18).
    3. Use "Save as" button in File menu to save the images as "seq" file.
    4. Open the "seq" file. When open or save a sequence of images, tick the option "open as sequence" or "save as series." For manual alignment, a reference point is selected that is identical in all the images generated. The images are then manually aligned to this reference point and a new image sequence is generated, based on successive logical addition of images, in which the movement trajectories of cells can be visualised. In the ImagePro package, this is done as follows.
    5. Click "Track objects" button in Macro menu and a dialog box named as Tracking data table displays.
    6. Choose "add track manually" button on the dialog. Keep clicking left mouse to mark object from the first to last frame. The marked object is used as reference for the following alignment.
    7. Choose the "added track" on the frame sequence.
    8. Click "align image" by selected track button in Macro menu to start alignment.
    9. Once the alignment finishes, a sequence of aligned images is generated.

10. To start track, click “firework” button in Macro menu. Based on aligned images, a new sequence of tracked images is generated using a running frame summation (i.e. Frame 2 = 1 + 2, Frame 3 = 1 + 2 + 3, Frame 4 = 1 + 2 + 3 + 4, etc.).
11. To colour the tracked objects, use “heads” and “tails” button in Macro menu. The objects are pseudo-coloured yellow throughout, with the last few frames in green. A dialog box is used to select the frame number that is displayed in green.
12. Use “Save as” button in File menu to save the sequence of tracked colour images as “tiff” and “avi” file.

For automatic alignment:

1. Click align image global button in Macro menu.
2. A dialog box displays for selecting area of interest (AOI) and a square bar appears on the image.
3. Move the square bar using the mouse to position AOI and adjust it to appropriate size. The positioned AOI is used as reference for the following alignment (see Note 19).
4. Press “continue” button in the dialog to start alignment. Once finished, a new sequence of aligned images is generated.

Alternatively, we performed successive logical addition of images with macros written with the Optimas VI imaging library:

1. Open Optimax software for tracing images. For image alignment, click Marco in menu. Enter preset mac C:/Marco/xycorrect.mac, which includes alignment and track functions as in Image-Pro.
2. Open a series of image sequences for alignment. Set images’ number and name to be processed.
3. Define tacking window sidelength, *X* and *Y* direction. Adjust intensity map to optimise contrast on images.
4. Click left mouse and a square bar will appear. Set reference by moving the square to the targeted region on the start image.
5. Press right mouse, define threshold range to reduce background florescent. Release the mouse and let alignment start. After it finishes, a series of aligned images is saved in distributed folder as “tiff” file.
6. For images tracing, click Utilities on menu and choose autotrack. Open images and set images’ number to be traced. Press “OK” to start trace. Once finished, traced images will be saved as a series of .tiff files in designated name.
7. Create an “avi” movie out of a series of “tiff” files using Image J or Quicktime Pro.

4. Notes

1. Only choose well-developed embryos for all micromanipulations.
2. When preparing agar–albumen medium, aliquot 2.5 ml of the mixture per Petri dish (diameter 3 cm). Do this gently and quickly to avoid introducing bubbles into the medium, which will be problematic during embryo development. The prepared medium should be stored in 4°C. Medium older than 1 week is not recommended.
3. In EC culture, air bubbles often form between the filter paper carrier and solid medium along the paper edge. Air bubbles can damage the embryo during incubation and/or obscure the view during time-lapse imaging. To reduce the formation of air bubbles, use tissue paper to remove any liquid attached along the filter paper edge before placing it on the medium.
4. To target the epiblast, DNA should be injected into the space between the epiblast and vitelline membrane. Poor injection will result in low electroporation efficiency.
5. DNA solution is injected as long as the foot control is pressed under manual mode. The optimal injection time (to inject 1–2 μl DNA solution) needs to be established empirically.
6. Commercial electrodes are available, but the electrode can also be made in house at low cost. The cathode and anode are made from two 10 × 0.35 mm platinum wire bent 45° approximately 3 mm from the end into an "L-shape." The cathode and anode are parallel to each other. The distance between them is maintained at approximately 2 mm distance.
7. The electroporation program we use results in efficient transfection and good embryonic survival rates in our hands. However, the conditions should be established empirically and may need to be adjusted for the specific equipment used and the area targeted. Usually low voltage and long interval times between pulses are applied. If the voltage is too high, this can lead to severe malformations or embryonic death.
8. The tungsten needles for grafting need to be very sharp. They can be sharpened with high temperature using a gas lighter.
9. Cell pellets vary in size and are between approximately 50–120 μm in diameter. Same-size pellets should be selected unless systematic size variation is desired.
10. Through careful positioning, it is possible to culture up to four explants on the area opaca of a single HH4 host.
11. It is important to do the operation rapidly and effectively. The longer the embryos stay at room temperature, the less likely they are to survive the manipulations.

12. The temperature-controlled chamber can be set up to fit 6-well, 12-well, or 24-well multi-dishes. In ovo imaging is also possible. The setting used depends on the type and size of the organism cultured.
13. In most imaging software, the intensity of light in brightfield and GFP channel can be adjusted manually and saved as default setting. Once one imaging cycle ends, the laser shutter for GFP excitation should be closed automatically. This prevents bleaching and damage to live organisms due to overexposure.
14. When setting up time-lapse imaging for the first time, a trial will help to determine how often images need to be captured to achieve optimal image quality.
15. It is also possible to image multiple area opaca explants (see Subheading 3.6) on a single HH4 host; this significantly increases the number of samples per over-night image experiment. The position of each one has to be defined as described in step 7.
16. We describe the techniques to prepare a whole embryo for in vitro culture and to image its ventral side. The method can also be modified for recording the dorsal side or potentially for in ovo imaging.
17. We use Axiovision software for time-lapse imaging and Image-Pro and Optimax for tracing. These methods can be adapted for use with other image processing software; however, the exact procedure will vary slightly from the one described here and has to be established for the specific software used.
18. When using Image-Pro, black–white images are required for tracking.
19. We align all images to a common reference point that is within the original grafting site in the host embryo.

References

1. Tickle, C. (2004) The contribution of chicken embryology to the understanding of vertebrate limb development. *Mech Dev* **121**, 1019–1029.
2. El-Ghali, N., Rabadi, M., Ezin, A. M. & De Bellard, M. E. (2010) New methods for chicken embryo manipulations. *Microsc Res Tech* **73**, 58–66.
3. Nakamura, H., Katahira, T., Sato, T., Watanabe, Y. & Funahashi, J. (2004) Gain- and loss-of-function in chick embryos by electroporation. *Mech Dev* **121**, 1137–1143
4. Chapman, S. C., Collignon, J., Schoenwolf, G. C. and Lumsden, A. (2001) Improved method for chick whole-embryo culture using a filter paper carrier. *Dev Dyn* **220**, 284–289.
5. Streit, A. (2008) EC culture: a method to culture early chick embryos. *Methods Mol Biol* **461**, 255–264.
6. Dormann, D. & Weijer, C. J. (2006) Imaging of cell migration. *Embo J* **25**, 3480–3493.
7. Zamir, E. A., Czirok, A., Cui, C., Little, C. D. and Rongish, B. J. (2006) Mesodermal cell displacements during avian gastrulation are due to both individual cell-autonomous and convective tissue movements. *Proc Natl Acad Sci USA* **103**, 19806–19811.

8. Hamburger, V. and Hamilton, H. L. (1992) A series of normal stages in the development of the chick embryo. 1951. *Dev Dyn* **195**, 231–272.
9. Brand, T. (2003) Heart development: molecular insights into cardiac specification and early morphogenesis. *Dev Biol* **258**, 1–19.
10. Redkar, A., Montgomery, M. and Litvin, J. (2001) Fate map of early avian cardiac progenitor cells. *Development* **128**, 2269–2279.
11. Yue, Q., Wagstaff, L., Yang, X., Weijer, C. and Münsterberg, A. (2008) Wnt3a-mediated chemorepulsion controls movement patterns of cardiac progenitors and requires RhoA function. *Development* **135**, 1029–1037.
12. Yang, X., Dormann, D., Munsterberg, A. E. and Weijer, C. J. (2002) Cell movement patterns during gastrulation in the chick are controlled by positive and negative chemotaxis mediated by FGF4 and FGF8. *Dev Cell* **3**, 425–437.
13. Mende, M., Christophorou, N. A. and Streit, A. (2008) Specific and effective gene knockdown in early chick embryos using morpholinos but not pRFPRNAi vectors. *Mech Dev* **125**, 947–962.
14. Yang, X., Chrisman, H. and Weijer, C. J. (2008) PDGF signalling controls the migration of mesoderm cells during chick gastrulation by regulating N-cadherin expression. *Development* **135**, 3521–3530.
15. Cui, C., Lansford, R., Filla, M. B., Little, C. D., Cheuvront, T. J. and Rongish, B. J. (2006) Electroporation and EGFP labeling of gastrulating quail embryos. *Dev Dyn* **235**, 2802–2810.
16. Lopez-Sanchez, C., Puelles, L., Garcia-Martinez, V. and Rodriguez-Gallardo, L. (2005) Morphological and molecular analysis of the early developing chick requires an expanded series of primitive streak stages. *J Morphol* **264**, 105–116
17. Alvares, L. E., Schubert, F. R., Thorpe, C., Mootoosamy, R. C., Cheng, L., Parkyn, G., Lumsden, A. and Dietrich, S. (2003) Intrinsic, Hox-dependent cues determine the fate of skeletal muscle precursors. *Dev Cell* **5**, 379–390.
18. Münsterberg, A. E., Kitajewski, J., Bumcrot, D. A., McMahon, A. P. and Lassar, A. B. (1995) Combinatorial signaling by Sonic hedgehog and Wnt family members induces myogenic bHLH gene expression in the somite. *Genes Dev* **9**, 2911–2922.

Chapter 25

Characterizing System Performance in Total Internal Reflection Fluorescence Microscopy

Juliane P. Schwarz, Ireen König, and Kurt I. Anderson

Abstract

Total internal reflection fluorescence microscopy (TIRF-M) has become an increasingly popular tool to study events in close proximity to the cell cortex, such as cell adhesion (Axelrod, J Cell Biol 89:141–145, 1981; Gingell et al., J Cell Biol 100:1334–1338, 1985; Patel et al., J Cell Sci 121:1159–1164, 2008), actin (Bretschneider et al., Curr Biol 14:1–10, 2004; Gerisch, Biophys J 87:3493–3503, 2004; Merrifield et al., Nat Cell Biol 4:691–698, 2002), and membrane dynamics (Oheim et al., Eur Biophys J 27:83–98, 1998; Steyer et al., Nature 388:474–478, 1997; Weisswange et al., J Cell Sci 118:4375–4380, 2005). In TIRF-M, dim fluorescence from cortical structures can be imaged with high contrast despite large cytoplasmic background from the bulk of the cell body. With any imaging method, standard samples are required to ensure correct alignment and monitor system performance over time. Here, we describe procedures for the production and use of a test sample to characterise and optimize TIRF system performance.

Key words: TIRF – total internal reflection, TIRF-CIF – TIRF contrast improvement factor, ICE – image contrast estimation, Critical angle, Fluorescent beads-fluorescein

1. Introduction

Total internal reflection fluorescence microscopy (TIRF-M) is an ideal tool for visualisation of cellular events taking place near the bottom surface of the cell. TIRF illumination is based on the phenomenon that when a light beam passes from a material of high-refractive index into a material with lower refractive index, part of it is reflected, while part of it is refracted (reviewed in (1, 2)). Increasing the angle of incidence reduces the refracted portion until, near the so called "critical angle", the refracted portion is negligible and the entire incident beam is reflected. This creates

Claire M. Wells and Maddy Parsons (eds.), *Cell Migration: Developmental Methods and Protocols*,
Methods in Molecular Biology, vol. 769, DOI 10.1007/978-1-61779-207-6_25,

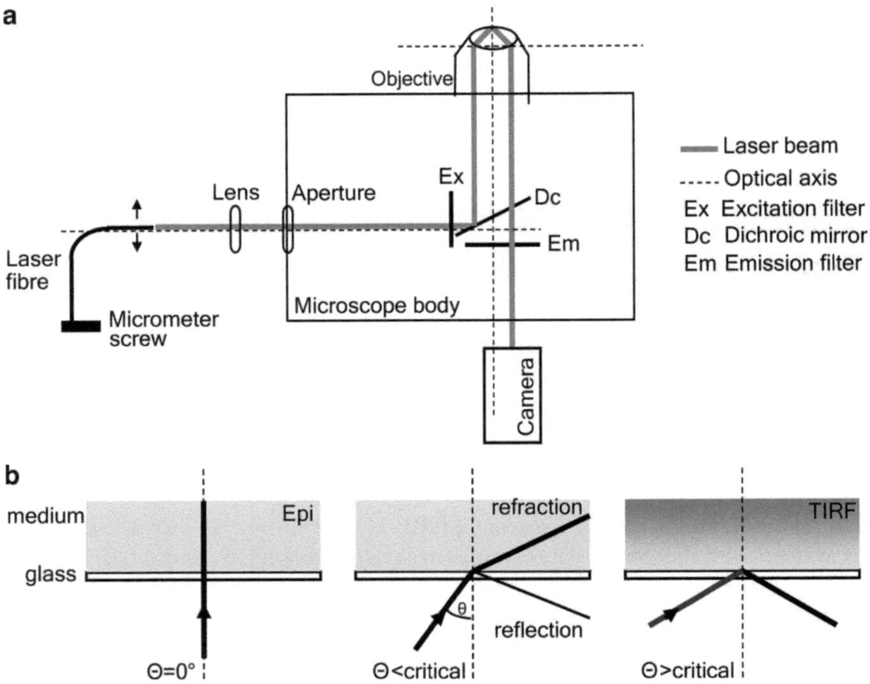

Fig. 1. TIRF microscopy. (**a**) Schematic of the microscope setup used to develop the experiments described. Laser light is introduced parallel to the optical axis of the microscope using a laser fiber that can be moved sideways (indicated by *arrows*) using a micrometer screw. Movement away from the optical axis shifts the focus of the laser beam to the edge of the back focal plane of the objective. This increases the angle of illumination resulting in the total internal reflection of the beam. A lens introduced between laser output and back aperture of the microscope allows the adjustment of focus when changing between objectives with different numerical apertures. Important optical axes are indicated by *dashed lines*. (**b**) Principles of TIRF illumination. In Epi fluorescence, when the angle of incidence is 0°, the light passes glass and medium nearly unobstructed (*left panel*). When the laser beam is tilted away from this optical axis, a portion of light is refracted, while a much smaller portion is reflected (*middle panel*). Increasing the angle of incidence increases the reflected portion until, when reaching the critical angle and above, the beam is completely reflected back into the objective, creating an evanescent wave that is most intense at the surface between the two media, but then quickly loses intensity (*right panel*).

an evanescent field which is most intense at the surface between the high- and low-refractive index materials but exponentially looses intensity further away from it (Fig. 1.b). In biological samples, the evanescent wave is created at the border of a high-refractive index coverslip and low-refractive index sample medium. Therefore, only features of the specimen that are close to the substrate-attached cell cortex can be visualized. TIRF-M can be used in negative or positive contrast modes. For negative contrast, non-fluorescent cells are bathed in fluorescent media, and regions of close contact between cell and substrate appear dark through exclusion of the dye (3, 4). For positive contrast, features of the cell are labelled using fluorophores such as membrane dyes (5, 6) or fluorescent proteins (7). Optical sectioning in TIRF has been estimated (5) and measured (8) in the range of 100–200 nm and is superior to that of confocal microscopy. However, use of a coherent illumination field makes it more prone to disturbances through interference with dirt on optical surfaces and in the immersion oil.

The samples and measurements described here provide a means of measuring key parameters of TIRF illumination, such as angle of incidence and depth of illumination. Furthermore, we describe an approach to estimate the improvement in image contrast (TIRF Contrast Improvement Factor, TIRF-CIF) achieved by TIRF-M over wide-field imaging, which can be used to assess the quality of the system setup and to control consistency of performance between different experiments.

2. Materials

2.1. TIRF Microscope

Most of the experiments described here can be performed with any TIRF microscope. In our case, we used a custom-built microscope allowing manual manipulation of the illumination angle (see Notes 1 and 2 and Fig. 1a). The laser safety features of some commercial systems may prevent determination of illumination angles, as described in Subheading 3.2, but will still allow the measurement of image contrast estimation (ICE) and TIRF-CIF using the beads sample (see Subheading 3.3) and the estimation of the illumination depth using the short focal length lens (see Subheading 3.4).

1. A TIRF microscope with inverted stand.
2. TIRF objective (100× magnification, 1.45NA, see Note 3).
3. Filter set consisting of a Z473/10 excitation filter, a Z488RDC dichroic mirror, and a ET525/50M emission filter (Chroma Technology Corp., USA) optimized for the detection of GFP using 473 nm excitation (use of other wavelengths, e.g. 488 nm, is also possible).

2.2. Measurement of Illumination Angles

1. 30 cm ruler.
2. Glass hemisphere (21.72 mm diameter) made from BSM16 glass ($n = 1.6204$, Melles-Griot).

2.3. Preparation of TIRF Sample

1. Fluorescein (in DMSO, Reference Dye Kit, Invitrogen) diluted to 2 μM in PBS/19% Glycerol (see Note 4).
2. FluoSpheres® Fluorescent microspheres, 100 nm yellow/green (505/515) fluorescent carboxylate-modified beads (Invitrogen, see Note 5). Prepare a 1 in 100 dilution of the beads in Millipore water.

3. High-vacuum grease (see Note 6).
4. Scotch® Magic™ tape (3 M), glass slides, glass coverslips (20 mm^2, No. 1½), ethanol, methanol, sonicating waterbath, tweezers, and rack for holding coverslips.

2.4. Illumination Depth Estimation

1. Small diameter, short focal length lens (01LPX421, r=2.004 mm, f=3.5 mm, Melles-Griot).
2. 2 μM Fluorescine/19% Glycerol in PBS.
3. Glass bottom dish (e.g. MatTek, 14 mm diameter).

2.5. Software for Image Analysis

1. Image J, Version 1.43 (www.rsbweb.nih.gov/ij/) (see Note 7).
2. Excel (Microsoft).

3. Methods

3.1. Adjusting the Laser and the Stage

1. Mark an X on the ceiling directly over the objective to serve as a target for the laser at Epi-Illumination. This can be estimated by eye or even determined using a plummet suspended from the ceiling. Try to be as accurate as possible (within centimetre range).
2. If using a stage insert, make sure the specimen holder is normal to the optical axis. Using a test sample (e.g. a slide with fixed cells), focus on one end of the sample, translate the stage to the other end of the sample, and check the focus. Adjust the set screws on the stage insert to ensure the sample remains in focus over 15–20 mm of lateral stage movement.

3.2. Measurement of Illumination Angles

1. With no sample in place, center the laser beam on the X marked above the microscope.
2. Place the glass hemisphere on the specimen holder and make oil contact with the objective. Adjust the stage position so that the laser beam strikes the X on the ceiling again and center using the back-reflections of the laser beam within the hemisphere.
3. Erect a ruler on the side of the microscope stage. Note the position of the bottom of the ruler relative to the bottom of the hemisphere (offset, equal to height of stage in Fig. 2b and 0.71 cm in Fig. 2c) and lateral distance of the ruler from the objective aperture. Make sure the ruler is in line with the beam. An example of this setup is shown in Fig. 2a.
4. Reduce the illumination angle until the laser beam strikes the ruler and record the height of the beam and the position of the beam, e.g. the micrometer screw reading or value obtained from the software used to control illumination angles.

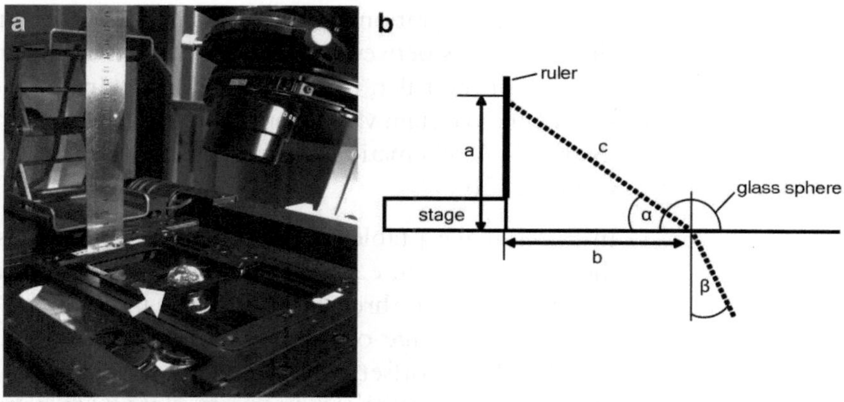

c

	A	B	C	D	E	F	G	H	I	J	K	L	M
1	beam adjust	ruler reading 1	ruler reading 2	ruler reading 3	average reading	a	c	sin alpha	alpha (rad)	alpha (deg)	beta	extrapolated angles	corrected angles
2	0										0	0.00	0.00
3	150	19.4	19.35	19.3	19.35	20.06	23.38	0.86	1.03	59.11	30.89	29.28	31.35
4	200	11.9	11.95	12	11.95	12.66	17.44	0.73	0.81	46.53	43.47	43.04	46.55
5	210	10.8	10.75	10.8	10.78	11.49	16.62	0.69	0.76	43.76	46.24	46.03	49.96
6	220	9.65	9.65	9.65	9.65	10.36	15.85	0.65	0.71	40.80	49.20	49.10	53.52
7	230	8.45	8.45	8.5	8.47	9.18	15.11	0.61	0.65	37.40	52.60	52.26	57.26
8	235											53.86	59.21
9	240	7.3	7.3	7.3	7.30	8.01	14.43	0.56	0.59	33.72	56.28	55.49	61.22
10	245	6.7	6.65	6.7	6.68	7.39	14.09	0.52	0.55	31.64	58.36	57.13	63.31
11	250	6.1	6.05	6.15	6.10	6.81	13.80	0.49	0.52	29.57	60.43	58.80	65.48
12	255	5.5	5.5	5.5	5.50	6.21	13.51	0.46	0.48	27.36	62.64	60.49	67.77
13	260											62.19	70.19
14	265											63.92	72.82

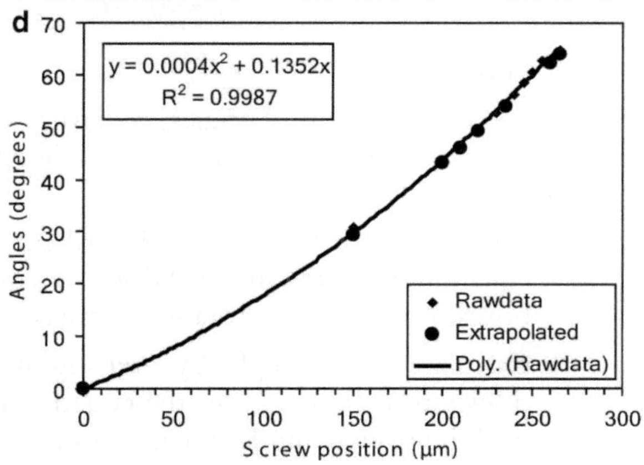

Fig. 2. Measurement of illumination angles. (**a**) Picture of the microscope stage with glass hemisphere (*white arrow*) and ruler used to measure illumination angles. Note spot from laser beam on ruler at approximately 10.5 cm. (**b**) Schematic representation of the microscope setup shown in (**a**) indicating the variables needed for the angle calculation. (**c**) Excel table illustrating the different steps to calculate the angles of incidence. A detailed description of the formulae and commands used can be found in Subheading 3.2.1. (**d**) Calibration curve created in Excel showing measured (raw data, *full squares*) and extrapolated (*full circles*) angles of incidence on a polynomial trendline.

5. Continue changing the angle of the laser beam in regular increments and record the ruler readout until the beam no longer comes out of the objective.
6. Repeat steps 4 and 5 three times.

3.2.1. Calculating the Angle(s) of Incidence Using Excel

Constructing a calibration curve in Excel allows determination of illumination angles between the measured points. Depending on the position of the ruler, it may also be necessary to extrapolate measurements to obtain values for the highest illumination angles (see Note 8). A schematic definition of the relevant variables is shown in Fig. 2b.

1. In Excel, make a table, as displayed in Fig. 2c, consisting of the beam position, e.g. micrometer screw or software value in column A, the three ruler readings in column B, C, and D, and the average of these readings in column E. In column F, add the offset between the image plane and the bottom of the ruler (if there is any, see Note 8) to get the value for a.
2. In column G, calculate c using the formula: $c=\sqrt{(a^2+b^2)}$, with b being the distance between the objective and the ruler.
3. In column H, calculate $\sin\alpha = a/c$ and in column I, calculate the value for α in radians (=$a\sin$(H)).
4. Convert radians into degrees in the next column: 1 radian = 57.29°, thus α in degrees = 57.29(I).
5. Finally calculate the incident angle β by:

$$\beta = 90° - \alpha$$

in column K

6. For extrapolation of angles, construct a calibration curve of beam position (column A) versus illumination angles (column K).
7. Add a trendline (polynomial fit, order 2) and tick the "display equation on chart" and the "display R squared value on chart" boxes. An example calibration curve is shown in Fig. 2d.
8. Calculate missing angles by inserting the required beam positions into column A. Put the equation given by the polynomial fit into column L and add in the beam positions for x.
9. Add the calculated values into your graph as a second data series to check how well they fit.
10. As the beam passes through a coverslip when imaging, rather than the glass hemisphere, the angles need to be corrected for the difference in the refractive index of the hemisphere ($n_2 = 1.620$) compared to the refractive index of the coverslip ($n_1 = 1.523$) using Snell's law:

$$n_1 \sin\beta = n_2 \sin\alpha,$$
$$\sin\beta = (n_2/n_1)\sin\alpha.$$

In column M, type the term = DEGREES ($a$sin(1.620/1.523) sin(RADIANS(L))) into the formula bar to get the illumination angles for the beads sample.

3.2.2. Calculating the Critical Angle

The critical angle (θ_c) is the angle of illumination at which the incident beam is entirely reflected. Increasing the angle of incidence above this value leads to the creation of the evanescent field. The critical angle is a theoretical value that is dependent on the refractive indices of the two optically different media. It is calculated according to the formula:

$$\theta_c = \arcsin\,(n_2/n_1),$$

with n_2 being the refractive index of the less dense material and n_1 the refractive index of the more dense material. For the beads sample described below, the critical angle is calculated as:

$$\theta_c = \arcsin\,(1.523/1.363) = 63.5°,$$

with n_2 (corning coverslip) = 1.523 and n_1 (fluorescein/19% glycerol) = 1.363.

3.3. Preparation of Beads Sample

3.3.1. Cleaning of Coverslips (See Note 9)

1. Rinse the coverslips individually in ethanol (using tweezers).
2. Collect them in a large glass beaker and rinse them once more with ethanol.
3. Discard most of the liquid and allow the remaining liquid to evaporate (see Note 10).
4. Rinse the coverslips once with Millipore water, put into a suitable holder, and sonicate them in a sonicating water bath (filled with Millipore water) for about 1–2 min.
5. Finally, rinse them again individually with ethanol (using tweezers) and leave to air-dry on a piece of tissue.

3.3.2. Assembly of the Beads Sample

1. Take a glass slide and stick two strips of Scotch® Magic™ tape along each long side so that the tape reaches about 5 mm into the slide. Cut off the remaining tape (see Fig. 3a).
2. Run a continuous bead of vacuum grease on the glass slide along the edges of the tape where the coverslip will be attached.
3. Suspend 0.3 μl FluoSpheres® (1 in 100 dilution in water) in 15 μl methanol. Mix well and sonicate briefly to break up clumps.
4. Pipette onto an ethanol-washed coverslip and leave to air-dry. Use a glass pipette or a plastic pipette tip to streak out possible drops that form during evaporation.
5. Lay the bead-coated coverslip down on top of the vacuum grease (bead surface on the inside of the chamber) and gently press the coverslip onto the grease to form a seal.
6. Fill the resulting cavity with 2 μM fluorescein/19% glycerol just before imaging. If desired, the ends of the chamber can be filled with vacuum grease to prevent evaporation.

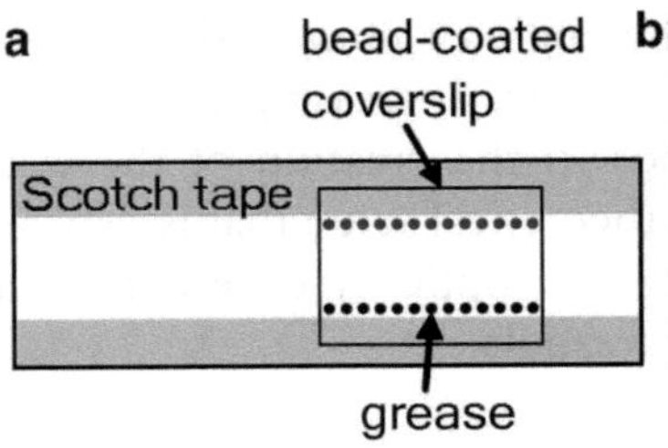

b

	A	B	C	D	E	F
1	angle degrees	beads average	bkg average	beads-bkg	ICE	CIF
2	0	5245.4	4707.8	537.6	0.1	1.0
3	57.3	8532.0	7106.1	1425.9	0.2	1.8
4	59.2	8954.6	6664.9	2289.7	0.3	3.0
5	61.2	8839.4	4595.3	4244.1	0.9	8.1
6	63.3	5655.4	244.4	5411.0	22.1	193.9
7	65.5	4651.2	160.2	4491.0	28.0	245.6
8	67.8	2513.8	83.7	2430.1	29.0	254.4
9	70.2	29.2	7.1	22.1	3.1	27.5

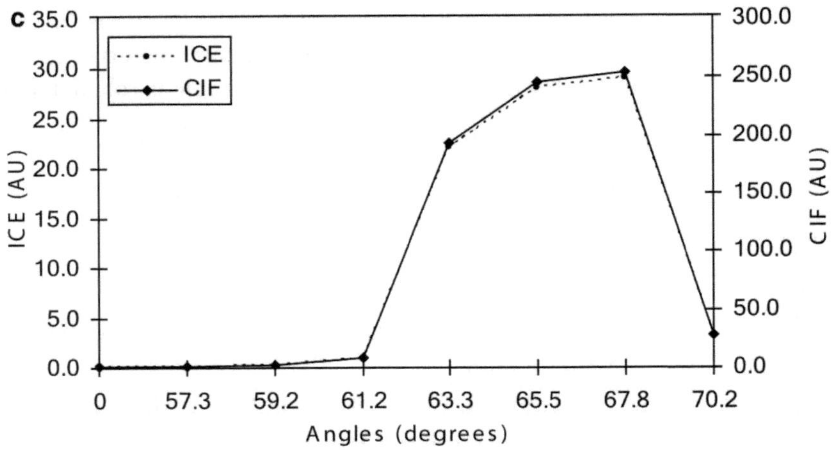

Fig. 3. ICE and TIRF-CIF. (**a**) Schematic of the beads sample described in Subheading 3.3.2. (**b**) Excel table illustrating the analysis of the intensity data of the beads sample acquired in Image J. (**c**) Curve depicting the change of ICE (*dashed line with circles*) and TIRF-Contrast Improvement Factor (CIF, *full line with diamond*) with increasing illumination angle.

3.3.3. Acquisition of the Angle Series

1. To set up the sample, focus on the beads in TIRF and adjust illumination intensity and exposure time so that the beads are not overexposed in TIRF (see Note 11).
2. Go to Epi-illumination and record a series of images (as a stack) at different illumination angles (see Note 12). Start with epi-illumination and then increase the angle approximately half way to the critical angle (around 57.3°). From there, increase the illumination angle gradually in suitable increments and take images until you lose the beads signal completely (at around 72.8°).
3. Repeat at least five times using different fields.

3.3.4. Image Analysis of the Angle Series Using Image J

1. Open an image stack; select a frame corresponding to an illumination angle near the critical angle.
2. Make sure individual beads can be seen clearly by selecting: Image > Adjust Brightness/Contrast > Reset.
3. Find the beads by: Process > Binary > Find Maxima. Select "Preview point selection," set the noise tolerance so that all beads in the field are located (see Note 13) and select point selection as output type.

4. Save the selection of the located beads by: Analyze > Tools > ROI manager > Add. Rename the point set with the image file name, e.g. Field01. Do not close the ROI manager window.
5. To record the background go to: Edit > Selection > Select All. In the ROI manager choose "Add [*t*]." Rename selection as Background, e.g. Field01-Bkg, (see Note 14).
6. To measure the intensities, highlight "Field01" and "Field01-Bkg" in the ROI Manager. Choose: More > Multi-Measure. Tick the boxes "measure all slices" and "one row per slice."
7. Cut the measurements displayed in the results window and paste them to an Excel workbook.
8. Repeat steps 1–7 for each field of beads acquired.
9. At the end of your Image J analysis, select all point sets in the ROI manager window, select More > Save and save them as a ROI set. This is a record of which beads were analyzed.

3.3.5. Calculating ICE and TIRF-CIF in Excel

1. In Excel, make a table with the relevant angle values (as calculated in Subheading 3.2.1) in column A; see Fig. 3b.
2. Average the bead and background intensities of your Image J measurements in column B and C.
3. In column D, calculate Beads-Background and in column E, divide (Bead-Background)/Background (=D/C) to obtain the ICE.
4. Normalize by dividing all ICE values of the angle series individually by ICE-Epi (=ICE at 0°) in column E to generate the TIRF-CIF for each illumination angle.
5. Plot ICE and CIF against the angles (using the lines on two axes option under custom types in the chart wizard) as shown in Fig. 3c.

3.4. Illumination Depth Estimation

1. With no sample on the microscope, center the laser beam on the X on the ceiling.
2. Place the short focal length lens with the curved surface down on a glass bottom dish.
3. Fill the dish with 2 μM fluorescine/19% glycerol until the lens is completely submerged.
4. Place the dish on the microscope and center the submerged lens over the objective by eye. Then turn on the laser and move the sample until the laser beam passes through the lens and strikes the ceiling X.
5. Using laser illumination at the same intensity used to record the angle series, focus on the interface between the curved lens and the coverslip. To begin with, close down the field diaphragm and adjust the focus until the edges of the field diaphragm appear in focus (see Note 15).

6. Open the field diaphragm again. With the field diaphragm fully open, the point of lens contact appears as a dark shadow in an otherwise bright image. The dark shadow is a negative contrast image of the lens surface, which is darkest in the center where the lens rests on the glass. It is important to center this shadow in the camera image so that inversion of the intensity scan will be symmetrical (see Subheading 3.4.1, step 4 and Note 16).
7. Gradually increase the illumination angle and adjust the focus until the illumination angle of maximum image contrast is reached (as previously described in Note 11).
8. Acquire an image.

3.4.1. Calculation of the Illumination Depth Using Image J and Excel

1. Draw a line ROI across the image through the center of the dark spot, perpendicular to the plane of laser incidence on the sample, i.e. if the laser impinges on the sample from left to right, draw the line from top to bottom. Ensure that the center of the linescan is aligned with the center of the dark spot in the image see Fig. 4a.
2. Increase the width of the ROI to between 10 and 20 pixels in order to smooth the subsequent intensity scan through averaging.
3. Use Menu > Analyze > Plot Profile to create a linescan, which should approximately resemble an inverted Bell curve. Click "Copy" in the plot window and paste the values (consisting of columns for pixel number and intensity) into Excel.
4. In Excel, paste your data into column A so that the pixel number, starting with 0, will be displayed in column A and the intensity values in column B. Paste the same values again into columns C and D and invert them by selecting: Data > Sort > by column C, descending so that the first intensity value in column D becomes the last value in the column.
5. In column E, average the values from B and D, and in column C, multiply the pixel number (in column A) by the appropriate calibration factor (μm/pixel for your imaging system) to convert pixel number into distance (μm). An example Excel table is shown in Fig. 4b.
6. Plot average intensity (*E*) against distance (*C*) and estimate the width of the curve at one half of the curve height (full-width at half maximum, FWHM) as shown in Fig. 4c.
7. The illumination depth (*d*) is calculated using basic trigonometry based on the FWHM and the radius of the lens as follows:

$$x = \frac{1}{2}\,\mathrm{FWHM}$$

$$r = 2{,}004\ \mu\mathrm{m}$$

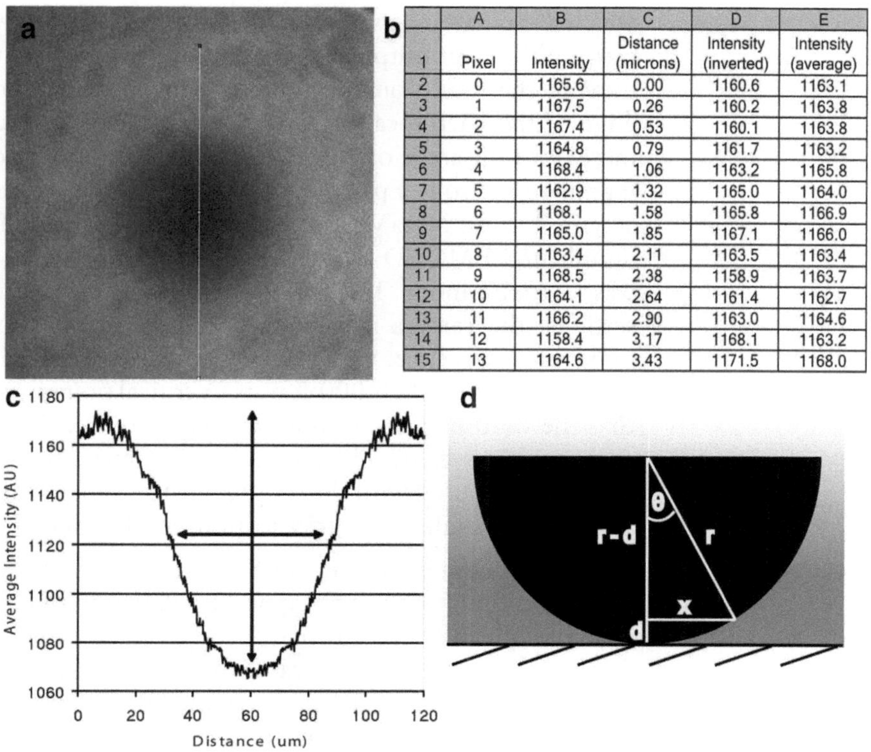

	A	B	C	D	E
1	Pixel	Intensity	Distance (microns)	Intensity (inverted)	Intensity (average)
2	0	1165.6	0.00	1160.6	1163.1
3	1	1167.5	0.26	1160.2	1163.8
4	2	1167.4	0.53	1160.1	1163.8
5	3	1164.8	0.79	1161.7	1163.2
6	4	1168.4	1.06	1163.2	1165.8
7	5	1162.9	1.32	1165.0	1164.0
8	6	1168.1	1.58	1165.8	1166.9
9	7	1165.0	1.85	1167.1	1166.0
10	8	1163.4	2.11	1163.5	1163.4
11	9	1168.5	2.38	1158.9	1163.7
12	10	1164.1	2.64	1161.4	1162.7
13	11	1166.2	2.90	1163.0	1164.6
14	12	1158.4	3.17	1168.1	1163.2
15	13	1164.6	3.43	1171.5	1168.0

Fig. 4. Illumination Depth Estimation. (**a**) TIRF-image of a glass lens submerged in fluorescein. A 15px wide linescan was used to acquire intensity data in Image J. (**b**) Section of an Excel table (data points 0–13 of 458) illustrating the calculation of average intensity and distance from the measurement depicted in (**a**). (**c**) Excel-based graph of intensity along the linescan. The height of the curve is approximately 100 AU. The width at one half of the curve height (FWHM) is approximately 55 μm. (**d**) Schematic diagram of the experimental setup, defining the variables used to calculate the penetration depth of the evanescent field. A detailed description of the formulae used can be found in Subheading 3.4.1. For the analysis described here, the illumination depth *d* was calculated as 188.6 nm.

$$r - d = \sqrt{(r^2 - x^2)}$$

$$d = r - (r - d)$$

Refer to the schematic in Fig. 4d for an explanation of the relevant variables.

4. Notes

1. The procedures described here were developed on a Nikon Eclipse TE2000-U inverted microscope, in which the Nikon Epi-fluorescence condenser is replaced with a custom condenser. Laser light (473 nm diode laser, Omicron GmbH, Germany) controlled by a DAC 2000 card is introduced parallel to the optical axis of the microscope directly from the output of a modified 3.5 μm optical fiber (IOM GmbH, Berlin). In Epi-fluorescence, the fiber output is centered on the optical axis to generate a collimated laser beam,

which illuminates the sample at 0°. A micrometer screw allows moving the fiber output in a conjugate image plane, thereby moving the focal point of the beam from the center to the edge of the back focal plane of the objective. The resulting increase of the angle of incidence at the coverslip surface shifts illumination from Epi-fluorescence into TIRF. Images are acquired using MetaMorph software and a Cascade II back-illuminated EMCCD camera (Roper Scientific). The camera was set to: FT mode, 1 MHz Digitizer, and Gain1. With these settings, the camera is capable of a 16 bit dynamic range; however, in practice, images were generally acquired having between 1.000 and 10.000 counts. A critical point is to ensure that the bead images are not over-exposed, i.e. the full-well capacity of the camera is not exceeded. Generally, any TIRF system can be used with the settings that seem most appropriate, but the results may differ from those described here.

2. Laser safety is an important issue for TIRF-M because parallel light is emitted from the objective. The light emitted from the objective during confocal microscopy rapidly loses intensity due to divergeance of the beam. In contrast, the intensity of parallel light used in TIRF-M is retained many yards from the objective. The increased hazard of parallel light is, however, mitigated by the low intensities generally used for live cell imaging, which are typically on the scale of a few milliwatts. For all procedures listed, ensure that the power of light emitted from the objective does not exceed 5 mW. Always be aware of the direction of laser emission from the objective and ensure that bystanders are protected from accidental exposure.
3. Using a TIRF objective with different properties than described here will lead to some variation in the data. Using lower magnification (e.g. 60×) or higher numerical aperture (e.g. 1.49NA) objectives will result in a broader range of illumination angles and different illumination power output at the objective.
4. Fluorescein was mixed with glycerol to match the refractive index of the solution to the refractive index of cytoplasm (9), which would normally surround the labelled bright structures, e.g. focal adhesions in cell samples. The refractive index of this solution is around 1.363, as measured using a refractometer (Ramsey, Cambridgeshire).
5. FluoSpheres® are sold according to their nominal size. The real size of the beads varies between batches and is displayed on the packaging. The 100 nm beads used to develop the sample described here had a real size of 110 nm.
6. High-vacuum grease was filled into a 5 ml plastic syringe (BD) with a cutoff pink luer lock needle serving as applicator.
7. Image J was run on a PC. The location of some menu commands may be slightly different on a Mac.

8. Ideally, the bottom of the ruler should be at the same height as the glass hemisphere. If the ruler is placed too high above the hemisphere, it may prohibit measurement of the largest illumination angles.
9. Washing the coverslips this way improves the surface properties of the glass, resulting in very even spreading and evaporation of the beads-containing solution.
10. The dried cover slips can have white residues from the alcohol, which will be removed in the next step.
11. Focus using live camera mode. Rather than focusing by eye, focus by maximizing bead intensities. This can be done most easily by viewing an image histogram in live imaging mode. For our experiments, an illumination intensity of 200 μW at an exposure time of 400 ms resulted in an average bead intensity of 5,000 counts in TIRF (at screw position 245 μm, corresponding to an angle of incidence of 63.3°).
12. For our experiments, the angle series was recorded at micrometer screw positions 0, 230, 235, 240, 245, 250, 255, and 260 μm corresponding to the following angles: 0°, 57.3°, 59.2°, 61.2°, 63.3°, 65.5°, 67.8°, 70.2°, and 72.8° (see Fig. 2c). Try not to refocus between individual images of the same field, as bead intensity values sensitively depend on focus. Additionally, if stage and sample are adjusted correctly, there should also be no need to refocus when changing between adjacent image fields.
13. To find a suitable value for the noise tolerance, consider what is the highest value which selects all beads? What is the lowest value which does not include non-beads? Set the value higher rather than lower, e.g. 500–1,000, to ensure discrimination for data where bead intensities are low.
14. In our experience, the average intensity of the entire image is accurate within ~0.01% to the intensity of the background alone. This is because the beads occupy only a very small percentage of the image area.
15. Dust or debris on the coverslip surface can be helpful to now find the focus at the lens/coverslip interface. If all else fails, fluorescent beads can be attached to the substrate. Care should be taken to keep the density of beads on the coverslip surface low. To attach beads to the glass surface of a dish, resuspend them in ethanol, sonicate, and spread out on the dish using a bent glass pipette or plastic pipette tip until the alcohol has evaporated, essentially as described in Subheading 3.3.2. The amount of ethanol needed is dependent on the size of the glass bottom dish used; the amount of beads should at least be ten times less than described for the beads sample as really just one fluorescent bead is needed on the dish to find the focal plane and beads in the vicinity of the lens interferes with the measurement.

16. It is important that the central spot does not completely fill the camera image, i.e. a region of background remains all the way around the spot. It is necessary to clearly recognize background level in the intensity scan in order to estimate the curve height. If the spot size is too large, it may be necessary to use a de-magnifying lens between the microscope and camera to image a larger field of view on the camera.

Acknowledgements

The authors would like to thank Brad Amos for his suggestion of using an inverted lens bathed in fluorescein to estimate the evanescent field penetration depth.

References

1. Toomre, D., and Manstein, D. J. (2001) Lighting up the cell surface with evanescent wave microscopy. *Trends Cell Biol* **11**, 298–303.
2. Simon, S. M. (2009) Partial internal reflections on total internal reflection fluorescent microscopy. *Trends Cell Biol* **19**, 661–8.*
3. Gingell, D., Todd, I., and Bailey, J. (1985) Topography of cell-glass apposition revealed by total internal reflection fluorescence of volume markers. *J Cell Biol* **100**, 1334–8.
4. Gerisch, G., Bretschneider, T., Muller-Taubenberger, A., Simmeth, E., Ecke, M., Diez, S., and Anderson, K. (2004) Mobile actin clusters and traveling waves in cells recovering from actin depolymerization. *Biophys J* **87**, 3493–503.
5. Axelrod, D. (1981) Cell-substrate contacts illuminated by total internal reflection fluorescence. *J Cell Biol* **89**, 141–5.
6. Weisswange, I., Bretschneider, T., and Anderson, K. I. (2005) The leading edge is a lipid diffusion barrier. *J Cell Sci* **118**, 4375–80.*
7. Bretschneider, T., Diez, S., Anderson, K., Heuser, J., Clarke, M., Muller-Taubenberger, A., Kohler, J., and Gerisch, G. (2004) Dynamic actin patterns and Arp2/3 assembly at the substrate-attached surface of motile cells. *Curr Biol* **14**, 1–10.
8. Mattheyses, A. L., and Axelrod, D. (2006) Direct measurement of the evanescent field profile produced by objective-based total internal reflection fluorescence. *J Biomed Opt* **11**, 014006.*
9. Curl, C. L., Bellair, C. J., Harris, T., Allman, B. E., Harris, P. J., Stewart, A. G., Roberts, A., Nugent, K. A., and Delbridge, L. M. (2005) Refractive index measurement in viable cells using quantitative phase-amplitude microscopy and confocal microscopy. *Cytometry A* **65**, 88–92.
10. Patel, H., Konig, I., Tsujioka, M., Frame, M. C., Anderson, K. I., and Brunton, V. G. (2008) The multi-FERM-domain-containing protein FrmA is required for turnover of paxillin-adhesion sites during cell migration of Dictyostelium. *J Cell Sci* **121**, 1159–1164.
11. Merrifield, C. J., Feldman, M. E., Wan, L., and Almers, W. (2002) Imaging actin and dynamin recruitment during invagination of single clathrin-coated pits. *Nat Cell Biol* **4**, 691–8.*
12. Oheim, M., Loerke, D., Stuhmer, W., and Chow, R. H. (1998) The last few milliseconds in the life of a secretory granule. Docking, dynamics and fusion visualized by total internal reflection fluorescence microscopy (TIRFM). *Eur Biophys J* **27**, 83–98.*
13. Steyer, J. A., Horstmann, H., and Almers, W. (1997) Transport, docking and exocytosis of single secretory granules in live chromaffin cells. *Nature* **388**, 474–8.

* These papers are especially recommended for readers that want to know more about biological applications of TIRF microscopy.

Chapter 26

Fluorescence Recovery After Photobleaching

Alex Carisey*, Matthew Stroud*, Ricky Tsang*, and Christoph Ballestrem

Abstract

This chapter describes the use of microscope-based fluorescence recovery after photobleaching (FRAP). To quantify the dynamics of proteins within a subcellular compartment, we first outline the general aspects of FRAP experiments and then provide a detailed protocol of how to measure and analyse the most important parameters of FRAP experiments such as mobile fraction and half-time of recovery.

Key words: Fluorescence recovery after photobleaching, Fluorescence microscopy, Green fluorescent protein, Half-time of recovery, Mobile fraction, Diffusion, Binding reaction kinetics, Focal adhesions, Vinculin

1. Introduction

1.1. Principles of Fluorescence Recovery After Photobleaching

Fluorescence recovery after photobleaching (FRAP) is a powerful, microscopy-based methodology for investigating molecular dynamics within living cells. Whereas traditional fluorescence microscopy yields information in a qualitative "yes or no" manner as to the localisation of molecules in cells, FRAP allows us to elucidate the dynamics of the protein of interest. In order to perform FRAP, the protein of interest must be tagged to a fluorophore. Green fluorescent protein (GFP) is a well-characterised fusion tag widely used for protein labelling in live cells. It was originally discovered in the jellyfish, *Aequorea victoria*, and has been subsequently modified to produce brighter and more photostable variants (1). In FRAP, fluorescent molecules that localise to a region of interest (ROI) are irreversibly bleached using a high-power laser illumination (see Fig. 1a). The fluorescence recovery over time within the ROI provides details about the dynamics of the protein of interest (2) (see Fig. 1a, b). The rate of fluorescence

*Alex Carisey, Matthew Stroud, and Ricky Tsang contributed equally to the manuscript.

Claire M. Wells and Maddy Parsons (eds.), *Cell Migration: Developmental Methods and Protocols*, Methods in Molecular Biology, vol. 769, DOI 10.1007/978-1-61779-207-6_26, © Springer Science+Business Media, LLC 2011

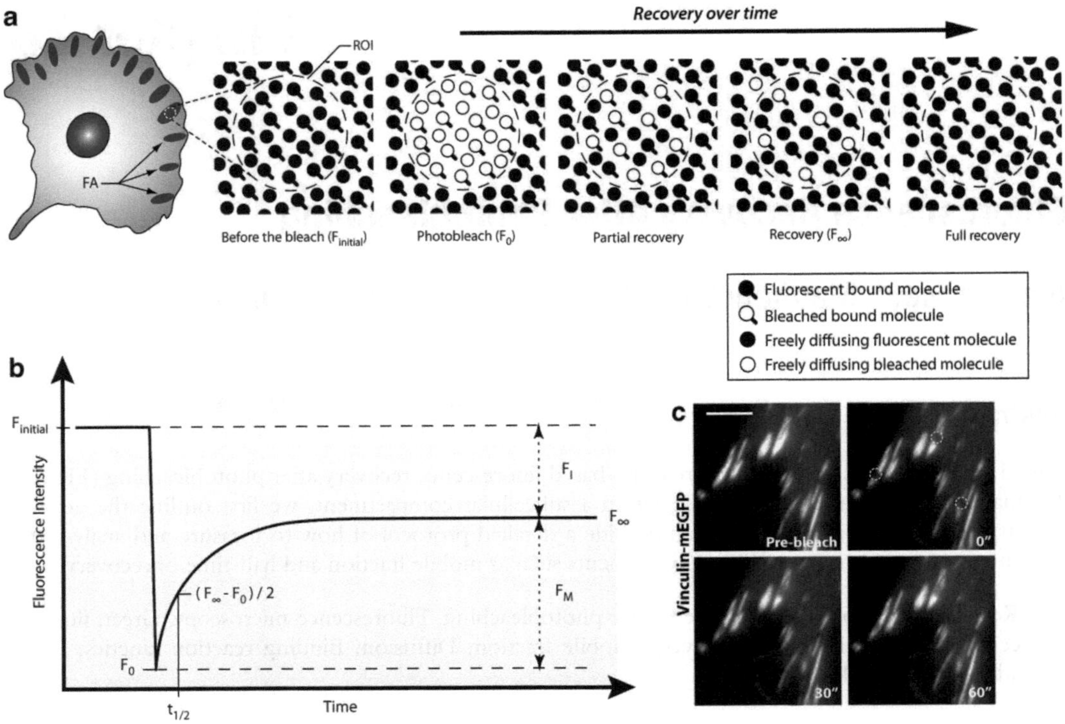

Fig. 1. Principle of FRAP experiment and example. (**a**) Scheme depicting the photobleaching and fluorescence recovery of a region of interest (ROI) within a focal adhesion (FA) in a cell. Before the bleach event, the fluorescent proteins are uniformly distributed within the structure and are in dynamic equilibrium, immediately after photobleaching, the equilibrium is lost and the fluorescence intensity recovers as fluorescent proteins move back into the ROI. This model represents a reaction-dominant recovery, in which diffusion is negligible. (**b**) Fluorescence intensity is recorded as a function of time; this enables the half-time of recovery ($t_{1/2}$), and the mobile and immobile fractions (F_M and F_I, respectively) to be measured. The fluorescence intensities before and immediately after bleaching are indicated ($F_{initial}$ and F_0, respectively) as well as the intensity after recovery (F_∞). (**c**) Example of a FRAP experiment on a focal adhesion plaque component; FRAP was performed on NIH 3T3 cells transiently expressing vinculin-mEGFP. Panels represent still frames taken at the indicated timepoints (in seconds), throughout the course of the experiment. *Dashed circles* indicate the bleached ROI, scale bar = 5 μm.

recovery is governed by two major events (3). The first is the diffusion of the fluorescently tagged protein within the localised environment, a fast process occurring over a few milliseconds. The second event is the binding between the fluorescently tagged protein and potential binding partners within the ROI. Most proteins in a cell undergo continuous turnover within complexes allowing bleached fluorescent proteins to be replaced by newly recruited fluorescent proteins, thus leading to the recovery of fluorescence (see Fig. 1a). Such recovery can be quantified by plotting the intensity of the fluorescence over time within the defined ROI, before, during, and after the bleaching event (see Fig. 1b). Further mathematical analysis including curve fitting then allows a detailed assessment of the behaviour of a fluorescently tagged protein within live cells.

This protocol focusses on FRAP as a method to assess the mobility of fluorescently labelled proteins within focal adhesions of a live cell (see Fig. 1c). We describe here the acquisition of time-lapse images, the bleaching and recovery of fluorescence with an emphasis on the compulsory controls, and cover some aspects of data processing. A complete methodology detailing the processing of raw data from FRAP experiments is described in refs. 4, 5.

1.2. Readouts of FRAP Experiments

Analysis of FRAP experiments yields information about the mobility of the fluorescently tagged protein. Two main parameters can be readily assessed, namely the mobile fraction and the half-time of recovery.

The mobile fraction (F_M) is the proportion of bleached proteins that are replaced by unbleached proteins during the monitoring of the recovery event. Thus, mobile fraction can be determined by calculating the ratio of fluorescence intensity between the end of the time-lapse recordings (F_∞) and the initial intensity before the bleaching event ($F_{initial}$), corrected by the experimental bleach value (F_0) (2) (see Fig. 1b). This ratio results in values between 0 and 1, or when expressed as a percentage, between 0% and 100%. Due to binding reactions within the ROI, the theoretical value of 100% of recovery is rarely achieved in practice. Therefore, the recovery reaches a plateau below this value by the end of the time-lapse (see Fig. 1b). Moreover, unintentional bleaching and dynamics of the overall structure during recovery limit the length of the time-lapse recording. In general, proteins that bind strongly to fixed components will have lower mobile fractions than those that interact weakly. Proteins that can freely diffuse show a maximal mobile fraction of 100%, as they do not interact with any partner.

The second parameter, the half-time of recovery, is the time that takes for the fluorescence to reach half of its maximal recovery intensity (termed $t_{1/2}$) (see Fig. 1b). The rate of the recovery depends on the size of the complexes and the stability of the interaction with large or fixed proteins (6). Under some circumstances, the recovery may exceed 100% of the initial fluorescence value. This observation indicates that the bleached structure undergoes a growth phase during the time frame of the experiment. The precise determination of F_M and $t_{1/2}$ is only possible in biological systems that have reached equilibrium before photobleaching so that the total amounts of both fluorescent protein and its binding sites remain constant over the course of the experiment. In this case, the only parameter changing after photobleaching is the equilibrium between the bleached and the unbleached molecules and the return to equilibrium provides the readout of the recovery curve. Finally, reliable measurements will only be obtained for proteins that are part of an immobile complex during the length of the monitored recovery.

1.3. Requirements of the Fluorescent Label

To study the dynamics of a protein in FRAP experiments, the protein of interest must be fused to a fluorophore suitable for expression and monitoring in live cells. To avoid artefacts, this fluorophore should have the following characteristics:

1. It should be bright enough to obtain a high signal-to-noise ratio.
2. It should be photostable to prevent excessive bleaching during time-lapse recordings; however, it should not resist bleaching with the high-power laser.
3. It should not be prone to photoreversible bleaching, an event whereby a proportion of the fluorophore is able to revert back to its original fluorescent state (see Note 1). Such an event can be tested to a certain extent by performing FRAP on fixed cells.
4. It should be monomeric to avoid any irrelevant association between the tagged proteins.

The monomeric enhanced GFP (mEGFP) is usually a good choice, since it has all these properties, to a greater extent than the original GFP and many other GFP derivatives. Also most confocal microscopes will be equipped with an argon laser line at 488 nm, which enables high-power photobleaching and low-power monitoring. The use of dyes emitting in the red spectrum is not preferred as bleaching is less efficient with these lasers (see Note 2).

Independent of the chosen tag, the fused protein should carry out the same function as the endogenous protein. It is therefore crucial to choose wisely where the fluorescent tag is placed on the protein of interest in order to avoid artefactual readouts (see Note 3).

1.4. Use of the Correct Microscope

FRAP experiments can be carried out on both confocal and wide-field microscopes equipped with the necessary laser, but one should be aware of the strengths and weaknesses of the respective systems.

Because of their design, wide-field microscopes capture the light emitted by the fluorescent molecules within a thicker optical section of the sample. On one hand, this allows the measurement of low fluorescent intensities that would otherwise be below the threshold for a confocal microscope. Such systems can be fitted with electron multiplying charge-coupled device cameras that have a higher sensitivity, a larger dynamic range, and a faster acquisition rate than most conventional confocal scanning heads. On the other hand, FRAP experiments should be performed on the assumption that the movement of fluorescently labelled proteins is recorded within a focal plane (5). To a certain extent, it is possible to overcome this problem on a wide-field system by the use of background subtraction methods. However, since confocal microscopes monitor a precise optical section of a defined thickness, in many cases this type of microscope can deliver more accurate results.

In our hands, results of FRAP experiments on adherent cells expressing vinculin-mEGFP (see Fig. 1c) are not significantly different when performed on a Leica SP5 confocal microscope or on a DeltaVision QLM wide-field microscope using the integrated analysis software.

1.5. Validation of Measurements and Data Fitting

When performing FRAP experiments on proteins expressed transiently, the variability of the expression levels of the fluorescent protein can lead to misinterpretation of the results. The level of expression dictates the proportion between the pool of proteins interacting with other molecules and the pool of freely diffusing proteins in the immediate environment. Thus, it is important to determine whether there is a dependence between the FRAP coefficients and the protein expression levels. FRAP coefficients over a wide range of expression levels must be calculated and plotted against the overall cell fluorescence intensity; the slope of the regression line through the data points should not be statistically different from 0, with 95% confidence (see Subheading 3.3.3).

Correct data fitting is essential for the interpretation of the results. There are many software packages available but we recommend using MATLAB in order to process large amounts of data. MATLAB allows easy automation of the analysis by fitting the recovery curve and extracting the FRAP coefficients such as F_{M} and $t_{1/2}$.

The fitting parameters depend on several factors, which have been outlined in great detail in ref. 5. Two major processes can influence the kinetics of the recovery: the diffusion of the fluorescently tagged protein and the binding to its partners. Most of the FRAP protocols used by biologists assume that the binding reaction is dominant in their system; however, this should be determined empirically on a protein-by-protein basis. To test whether the binding reaction is dominant, it is important to compare the different fluorescence recovery curves obtained by changing the diameter of the region bleached by the FRAP laser. If the fluorescence recovery curves are similar and give the same rates of recovery independently of the diameter of the bleached spot, then diffusion is negligible.

This is referred to as a "reaction-dominant model" (5) in which diffusion occurs a lot faster than the binding reaction. Therefore, the recovery of fluorescence is driven by the following binding equation:

$$F + S \underset{k_{off}}{\overset{k_{on}}{\rightleftarrows}} FS$$

where F represents free proteins, S represents unoccupied binding sites for the tagged protein, and FS represents the bound complex. In the case of a single binding state, the predicted FRAP recovery curve is the inverse of an exponential decay. The FRAP data can then be fitted with the following equation to obtain the off-rate constant of binding (k_{off}):

$$F(t) = 1 - Ae^{-\mathrm{koff}(t)}$$

The parameter A can be used to calculate the association rate (k_{on}), using:

$$A = k_{on}/(k_{on} + k_{off})$$

The mobile fraction is given directly by:

$$F_M = F_\infty - F_0/F_{initial} - F_0$$

The half-time of recovery can be determined from the curve fitting equation with $F(t_{1/2}) = (F_\infty - F_0)/2$:

$$t_{1/2} = \ln 2/k_{off}$$

2. Materials

2.1. Cell Culture and Transfection: Equipment and Reagents

1. Laminar flow, tissue culture cabinet.
2. Temperature-regulated, humidified incubator set at 37°C and 5% CO_2 gas for the maintenance of mammalian cells.
3. Cell culture medium (Dulbecco's modified Eagle's medium, DMEM) supplemented with 10% foetal calf serum (FCS), glutamine, and antibiotics.
4. Trypsin–ethylene diamine tetraacetic acid (EDTA) dissociation solution.
5. Sterile phosphate-buffered saline (PBS) without Mg^{2+} and Ca^{2+}.
6. Cultured adherent cells (e.g. NIH 3T3 mouse fibroblasts) expressing the protein of interest as a fusion protein with mEGFP (see Note 4).
7. Flasks or Petri dishes treated for cell culture.
8. Six-well tissue culture plates.
9. Transfection kit suitable for the cell type used. We use Lipofectamine Plus reagent (Life Technologies) according to manufacturer's instructions.
10. mEGFP expression vectors (Clontech).

2.2. Microscope Setup and Reagents for Imaging

The microscope setup described here is currently used in our laboratory; however, any laser scanning confocal microscope with an equivalent setup is suitable.

1. Leica TCS SP5 AOBS inverted laser scanning confocal microscope (Leica microsystems) (see Notes 5 and 6).
2. Leica HCX PL Apo 63x (lambda blue and NA iris corrected) objective (NA=1.4) or another high numerical aperture oil immersion objective.
3. An argon 488 nm laser line (see Note 6).
4. Leica LAS AF software for image acquisition and FRAP procedure (supplied with the microscope) or another image acquisition software package.
5. Heated chamber enclosing the microscope stage (see Note 7).
6. Glass bottom sterile culture containers (e.g. glass bottom dishes from MatTek or CellView from Greiner BioOne) for live imaging of cells (see Note 8).
7. Fibronectin from bovine plasma (see Note 9).
8. Ham's F12 medium supplemented with a final concentration of 25 mM HEPES and pH adjusted to 7.3 using a 1 M NaOH solution. After sterile filtration, supplement with glutamine and antibiotics (see Notes 7 and 10).
9. 4% paraformaldehyde (PFA) solution in PBS (w/v) (see Note 11).

2.3. Software for Analysis

1. Software to extract the fluorescence measurements from the time-lapse images generated by the microscope system. Many software suites controlling microscopes are more or less suitable to perform the complete analysis through simplified wizards (e.g. Leica LAS AF software from Leica Microsystems, softWoRx Suite from Applied Precision). For a better controlled and a more accurate analysis, we prefer to use an open-source image-processing program such as ImageJ (NIH, freely available at http://rsbweb.nih.gov/ij/).
2. Software to perform the data analysis. Any basic spreadsheet software is suitable; however, we recommend the use of a more advanced software package such as MATLAB (The MathWorks), enabling rapid automation in the analysis of large datasets (see Note 12).

3. Methods

3.1. Transfection and Preparation of Cells

This protocol is based on conditions optimised for our measurements of FRAP coefficients of focal adhesion plaque and cytoskeleton-associated proteins.

1. Prepare the mEGFP fusion proteins using standard molecular biology techniques.

2. Transfect cells with construct encoding the fusion protein in a six-well plate using Lipofectamine Plus reagent according to the manufacturer's instructions (see Note 4).
3. Coat the glass bottom dishes for 1 h at room temperature with 10 μg/mL of fibronectin (see Note 9).
4. After 4 h of incubation, wash the cells twice with sterile PBS.
5. Trypsinize the cells and wash them twice in DMEM supplemented with FCS and plate 5×10^4 cells per 35-mm glass bottom dish (see Note 8).
6. Incubate the cells for 12–48 h in an incubator until the expression of the transfected plasmid reaches its maximum (see Note 13).
7. Replace the DMEM with pre-warmed Ham's F12 medium (see Note 10).
8. Place the dish in the pre-warmed microscope chamber at least 1 h prior to the FRAP experiment to allow the medium to equilibrate.

3.2. Data Acquisition

3.2.1. Protocol

Three separate phases of the experiment can be distinguished. Firstly, the ROI must be defined and the fluorescence intensity of the GFP fusion protein within this ROI monitored using low-power laser settings. Secondly, a high-intensity laser pulse is emitted to irradiate the ROI and eliminate the fluorescence within this region. Thirdly, immediately after bleaching, the sample is monitored using the same low-intensity imaging settings as before the bleach. Images are collected at regular time intervals until the intensity in the bleached ROI reaches a plateau. The microscope can be fitted with a real-time autofocusing system to facilitate the acquisition of in-focus images over a long-time period and prevent focus drift due to temperature variations.

1. Select cells expressing a reasonable level of fluorescent protein in the appropriate cellular compartment using the eyepiece of the microscope (see Note 4).
2. Set the parameters following the guidelines provided for monitoring the fluorescence. The individual parameters must be adjusted empirically to obtain high-quality data while achieving three objectives: firstly, obtaining an almost complete bleach in the specified ROI, secondly, monitoring over a sufficient period of time until the fluorescence recovery becomes stable, and finally, keeping the photobleaching and phototoxicity to a minimum (see Note 14). Additionally, the intensity measured must always be confined to the dynamic range of the scan head (see Note 15).

The settings we use on the confocal microscope are:

- Confocal scan head settings: scan rate of 400 Hz for an intermediate image size of 512×512 pixels, using bi-directional scanning (see Note 16).
- Laser power of the 488 nm argon laser line is set to 20% and the transmission of the AOBS to 20% during the monitoring. For the bleaching pulse, the AOBS setting is increased to 100%.
- Pinhole setting: 1 Airy Unit (see Note 17).
- Zoom factor: set between 1.6 and 2 to fit the ROI for the bleaching event only.
- Gain adjustment: 900 V (see Note 18).
- Offset value: −0.50 (as close to zero as possible until a few zero intensity pixels appears on screen).
- Beam size (or ROI): set to 1 μm diameter (see Note 19).
- Iterations: three to five iterations to achieve full bleach (see Note 14).
- Time-lapse (see Note 20):
  - One frame every 10 s for three frames before bleach.
  - One frame every 10 s for 10 min after bleach, starting straight after the bleaching event.

3. Adjust precisely the focus and select your desired ROI (see Note 21).
4. Start your time-lapse.
5. Proceed with another cell. Acquire multiple cells showing a broad range of expression levels, as this will give useful information for the analysis (see Subheadings 1.5 and 3.3.3).

3.2.2. Controls

Using this detailed protocol and the same settings as for the experimental samples, several controls must be performed at the same time to accurately interpret the results of a FRAP experiment:

1. Carry out initial experiments to optimise the ideal time course and image acquisition number to ensure that the recovery of fluorescence is complete and photoreversible bleaching is minimised.
2. Bleach several ROI with different spot sizes to test whether diffusion has a major role in the fluorescence recovery. This will allow the user to apply the appropriate mathematical model for determining the protein dynamics according to ref. 5.
3. Ensure that the overall fluorescence intensity of the entire cell is not significantly different before and after photobleaching. Bleached spots must comprise a relatively small proportion of

the cellular pool of fluorescent proteins. To do this, one should perform several FRAP experiments with increasing size and/or number of bleached ROI to determine the limit for the bleach area. Slower recovery curves in experiments with an overall larger bleach area may indicate that recovery of fluorescence is affected by a depletion of the pool of fluorescently labelled proteins.

3.3. FRAP Data Analysis

3.3.1. Intensity Measurements

1. Collect a complete set of FRAP experiments (20–30 cells per condition) and transfer the images to the analysis workstation.
2. Correct any noticeable stage drift by realigning the individual images, e.g. ImageJ plug-in StackReg (7) (see Note 22).
3. Redraw the boundaries of the bleached ROI using the circular selection tool (see Note 19). The ROI position and diameter are likely to be found in the metadata recorded by the microscope software during acquisition.
4. Extract the fluorescence intensity values for:
   - The bleached ROI.
   - Unbleached areas representative of the background fluorescence.
   - Unbleached structures similar to the bleached target (see Note 23).
5. Label the data in a fully comprehensive manner and store them in a backed-up text file or spreadsheet.

3.3.2. Normalisation and Curve Fitting

Several calculations must be carried out to allow the comparison of the FRAP readouts between different biological samples or FRAP experiments.

1. Plot all the raw data from each ROI separately to validate the integrity of the measurements (see Fig. 2a and Note 24).
2. Obtain the background value ($F_{\mathrm{bgd}}[t]$) by averaging the fluorescence intensity from unbleached areas representative of the background and subtract this background value from each measurement of a photobleached ROI ($F_{\mathrm{bleach}}[t]$):

$$F[t] = \left(F_{\mathrm{bleach}}[t] - F_{\mathrm{bgd}}[t]\right)$$

3. Obtain the control value ($F_{\mathrm{ctr}}[t]$) by averaging the fluorescence intensity from several unbleached structures similar to the bleached target (see Note 23) in order to compensate for the overall loss of fluorescence.

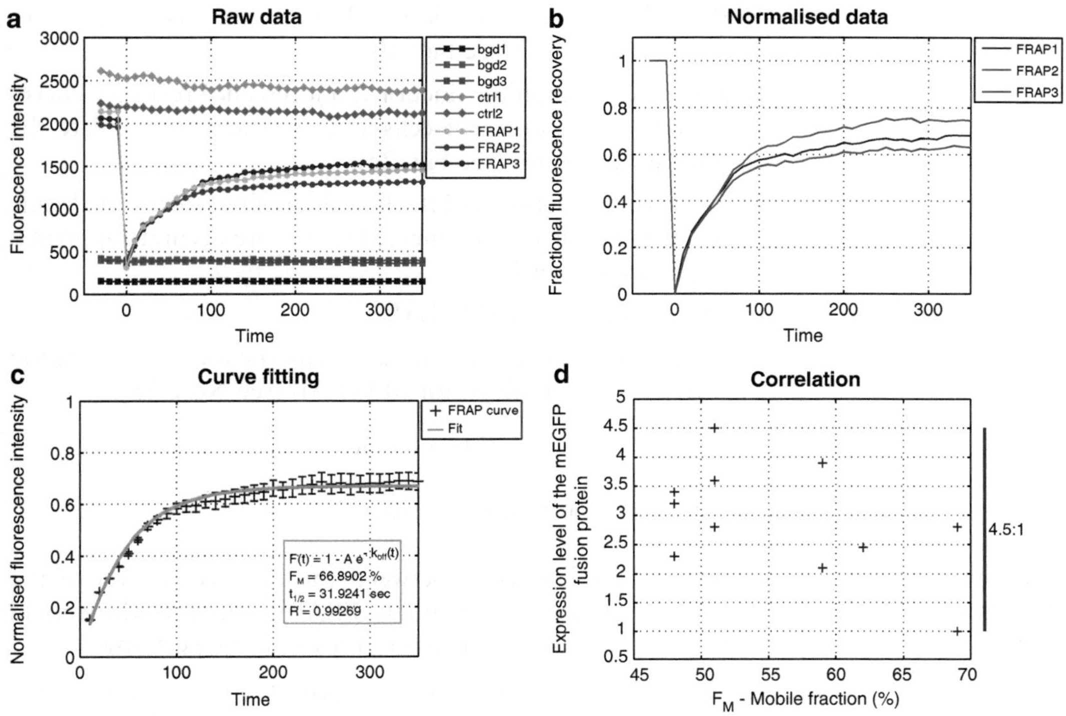

Fig. 2. Workflow of analysis of FRAP data. (**a**) Fluorescence intensities extracted from the time-lapse images are individually plotted over time to visualise any abnormal behaviour. The intensities recorded within several bleached ROI are termed FRAP1, FRAP2, and FRAP3. The background readings are termed bgd1, bgd2, and bgd3; and finally, the recordings for the unintentional bleaching are labelled as ctrl1 and ctrl2. (**b**) After background subtraction and correction for the postbleach intensity, the fractional fluorescence recovery of the three bleached ROI are plotted over time. The intensity of the monitored fluorescence now varies between 1 (prebleached value) and 0 (immediately after the bleach event). (**c**) Multiple fractional fluorescence recovery curves acquired in the same cell are averaged and plotted with their standard deviation of the mean. The curve is the best fit obtained from the experimental data. The main parameters discussed in the text have been extracted for these data. (**d**) A scatter plot is shown here to illustrate the absence of correlation between the overall fluorescence intensity of the cell, e.g. the expression level of the tagged protein, and the mobile fraction over a 4.5-fold range. A similar graph should be plotted with the half-time of recovery and the overall fluorescence intensity (see Subheading 3.3.3).

4. Correct each measurement with the background-subtracted control value from an unbleached ROI to obtain the experimental recovery ($R_{\text{norm}}[t]$) as follows:

$$R_{\text{norm}}[t] = F[t] / \left(F_{\text{ctr}}[t] - F_{\text{bgd}}[t]\right)$$

5. Correct the experimental recovery with the first postbleach fluorescence intensity set to zero (F_0) to obtain the fractional recovery measurement ($R_{\text{frac}}[t]$):

$$R_{\text{frac}}[t] = \left(R_{\text{norm}}[t] - F_0\right) / \left(1 - F_0\right)$$

6. Average the multiple $R_{\text{frac}}[t]$ obtained within the same cell and over the same time course.

7. Plot the average fractional fluorescence recovery curve ($R_{frac}[t]$) obtained (see Fig. 2b).
8. Superimpose multiple fractional fluorescence recovery curves obtained from different cells and use them to calculate the standard deviation (see Fig. 2c).
9. The mobile fraction and the half-time of recovery can already be estimated from this curve, but a more accurate approach requires curve fitting.
10. Export the data to MATLAB.
11. Fit the plotted data (see Fig. 2c) using the following equation with the ezyfit toolbox for MATLAB (see Note 25):

$$F(t) = 1 - Ae^{-\mathrm{koff}(t)}$$

12. Obtain the FRAP coefficients.

3.3.3. Evaluation of the Relationship Between Expression Level and FRAP Coefficients

To assess any relationship between the expression level of the fluorescently tagged protein and the FRAP coefficients which could invalidate the results, a correlation study must be performed:

1. With the help of the selection tool in ImageJ, measure the intensity of the fluorescence in the cell before the bleaching event.
2. Plot the overall intensity value against its estimated mobile fraction in a scatter plot (see Fig. 2d); in a second figure, plot the same value against the estimated half-time of recovery.
3. Calculate the slope of the regression line through the data points. It must not be statistically different from 0 at 95% confidence.

4. Notes

1. Photoreversible bleaching can pose a problem during the quantitative analysis of FRAP data and lead to an artefactual measurement of protein dynamics. Furthermore, different fluorescent proteins show different rates of photoreversible bleaching; therefore, comparisons between GFP variants within an experiment may give misleading results. Photoreversible bleaching can occur in the GFP species on a millisecond time scale (8). It is known that YFP species also exhibit photoreversible bleaching, which occurs in the timescale of seconds and is dependent on complex protonation reactions (9). The longer time scale associated with YFP may become apparent in experiments with short time intervals, therefore GFP is usually the preferred option.

2. Being able to completely bleach the fluorescence emitted by the tagged protein is crucial. Theoretically, bleaching using a high-power laser beam can be achieved on any fluorescent protein, depending on the availability of the appropriate laser line on the microscope system; this should match the excitation peak of the targeted molecule. For example, the red fluorescent protein (RFP) or its more photostable equivalent mCherry (10) may be used for FRAP; however, the fluorescence quantum yield is significantly lower than that of EGFP. Also, suitable lasers (such as the Helium-Neon Orange 594 nm) are known to be weak and may not achieve full bleach, therefore the GFP variant is preferred.
3. Ideally, the fluorescent tag should be inserted in at least two independent positions, for example on the N- or C-terminus end of the protein. The localisation of the overexpressed construct should be similar to that of the endogenous protein, and issues about a potential perturbation of function must be considered if the FRAP coefficients obtained are significantly different between the two constructs.
4. In most cases, either transient or stable expression of the fusion protein in cells can be performed. Using a stable cell line for FRAP ensures that the measurements are more consistent by reducing the variability of expression levels (see also Subheading 3.3.3).
5. The Acousto-Optical Beam Splitter (AOBS) is a common feature on more recent confocal microscopes allowing the physical separation of emission and excitation light paths with high transmission efficiency. An Acousto-Optical Tunable Filter (AOTF) allows the rapid alternation between the lasers and the adjustment of their repective intensities by modifying the transmission coefficient, as opposed to changing the power of the laser emission itself. This component is essential to perform automated monitoring and reduce bleaching of the sample during the recovery period.
6. The use of laser-based microscopes is restricted to persons who have been adequately trained and are familiar with the local safety regulations.
7. We use a temperature- and CO_2-controlled chamber to provide the optimal conditions for live imaging. It is possible to avoid the need for CO_2 by using a CO_2-independent buffered system such as HEPES.
8. Plastic culture dishes and plates are poorly suited for fluorescence studies as they exhibit non-stable auto fluorescence (11), therefore cells must be transferred to a glass bottom dish after transfection.

9. Other extracellular matrix proteins such as laminin, vitronectin, or collagen may be required to ensure the spreading of some cell lines.
10. In order to reduce background fluorescence due to the presence of riboflavin and phenol red in DMEM, Ham's F12 medium is preferred. Riboflavin exhibits high autofluorescence in the excitation range of 450–490 nm and emission range of 500–560 nm. Phenol red has a maximum excitation peak around 555 nm. To further minimise the autofluorescence, FRAP experiments are preferably performed in low-serum or serum-free medium.
11. PFA is a toxic powder that must be handled with care using gloves and a respiratory mask in an isolated environment. A 4% PFA solution in PBS (w/v) will dissolve overnight with constant stirring at 50°C. The stock solution can be aliquoted and stored at −80°C for a year. Thawed aliquots must be kept at 4°C and used within a week. Fixation of cells is performed at room temperature for 15 min followed by three washes in PBS.
12. MATLAB can be linked to ImageJ through an additional Java package created by Daniel Sage and Dimiter Prodanov (Biomedical Imaging Group, Swiss Federal Institute of Technology, Lausanne, Switzerland, http://bigwww.epfl.ch/sage/soft/mij/).
13. The expression of the construct in a suitable cell line for the study must be monitored by fixing the cells in PFA at various time points after transfection as the expression level and maturation speed of the fluorescently tagged protein vary considerably depending on the tag and the cell line. Moreover, the delivery of the overexpressed protein in the appropriate cellular compartment is another feature that varies, as it is dependant on the renewal rate of the endogenous protein.
14. Despite the fact that the parameters will be used to monitor living cells, it is more convenient to optimise the settings using a fixed sample maintained in the same conditions as for the live cells. The laser settings for the bleaching event should be set to achieve a sufficient decrease in fluorescence within the ROI. As a guideline, we consider that the first reading after bleach should be less than 20% of the measurement before the bleaching event. Although this value is subject to discussion as the bleaching illumination must be kept as low as possible, for the following two reasons: firstly, to avoid excessive phototoxicity and secondly, because the beginning of fluorescence recovery must not overlap with the last bleaching iteration. Indeed, the bleaching duration must be confined to a limited space (i.e. the ROI) and time (ideally instantaneous) to allow accurate measurement. Unfortunately,

there is no general rule and it is more a case of trial and error for the user to obtain the optimal bleaching and imaging parameters.

15. It is crucial to adjust the acquisition parameters of the photomultiplier tube (PMT) in order to remain in the dynamic range of intensity acquisition. Saturation of fluorescence intensity during the time-lapse will result in inaccurate measurements of FRAP coefficients.
16. On the majority of line scanning confocal microscopes, an option is available for bi-directional scanning; however, the user must bear in mind that phase correction may need to be adjusted. This allows the scan head to image the sample in both directions along the *x*-axis and inevitably increase the acquisition rate. Additionally, the frame averaging function should be avoided while doing live measurements as it leads to a loss of time resolution.
17. The pinhole value, determined by the numerical aperture of the objective and the light wavelength, should ideally be adjustable. For live cell work, the pinhole should be opened wide enough to acquire an adequate signal, while keeping the laser intensity low to minimise photobleaching and phototoxicity. Meanwhile, opening the pinhole affects the resolution by increasing the thickness of the optical section.
18. The PMT gain value should be around 900 V. A value below 600 V means that the laser line intensity can be lowered; conversely, it needs to be increased if the gain setting is above 1,100 V.
19. If these experiments will be analysed according to the method detailed by Sprague et al. (5), the ROI must be circular.
20. It is necessary to record three to five frames before the bleaching step to obtain the value of the initial maximum intensity within the ROI.
21. It is recommended to bleach multiple ROI of identical size for each time-lapse. Recovery curves obtained from multiple ROI within the same cell should be averaged after normalisation to obtain smooth data (see Subheading 3.3.2 and Fig. 2c). Pooling data is acceptable under these circumstances as the level of expressed fusion proteins (i.e. both fractions of bound and free molecules) is the same between the different ROI.
22. Plug-in created by Philippe Thévenaz (Biomedical Imaging Group, Swiss Federal Institute of Technology, Lausanne, Switzerland, http://bigwww.epfl.ch/thevenaz/stackreg/).
23. The data have to be corrected for unintentional bleaching that occurs during the monitoring phases. Two techniques can be used: the first consists of recording the fluorescence

intensity within several identical unbleached regions to the one displayed in the same cell as the bleached ROI. The second technique involves the recording of the averaged intensity of the cell.

24. Discard any sets of data that exhibit any of the following defects: inconsistent background fluorescence intensities, unstable prebleach intensities, and recovery curves that do not exhibit a plateau.
25. Ezyfit toolbox for MATLAB created by Frederic Moisy (Laboratory FAST, University Paris Sud, Paris, France, http://www.fast.u-psud.fr/ezyfit/).

Acknowledgments

We would like to thank Dr. Janet Askari for critical reading of the manuscript. CB acknowledges BBSRC (BB/G004552/1). The Bioimaging Facility microscopes used in this study were purchased with grants from BBSRC, Wellcome Trust and the University of Manchester Strategic Fund.

References

1. Tsien, R. Y. (1998) The green fluorescent protein, *Annu Rev Biochem* 67, 509–544.
2. Axelrod, D., Koppel, D. E., Schlessinger, J., Elson, E., and Webb, W. W. (1976) Mobility measurement by analysis of fluorescence photobleaching recovery kinetics, *Biophys J* 16, 1055–1069.
3. Sprague, B. L., and McNally, J. G. (2005) FRAP analysis of binding: proper and fitting, *Trends Cell Biol* 15, 84–91.
4. Sprague, B. L., Muller, F., Pego, R. L., Bungay, P. M., Stavreva, D. A., and McNally, J. G. (2006) Analysis of binding at a single spatially localized cluster of binding sites by fluorescence recovery after photobleaching, *Biophys J* 91, 1169–1191.
5. Sprague, B. L., Pego, R. L., Stavreva, D. A., and McNally, J. G. (2004) Analysis of binding reactions by fluorescence recovery after photobleaching, *Biophys J* 86, 3473–3495.
6. Lippincott-Schwartz, J., Snapp, E., and Kenworthy, A. (2001) Studying protein dynamics in living cells, *Nat Rev Mol Cell Biol* 2, 444–456.
7. Thevenaz, P., Ruttimann, U. E., and Unser, M. (1998) A pyramid approach to subpixel registration based on intensity, *IEEE Trans Image Process* 7, 27–41.
8. Dickson, R. M., Cubitt, A. B., Tsien, R. Y., and Moerner, W. E. (1997) On/off blinking and switching behaviour of single molecules of green fluorescent protein, *Nature* 388, 355–358.
9. McAnaney, T. B., Zeng, W., Doe, C. F., Bhanji, N., Wakelin, S., Pearson, D. S., Abbyad, P., Shi, X., Boxer, S. G., and Bagshaw, C. R. (2005) Protonation, photobleaching, and photoactivation of yellow fluorescent protein (YFP 10 C): a unifying mechanism, *Biochemistry* 44, 5510–5524.
10. Shaner, N. C., Campbell, R. E., Steinbach, P. A., Giepmans, B. N. G., Palmer, A. E., and Tsien, R. Y. (2004) Improved monomeric red, orange and yellow fluorescent proteins derived from Discosoma sp. red fluorescent protein, *Nat Biotechnol* 22, 1567–1572.
11. Piruska, A., Nikcevic, I., Lee, S. H., Ahn, C., Heineman, W. R., Limbach, P. A., and Seliskar, C. J. (2005) The autofluorescence of plastic materials and chips measured under laser irradiation, *Lab Chip* 5, 1348–1354.

Chapter 27

Measuring FRET Using Time-Resolved FLIM

Penny E. Morton and Maddy Parsons

Abstract

Cell migration is a process that is controlled by the formation and correct localization of protein complexes and by post-translational modification of individual proteins. Forster or fluorescent resonance energy transfer (FRET) detected using fluorescence lifetime imaging microscopy (FLIM) provides a method by which protein–protein interactions may be detected and spatially localized within a cell. This technique can be used to map protein activation states and the formation and dissolution of protein complexes that control movement of a cell. This chapter describes a protocol for detecting FRET between GFP- and mRFP1-tagged proteins in fixed adherent cells. A background to both FRET and FLIM is provided followed by an overview of the method and a full protocol for sample preparation, data acquisition, and analysis.

Key words: FRET, FLIM, Forster resonance energy transfer, Fluorescence lifetime imaging microscopy, eGFP, mRFP1

1. Introduction

The process of cell migration is controlled by diverse biochemical events such as protein interaction and post-translational modification. These dynamic events control the localization, conformation, and functional activity of proteins involved in the movement of a cell. Understanding how protein complexes and post-translational modifications occur as well as where and when these events happen in the cell is crucial to understanding the processes that control cell migration. Over recent years, the use of Forster or fluorescent resonance energy transfer (FRET) in cell biology to measure distances between proteins and to localize protein modifications and complexes has yielded several interesting discoveries and has become a useful and powerful tool in the field of cell adhesion and

Claire M. Wells and Maddy Parsons (eds.), *Cell Migration: Developmental Methods and Protocols*,
Methods in Molecular Biology, vol. 769, DOI 10.1007/978-1-61779-207-6_27, © Springer Science+Business Media, LLC 2011

migration (1–3). FRET is a non-radiative process where energy is transferred from an excited molecular donor fluorophore to an acceptor fluorophore by means of intermolecular dipole–dipole coupling (4). The rate of energy transfer between the donor and the acceptor molecules or FRET efficiency (E) is dependent on the distance between the two participating fluorophores and is described by the following equation:

$$E = R_0^6 / \left(R_0^6 + r^6\right)$$

where r (nm) is the distance between the two fluorophores and R_0 (nm) is the distance at which the energy transfer is 50% efficient. The distance over which this energy transfer can occur is dependent on the fluorophores involved but it typically can occur between fluorophores which are up to 11 nm apart. These distances are similar to protein diameters and therefore the distance dependency of FRET allows measurement of direct protein–protein interactions as well as changes in protein conformation or post-translational modifications.

1.1. Microscopical Methods for Measuring FRET in Biological Samples

FRET can be measured using either steady-state or time-resolved methods. In steady-state measurements, the sample is illuminated with a continuous beam of light typically for milliseconds or seconds and the intensity or emission spectra is recorded. The measured fluorescence is dependent on the concentration, quantum yield, and fluorescence decay kinetics of the probe. FRET can be measured using steady-state fluorescence because in FRET the presence of an acceptor within a FRET pair leads to quenching of donor emission and an increased or sensitized acceptor emission. Intensity-based FRET detection techniques make use of these properties. However, intensity-based methods can be highly problematic in cases where the concentration of individual donor and acceptor molecules cannot be controlled and are therefore most suited to experiments using chimeric constructs where donor and acceptor are attached to the same molecule. In contrast, fluorescence lifetime imaging microscopy (FLIM) directly measures the concentration-independent fluorescence decay kinetics of the fluorophore (4). Consequently, FLIM measurements do not require subsequent correction for donor and acceptor concentration. Instead, FLIM measures the inherent fluorescence lifetime of the donor fluorophore. Fluorescence lifetime refers to the average time that the molecule stays in the excited state before emitting a photon. The main drawback of FLIM has traditionally been the high cost and complexity of the equipment required to permit lifetime acquisition. However, the recent launch of commercial FLIM systems or modules to add onto existing microscopes has made this technique for measuring FRET more widely accessible to the cell biology community.

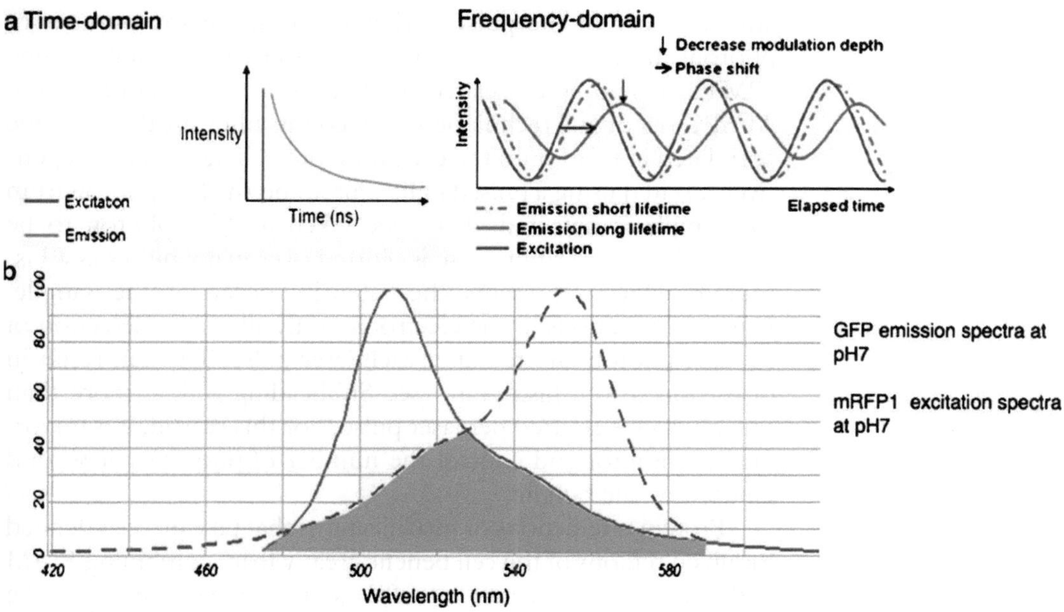

Fig. 1. (**a**) Frequency-domain and time-domain FLIM. The *left-hand panel* represents a typical fluorophore excitation event and the subsequent fluorescence decay curve. In time-domain FLIM, this curve is measured and used to obtain an average lifetime for the fluorophore. The *right-hand panel* shows typical frequencies for fluorophores with long (unFRETing) and short (FRETing) lifetimes and the consequent effect on the phase and modulation of the curve. This change in phase is then measured compared with a sample of known lifetime. (**b**) The spectral overlap between the excitation spectrum for mRFP1 and the emission spectrum for GFP. This shows that GFP and mRFP1 are suitable as a FRET pair.

There are two methods to measure fluorescence lifetime.

1. In *time domain FLIM* measurements, the sample is excited with a pulse of light where the pulse width is much shorter than the decay time (or lifetime) of the fluorophore. The time-dependent intensity is measured after a pulse and the decay time is calculated.
2. In *frequency domain FLIM*, the sample is excited with intensity-modulated light. The intensity of the incident light is varied at a high frequency and the emission responds at the same modulation frequency. However, the emission is delayed in time relative to the lifetime of the fluorophore. This delay can be measured as a phase shift and then used to calculate the decay time (see Fig. 1a for a diagrammatic representation).

1.2. Time Domain FLIM (TCSPC)

In time-domain lifetime measurements, the sample is excited with a brief pulse of light. The duration of the pulse is significantly shorter than the fluorescence lifetime of the fluorophore and is typically in the order of tens to hundreds of femtoseconds. The pulse excitation results in a population of fluorophores which are in the excited state and these will relax to the ground state producing a fluorescence decay profile. In a time-domain FLIM instrument, the exponential decay in intensity is directly measured after delivery of the pulse, either by counting the number of photons in time bins using gated detection photomultiplier tube (PMT) or by time-correlated single photon counting (TCSPC). TCSPC is a digital technique which counts photons that are time correlated in relation to the excitation pulse. The sample is repetitively excited using a pulsed light source and the first photon from the sample is detected. The time taken for this photon to be detected after the light pulse is plotted over many pulses as a histogram which represents the intensity decay of the sample. Current limitations in the electronics only allow the detection of the first photon emitted after each laser pulse. This can result in distortion of the histogram (see Subheading 3.3) if more than one photon is emitted per laser pulse. For this reason, it is important to monitor and control the number of photons per second emitted by the sample.

Protein interactions or modifications that take place at defined localized regions of the cell benefit greatly from combining FLIM with two-photon excitation (2PE) sources which enhances the spatial detection of FRET by removing the contribution of non-interacting species and out-of-plane regions. In almost all fluorescence experiments, excitation is due to absorption of a single photon by each fluorophore; however, it is also possible for a fluorophore to absorb two or more long-wavelength photons to reach the same excited state. This process is called 2PE or multiphoton excitation (5). 2PE occurs by the simultaneous absorption of two low-energy photons whose total energy equals the energy required for one-photon excitation. Using a laser illumination source allows this excitation process to place in the infrared spectral range. This is particularly useful when using 2PE to excite a donor fluorophore in FRET experiments since using such a long wavelength light to excite the donor prevents unintentional excitation of the acceptor fluorophore. The most important benefit of 2PE is the increase in resolution. In 2PE, the excitation region is limited to within a sub-femtolitre volume. This means that the emission region is intrinsically confocal. Molecules away from the focal region of the objective lens do not contribute to the image formation process and are not affected by photobleaching or phototoxicity (5) which makes this technique particularly useful for live cell imaging (4).

1.3. Fluorophores Used in FLIM Measurements of FRET

The first consideration when choosing fluorophores to use as FRET pairs is that the donor fluorophore emission spectra must overlap with the excitation spectra of the acceptor fluorophore in order for the energy transfer to occur (see Fig. 1b for diagrammatic representation of this overlap). Various pairs of fluorophores with overlapping emission and excitation spectra have been used in FRET experiments such as CFP/YFP, BFP/EGFP, and more recently EGFP/mRFP1 (4) and mTFP/mVenus (6). There are advantages and disadvantages to using these pairs in FRET/FLIM measurements (reviewed in (7, 8)). CFP/YFP is the most commonly used FRET pair, particularly in intensity-based FRET measurements. However, since analysis of FLIM within a sample involves the fitting of the donor fluorescent lifetime to a mono- or bi-exponential decay curve, the inherent bi-exponential nature of the CFP fluorescence lifetime (9) makes quantitative analysis of CFP/YFP FRET using FLIM impractical. For FLIM, only two donor fluorophores are currently available that have high quantum efficiency, are photostable, exhibit fast maturation in cells, and have a monoexponential lifetime decay. These are eGFP (10) and mTFP (11). eGFP paired with either mRFP1 (4) or mCherry (12) has been extensively used for FLIM measurements of FRET and shown to be an efficient FRET pair although GFP/mCherry has been shown to have a lower FRET efficiency than GFP/mRFP1 making mRFP1 the preferred acceptor for GFP. mTFP has been shown to be a good donor for FLIM when paired with either mVenus or EYFP and is being established as a good alternative FRET pair in FRET/FLIM experiments (6).

2. Materials

2.1. Sample Preparation

1. eGFP vector can be obtained from Clontech (Palo Alto, CA; www.clontech.com) and mRFP1 was obtained from Roger Tsien and cloned into pcDNA3.1 (Invitrogen Carlsbad, CA).
2. MCF-7 cells are available from ATCC.
3. For transfection Fugene 6 (Roche, Indianapolis, USA) was found to have good transfection efficiency without cell toxicity (see Note 1 for other transfection methods plus microinjection).
4. Growth media: MCF-7 cells were grown in Dulbecco's modified Eagle's media (DMEM) supplemented with 10% fetal calf serum, penicillin/streptomycin, and glutamine.
5. Transfection media: For transfection serum-free Optimem medium (Gibco, Invitrogen) should be used.
6. Cell fixation: For fixation, 4% paraformaldehyde (PFA) is dissolved in phosphate-bufferred saline (PBS) using NaOH then corrected to pH 7.4. Four percent PFA-PBS can then be frozen in aliquots until required. Sodium borohydride solution must be prepared immediately prior to use.
7. Mounting medium: Fluorsave Reagent (Calbiochem) is used to mount the coverslips (see Note 2).

2.2. Image Acquisition Using Time-Domain Multiphoton FLIM

Time domain FLIM is performed with a mutiphoton microscopy system which is based on a Mira Ti:Sapphire laser (Coherent, Santa Clara, CA, USA), a TE2000 microscope body (Nikon), an in-house developed scan head, SPC830 single-photon counting electronics (Becker and Hickl) and a temperature controlled enclosure. Non-descanned detection is afforded by the use of fast single-photon response photomultiplier tubes (7400 series, Hamamatsu Ltd, Japan) situated in the re-projected pupil plan of the objective.

This system was custom made in-house; however, various companies offer time-domain and frequency-domain FLIM systems which are compatible with many commercial confocal or widefield microscopes (see Note 3).

2.3. Analysis of FLIM Data

All analysis of FLIM data presented in this chapter was performed using Time Resolved Imaging 2 (TRI2) which was developed by P.R. Barber, R.Locke, R.Edens, S. Ameer-Beg, B.Vojnovic, and J. Gilbey at the Gray Cancer Institute. Commercial software is also available (see Note 4). Further analysis to produce FRET efficiency histograms was performed in Microsoft Excel.

3. Methods

3.1. Preparation of Samples

1. Seed cells of interest onto coverlips the day before transfection in growth media. Cells should be between 30 and 60% confluent on the day of transfection.
2. Prepare plasmid DNA of donor (GFP) and acceptor (RFP or mCherry) fusion proteins of interest for transfection.
3. Transfection: Co-transfect cells with plasmids encoding the donor and acceptor fusion proteins using Fugene6 or similar reagent according to the manufacturer's instructions. It is important that the acceptor concentration exceeds that of the donor therefore DNA transfection of at least a 2:1 ratio is recommended in order to ensure that excess acceptor, which is required for accurate FRET efficiency determination, is present. For example, when transfecting a 35 mm dish of cells 1 μg of donor and 2 μg of acceptor should be used. In addition to this, a "donor-alone" sample must be prepared with cells transfected only with the donor construct. If investigating a stimulus-induced interaction, then one "donor-alone" sample must be prepared per condition. The GFP-alone sample will be used to determine the control GFP lifetime in the absence of acceptor (see Note 5).
4. 36–48 h post-transfection apply any required stimulus, wash cells with PBS and fix using 4% PFA for 15 min at room

temperature. Prepare solution of 1% $NaBH_4$ in PBS (see Note 6). This should always be prepared just before use. Wash the fixed samples with PBS three times and then cover the samples with the $NaBH_4$ solution for 10 min. Wash again with PBS three times before mounting the coverslips onto glass microscope slides using an anti-fade-containing mounting medium, taking care not to create bubbles in the mounting medium (see Note 8). Allow the mountant to dry and harden completely before imaging samples.

3.2. Image Acquisition

1. All images should be acquired using a time-domain multiphoton FLIM system (described in Subheading 2). For eGFP imaging, 890 nm laser excitation wavelength is used. This is selected by adjusting the wavelength selector prism and provides a good compromise between oscillation efficiency (of GFP) and laser power/stability; however, 2PE of GFP can occur using between 720 and 920 nm. Photons are detected using photomultiplier tubes which are the most widely used time-resolved detectors. The laser power is adjusted to give average photon counting rates of the order 10^4–10^5 photons/s, below the maximum counting rate afforded by the TCSPC electronics to avoid pulse pile-up. It is important to show the localization of the acceptor fluorophore, therefore an image should be acquired of the mRFP-tagged molecule immediately prior to the FLIM acquisition using widefield illumination, a standard cy3 filter cube (Chroma) and CCD camera detection using a Hamamatsu Orca-ER. When using a multiphoton illumination source, it is much easier to acquire the acceptor image using a widefield camera detection; however, in cases where illumination is widefield, for example with frequency domain system, it is just as easy to acquire acceptor images in the same way as the lifetime data.
2. For FLIM, the photon arrival times, with respect to the approximately 80 MHz repetitive laser pulses, are binned into 256 time windows over a total measurement period of 10 ns. Images are captured with a 40× objective lens (Plan Fluor 40×/1.3 oil, Nikon) at 256×256 pixels and saved as .ics files for later analysis (see Note 4).
3. It is important to know what the instrument response time is in order to be able to fit the data. The instrument responses are measured from the hyper-Rayleigh scattering of highly attenuated excitation in a suspension of 20 nm colloidal gold (G-1652, Sigma-Aldrich Company Ltd, Dorest, UK; Habenicht et al. 2002) these measured responses were used in the fitting of data. Photons are collected at 500 nm (filter 35–5040 Coherent) (see Note 7).

3.3. Data Analysis

The analysis described here uses TRI2 which was developed in-house; however, there are various commercial software systems available and the process is the same.

1. The file containing the acquired data is imported into the software and TRI2 fits this data using a Non-linear Least Squares fit with the Levenburg–Marquardt (LM) algorithm to minimize the Chi square value. Initial estimates are provided by an algorithm based on the "rapid lifetime determination" technique. Iterations of the LM continue until a target Chi square is reached or there is no significant change in that value. At this point, the data can also be binned to reduce noise in the resulting image. Binning combines a specified number of pixels into one new pixel. This can increase the signal to noise ratio of the image but can also reduce the resolution of the image. The data are 3D with the third dimension being time. The time course of each pixel (the transient) at each pixel or binned area of pixels is fitted with the equation below and the best fit parameters are recorded, i.e. Z, A, and tau. These parameter values can be used to create parameter maps when arranged in the (x, y) formation as in the original image (see Fig. 2 for an examples of these).

$$\text{fit function}\, f(t) = Z + A\exp\left(-\frac{t}{\tau}\right)$$

where Z is the background obtained from the instrument response, t is the time variable, A is the peak strength of the exponential, and τ is the eGFP lifetime.

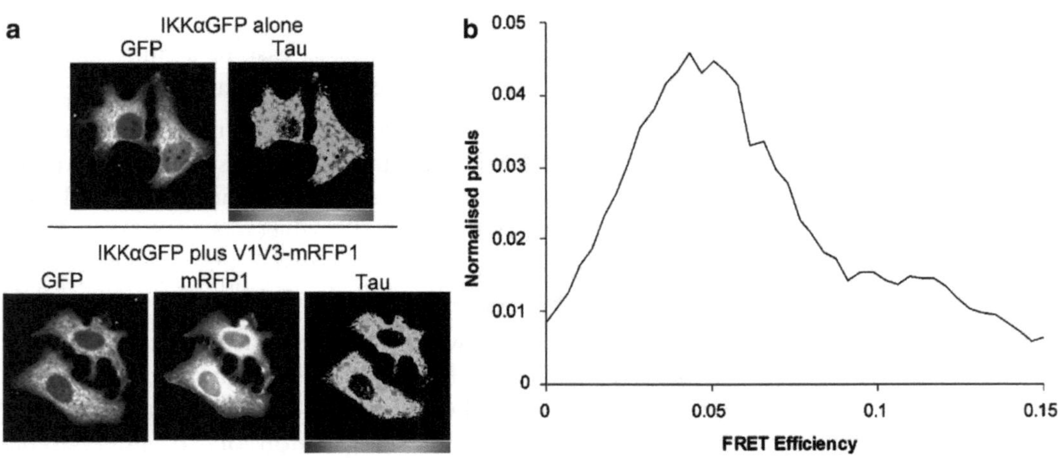

Fig. 2. Inhibitor of κB kinase α (IKKα) interacts with the regulatory domain of Protein Kinase C α (PKCα). (**a**) Multiphoton FLIM was undertaken to determine the extent of FRET between the GFP-IKKα donor and V1V3-mRFP1 (PKCα regulatory domain) acceptor in cells expressing both constructs. Transfected MCF-7 cells were fixed without prior treatment. GFP images relate to the GFP intensity image obtained from the FLIM whereas mRFP1 relates to mRFP1 images acquired using a camera. (**b**) Data from 5 cells per condition were accumulated and presented as a scatter plot to show the average FRET.

2. Lifetime images are produced from the lifetime data which show the donor lifetime at each pixel of the image using a pseudocolour display (see Fig. 2 for an example of these).
3. This image is then used in two ways. Firstly, the image itself is presented to show localization of the GFP-tagged protein as well as the spatial distribution of GFP lifetime values (see Fig. 2). Secondly, histograms generated from the lifetime images containing information about the number of pixels with each value of the fluorescence lifetime (tau) are saved as text files before importing into Microsoft Excel (or similar software) for further analysis.
4. The text files contain information corresponding to tau and the number of pixels at each tau value. The values of tau are converted to FRET efficiency values using the following equation and plotted using microsoft excel (see Fig. 2).

$$\text{FRET efficiency} = 1 - \tau_{\text{fret}}/\tau_D$$

where τ_D is the lifetime of the donor, and τ_{fret} is the donor lifetime in the presence of the acceptor.
5. Fitting the lifetime decay data to a monoexponential curve provides information about the average lifetime at each pixel of the image. In most cases, this is sufficient; however, in some cases, it is desirable to fit the data to a biexponential curve and examine both FRETing and non-FRETing populations separately (see Note 9).

4. Notes

1. Other methods such as electroporation can be used where appropriate and should be optimized for the cell type. In addition, alternative transfections reagents such as Lipofectamine 2000 can be used; however, in our hands, Fugene6 give good transfection efficiencies with low cell toxicity in most cell types. Microinjecting cells with DNA is also an option; this can be useful if the molecules of interest are toxic to the cell or when transfection does not produce the required 1:2 donor:acceptor ratio. Microinjecting 1 μg GFP donor plasmid and 2 μg mRFP1 acceptor plasmid produces suitable expression levels in 4–6 h. Alternatively, microinjecting 0.3 μg GFP donor plasmid and 0.6 μg mRFP1 acceptor plasmid can lead to good expression levels in around 16 h. However, expression times and levels can also vary depending on the size of the protein(s) being expressed.

2. Samples prepared for FLIM need to be mounted in media which are not non-absorbing, containing no autofluorescence, or light scattering, but also have an "anti-fade" agent which is capable of reducing light-induced fading (photobleaching) of the fluorophore. Fluorsave Reagent (Calbiochem) is a suitable mounting medium and contains an anti-fade; however, there are various other suitable media available. For example, a solution of 10% Mowiol (Calbiochem) containing 100 mM Tris, pH 8.5, and 25% glycerol (incubate for at least 3 h at 50°C for Mowiol to dissolve) is also a suitable mounting media if used with an anti-fade such as DABCO (Sigma) which should be added at 2.5% prior to use.
3. Commercial frequency domain and time domain FLIM systems. Lambert Instruments can provide frequency domain FLIM systems for combining with confocal or wide-field microscopes from Leica, Zeiss, Nikon, and Lambert Instruments (www.lambert-instruments.com). Becker and Hickl can provide time domain FLIM systems which are compatible with both Leica and Zeiss confocal microscopes which utilize their TCSPC electronics.
4. Commercial software for FLIM detection and analysis. Both Lambert Instruments and Becker and Hickl sell software either alone or with their FLIM systems which can be used in a similar way to TRI2.TRI2 generates lifetime files in .ics format which are a generic filetype, however, different commercial software systems may use other file types.
5. It is necessary to acquire a control "GFP-alone" lifetime measurement in order to accurately analyze the FRET efficiency in the presence of acceptor. The FRET efficiency is a comparison between the control GFP lifetime and the GFP lifetime in the presence of acceptor. The fluorescence lifetime of eGFP is typically in the region of 2.1–2.2 ns; however, it is necessary to acquire a control GFP lifetime in every experiment due to environmental factor which can slightly alter the fluorescence lifetime of GFP. These include pH, ion concentration, and the age of the sample.
6. Sodium borohydride is used here to remove autofluoresence caused by disulphide bonds created during the fixation process.
7. It is necessary to use a filter here which is matched to the fluorescence emission of interest. This reduces cross-talk and also acts to remove scattered/reflected excitation light which would otherwise swamp the detector.
8. If bubbles occur, it is possible to remove them by gently rolling a yellow tip over the cover slip to expel the air bubbles.

9. Bi-exponential data analysis: For intermolecular FRET, it is sometimes desirable to determine the source of the observed lifetime reduction. To do this, a bi-exponential fluorescence decay model is applied to the data to determine the fluorescence lifetime of noninteracting and interacting subpopulations. The data may be fit by iterative re-convolution to

$$I(t)=\int_{-\infty}^{\infty} I_{\text{instr}}(t)\left\{Z+\alpha_1\text{Exp}(-t/\tau_1)+\alpha_2\text{Exp}(-t/t_2)\right\}\text{d}t,$$

where I_{instr} is the instrumental response, Z is the baseline offset, τ_1 and τ_2 are lifetimes of the interacting/noninteracting populations, and α_1 and α_2 are the pre-exponential factors relating to absolute species concentration (13). This equation is embedded into the TRI2 software.

References

1. Killock, D. J., Parsons, M., Zarrouk, M., Ameer-Beg, S. M., Ridley, A. J., Haskard, D. O., Zvelebil, M., and Ivetic, A. (2009) In Vitro and in Vivo Characterization of Molecular Interactions between Calmodulin, Ezrin/Radixin/Moesin, and L-selectin, *J Biol Chem* **284**, 8833–8845.
2. Worth, D. C., Hodivala-Dilke, K., Robinson, S. D., King, S. J., Morton, P. E., Gertler, F. B., Humphries, M. J., and Parsons, M. (2010) Alpha v beta3 integrin spatially regulates VASP and RIAM to control adhesion dynamics and migration, *J Cell Biol* **189**, 369–383.
3. Makrogianneli, K., Carlin, L. M., Keppler, M. D., Matthews, D. R., Ofo, E., Coolen, A., Ameer-Beg, S. M., Barber, P. R., Vojnovic, B., and Ng, T. (2009) Integrating receptor signal inputs that influence small Rho GTPase activation dynamics at the immunological synapse, *Mol Cell Biol* **29**, 2997–3006.
4. Peter, M., Ameer-Beg, S. M., Hughes, M. K. Y., Keppler, M. D., Prag, S., Marsh, M., Vojnovic, B., and Ng, T. (2005) Multiphoton-FLIM quantification of the EGFP-mRFP1 FRET pair for localization of membrane receptor-kinase interactions, *Biophys J* **88**, 1224–1237.
5. Diaspro, A., Chirico, G., and Collini, M. (2005) Two-photon fluorescence excitation and related techniques in biological microscopy, *Q Rev Biophys* **38**, 97–166.
6. Grashoff C, H. B., Brenner MD, Zhou R, Parsons M, Yang MT, McLean MA, Sligar SG, Chen CS, Ha T, Schwartz MA. (2010) Measuring mechanical tension across vinculin reveals regulation of focal adhesion dynamics., *Nature **In press***.
7. Pollok, B. A., and Heim, R. (1999) Using GFP in FRET-based applications, *Trends Cell Biol* **9**, 57–60.
8. Selvin, P. R. (2000) The renaissance of fluorescence resonance energy transfer, *Nat Struct Biol* **7**, 730–734.
9. Tramier, M., Gautier, I., Piolot, T., Ravalet, S., Kemnitz, K., Coppey, J., Durieux, C., Mignotte, V., and Coppey-Moisan, M. (2002) Picosecond-hetero-FRET microscopy to probe protein-protein interactions in live cells, *Biophys J* **83**, 3570–3577.
10. Tramier, M., Zahid, M., Mevel, J. C., Masse, M. J., and Coppey-Moisan, M. (2006) Sensitivity of CFP/YFP and GFP/mCherry pairs to donor photobleaching on FRET determination by fluorescence lifetime imaging microscopy in living cells, *Microsc Res Tech* **69**, 933–939.
11. Ai, H. W., Henderson, J. N., Remington, S. J., and Campbell, R. E. (2006) Directed evolution of a monomeric, bright and photostable version of Clavularia cyan fluorescent protein: structural characterization and applications in fluorescence imaging, *Biochem J* **400**, 531–540.
12. Albertazzi, L., Arosio, D., Marchetti, L., Ricci, F., and Beltram, F. (2009) Quantitative FRET analysis with the EGFP-mCherry fluorescent protein pair, *Photochem Photobiol* **85**, 287–297.
13. Barber, P. R., S. M. Ameer-Beg, J. Gilbey, L. M. Carlin,, and M. Keppler, T. C. N. a. B. V. (2009) Multiphoton time-domain fluorescence lifetime imaging microscopy: practical application to protein–protein interactions using global analysis, *J. R. Soc. Interface* **6**, S93–S105.

Chapter 28

Cell Migration in Confinement: A Micro-Channel-Based Assay

Mélina L. Heuzé, Olivier Collin, Emmanuel Terriac, Ana-Maria Lennon-Duménil, and Matthieu Piel

Abstract

This chapter describes a method to study cells migrating in micro-channels, a confining environment of well-defined geometry. This assay is a complement to more complex 3D migration systems and provides several advantages even if it does not recapitulate the full complexity of 3D migration. Important parameters such as degree of adhesion, degree of confinement, mechanical properties, and geometry can be varied independently of each other. The device is fully compatible with almost any type of light microscopy and the simple geometry makes automated analysis very easy to perform, which allows screening strategy.

The chapters is divided into five parts describing the design of different types of migration chambers, the fabrication of a mold by photolithography, the assembly of the chamber, the loading of cells, and finally the imaging on live or fixed cells.

Key words: Micro-fabrication, Micro-channels, Cell migration

1. Introduction

Although mechanisms controlling cell motility on 2D flat surfaces have been largely addressed, mechanisms allowing cells from vertebrate organisms to move in the confined environments of tissues remain largely unknown. This type of motility is relevant to multiple cell types and physiological processes, such as development, immunity, and cancer (1–3).

During in vivo cell migration, several parameters influence not only the type of locomotion used by cells, such as the presence of attractant molecules such as chemokines, or the cell capacity to degrade the extracellular matrix, but also the physical and

Claire M. Wells and Maddy Parsons (eds.), *Cell Migration: Developmental Methods and Protocols*,
Methods in Molecular Biology, vol. 769, DOI 10.1007/978-1-61779-207-6_28, © Springer Science+Business Media, LLC 2011

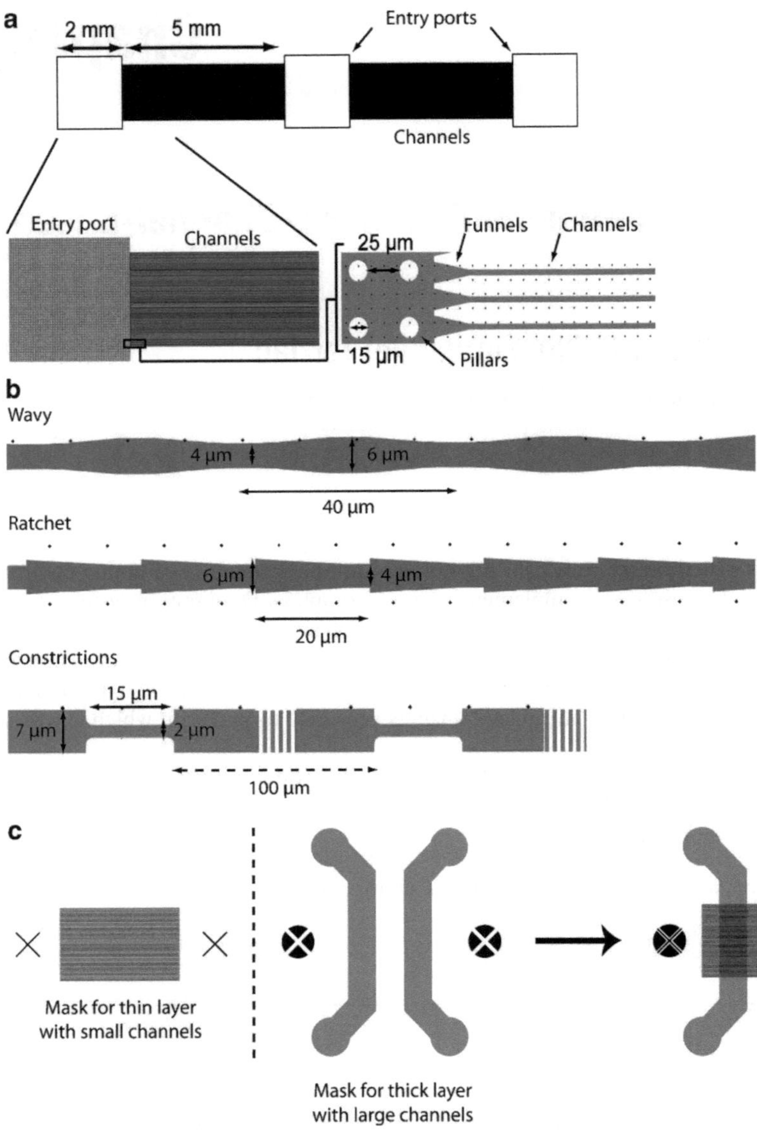

Fig. 1. Examples of micro-channel designs for migration of dendritic cells. (**a**). Multiscale drawing of an entire channel with three entry ports initially designed with 15-µm diameter pillars, linked by two rows of 4-µm large micro-channels. Note the funnel shaped entry of the channels. (**b**). Three different examples of micro-channels used in migration assays. (**c**) Design for a two-layer micro-channel chamber for immunostaining and drug delivery experiment. The crosses on the side allow the alignment of the two masks during the lithographic process.

biochemical nature of the space wherein cells move. Because of this complexity, it has been difficult to define how these parameters individually contribute to the type of locomotion used by migrating cells.

Systems such as collagen gels are commonly used as simplified models to study cell migration in 3D environments (4). However, even this very simplified system remains highly complex and results are thus difficult to interpret: many physical parameters act simultaneously and affect the motility of cells embedded in 3D collagen gels. These parameters include not only geometrical confinement, which will depend on gel density, but also local constrictions (due to irregular mesh size of the gel), which will force cells to squeeze their nucleus temporally to pass through narrow spaces, as well as gel elasticity (ability to deform), and the capacity of cells to reorganize the gel; in collagen gels, these parameters are all interwoven and their specific role on motility cannot be extracted.

Here, we propose a versatile and simple cell migration assay based on micro-channels, in which these parameters can be controlled and varied independently of each other. In its simplest version – straight channels made of silicone rubber (here polydimethylsiloxane, PDMS) – such an assay already proved very useful to assess the migration properties of cells under confinement, i.e., in conditions in which cells have a constrained geometry (channels of subcellular dimensions, between 2 and 10 μm in width and height depending on cell type). Such an approach has proved to be relevant for both cancer cells (5, 6) and immune cells (7, 8), and neurons (9). Such geometry also makes automated analysis very easy to perform, which allows screening strategy.

Channels can then be modulated to address specific biological questions: for example, trans-migration capacity can be assayed by introducing constrictions of a well-defined geometry (see Figs. 1b and 4B.b), or the influence of substrate elasticity by varying the material in which the channels are made (this aspect is not described in this paper). The geometry can also be complexified, to allow cells to take turns and explore the space in 2D while maintaining confinement (creating maze instead of simple straight channels). Once the basic micro-fabrication technology is implemented, designing new devices is fast and straightforward.

2. Materials

Materials needed for this assay can be subdivided into two groups: common chemicals and equipment found in biology laboratories, and specialized equipment more common in micro-fabrication facilities (clean room). Two strategies are possible for a biology laboratory wanting to implement such an assay: (1) find a micro-fabrication collaborator for fabrication of a mold for the chambers. Once the mold is done, almost no specialized equipment is needed. Molds can also be duplicated with no specialized equipment (see Note 1). The only limitation if this option is chosen is the design and fabrication of new types of chambers which might be slowed down. (2) Assemble a small micro-fabrication setup, which can simply be done under a chemical hood plus a small bench (a "gray room"). The only two specialized pieces of equipment needed are a collimated UV lamp and a spincoater. This solution is recommended to allow free development of new assays, but will require a specific training to implement fabrication processes efficiently.

2.1. Mask Design

1. Software: Several pieces of software will allow designing mask features, depending on the requirements of the mask-producing company. In the simplest case, drawing software with indication of sizes can be used, and the mask manufacturer will convert it into a proper file format (this will, of course, have a cost). A common file format used by manufacturers is GDS II (other formats are CIF or DXF). Any software that can produce such file will work. Some pieces of software are specifically meant to design masks, they are not only very convenient, but are also often expansive and will require a short learning phase (e.g., L-Edit, Clewin, and AutoCAD).
2. Photomask: Examples of companies to which we have ordered photomasks of good quality are Delta Mask (the Netherland), Toppan photomasks (present in many countries), and Microtronics Photomasks (USA). Size of features is limited by two factors: the resolution of the photomask, which can go down to a fraction of microns for the most expansive ones and will be around one micron for regular ones. The second factor is the quality of the contact between the substrate and the photomask. Features of 1 μm are possible to obtain with care, and features of a few microns are easy to obtain (see Note 2).

2.2. Wafer Fabrication

1. Silicon wafers, [100] orientation and 500-μm thickness (Siltronix).
2. Epoxy-based negative photoresist (SU-8 2005 and 2050, MicroChem).
3. Primer solution (Omnicoat, Microchem).
4. Developer solution (SU-8 developer, Microchem).
5. Negative chromium mask (for structure larger than 30 μm, a high-resolution printed transparency mask is enough).
6. High-grade isopropanol (Propan-2-ol).
7. Crystallizing pan.
8. Two hotplates (Super Nuova, Thermo Scientific, Model HP 1317020 33Q).
9. Spincoater (Model WS-650SZ-6NPP/Lite, Laurell Technologies Corporation).
10. Collimated 365-nm UV lamp controlled by a timer (OAI, Model 30 Enhanced Light Source).

2.3. Chamber Fabrication and Assembly

1. PDMS chip fabrication
   (a) Curable silicone rubber (e.g., Sylgard 184, Dow Corning and RTV 615, GE Silicones).
   (b) Mold. cf. 2.2 and 3.2.
   (c) Vacuum bell jar.
   (d) Oven.
2. Chip assembly
   (a) Surgical blade.
   (b) Hole driller: core sample cutter (e.g., Harris Uni-Core, tip diameter 2.0 mm, Ted Pella Int.) or a sharp needle.
   (c) Adhesive tape.
   (d) Ultrasonic bath.
   (e) Pure ethanol.
   (f) Air or Oxygen Plasma cleaner (Plasma cleaner PDC 32 G, Harrick Plasma).
   (g) Glass slide.
3. Chip assembly
   (a) Air or Oxygen Plasma cleaner (Harrick PDC 32 G).
   (b) PBS.
   (c) Fibronectin from bovine plasma (fibronectin from bovine plasma 0.1% solution). For poly(ethylene glycol) (PEG, which prevents cell adhesion) coating.
   (d) 2-(N-Morpholino)ethanesulfonic acid, 4-morpholineethanesulfonic acid (MES; low moisture content).
   (e) N-(3-Dimethylaminopropyl)-N′-ethylcarbodiimide hydrochloride (EDC).
   (f) Poly-(L-lysine)(20)-grafted[3.5]-poly(ethylene glycol)(2) (PLL-g-PEG, SuSos Product Name PLL(20)-g[3.5]-PEG(2)).
   (g) HEPES buffer 10 mM, pH 8.6.

2.4. Cell Culture

Use the regular growth medium and cell culture materials adapted to the cells you want to assay.

2.5. Live Cell Microscopy and Image Analysis

This device can be adapted to any type of light microscopy. The bottom coverslip is a regular microscopy coverslip. The device can also be bound to the bottom of any sort of dish (glass or plastic bottom).

3. Methods

3.1. Designing a Chamber for a Migration Assay

The general behavior and motility of cells in micro-channels highly depend on the geometry of the channels. It is thus important to design channels with a geometry adapted to a given cell type and a given question. We give below a few design principles and several examples that worked for our assays with bone marrow-derived dendritic cells.

1. Micro-channel designs are made using specialized software. The chamber design presented in Fig. 1 includes three entry ports (Fig. 1a) containing micro-pillars of 15-μm diameter positioned every 25 μm. These micro-pillars ensure that the roof of the entry port will not collapse on the glass-bottom coverslip when assembling the chamber. The distance between each pillar must be large enough for cells to migrate between them and reach the entrance of the micro-channels. Each entry port is linked to the next one by hundreds of parallel micro-channels, whose geometry depends on the purpose of the experiment. Figure 1b presents three examples of micro-channels with "wavy" and "ratchet" structures used in migration assays, and narrow constrictions of dimensions smaller than the cell nucleus, used to study diapedesis in conditions in which nuclear deformation becomes the limiting factor for migration (for BMDCs, around 2 μm in width). Each micro-channel entry is designed with a funnel structure (Fig. 1a) that help cells find the channel entry and also makes it possible to inject cells in channels without damaging them, when needed (nonadhesive channels, for example).
2. Once a design is finalized, the corresponding photomask is ordered. The manufacturer usually requires a specific file format (e.g., DGSII). Note that the manufacturer can often help with the mask design. It is often possible to simply provide a drawing with dimensions and the manufacturer will finalize the chamber design.
3. For immunostaining assay or drug delivery experiments, it is important to be able to introduce chemicals rapidly inside the micro-channels. Due to the confined geometry of the channels, and the slow diffusion rate inside small and long capillaries, it can be useful to design a two-layer chamber with short micro-channels, to allow rapid diffusion without having a direct flow over the cells (Fig. 1c), (8, 9). The micro-channels of desired geometry and height are connecting larger channels with entry ports used for medium exchange. This device requires a two-layer photolithography process (see Subheading 3.2). The photomask used for the small channels is

the same, previously designed for one-layer micro-channel design. A second mask with the design of the large channels (e.g., 1 mm large and 50-μm-high channels) has to be designed and aligned to the first one during the lithography step. Crosses can be placed on the two masks to help with alignment, but a nearly perpendicular alignment of large channels with a large bundle of small channels is relatively easy to achieve and does not have to be very precise. Large channels are usually made using a simple transparency photomask.

3.2. Wafer Fabrication by Photolithography

This step is the most demanding in terms of specialized equipment and specialized training. It can be performed by collaborators, or after a significant training, either in a clean room or using minimal equipment under a hood. It is also possible to have customized molds made by some specialized companies (e.g., Biotray, Lyon, France, or Amo, Aachen, Germany). Some assembled micro-fabricated devices can also be found commercially (e.g., Ibidi, Martinsried, Germany). Note that the protocol below is a stereotypical photolithography process. A large literature exists on such fabrication processes, and documentation is also provided with the photoresist, containing tables and a protocol for fabrication of features of different sizes. Due to variation in UV lamps used, protocols always require a fine tuning. Once a precise protocol is found, it is usually very reproducible. The most important parameters to adjust are the illumination time and the development time.

1. All chemicals have to be at room temperature (place photoresist at room temperature at least 30 min before experiments).
2. Place the wafer on a hotplate at 95°C for 5 min to let water evaporate. A higher temperature could cause oxidization of the silicon wafer, which would alter adhesion between the photoresist and the wafer.
3. Pour omnicoat solution on the wafer to cover a large surface. Typically, 3 mL is sufficient for a 4-in. wafer.
4. Typical spincoating parameters for primer are 500 rpm for 5 s (acceleration of 100 rpm/s) and 3,000 rpm for 30 s (acceleration of 300 rpm/s).
5. Bake the coated wafer for 1 min at 200°C on a hotplate. To avoid thermal shock, approach the wafer slowly when placing it on the hotplate, or use a temperature ramp. For the same reason, hold the wafer until it has cooled down before placing it on a room temperature surface after baking.
6. Pour gently (i.e., try to avoid making bubbles) 5 mL of photoresist on the wafer.

7. Typical spincoating parameters for photoresist are 500 rpm for 5 s (acceleration of 100 rpm/s) and 30 s at a speed proportional to the thickness wanted (e.g., 5 μm with SU-8 2005 requires a final speed of 3,000 rpm, see application notes given by the supplier).
8. Softbake step: place the wafer on the hotplate at 65°C for 1 min and then for 2 min at 95°C.
9. Place the mask on the baked layer of photoresist. The mask must be very clean in order to ensure a good contact between the mask and the photoresist. Use an air gun to remove dust. Press the mask against the wafer to ensure a good contact. If no vacuum holder is available, you can place a weight on the mask.
10. Insolate the photoresist through the mask according to the thickness of the layer. For example, for 5 μm, photoresist needs 50 mJ/cm^2 to be fully activated, while a 30-μm-thick layer requires 150 mJ/cm^2. This step is the most important one as an over- or an underexposure will strongly affect the shape of the channels. It is common to make a few try on pieces of wafers to calibrate the timing correctly.
11. A postexposure bake step ensures the cross-linking of the activated region obtained through illumination. Duration of the bake depends on the thickness of the layer. In any case, this must be done in three steps to avoid cracking of the photoresist. Put the wafer on a hotplate set at 60°C for 1 min, then on another hotplate set at 95°C (3 min for 5 μm, 5 min for 30 μm), and finally put it back on the 60°C hotplate for 1 min. After this step, structures appear. When the photoresist layer is thick (several tens of microns), it might be necessary to shut down the hotplate and let it cool down slowly down to about 50°C (in about 30 min) to avoid storing stress in the structures, which could lead to detachment of the photoresist layer in later steps of the protocol (in particular, when using the wafer as a mold for PDMS). At this stage, a second layer of photoresist can be applied to make larger channels (see Note 3).
12. Once the wafer is back at room temperature, place it in a crystallizing pan containing the developer solution. Agitate gently all along the development process. For small thickness of 5 μm, the wafer must not stay longer than 1 min in the solution, while for 30 μm, this may take up to 5 min.
13. Rinse the wafer with propan-2-ol. If unexposed photoresist remains on the wafer, white traces will appear. Repeat the development and rinsing process until no traces appear. It may be useful to have some development solution in a wash bottle to rinse the wafer when developing very small structures.

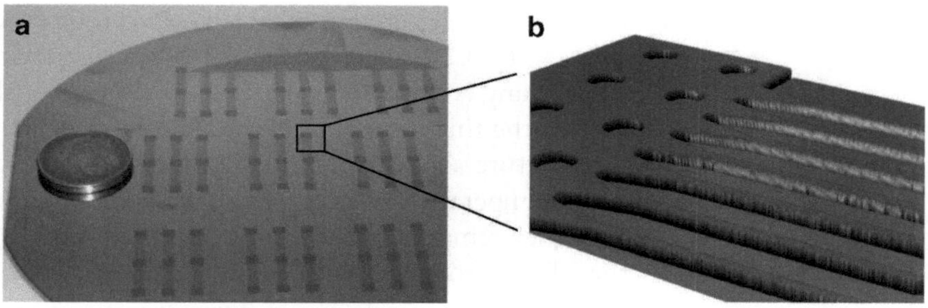

Fig. 2. Example of a silicon wafer obtained by photolithography. (**a**). Wafer obtained by photolithography (a 1 cent coin is shown to compare sizes). (**b**). Image of features made of photoresist, obtained by optical interferometry. Color code is set to *light grey* (ground) to *darker grey* (5 μm) (see online version for color).

14. The last step, called hard bake, is optional but appears to be useful when a wafer is frequently used as a mold. This step avoids photoresist detachment after several chip fabrication. The wafer is placed on a hotplate at 150°C for 10 min. It is best to use a ramp before reaching such a high temperature and cooling down must also be slow.
15. As vertical dimensions obtained can be hard to assess with a microscope, an optical interferometer can be used for a precise measure of features.

The mold (or wafer) resulting from this step is shown in Fig. 2.

3.3. Chamber Fabrication, Assembly, and Coating

From this point on, everything can be performed in a regular biology laboratory. The only specialized equipment needed is a plasma cleaner, which is small and not expensive.

1. PDMS chip fabrication
   (a) Prepare a mix of uncured silicon rubber and curing agent at a ratio of 10:1 by weight. Mix the two parts thoroughly. Mixing will induce air bubbles, which may help to assess the homogeneity of the mix.
   (b) Bubbles can be removed either in a vacuum bell jar or by centrifugation of the mix. Put the solution in a tube and centrifuge for 1 min at a speed of 1000 × *g*. Centrifugation works best and even small bubbles will be removed, such bubbles might be a problem when using the device for microscopy studies.
   (c) Pour the solution on the mold. Thickness of the PDMS should be between 5 mm and 1 cm.
   (d) Put the mold and PDMS layer in a vacuum bell jar for ca. 30 min. Even if air bubbles were removed by centrifugation, some may form when pouring the solution on the mold, especially when there are small structures on the mold.

(e) Once all bubbles are removed, PDMS can be cured in an oven. At 65°C, a 5-mm layer of PDMS is cured within 2 h. This timing is dependent on the temperature of the oven and on the thickness of the layer. Note that when using a temperature-sensitive mold, PDMS can also be cured at room temperature over 24 h; a short baking of the PDMS at higher temperature is then performed after removing the PDMS from the mold.

(f) From a single mold, many replicas can be made in very short time. This will protect the original mold (see Note 1 to use a PDMS counter-mold to make an epoxy replica of the silicon wafer).

2. Chip assembly

(a) Cut the cured PDMS with a surgical blade, leaving enough space (about 2 mm) around the structures to allow a good binding to the bottom glass substrate.

(b) Drill holes in designed entry ports, to access channels after binding.

(c) Clean PDMS chip by sticking and peeling adhesive tape on the structures side.

(d) Place PDMS in a beaker containing pure ethanol and sonicate for 30 s. PDMS is a porous elastomer and a solvent like ethanol may diffuse inside and be released later causing cell death, so cleaning with ethanol should not exceed a few minutes.

(e) Dry the PDMS with an air gun equipped with a filter.

(f) Place the glass substrate and the PDMS piece with the channels side up in the plasma cleaner. Treat the surfaces at the highest power for 30 s at 300 mTorr (if no barometer is available, check the color of the plasma and adjust the air influx to get a bright purple). This will activate the surfaces, making them able to stick together. Note that surfaces will remain activated only for a short amount of time, so binding should be performed immediately after treatment. See Fig. 3.

(g) Stick the glass and the PDMS together. One hour incubation at 60°C will strengthen the binding. PDMS device can be assembled on plastic substrate (see Note 4).

3. Chip fictionalization

(a) Activate the upper surface of the chip for 30 s at 300 mTorr in the plasma cleaner to make inlets hydrophilic. After the 30 s, turn the power off but leave the chip at 300 mTorr for another 5 min to lower the air pressure inside the PDMS. This will later help getting liquid inside the channels and removing bubbles.

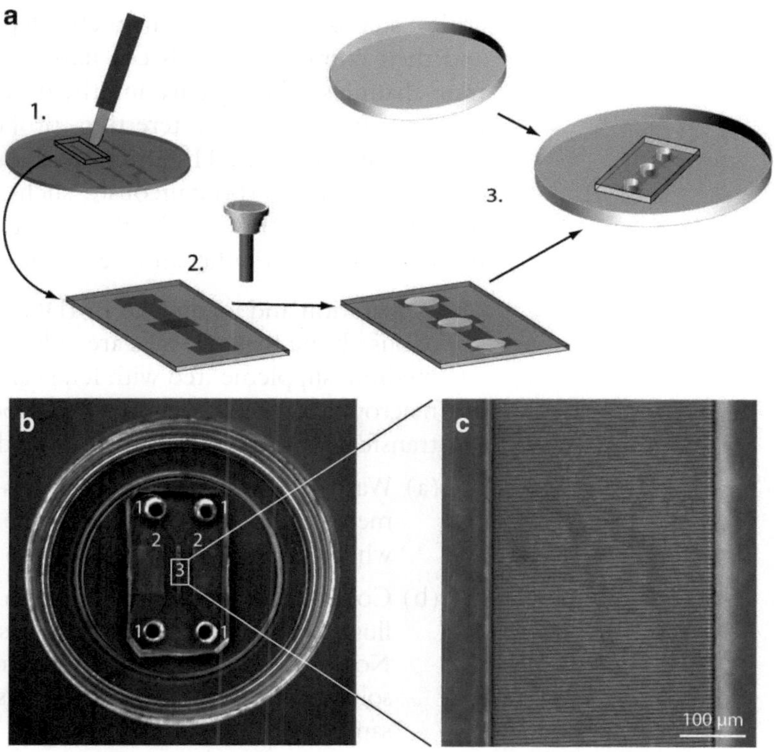

Fig. 3. Assembly of the cell migration chamber. (**a**). Schematic view of a device assembly. 1. Chip is cut from a large piece of PDMS cured on the mold. 2. Holes are drilled to create connections between the channels and the exterior of the chip. Note that drilling starts from the engraved part to the exterior of the PDMS chip. 3. Drilled chip and glass-bottom Petri dish are plasma activated and bound together. (**b**). Overview of a "dual channels device." The PDMS chip is stuck in a 35-mm glass-bottom Petri dish. It comprises four injection holes (label 1), two large channels 50 μm high (label 2), and a series of smaller channel in which migration is recorded (label 3). (**c**). Zoom on the central region where small channels of 4 μm are connecting two large channels.

(b) Fill inlets with a 25 mg/mL fibronectin solution in PBS. The low pressure inside PDMS will suck the solution inside small channels. If inlets are small (typically 2 mm in diameter), bubbles can form and prevent the solution from reaching the channels. Such bubbles can be removed with a thin needle. Incubate for 1 h at room temperature.

(c) Rinse the channels with PBS by aspiration from one of the inlets.

(d) The system can be used immediately or stored up to 3 days at 4°C.

An alternative coating protocol can be used to prevent cell adhesion inside the channels or enable covalent binding of proteins (see Note 5).

3.4. Cell Culture in Micro-Channels

One major advantage of the micro-channels technology is the small amount of cells needed to run an experiment. However, the critical step is entry in the channels. Depending on their motility and their contractility, cells can have different capacity to penetrate channels. The capacity and the time required to enter channels are by itself an interesting property to test using the micro-channels assay. Highly mobile and deformable cells will enter the channels spontaneously, such as bone marrow-derived dendritic cells (BMDCs) (7, 10); in that case, simply placing cells in the entry well is sufficient (see Fig. 4A).

1. Preparation and loading of BMDCs in the simple device
   Mouse bone marrow cells are cultured during 10–12 days in medium supplemented with fetal calf serum and granulocyte-macrophage colony stimulating supernatant obtained from transfected J558 cells, as previously described (12).
   (a) Wash twice the micro-channels chip with cell culture medium (see Note 6). Keep the device in the incubator while collecting cells.
   (b) Collect cells: replace the medium by PBS and then gently flush to collect semi-adherent cells in a centrifuge tube (see Note 7). Add a 37°C pre-warmed PBS + EDTA 5 mM solution for 5 min. Flush the cells and collect them in the same centrifuge tube.
   (c) Centrifuge the collected cells for 5 min at 1,000 × g. Resuspend the cell pellet in culture medium to get a concentration of 12 millions cells per mL.
   (d) Gently empty the entry well of the micro-channel chip. Load 5 μL of cell suspension (about 60,000 cells) in the entry well. Avoid making bubbles (see Note 8). Let cells fall at the bottom of the well and then completely fill the entry well with culture medium to avoid drying. BMDCs will take several hours to reach and penetrate the channels massively (see Note 9).
2. Loading of BMDCs in the two-layer device
   (a) Repeat steps 1a and 1b in Subheading 3.4.
   (b) After centrifugation, resuspend cells in culture medium at a concentration of 30 million cells per mL.
   (c) Gently empty the four large entry wells. Load 5 μL of cell suspension only in the two entry wells of one of the two large flow channels (see Fig. 1c).
   (d) Wait for 5–10 s and fill the two remaining entry wells with culture medium.

   This procedure will create a slight fluid flow that will gently push cells against the entry of the micro-channels connecting the two larger flow channels.
3. Alternative ways to introduce cells in the channels

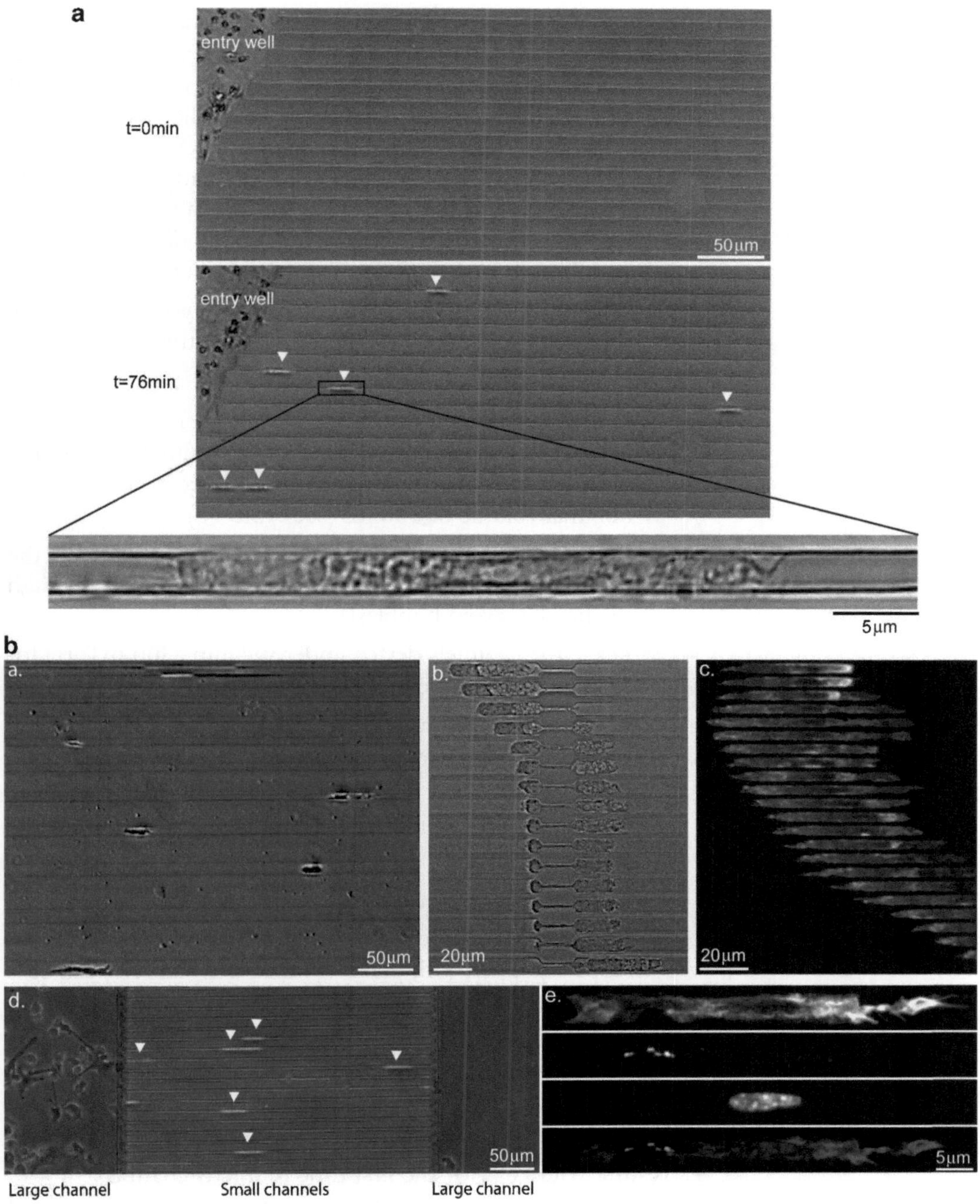

Fig. 4. Examples of applications of the micro-channel migration assay. (**A**). Low-magnification phase contrast images of dendritic cells migrating in channels, at two different time points. BMDCs were loaded in the entry well connected to 4-μm width channels coated with fibronectin. *Insert*: High-magnification phase contrast image of a dendritic cell migrating in a channel. (**B**). (*a*). Phase contrast image of BMDCs migrating in channels from 1- to 13-μm width. (*b*). Phase contrast image sequence of a dendritic cell passing through a constriction of 2-μm width and 20-μm length, in a 8-μm width channel (total time 15 min, time lapse 1 min). (*c*). Fluorescent live cell image sequence of a dendritic cell overexpressing the Lifeact peptide fused to mCherry (11), in a 4-μm width channel (total time 10 min, time lapse 30 s). (*d*). Phase contrast image of BMDCs migrating in the "dual channels device" (with small channels 350 μm long and 4 μm in width). (*e*). Immunostaining of a dendritic cell fixed inside a channel, after a pulse-chase with Lucifer Yellow to visualize macropinocytosis. From *top* to *bottom*: F-actin (*red* in the merge panel at the *bottom*), Lucifer Yellow (*green*), and DAPI (*blue*) (see online version for color).

If cells of interest are not able to enter spontaneously in the channels, two alternative ways of loading are proposed. If cells express chemokine receptors (it is the case of activated BMDCs), a chemokine gradient can be applied in the simple channels device by drilling a second well, opposite to the entry well, in which chemokines will be loaded. For a more controlled gradient, the two-layer device can be improved by adding short micro-channels (8). A more direct way is to inject cells with a pressure gradient as described below.

(a) When assembling the device, make sure you drill holes of the right diameter to fit with the tubing used for injection.

(b) Prepare cells as described before.

(c) After centrifugation, resuspend cells in culture medium at a concentration of two to four million cells per mL, in a minimal volume of 500 μL (see Note 10).

(d) Prepare a 1-mL syringe with a needle that fits with the diameter of the tubing. Aspirate the cell suspension through the needle (avoid bubbles!).

(e) Leave the channels device under vacuum (300 mTorr) for 15 min to lower the pressure inside PDMS.

(f) Immediately load a small drop of cells in the entry well with the needle, it will be sucked inside due to the depression in the PDMS.

(g) Connect a piece of tubing to the syringe and fill it with the cell suspension.

(h) Plug tubing to the entry well.

(i) Place the device under a simple cell culture inverted microscope equipped with phase contrast and start to push cells gently inside the device.

(j) Stop injection as soon as the first cells have entered the micro-channels. The goal is to obtain a concentrated front of cells at the entry of the channels, but pushing them further would damage them.

4. Testing your cells for the first time in micro-channels
When using such an assay for the first time, it is recommended to find the right channel dimension for your cells. A good solution is to use a device that has a series of channels of various sizes connected to the same entry well (ranging from 2 to 40 μm large, for example). Load the cells and check from a few hours up to 48 h in which channels they penetrate, migrate, and survive (Fig. 4B.a). Finally, the fibronectin concentration can turn out to be determinant for entry and/or migration of cells in the channels. Thus, it can also be informative to perform a migration test with a range of fibronectin concentrations (or other coating protein you might want to use).

3.5. Imaging in Micro-Channels

1. Live imaging in micro-channels (see Fig. 4)
   The micro-channel assay is compatible with any kind of microscopy, as the bottom of the device is a classical glass coverslip. The timescale of acquisition and the magnification will vary depending on the parameters you want to analyze. Thus, you can easily image the dynamics of fluorescent proteins, using either widefield or confocal microscopy. Moreover, the geometry of the channels greatly facilitates the collection of data on cell motility through the generation of kymographs. This tool can be applied on phase contrast images, even at low magnification (which enables to image a large number of cells in different conditions). It is also possible to automate image analysis fully (7). Many different parameters can be extracted from kymographs. For each line of the kymograph, the two ends of each cell in the channel, and the center of mass can be extracted. From these measurement, once the tracking is performed (identifying cells from one line to the next), cell size, cell speed, persistence, cell–cell contacts, etc., can be easily measured. Note that phase contrast images at low magnification will be affected by the channels walls, almost completely masking the cells. There is an easy image cleaning solution for this: make sure images in the movie are correctly registered (multi-position acquisition often induces slight errors in repositioning and a giggling of the movie), and use a registration routine if it is not the case – several are available in ImageJ, for example. Then produce a time-averaged image from the registered movie and subtract it from all images (or use a routine which does all of that, which is basically cleaning all the nonmoving parts of the movie).

   A second advantage of the micro-channel assay is the possibility to combine it with micro-fluidic devices of diverse complexity. The simplest one consists in changing the fluid in the micro-channels during the experiment. For a fast exchange of fluid, the most adapted device is the dual channels device (with large and small channels, see Figs. 1c, 3b, and 4B.d) as diffusion in the short channels will rapidly lead to a uniform concentration.

2. Immunocytochemistry inside micro-channels
   Immunocytochemistry is a useful complementary approach to live imaging for the study of molecules involved in migration and the quantification of specific expression patterns. This approach can be adapted to micro-channel devices (see Fig. 4B.e).
   (a) Perform your experiment as described in previous parts, using the dual channel device (see Note 11).
   (b) When desired, gently remove the medium from entry wells (see Note 12).

(c) Immediately fill the wells with 4% PFA. Fix the cells for 15 min at room temperature.

(d) Wash three times with PBS (empty wells and fill with PBS).

(e) Proceed with your favorite protocol for immunostaining (e.g., permeabilization and antibody fixation). See Notes 13 and 14.

(f) At the end of the staining, remove all the liquid contained in the large channels, sucking from the entry wells with a vacuum aspirator.

(g) Fill the device with fluoromount, as slowly as possible, to avoid the formation of bubbles. Also completely fill the entry wells, leaving an excess of fluoromount on top of them.

(h) Leave the devices overnight at 37°C, in a dry chamber.

4. Notes

1. Mold replication.

   Material: surgical blade, vacuum bell jar, adhesive tape, epoxy resin (Soloplast, packages of epoxy resin R123 and hardener R614).

   Method:

   (a) Mould a PDMS chip from the original wafer as in part 3.3.1 (PDMS chip fabrication) PDMS replica should not be cut too close to the edges of the structure. Typically, a frame of around 5 mm around the channels must be kept. Clean the PDMS by sticking and peeling adhesive tape on the structure's side. It can also be sonicated in ethanol for further cleaning.

   (b) Remove air from the PDMS replica in a vacuum bell jar for 15 min.

   (c) Prepare epoxy resin mix: two-thirds of resin and one-third of hardener (w/w). As the viscosity of the two components are different, be careful when mixing both to obtain a homogenous solution.

   (d) Pour 1 cm high of the mix in a dish. As an example, a Petri dish of 145-mm diameter requires 90 g.

   (e) Place PDMS replica on top of the mix, with structures down. Small bubbles may appear at the interface between PDMS and epoxy. Since PDMS was placed under vacuum, bubbles will be sucked inside the PDMS.

   (f) Let epoxy harden for 24 h at room temperature. Remove PDMS pieces from the mold. The epoxy replica is now ready to be used as a mold for PDMS. It does not need any further treatment as it does not stick to PDMS. Epoxy will be fully hardened only after 2 weeks. In the mean time, it can be used as a mold, but it might deform slightly if placed at too high temperature during PDMS curing. Let it cool down at room temperature after curing before peeling out the cured PDMS.

2. Fabrication of the photomask mainly depends on the resolution and size of the structures to design. For large structures, it is possible to use a faster, cheaper but nonetheless very accurate method using transparency photomasks. For structures larger than 30 μm, the drawing can be made with regular software (e.g., Adobe Illustrator) and can be printed on a transparency mask using high-resolution printers (at least 3,600 dpi). Some companies also provide high-resolution printing on transparencies for a low price (e.g., the Asseco company).
3. Two-layer photolithography. It may often be interesting to have different sizes of structures on the same mold, as, for example, series of small channels bridging larger ones in which fluids can be flown without imposing any flow on cells inside small channels (6, 8, and 9). For such a fabrication, two masks are needed, one for small features (high-resolution mask) and one for large features (a transparency is enough). Some large masks might also be added (e.g., complementary crosses) positioned at the same place on both masks to make alignment of the two sorts of channels easier. When using large series of small channels, they will be visible on the resist after the post-bake step, so it will be very easy to place the large channels at 90° and to align with the small ones. The small channels have to be made longer than the distance between the two large channels for an easy alignment without a mask aligner. First make the thin layer with small channels until step 11, then spread the second thicker layer on top of the first layer, adjusting the speed to get the right thickness, all other steps are then performed as for the first layer.
4. It is possible to stick PDMS on plastic surfaces (usually polystyrene). Leave the plastic substrate overnight in a sealed box containing 100 μL of dimethyldichlorosilane. Silane will form a thin layer on the plastic. Then activate by plasma, like for glass, and bind PDMS immediately.
5. PEG coating to prevent cell adhesion inside the channels (the method below can also be applied to bind proteins covalently to the channel walls)
   (a) Prepare a buffer solution containing 0.05 M MES and 0.5 M NaCl at pH 6.
   (b) For 1 mL of buffer solution, dissolve 17.2 mg of Sulfo-NHS and 40 mg of EDC.

(c) Prepare a solution of 0.5 mg PLL-g-PEG in HEPES 10 mM, pH 8.6.

(d) Plasma treat the assembled device for 30 s at 300 mTorr.

(e) After treatment, let the system at low pressure for 5–10 min.

(f) Flow EDC/Sulfo-NHS solution in the device. This step must be done quite rapidly to suck the solution efficiently inside the small channels (PDMS treated at low pressure will inflate for a few minutes when placed back at normal pressure, inducing a suction inside channels). Always avoid bubbles in the entry wells. Completely fill entry wells with the solution.

(g) Incubate for 30 min and then empty entry wells.

(h) Flow PBS through the channels.

(i) Fill the channels with the PLL-g-PEG solution and incubate for at least 3 h (up to 12 h) at room temperature.

(j) Flow PBS through the channels.

(k) The system can be stored up to 1 week at 4°C.

6. In case channels directly end into the dish, pay attention to fill the dish entirely with medium to avoid drying or liquid flows.

7. To prevent activation of BMDCs, be careful not to flush them too hard when harvesting. Keep in mind that even mechanical stress can induce a pre-activation of immature BMDCs.

8. This protocol is designed for an entry well of 2.5-mm diameter. It can be adapted to smaller wells in order to load a smaller amount of cells (but in smaller wells, bubbles are more difficult to avoid). The amount of cells loaded can also vary from one cell type to another, depending on their size and their motility.

9. Entry of BMDCs in the channels is random (in the simplest setup, they are not attracted by channels). Only those that stand close to the channels will get inside. High concentration in the well will increase the probability of a cell to find a channel entry.

10. The minimal volume required is 500 μL for easy manipulation (to fill half of a 1-mL syringe). Therefore, at least one million cells are needed for this experiment.

11. For such an experiment, the amount of cells per channel can be critical. Indeed, during the staining process, reagents will reach the cells by diffusion inside the small channels: too many cells will obstruct the channel and prevent the reagents from staining all the cells.

12. During the whole procedure of fixation and staining, it is important to remove the medium only in entry wells and to leave the large channels full of fluid, to avoid bubbles in the next steps.
13. A better result is obtained with a one-step staining, using coupled antibodies or fluorescent phalloïdin, for example. With a two-step protocol (unstained primary antibody and stained secondary antibody), the quality of the staining might be more variable.
14. For a first try, perform your usual immunostaining protocol. If the staining is weak, you can eventually increase the concentration of antibodies. In case the staining appears dirty (nonspecific signal or spots appearing on the sides of the channels), try to increase the number and/or the duration of the washes, and to increase the BSA concentration in the buffer.

Acknowledgments

This work was supported by ANR Grant ANR-09-PIRI-0027 and by a grant of the Fondation InNaBioSanté, to M.P. and A.M.L. M.L.H was supported by Association pour la Recherche sur le Cancer (ARC).

References

1. Friedl P, Wolf K. Plasticity of cell migration: a multiscale tuning model. J Cell Biol. 2010 Jan 11;188(1):11–9.
2. Nourshargh S, Hordijk PL, Sixt M. Breaching multiple barriers: leukocyte motility through venular walls and the interstitium. Nat Rev Mol Cell Biol. 2010 May;11(5):366–78.
3. Richardson BE, Lehmann R. Mechanisms guiding primordial germ cell migration: strategies from different organisms. Nat Rev Mol Cell Biol. 2010 Jan;11(1):37–49.
4. Wolf K, Alexander S, Schacht V, Coussens LM, von Andrian UH, van Rheenen J, Deryugina E, Friedl P. Collagen-based cell migration models in vitro and in vivo. Semin Cell Dev Biol. 2009 Oct;20(8):931–41.
5. Rolli CG, Seufferlein T, Kemkemer R, Spatz JP. Impact of tumor cell cytoskeleton organization on invasiveness and migration: a microchannel-based approach. PLoS One. 2010 Jan 15;5(1):e8726.
6. Irimia D, Toner M. Spontaneous migration of cancer cells under conditions of mechanical confinement. Integr Biol (Camb). 2009 Sep;1(8-9):506–12.
7. Faure-André G, Vargas P, Yuseff MI, Heuzé M, Diaz J, Lankar D, Steri V, Manry J, Hugues S, Vascotto F, Boulanger J, Raposo G, Bono MR, Rosemblatt M, Piel M, Lennon-Duménil AM. Regulation of dendritic cell migration by CD74, the MHC class II-associated invariant chain. Science. 2008 Dec 12;322(5908):1705–10.
8. Irimia D, Charras G, Agrawal N, Mitchison T, Toner M. Polar stimulation and constrained cell migration in microfluidic channels. Lab Chip. 2007 Dec;7(12):1783–90.
9. Taylor AM, Blurton-Jones M, Rhee SW, Cribbs DH, Cotman CW, Jeon NL. A microfluidic culture platform for CNS axonal injury, regeneration and transport. Nat Methods. 2005 Aug;2(8):599–605.

10. Lautenschläger F, Paschke S, Schinkinger S, Bruel A, Beil M, Guck J. The regulatory role of cell mechanics for migration of differentiating myeloid cells. Proc Natl Acad Sci USA. 2009 Sep 15;106(37):15696–701.
11. Riedl J, Crevenna AH, Kessenbrock K, Yu JH, Neukirchen D, Bista M, Bradke F, Jenne D, Holak TA, Werb Z, Sixt M, Wedlich-Soldner R. Lifeact: a versatile marker to visualize F-actin. Nat Methods. 2008 Jul;5(7):605–7.
12. Winzler C, Rovere P, Rescigno M, Granucci F, Penna G, Adorini L, Zimmermann VS, Davoust J, Ricciardi-Castagnoli P. Maturation stages of mouse dendritic cells in growth factor-dependent long-term cultures. J Exp Med. 1997 Jan 20;185(2):317–28.

Chapter 29

Functional Screening with a Live Cell Imaging-Based Random Cell Migration Assay

Wies van Roosmalen*, Sylvia E. Le Dévédec*, Sandra Zovko, Hans de Bont, and Bob van de Water

Abstract

Cell migration, essential in cancer progression, is a complex process comprising a number of spatiotemporally regulated and well-coordinated mechanisms. In order to study (random) cell migration in the context of responses to various external cues (such as growth factors) or intrinsic cell signaling, a number of different tools and approaches have been developed. In order to unravel the key pathways and players involved in the regulation of (cancer) cell migration, a systematical mapping of the players/pathways is required. For this purpose, we developed a cell migration assay based on automatic high-throughput microscopy screen. This approach allows for screening of hundreds of genes, e.g., those encoding various kinases and phosphatases but can also be used for screening of drugs libraries. Moreover, we have developed an automatic analysis pipeline comprising of (a) automatic data acquisition (movie) and (b) automatic analysis of the acquired movies of the migrating cells. Here, we describe various facets of this approach. Since cell migration is essential in progression of cancer metastasis, we describe two examples of experiments performed on highly motile (metastatic) cancer cells.

Key words: Random cell migration, Functional genomics, High-throughput screening, Quantitative image analysis, Automation

1. Introduction

Cell migration is essential for various biological processes, such as embryonic development, immune responses, and tissue remodeling as well as in pathologic conditions, such as cancer progression and metastasis (1). The complexity of the cell migration as a process can be ascribed to the involvement of many regulating genes, signal transduction pathways, and numerous external stimuli such as growth factors and chemokines. Moreover, the integration and

*Equal contribution (Wies van Roosmalen and Sylvia E. Le Dévédec).

Claire M. Wells and Maddy Parsons (eds.), *Cell Migration: Developmental Methods and Protocols*,
Methods in Molecular Biology, vol. 769, DOI 10.1007/978-1-61779-207-6_29, © Springer Science+Business Media, LLC 2011

spatiotemporal coordination of the processes underlying the cell migration add to its complexity (2–4). Thus, it is a major challenge to develop tools and approaches to study these individual processes that make up cell migration. Additionally, in the context of cancer progression and metastasis, a deeper understanding of crucial factors in the migratory behavior of cells is necessary in order to develop targeted anticancer therapies.

We and others have developed various approaches to unravel the key players in the mechanisms underlying cell migration (5). One of them is the RNAi-based gene silencing high-throughput screening approach (6–9). The advantage of applying high-throughput screening is that multiple genes or conditions can be tested at once, allowing a more systematic approach toward understanding the mechanisms underlying cell migration. We present here a method that combines high-throughput screening with a live cell imaging-based random cell migration assay. This allows for screening of hundreds of genes, e.g., those encoding various kinases and phosphatases. Moreover, we have developed an automatic analysis pipeline comprising (a) automatic data acquisition (movie) and (b) automatic analysis of the acquired movies of the migrating cells. This protocol can be used in combination with functional genomics as well as chemical compound screens.

2. Materials

2.1. Cell Culture

1. Minimum Essential Medium (MEM) alpha medium (1×), liquid, with L-glutamine, without phenol red supplemented with 5% fetal bovine serum (FBS) (see Note 1) for the culture of MTLn3 cells (10).
2. RPMI 1640 medium (1×), liquid, with L-glutamine, without phenol red supplemented with 10% FBS for the culture of H1299 cells.
3. Imaging plate CG 96 well, glass bottom, surface treated (PAA Laboratories GmbH).
4. Collagen, rat tail, type I is diluted in PBS without Ca/Mg at 30 μg/mL (for assay with MTLn3).
5. Fibronectin from bovine plasma is diluted in PBS without Ca/Mg at 10 μg/mL. Once diluted, use directly (for assay with H1299).
6. Epidermal Growth Factor is dissolved at 100 μg/mL in 10 mM acetic acid containing 0.1% bovine serum albumin (BSA) and aliquots are stored at −80°C (see Note 3).
7. DharmaFECT2 transfection reagent and siGENOME smartpool targeting GFP, PTEN, and RAC1 (Dharmacon, Thermo Fisher Scientific) (see Note 4).

2.2. Automated High-Throughput Imaging

1. Nikon Eclipse TI with fluorescent lamp and 20× objective (Pan Apo, dry, NA 0.75, WD 1.0) equipped with an automated stage control and a Perfect Focus System (PFS).
2. Temperature- and CO_2-controlled imaging chamber (custom design).
3. Nis Elements AR version 3.10 SP3 build 634.
4. In Nis Elements implemented Macro: Wellplate2.mac (Nikon).

2.3. Automated Quantitative Image Analysis

1. The conversion and analysis macros are written with Image Pro Plus software (Media Cybernetics, Inc. Bethesda, MD) and are available on request.
2. Spreadsheet software (Excel, Microsoft Corporation).

3. Methods

3.1. Preparing Cells for Imaging

Cell migration can be investigated in various experimental settings. Here we describe two different example experiments illustrating the type of assays that can be performed using the following protocols. In these particular examples we use the MTLn3 cell line, a rat mammary carcinoma cell line, overexpressing ErbB1 (11) and the H1299 cell line (12), a human lung carcinoma cell line; both cell-lines express ectopically the Green Fluorescent Protein (GFP). However, the protocols can be adapted to various other mammalian cell culture systems and assays.

All following steps should be performed in a sterile flow cabinet and with sterile reagents.

3.1.1. Exposure of MTLn3-ErbB1 Cells to Epidermal Growth Factor

1. Dilute collagen type I in PBS without Ca/Mg to a concentration of 30 μg/mL. Then coat the 96-well plate by adding 50 μL of this solution to each well.
2. Incubate the plate for 1 h at room temperature, then aspirate the solution and let the plate air-dry for approximately 30 min.
3. Plate 10.000 cells per well (end volume is 100 μL per well) in phenol red free medium (MEMα) containing 5% FBS (see Notes 1, 2, and 5).
4. The next day, 3–4 h before start of the imaging, starve the cells by replacing the medium on the cells with medium without FBS.
5. Dilute EGF in MEMα without FBS to 5 and 10 nM and expose the cells by replacing the medium covering the cells with medium containing EGF (see Note 3).
6. The cells should now be placed on the microscope as soon as possible (see Note 6).

3.1.2. RNAi-Mediated Gene Silencing in H1299-GFP

1. Dilute fibronectin in PBS without Ca/Mg to a concentration of 10 μg/mL. Then coat the 96-well plate by adding 50 μL of this solution to each well.
2. Incubate the plate for 1 h at 37°C and aspirate the solution prior to plating cells. Do not dry the plate.
3. Prepare a 96-well V-bottom plate containing 50 nM siRNA per well by adding 0.25 μL/well from a 20-μM siRNA stock solution to 9.75 μL/well medium (RPMI) without serum (see Note 4).
4. In an eppendorf tube, dilute DharmaFECT 2 in serum-free medium. For one well, add 0.2 μL DharmaFECT 2 to 9.8 μL medium without serum. Multiply this for the total amount of wells, and add a few spare. Incubate at room temperature for 5 min.
5. Add 10 μL of the diluted DharmaFECT2 to each well, gently pipette up and down and leave the mix at room temperature for 20 min so that the siRNA complexes can form.
6. Add 3,000 cells per well in a volume of 80 μL complete medium (RPMI) containing 10% FBS to the siRNA mix in the V-bottom plate and gently pipette up and down to mix. Then transfer the total volume (100 μL) from the V-bottom plate to the coated glass bottom plate and place in the incubator.
7. After 16 h, replace the medium covering the cells with phenol red-free medium containing 10% FBS.
8. 72 h after transfection of the cells, replace the medium one more time to get rid of floating cells and start imaging.

3.2. Automated High-Throughput Imaging

This automated imaging protocol is developed for application on a NIKON inverted fluorescent microscope and differential interference contrast (DIC) imaging with automated stage control, but can be adapted to similar high-throughput imaging systems.

1. Prior to imaging, start heating the climate control chamber till 37°C.
2. Start the microscope including the camera, transmitted light (for DIC imaging), and/or mercury lamp (for fluorescent imaging).
3. Start the Nis Elements AR software. The “ND acquisition control panel” will pop up.
4. Place the 96-well plate in the plate holder on the motorized stage and open the CO_2 supply.
5. In the “ND acquisition control panel,” define the exposure time and acquire one live image of your cells in a random well.
6. Open the macro WellPlate2.mac and run it. The “WellPlate” window will pop up.

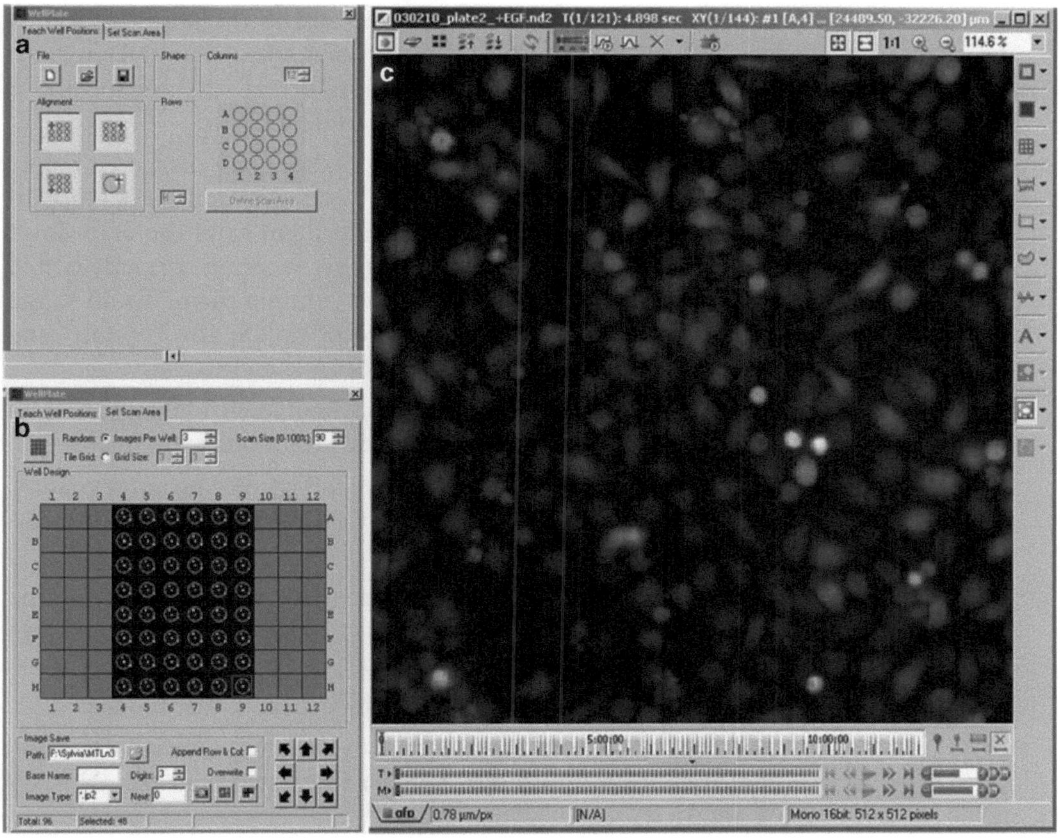

Fig. 1. Essential steps in the Wellplate2.mac macro. (**a**) In this control panel, define the plate, by moving the stage to well A1, A12, H1, and outside a well, respectively. Click on the corresponding buttons to align the plate. (**b**) In this control panel, define the region of wells you want to image and number of positions per well. The imaging position will be displayed as *white dots* in the plate scheme. (**c**) This window shows the final .nd2 file, including images of all positions and all time frames. Movies can be viewed directly by pressing the "play" button.

7. Position the well A1 in the center and click on "Alignment" (see Note 7 and Fig. 1a).
8. Position the well A12 in the center and click on "Alignment."
9. Position the well H1 in the center and click on "Alignment."
10. Position the outside of any well and click on "Alignment."
11. Choose for the correct number of rows (in letter, A–H) and columns (numbers, 1–12).
12. Click on the button "Define scan area." The tab "Set Scan Area" will become active (see Fig. 1b).
13. Select the region of wells you want to image by dragging your mouse over the wells.
14. Select at least three images (position) per well.
15. Select a scan size of 80% so that you do not image the extreme periphery of the well.

16. Click on the "well design" icon and all the positions will be directly given in the "*XY* position" tab of the "ND acquisition control panel" window. Important: do not include *Z* (checkbox) and keep PFS ON while moving.
17. In tab "Time" and then "Advanced" execute before Multipoint the macro "EpiOn" and execute after Multipoint the macro "EpiOff" by choosing them from the drag-down menu. Test if these two macros are working correctly before start imaging the entire plate by running them. To do so, go to "Macro" in the toolbar, select "EpiOn" or "EpiOff" and press run. The shutter should open and close.
18. Define camera settings and keep them unchanged when imaging multiple wells and/or plates within one experiment.
19. Right click on the button with the correct imaging channel ("GFP" or "DIC") in the upper toolbar and assign the current settings to this channel.
20. In tab "Time," define the imaging interval (e.g., 6 min) and duration of the experiment (e.g., 12 h). The number of loops will be calculated (see Note 8).
21. Give the experiment name and location to safe the file.
22. Click on "1 time loop" to test the imaging time of one round or click on "run now" to start the experiment (see Notes 9 and 10).
23. Finally, one .nd2 file is generated for all the wells and the entire imaging period (see Note 11 and Fig. 1c).

3.3. Automated Quantitative Image Analysis

The following protocol describes the automated quantitative image analysis that is designed to quantitatively analyze migratory tracks of all cells within each frame of a live cell movie. First, a number of parameters are set on a "test image/frame" followed by applying the analysis to the whole range of data (movies) recorded during an experiment.

The macros are written in Image Pro Plus Software and are available on request. The macro "TRACKDIC" should be applied when analyzing movies recorded using DIC microscopy, whereas the macro "TRACKFLUOR" should be used when analyzing the movies of cells containing fluorescent probes, such as GFP. Examples of the result of both macros are depicted in Fig. 2. In this particular example, MTLn3-ErbB1 cells are exposed to EGF and imaged with DIC. The result of the analysis of those movies is shown in Fig. 3. In the H1299-GFP cells, the expression of two proteins (Rac1 and PTEN) is silenced using siRNA. These cells were used to acquire fluorescent movies. The result of the analysis of these movies is shown in Fig. 4.

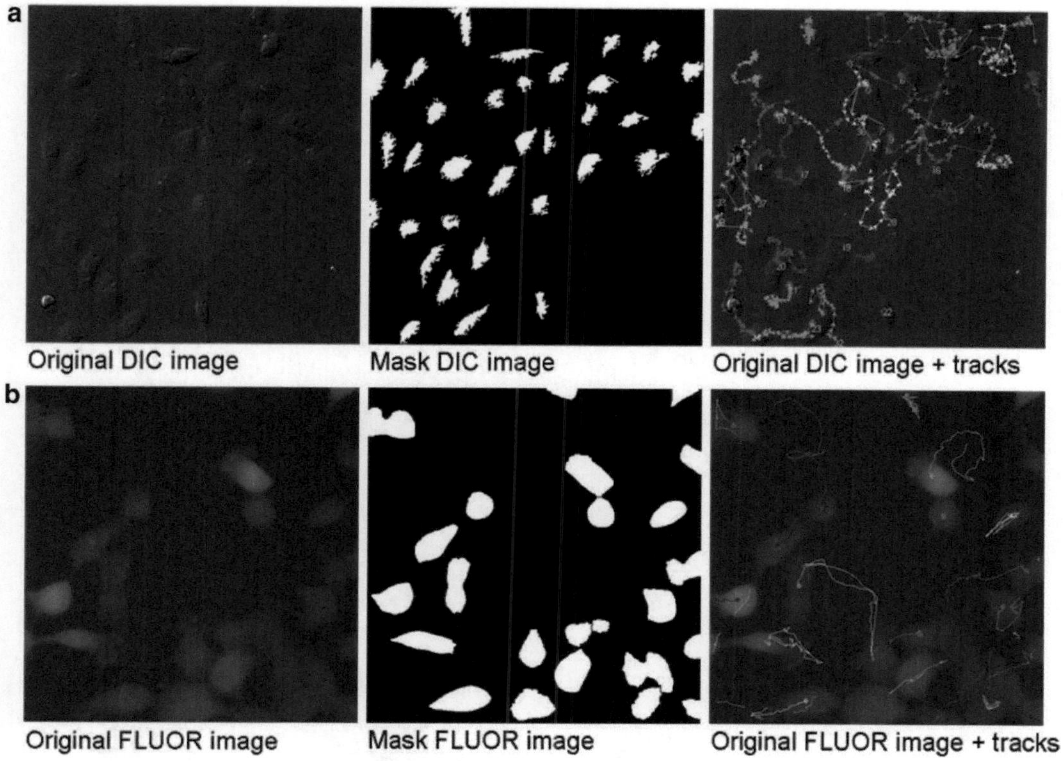

Fig. 2. Steps in image processing. (**a**) An example of DIC image is shown. After applying several filters on the original image, a mask is defining the cell area and in time, the trajectory of the cells is calculated and displayed on the original image. (**b**) As above, but now an example of FLUOR image is shown.

3.3.1. Conversion of .nd2 to avi

1. Within NIS Elements, export the file to tagged image file (tif). You will get one folder that contains all the .tif files of the experiment and that are automatically named *namet01xy01z01c01.tif* where *t* represents time, *xy* the position, *z* the focus plane, and *c* the channel.
2. Start ImagePro software and run the macro "TIFfolderAVI." A user dialog window is created and will pop up.
3. Give the number of imaged color channels (in these examples this value is 1), *Z*-stacks and positions. When imaging a full 96-well plate with three positions per well, this number is 288. Select "auto contrast on."
4. Select the first .tif file in your folder. The macro will automatically start to generate a time lapse series for each position over the entire imaging period.
5. Wait patiently until all the .tif files are converted to .avi files. These .avi files are saved in the same folder and are consequently named *namet01xy01z01c01.avi* for each position.

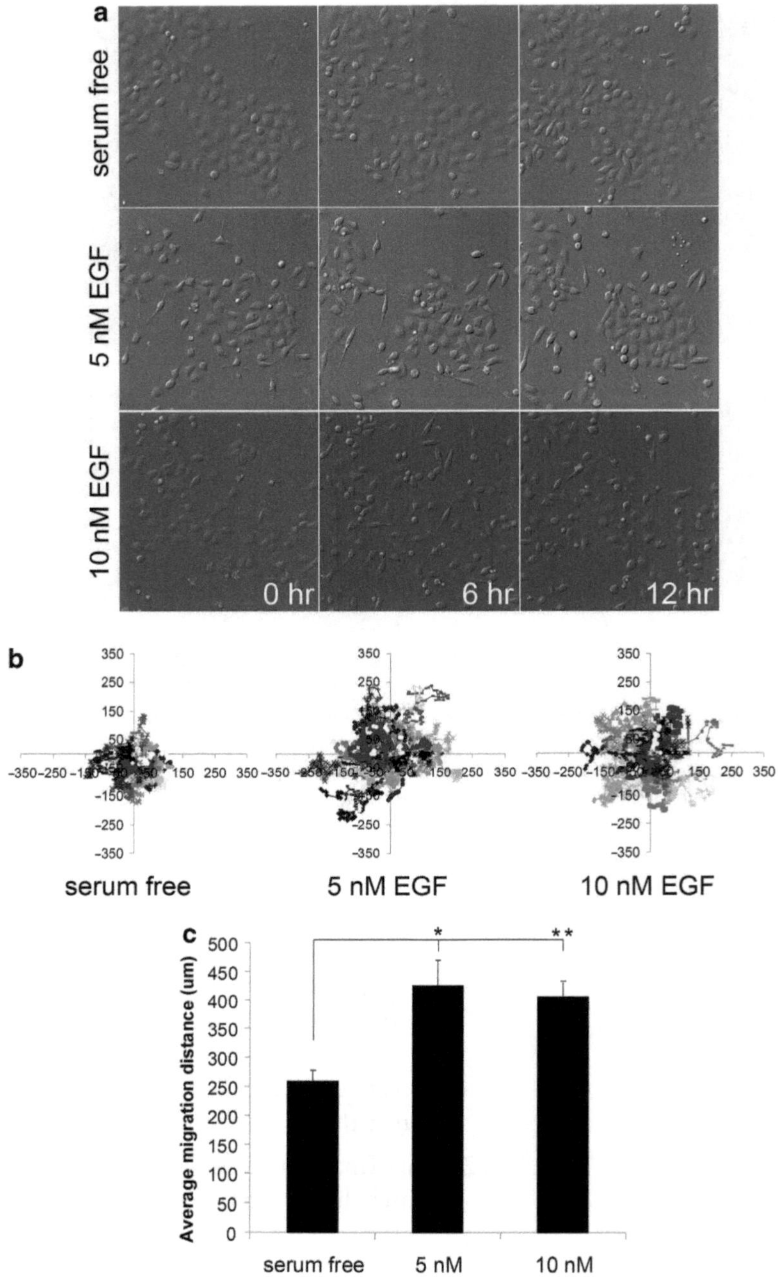

Fig. 3. Exposure of MTLn3-ErbB1 cells to EGF is triggering cell migration. MTLn3-ErbB1 cells were starved for 3 h and then exposed to 0, 5, and 10 nM EGF. (**a**) Snapshots of the entire imaging period of 12 h for different conditions. (**b**) Representation of individual tracks within one movie. Exposure to 0, 5, or 10 nM EGF is increasing the track length. (**c**) The calculated average migration distance of cells exposed to EGF. Exposure to EGF is significantly increasing the migration distance (*$P<0.05$, **$P<0.01$). Note that there is no additional increase when comparing 5 and 10 nM due to receptor saturation at the lower concentration.

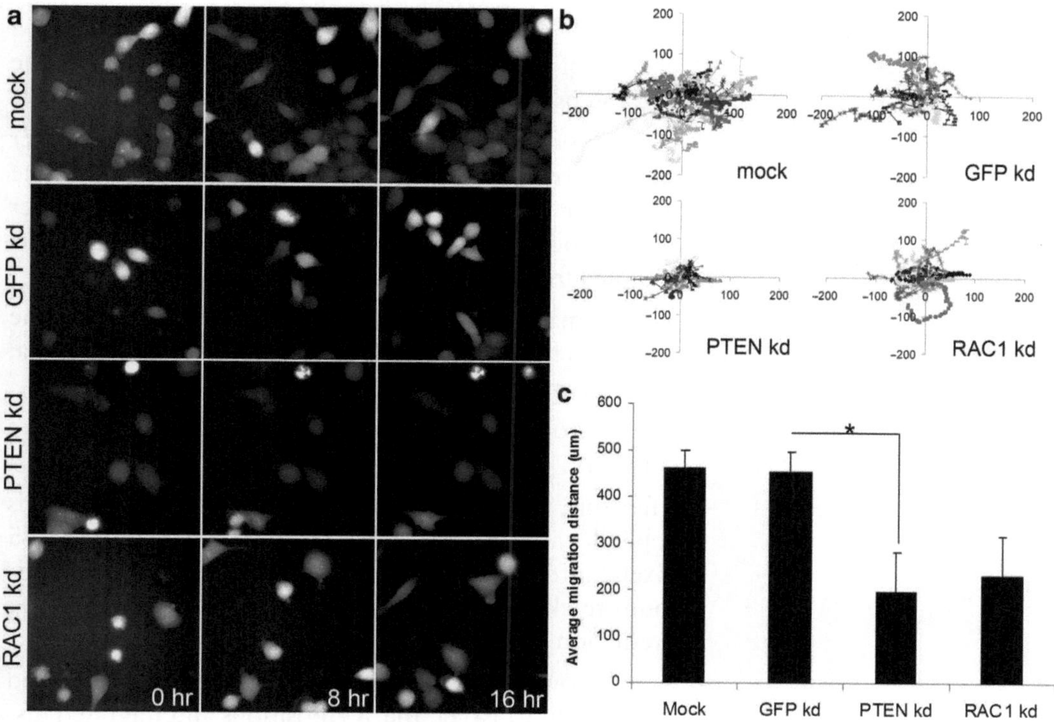

Fig. 4. Knockdown of PTEN in H1299-GFP is suppressing cell migration. The effect of gene knockdown on cell migration was examined 72 h after transfection. (**a**) Snapshots of the entire imaging period of 16 h for different genes. (**b**) Representation of individual tracks within one movie. Knockdown of PTEN is decreasing the track length, where RAC1 knockdown partially decreases track length. (**c**) The calculated average migration distance of cells after knockdown. Knockdown of PTEN is significantly reducing the migration distance ($P < 0.05$). Knockdown of RAC1 also seems to reduce the migration distance, but this is not significant.

3.3.2. Analysis with TRACKDIC

1. Start ImagePro software and run the macro "TRACKDIC." A user dialog window is created and will pop up. In this window, files or folders can be selected, minimum cell areas can be set, threshold and filter options can be set, and a status/instruction window is displayed.
2. Start Excel.
3. Select the first avi file. An avi file with the time lapse-sequence of cells recorded as described in Subheading 3.2 is opened. With this file the filters and settings can be calculated.
4. Determine manually the intensity threshold by moving the threshold ruler until the cells are red and the background is black. Determine the minimum area of 1 cell. All objects with high intensity and which are of the size equal to or larger than the set minimum area will be classified as cells. Low intensity pixels and small areas will be discarded as noise.
5. When the settings are correct, choose "select folder" and then the first avi in that folder. All avi files in the folder will be automatically analyzed.

6. Press "calculate" to start the analysis.
7. Calculation of the track is automatically done through the following steps:
   - First two filters are applied to the movie. A flatten filter which corrects for uneven background and a large edge plus filter to reduce background. The edge plus filter extracts positive edges (bright features on a dark background) from an image.
   - Then a mask (binary image) sequence is made with the option to fill holes in the segmentation steps (see Fig. 2a, middle panel).
   - Then an open filter is applied which will smooth object contours.
   - After this, the standard Image Pro Auto track function is applied with the following options: velocity limit = 75 pixels, auto acceleration limit, chaotic motion type, minimum track length = 10% of total movie length, tracking prediction depth = 3 (see Note 12).
   - The following parameters are automatically calculated and written to Excel: the *XY* positions and migrated distance per cell from frame to frame. Also averages are calculated for all cells for each frame and averages for all cells in the total movie.
8. A picture of the original first image with the track annotated on it is written to Excel (see Fig. 2a, right panel).

3.3.3. Analysis with TRACKFLUOR

See steps 1–6 described in Subheading 3.3.2, except run the macro "TRACKFLUOR."

1. Calculation of the track is automatically done through the following steps:
   - First six filters are applied to the movie. A flatten filter which corrects for uneven background, a gauss filer which reduces high intensity noise. Contrast of the movies is optimized. The large edge plus filter is reducing background. The erode filter will separate touching cells and the open filter will smooth object contours.
   - After the filters, a standard Image Pro Auto track function is applied with the following options: velocity limit = 75 pixels, auto acceleration limit, chaotic motion type, minimum track length = 10% of total movie length, tracking prediction depth = 3 (see Note 12).
   - The following parameters are automatically calculated and written to Excel: the *XY* positions and migrated distance per cell from frame to frame. Also averages are cal-

culated for all cells for each frame and averages for all cells in the total movie.

2. A picture of the original first image with the track annotated on it is written to Excel (see Fig. 2b, right panel).

3.3.4. Processing the Tracking Data for Representation

The TRACKDIC and TRACKFLUOR macros in ImagePro write results to Excel, but also generate .trc files within the same folder. These files can be used to generate graphs displaying all individual tracks within one well, with all tracks starting at the center of the graph.

1. Start ImagePro software and run the macro "RelativeTracks."
2. Open the first .avi file in the folder which also contains the .trc files. All other .avi files will be calculated automatically.
3. Give the calibration factor for calculation of the correct distances. This is the size of 1 pixel in μM and this value is displayed in NIS Elements at the bottom left corner of the final imaging panel (see Fig. 1c).
4. Start Excel and open "makegraph.xls." This excel file is part of the macro and available on request.
5. Go back to ImagePro and press "ctrl" and "G" on your keyboard. A new window will pop up.
6. Select the maximum number of tracks to be displayed in each graph, e.g., 50.
7. Enter the maximum scale in μM, e.g., 200. This value needs to be enlarged for highly migratory cells and can be decreased for cells that are less migratory.
8. Enter the number of imaged positions, corresponding with the number of .avi files within your folder. The graphs will automatically be generated for all positions and written to the active excel file (see examples in Figs. 3b and 4b).

4. Notes

1. Since phenol red exhibits auto fluorescence (it is excited at 488 nm and emits over a wide spectrum), it is highly recommended to use phenol red-free medium while imaging in order to reduce background signal. This is in particular important when imaging GFP signal.
2. When adapting this protocol for a different cell line, the cell density should be adjusted. The required confluency at time of imaging is ~70%, so that cells have enough free space to migrate.

3. Cancer cell migration is dependent on growth factors, such as EGF. For that reason, stimulation of cells with EGF during imaging is mimicking the in vivo microenvironment. Exposure to other growth factors or chemokines can also be combined with this assay. Also, exposure to growth factors can be combined with RNAi screening.
4. When performing siRNA experiments, it is essential to include the proper controls. In this case, siRNA targeting GFP is chosen as a negative control since it should not affect the cell migration. In this example, two specific genes are chosen, but this method can be applied to whole genome siRNA libraries or sub libraries. In addition, the optimal cell density and transfection reagent is cell type specific. A guideline is available at http://www.dharmacon.com/uploadedFiles/Home/Support_Center/Selection_Guides/dharmafect-cell-type-guide-dharmacon-products.pdf.
5. This protocol can be further adapted for full automation by applying robotics for automated liquid handling for, e.g., cell seeding and drug exposure.
6. Alternatively, first place the 96-well plate on the microscope and then carefully add the exposure medium. In this case, the concentration of EGF should be doubled, and a volume of 1:1 should be added to the medium already in the plate. Prior to placing the cells on the microscope, make sure the temperature and CO_2 levels in the chamber are optimal in order to allow the natural environment of the cells and to minimize stress levels.
7. The center of the well is easily determined by opening the shutter so that a beam of light hits the well. The beam can then be positioned at the center of the well.
8. When adapting this protocol for a different cell line, the optimal time interval should be determined. For fast moving cells such as the MTLn3, the maximum time interval is 6 min. In general, it is advised to make the time interval between imaging frames as short as possible, to ensure a proper tracking of the cells. If the interval is too long, the cell "jumps" between the ascending frames, making it difficult for the analysis programs to assign the track to the correct cell.
9. When imaging 96 wells and three positions per well, the total imaging time of one time loop is about 5 min.
10. Take note of direction of imaging (e.g., left to right or up to down) since the movies will be numbered accordingly.
11. The .nd2 file that is created at the end of the imaging period is, depending on the amount of positions imaged, ranging from 7 to 21GB in size when individual images are 512×512 in size. Make sure there is enough free space on the hard disk of your computer.

12. The velocity limit (search radius) defines maximum distance in calibrated units between object positions on the neighboring frames. The parameter is used with automatic tracking to find tracks on the first two frames (on all frames with chaotic movement type). If no object is found in the search radius, you will be prompted to click on the expected position manually. In automatic mode the track is closed at this point. Tracking prediction depth defines the number of frames ahead of the current, which are used to find most suitable track for the given object (default is 2). This parameter also defines the maximum number of consecutive occlusions on track. The occlusions number is less than the prediction depth by 1. For example, if the prediction depth is 3, the maximum number of consecutive object occlusions is 2, it means that if the tracking object in sequence disappears (or gets hidden) for 1 or 2 consecutive frames it will still be detected as a part of the track. The estimated position will be used for missed objects. The track points of missed objects do not have outlines and count/size measurement values. High prediction depth values increase computation time significantly.

Acknowledgments

The MTLn3-ErbB1 cells are a kind gift from Jeffrey Segall and the H1299-GFP from Benjamin Geiger. This work is financially supported by the Dutch Cancer Society (UL 2007–3860) and the EU FP7 Health Program Metafight (EU201862).

References

1. Chambers, A. F., Groom, A. C., and MacDonald, I. C. (2002) Dissemination and growth of cancer cells in metastatic sites. *Nat Rev Cancer.* **2**:563–572.
2. Carragher, N. O. and Frame, M. C. (2004) Focal adhesion and actin dynamics: a place where kinases and proteases meet to promote invasion. *Trends Cell Biol.* **14**: 241–249.
3. Ridley, A. J., Schwartz, M. A., Burridge, K., Firtel, R. A., Ginsberg, M. H., Borisy, G., Parsons, J. T., and Horwitz, A. R. (2003) Cell migration: integrating signals from front to back. *Science.* **302**:1704–1709.
4. Webb, D. J., Parsons, J. T., and Horwitz, A. F. (2002) Adhesion assembly, disassembly and turnover in migrating cells – over and over and over again. *Nat. Cell Biol.* **4**:E97-100.
5. Le Dévédec, S. E., Yan, K., de Bont, H., Ghotra, V., Truong, H., Danen, E. H., Verbeek, F., and van de Water, B. (2010) Systems microscopy approaches to understand cancer cell migration and metastasis. *Cell Mol Life Sci.* **67**:3219–3240.
6. Neumann, B., Held, M., Liebel, U., Erfle, H., Rogers, P., Pepperkok, R., and Ellenberg, J. (2006) High-throughput RNAi screening by time-lapse imaging of live human cells. *Nat Methods.* **3**:385–390.
7. Niederlein, A., Meyenhofer, F., White, D., and Bickle, M. (2009) Image analysis in high-content screening. *Comb. Chem High Throughput. Screen.* **12**:899–907.
8. Pepperkok, R. and Ellenberg, J. (2006) High-throughput fluorescence microscopy for systems biology. *Nat Rev Mol Cell Biol.* 7:690–696.

9. Zanella, F., Lorens, J. B., and Link, W. (2010) High content screening: seeing is believing. *Trends Biotechnol.* **28**: 237–245.

10. Neri, A., Welch, D., Kawaguchi, T., and Nicolson, G. L. (1982) Development and biologic properties of malignant cell sublines and clones of a spontaneously metastasizing rat mammary adenocarcinoma. *J. Natl. Cancer Inst.* **68**:507–517.

11. Xue, C., Wyckoff, J., Liang, F., Sidani, M., Violini, S., Tsai, K. L., Zhang, Z. Y., Sahai, E., Condeelis, J., and Segall, J. E. (2006) Epidermal growth factor receptor overexpression results in increased tumor cell motility in vivo coordinately with enhanced intravasation and metastasis. *Cancer Res.* **66**:192–197.

12. Ciernik, I. F., Berzofsky, J. A., and Carbone, D. P. (1996) Human lung cancer cells endogenously expressing mutant p53 process and present the mutant epitope and are lysed by mutant-specific cytotoxic T lymphocytes. *Clin Cancer Res.* **2**:877–882.

Chapter 30

Beads on the Run: Beads as Alternative Tools for Chemotaxis Assays

Eric Theveneau and Roberto Mayor

Abstract

Cell migration is required for a wide variety of processes from bacteria seeking for food to correct patterning of neuronal networks. The ability to sense external cues is critical for cells to get directions and reach their goals. So far, studies on chemotaxis have mainly focused their attention on individual cells and therefore available tools are designed to monitor cell behavior at the single cell level. However, as collective cell migration is now widely accepted as a main mode of cell migration from development to cancer, the question of how chemotaxis is achieved has also to be asked on a bigger scale. Here, we present two chemotaxis assays suitable for single cells, cell sheets, and cell explants. Using a simple combination of heparin-coated beads and high vacuum silicone grease, these techniques can be adapted to a wide variety of culture conditions. They allow time-lapse study, high-resolution microscopy, and can be set up at no extra cost.

Key words: Chemotaxis, Chemokine, Growth factors, Collective cell migration, Cell culture, Neural crest, Mesoderm, Growth cone, *Xenopus*

1. Introduction

The ability of cells to sense external cues plays a major role in driving directional cell migration and targeting cells to precise locations in various situations such as axon path finding (1–5), germ cell migration (6–8), homing of immune cells (9, 10), and cancer metastasis (11, 12). Several systems such as the Boyden Chamber (transwell assay) (13), Zigmond Chamber (under-agarose assay) (14), Dunn Chamber (15), and the Pipette Assay (3, 4, 16), to name a few, have been designed to study the ability of a given molecule to act as a chemoattractant. Despite being best suited to study chemotaxis at the single cell level, some of these techniques allow cell populations to be monitored. For example, Zigmond

Claire M. Wells and Maddy Parsons (eds.), *Cell Migration: Developmental Methods and Protocols*,
Methods in Molecular Biology, vol. 769, DOI 10.1007/978-1-61779-207-6_30, © Springer Science+Business Media, LLC 2011

and Dunn chambers allow migration of monolayers and fascicular networks of cells to be analyzed. However, none of them makes it possible to look at cell groups as they can be found in collective cell migration, a recognized migratory behavior (17–20). We describe here a new method (21) that can be used for single or collective cell migration, which offers several advantages when compared to more commonly used methods (see Table 1).

The Boyden chamber is a simple and easy method that is widely used for testing putative chemoattractants but cells can obstruct the pores and locally block protein diffusion. Moreover, cell movements cannot be monitored and therefore chemotaxis and chemokinesis can only be distinguished by using a tedious checkerboard analysis (22, 23), which is a major disadvantage of the technique. These limitations led Zigmond and colleagues to design an under-agarose assay that allows cells to be observed during migration (14) but its chamber, made of 3 mm thick glass, has limiting optical properties and the final experimental conditions, including the diffusion of the tested molecule from one compartment to the other, are difficult to reproduce (discussed in ref. 15). For these reasons, Dunn and colleagues worked on a new direct-viewing chamber solving the optical issues of the Zigmond chamber and providing a more reliable setup to allow reproducible experimental conditions (15). The pipette assay, originally designed to look at growth cone behavior (16), in which a pipette is set to release slowly a solution containing the molecule of interest into the culture medium, offers a different alternative. The system is directly adapted to your usual culture conditions and is therefore suitable for high-quality microscopy. It also gives the possibility to move the source of the putative chemoattractant and monitors the response of the cells according to the modifications done in the course of the experiment. However, the dish has to be kept open, it is nearly impossible to set up different conditions in parallel under the same microscope and its short-range gradient (100 μm) (4, 24, 25) is not adapted to cell group analysis. All these techniques are very powerful tools to study chemotaxis. They offer easy solutions for wide screening of putative chemoattractants (Boyden chamber), competition between putative attractants (Zigmond chamber) and precise analysis of cell movement (Dunn chamber), and the dynamics of cell response (Pipette assay). Unfortunately, these methods cannot be used to look at large multilayered cell groups undergoing collective cell migration. The Boyden chamber, for example, cannot be used without dissociating the cells beforehand, while the extra thin culture area of the Dunn chamber can only support the migration of individual cells or cell sheets. As collective migration is now accepted as a major mode of cell migration in normal and pathological situations (17–20), there is a need for methods allowing large groups of cells to be monitored.

Table 1
Comparison between Boyden, Zigmond, and Dunn chambers and Pipette Assay with the Fixed/Free bead assays

Technique	Time-lapse movie	Microscope	Multiple positions	Multiple experimental conditions	Cell culture	High-resolution microscopy
Boyden Chamber (transwell assay)	No	NA	NA	No	SC[a]	NA
Zigmond Chamber (under-agarose assay)	Yes	CM	Yes	No	SC, ML	No
Dunn Chamber (direct-viewing chamber)	Yes	CM	Yes	No	SC, ML	Yes
Pipette Assay	Yes	CM	No	No	SC, CC, ML[b]	Yes
Fixed Bead Assay	Yes	CM	Yes	Yes	SC, CC, ML	Yes
Free Bead Assay	Yes	SM	No[c]	Yes	SC, CC, ML	No

CM compound microscope, *SM* stereomicroscope, *CC* cell cluster (more than one cell layer), *ML* monolayer, *SC* single cells

[a] Cell dissociation is a prerequisite for the Boyden chamber

[b] In theory it would be possible to use the pipette assay with monolayer of cells or cell clusters but this technique usually gives a gradient around 100 μm (4, 24, 25) which makes it difficult to use with large cell population

[c] As the beads are not fixed to the substratum it is not possible to set multiple positions on a motorized stage, but this technique is better off on a dissecting microscope as the low magnification allows several beads and explants to be monitored at once

Here, we provide two easy methods for testing putative chemoattractants that can be used with individual cells, cell sheets, or multilayered cell clusters; they allow time-lapse study and high-resolution microscopy; they can be adapted to a wide variety of cell culture conditions; and importantly, they can be easily set up with routine bench materials at no extra cost (a comparison with other typical methods is summarized in Table 1). Briefly, cells are cultured on nontreated plastic dishes covered by extracellular matrix and heparin-coated beads are used to deliver the molecule of interest. Beads can be immobilized on the dish using silicon grease (fixed bead assay, Fig. 1) or broken in pieces and placed at the bottom of the dish next to the cells (free bead assay, Fig. 2). A gradient of diffusion is quickly generated around the bead (Fig. 3) and cell movement can be easily analyzed using open access software (ImageJ, Manual Tracking and Chemotaxis Tool plug-ins) if cells have distinguishable properties or are labeled beforehand. We usually use nuclear fluorescent molecules in order to track the cells (see Fig. 4 for typical cell tracking conditions). The fixed bead assay can be used to expose cells to different conditions. For example, beads can be soaked in different molecules or different

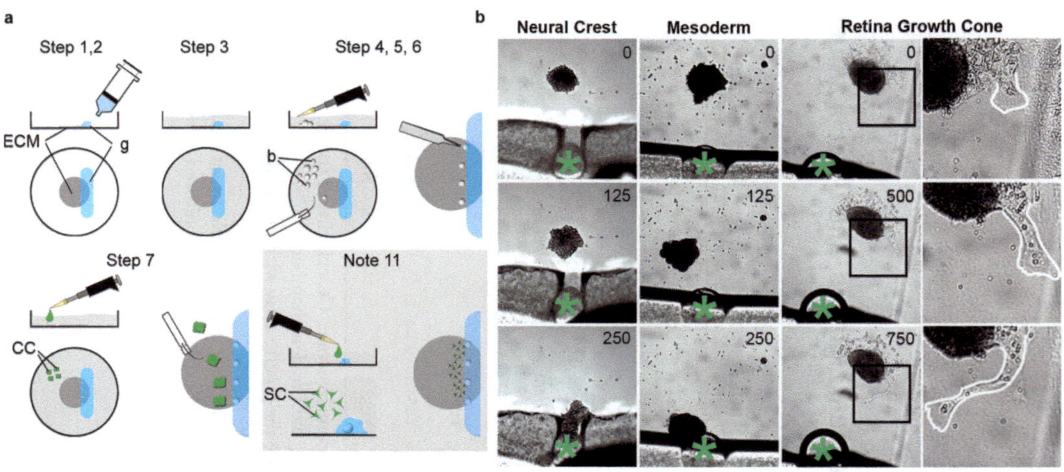

Fig. 1. Protocol for the fixed bead assay. (**a**) Steps for setting up a fixed bead assay: 1, remove the PBS from the last wash of matrix preparation; 2, using a syringe place a line of grease (g) in the extracellular matrix area; 3, fill the dish with culture medium; 4, place the beads (**b**) inside the dish; 5, select beads of equivalent size and place them in front of the grease using an eyebrow knife or tweezers; 6, push them inside the grease using tweezers; 7, place the cell clusters (CC) in front of the beads and wait for attachment. See Note 11 for variations regarding single cells (SC). (**b**) Examples of fixed bead assay with *Xenopus* cells endogenously expressing Cxcr-4 exposed to beads soaked in purified Stromal-cell derived factor 1. *Black frames* in the third column indicate regions of high magnifications shown in the fourth column. Time is in minutes. *Asterisks* indicate the positions of the beads.

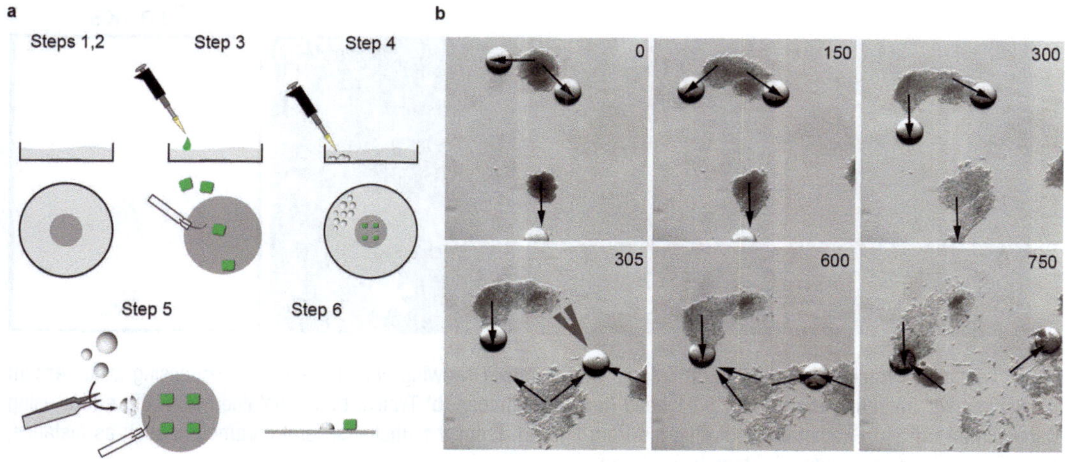

Fig. 2. Protocol for the free bead assay. (**a**) Steps for setting up a free bead assay: 1, remove the PBS from the last wash of matrix preparation; 2, fill up the dish with culture medium; 3, place the cell clusters as desired and wait for attachment; 4, place the beads in the dish; 5, break a few beads in pieces; 6, place the pieces on their flat side. See Notes 13 and 14 for the limitations of the free bead assay. (**b**) Example of a free bead assay with *Xenopus* neural crest cells exposed to beads soaked in purified Stromal-cell derived factor 1. Time is in minutes. *Arrows* indicate the movement of the explants. *Gray arrowhead* indicates that one of the beads "jumped" after being pushed by one of the explant.

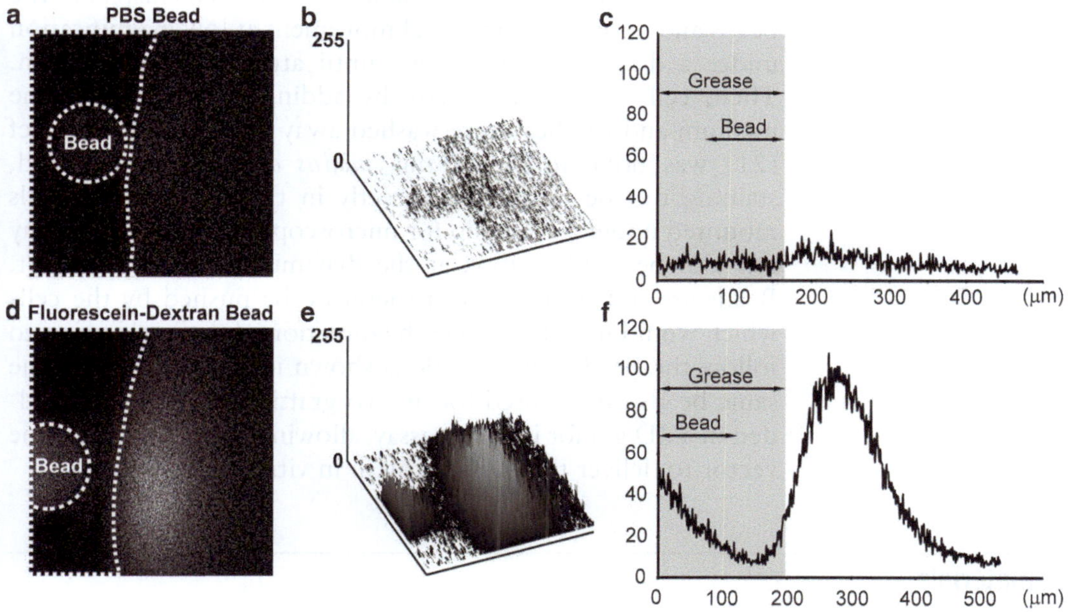

Fig. 3. A steep gradient is generated by diffusion from the bead. (**a**) Confocal image of a control PBS bead embedded in silicone grease. (**b**) 2.5D projection of the fluorescence intensity. (**c**) 2D projection of the fluorescence intensity. (**d**) Confocal image of a bead soaked in Fluorescein-Dextran (FDx) embedded in silicone grease. (**e**) 2.5D projection of the fluorescence intensity. (**f**) 2D projection of the fluorescence intensity. Diffusion of FDx from the bead generates a steep gradient spanning over approximately 300 μm.

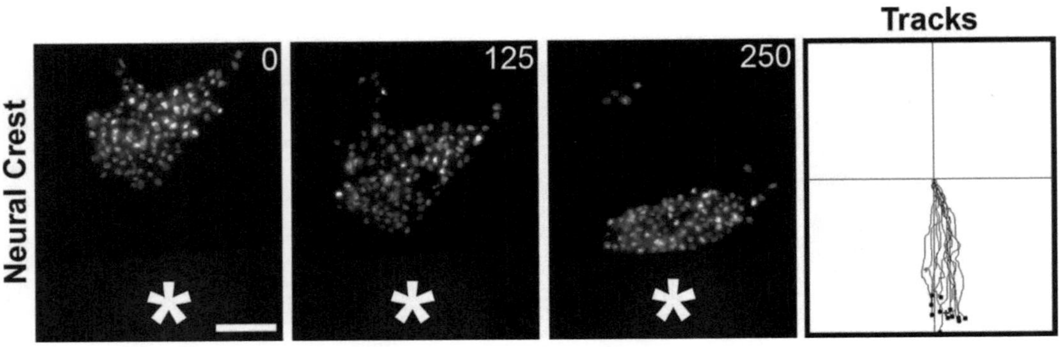

Fig. 4. Typical cell tracking conditions. (**a**) Frames from a movie showing neural crest cells expressing a fluorescent nuclear marker migrating towards an Sdf1 bead (time in minutes). (**b**) Typical tracks obtained from the movie using Manual Tracking and Chemotaxis Tool plug-ins from ImageJ. From the tracks, several parameters such as distance, velocity, and chemotaxis index can then be retrieved.

concentrations of the same molecule. In addition, two types of cells can be placed in front of the same bead to run a competition assay. Moreover, by using multiwell dishes the fixed bead assay can be coupled with the use of chemical inhibitor/activator added in the culture medium. The free bead assay, on the other hand, can be used to deliver the protein in a transient manner. For instance, it is possible to expose cell clusters to the beads and monitor their global movement at low magnification under a dissecting microscope until attraction can be seen. Then, cells are fixed directly by adding the fixative to the medium and the beads are washed away (FRET analysis in ref (21) was done following this *modus operandi*). Afterward, staining can be performed directly in the dish and the cells mounted under a coverslip for microscopy. The free bead assay can also be used to look at the dynamics of cell movement. Beads are not fixed and can therefore be pushed by the cells which would have to change the direction of their migration to follow the beads (an example is shown in Fig. 2). Finally, the same beads can be used for in vivo grafting studies or embedded in a 3D gel for invasion assay, allowing you to use the same vector to deliver the tested protein in vitro and in vivo.

2. Materials

2.1. Bead Preparation

1. Heparin beads (Sigma, H5263; Adar Biotech 6024-1, see Note 1).
2. PBS 1× (1 L): 8 g of NaCl, 0.2 g of KCl, 1.44 g of Na_2HPO_4, 0.24 g of KH_2PO_4. Adjust pH to 7.4. Adjust volume to 1 L

with additional distilled H_2O. Sterilize by autoclaving. Store at room temperature.

3. Protein to be tested. Aliquot at 10× and store at −80°C (see Note 2).

2.2. Extracellular Matrix Preparation

1. Commercially available heating block/plates set at 37°C.
2. Plastic dishes (see Note 3).
3. Fibronectin. Aliquot at 10× (100 μg/mL) and store at −20°C (see Note 4).
4. PBS 1×.
5. PBS 1×/0.1% BSA. Aliquot and store at −20°C.

2.3. Set Up the Dish for Fixed and Free Bead Chemotaxis Assay

1. Eyebrow knife (an eyebrow hair glued into a glass pipette using wax).
2. Tweezers.
3. Extracellular matrix-coated dish.
4. Beads (incubated with the molecule of interest).
5. High vacuum grease for fixed bead assay only.
6. Culture medium. In our case, we cultured all *Xenopus* cells in modified Danilchick's solution: NaCl 53 mM, Na_2CO_3 5 mM, K gluconate 4.5 mM, Na gluconate 35 mM, $MgSO_4(7H_2O)$ 1 mM, $CaCl_2$ 1 mM, BSA 0.1%. Adjust at pH 8 using bicine 1 M.

3. Methods

3.1. Beads Preparation

1. Dilute an aliquot of the protein to be tested directly by adding sterile PBS containing beads from the beads stock solution (see Note 5).
2. Protect from light and incubate at room temperature for 2 h on a rotator.

3.2. Extracellular Matrix Preparation

1. Dissolve the fibronectin to 10 μg/mL in PBS 1×.
2. Place the fibronectin solution at the center of the plastic dish with a pipette.
3. Incubate at 37° on a heating block for an hour with the lid on (see Note 6).
4. Remove the fibronectin solution and wash twice with PBS 1×.
5. Replace with PBS 1×/0.1% BSA and incubate at 37° on a heating block for 30 min with the lid on.
6. Wash once with PBS 1×. Replace with clean PBS 1×. The dish can be stored at 4°C for a few hours before use.

3.3. Set Up the Dish for a Fixed Bead Chemotaxis Assay (Fig. 1)

1. Remove liquid and allow to dry for 2 min at room temperature.
2. Make a line of high vacuum grease on the dish using a syringe.
3. Fill the dish with culture medium.
4. Take some of the beads with a pipette and place them inside the dish outside the fibronectin region (see Note 7).
5. Select beads of similar size (see Note 8) and gently place them along the grease (see Note 9) using an eyebrow knife (see Note 10) or tweezers. Avoid scratching the fibronectin.
6. Carefully push the beads one by one inside the grease using tweezers.
7. Place cells in front of the beads and wait for attachment (see Note 11).
8. Fill the dish with culture medium, close it (see Note 12), and start time-lapse recording.

3.4. Set Up the Dish for a Free Bead Chemotaxis Assay (Fig. 2)

1. Remove PBS from the last wash (see Subheading 3.2, step 6) and fill the dish with the culture medium.
2. Place your cells or explants as desired and wait for attachment.
3. Place the dish under a stereomicroscope.
4. Take some of the incubated beads with a pipette and place them inside the dish outside the fibronectin region.
5. Break some of the beads in pieces using tweezers (see Note 13).
6. Select some pieces of beads and place them next to the cells/explants and start time lapse recording (see Note 14).

4. Notes

1. We use heparin-acrylic and heparin-sepharose beads ranging from 75 to 300 μm with equivalent results. We found that plastic beads (such as BioRad Affi-gel blue beads) are more difficult to manipulate, do not stick to the grease and give poor results.
2. Use fresh aliquots each time. The most convenient stock concentration has to be determined for each protein. For example, with Sdf-1 and *Xenopus* neural crest (NC) cells we have tried the following concentrations: 0.1, 0.2, 0.5, and 1 μg/mL. Beads incubated in 0.1 μg/mL do not attract NC cells, while beads soaked in 0.2, 0.5, and 1 μg/mL attract around 50, 80, and 90% of the explants, respectively.

3. Several types of dishes can be used. The most suitable for each specific microscopic equipment or experimental condition will have to be determined. For the fixed bead assay, we routinely use the following alternatives: tight plastic dishes (Falcon 351006) completely filled with culture medium placed upside down and monitored with dry lenses under an upright microscope; multiwell dishes (Nunc 179820), with each well filled with a different culture medium and sealed with a coverslip, placed upside down and monitored with dry lenses under an upright microscope; any plastic dish left open for the use of water immersion lenses. Although we successfully use this technique on an upright microscope, the use of an inverted microscope could make it even easier as you would not have to seal the dish to use a dry lens. For the free bead assay under an upright microscope, it is possible to close the dish to prevent evaporation but it generally causes the beads to move.
4. In our hands fibronectin solution is not stable at 4°C. Freeze and store single-use aliquots from the stock solution. Any extra freeze/thaw cycle dramatically damages the fibronectin.
5. Sigma H5263 beads have to be washed thoroughly in PBS 1× before use. They can be kept in sterile PBS for months in the fridge. Adar Biotech Beads 6024-1 can be used directly from the stock bottle to dilute the protein.
6. For large Petri dishes (90 mm) place a piece of wet paper inside the dish while incubating at 37°C to maintain humidity.
7. The extracellular matrix components are known to bind a wide range of molecules (26). Therefore, to avoid creating any hot-spots of protein in the matrix, place the beads outside the matrix-coated region while selecting the beads before placing them along the grease.
8. To make the process easier, beads of similar size can be selected from the stock before incubation or even sorted out and stored in PBS according to their diameter. If the effect of different concentrations of the same molecule needs to be compared. We recommend using beads of the same size to reduce any bias related to the amount of protein released by the beads.
9. To make sure that each bead represents an independent experiment we usually place the beads apart from each other (around 2 mm between two consecutive beads).
10. To move the beads around the dish we preferentially use an eyebrow knife, which is made simply by sticking an eyebrow hair into a glass pipette with a bit of wax. However, this step can also be done using tweezers.

11. We place cell clusters around 250 and 500 μm from the bead. The average distance between the border of the bead and the border of the explant is 300 μm. It is possible to determine the diffusion of a protein based on its size and shape (27, 28) but this remains very difficult to predict and the best concentration and best distance between the cells and the beads would still have to be determined experimentally. It is therefore important to start with small distances between the cells and the beads (100–200 μm) and to try different concentrations of the tested molecules until a response can be obtained, if any. In addition, cell clusters can easily be placed one by one but single cells cannot be placed individually. The best alternative is to have cells in suspension. Then a drop of the cell suspension can be deposited on top of the bead using a pipette. Cells will sediment to the bottom of the dish and eventually attach to the matrix. This gives a uniform distribution of cells around the beads. Too many cells may accumulate next to the bead. If this happens, it is possible to remove mechanically the cells that are deemed to be too close using an eyebrow knife, generating a free space between the bead and the cells. Cells can then be tracked individually and their behavior analyzed according to their distance from the bead. Finally, if cells need to be cultured for a long period before running the chemotaxis assay, it is possible to apply the grease, fill the dish with the culture medium, plate the cells, and culture them until they are ready. Then, place the beads in the grease according to the position of the cells you wish to monitor. That is how we performed the growth cone turning assay (Fig. 1). Retinal explants were cultured overnight in a dish containing grease. Beads were added in front of the growing growth cones the day after.
12. Tight Petri dishes can be closed by adding an excess of liquid and quickly putting the lid back on. No air bubbles should remain as they may damage the cells while turning the dish upside down. If air bubbles are present, the dish can be kept in its original orientation if the depth of field of the required lens allows it.
13. The free bead assay can be done without breaking the beads into pieces. However, beads tend to move because of Brownian movements in the liquid. To avoid excessive movements of the beads, break them and position the pieces with their flat edge on the bottom of the dish.
14. In a free bead assay, closing the dish usually causes the beads to move. Therefore, using inverted microscopes or water immersion lenses is preferred, but micromanipulations at high magnification are very challenging. In addition, it is not recommended to set a motorized stage to monitor several positions

as it will also cause beads to move. Consequently, positions should be monitored one at a time or at low magnification, for example, with a stereo microscope to allow several beads to be seen at once.

Acknowledgments

This investigation was supported by grants from MRC, BBSRC, and Wellcome Trust. We thank Rachel Moore for corrections and comments on the manuscript.

References

1. Serafini, T., Colamarino, S. A., Leonardo, E. D., Wang, H., Beddington, R., Skarnes, W. C., and Tessier-Lavigne, M. (1996) Netrin-1 is required for commissural axon guidance in the developing vertebrate nervous system, *Cell* **87**, 1001–1014.
2. Tessier-Lavigne, M., and Goodman, C. S. (1996) The molecular biology of axon guidance, *Science* **274**, 1123–1133.
3. Ming, G. L., Song, H. J., Berninger, B., Holt, C. E., Tessier-Lavigne, M., and Poo, M. M. (1997) cAMP-dependent growth cone guidance by netrin-1, *Neuron* **19**, 1225–1235.
4. Song, H. J., Ming, G. L., and Poo, M. M. (1997) cAMP-induced switching in turning direction of nerve growth cones, *Nature* **388**, 275–279.
5. Fernandis, A. Z., and Ganju, R. K. (2001) Slit: a roadblock for chemotaxis, *Sci STKE* **2001**, pe1.
6. Boldajipour, B., Mahabaleshwar, H., Kardash, E., Reichman-Fried, M., Blaser, H., Minina, S., Wilson, D., Xu, Q., and Raz, E. (2008) Control of chemokine-guided cell migration by ligand sequestration, *Cell* **132**, 463–473.
7. Doitsidou, M., Reichman-Fried, M., Stebler, J., Koprunner, M., Dorries, J., Meyer, D., Esguerra, C. V., Leung, T., and Raz, E. (2002) Guidance of primordial germ cell migration by the chemokine SDF-1, *Cell* **111**, 647–659.
8. Richardson, B. E., and Lehmann, R. (2010) Mechanisms guiding primordial germ cell migration: strategies from different organisms, *Nat Rev Mol Cell Biol* **11**, 37–49.
9. Chen, A., Ganor, Y., Rahimipour, S., Ben-Aroya, N., Koch, Y., and Levite, M. (2002) The neuropeptides GnRH-II and GnRH-I are produced by human T cells and trigger laminin receptor gene expression, adhesion, chemotaxis and homing to specific organs, *Nat Med* **8**, 1421–1426.
10. Butcher, E. C., and Picker, L. J. (1996) Lymphocyte homing and homeostasis, *Science* **272**, 60–66.
11. Le, Y., Zhou, Y., Iribarren, P., and Wang, J. (2004) Chemokines and chemokine receptors: their manifold roles in homeostasis and disease, *Cell Mol Immunol* **1**, 95–104.
12. Pals, S. T., de Gorter, D. J., and Spaargaren, M. (2007) Lymphoma dissemination: the other face of lymphocyte homing, *Blood* **110**, 3102–3111.
13. Boyden, S. (1962) The chemotactic effect of mixtures of antibody and antigen on polymorphonuclear leucocytes, *J Exp Med* **115**, 453–466.
14. Lauffenburger, D., Rothman, C., and Zigmond, S. H. (1983) Measurement of leukocyte motility and chemotaxis parameters with a linear under-agarose migration assay, *J Immunol* **131**, 940–947.
15. Zicha, D., Dunn, G. A., and Brown, A. F. (1991) A new direct-viewing chemotaxis chamber, *J Cell Sci* **99** (Pt 4), 769–775.
16. Gundersen, R. W., and Barrett, J. N. (1979) Neuronal chemotaxis: chick dorsal-root axons turn toward high concentrations of nerve growth factor, *Science* **206**, 1079–1080.
17. Bianco, A., Poukkula, M., Cliffe, A., Mathieu, J., Luque, C. M., Fulga, T. A., and Rorth, P. (2007) Two distinct modes of guidance signalling during collective migration of border cells, *Nature* **448**, 362–365.
18. Deisboeck, T. S., and Couzin, I. D. (2009) Collective behavior in cancer cell populations, *Bioessays* **31**, 190–197.
19. Friedl, P., and Gilmour, D. (2009) Collective cell migration in morphogenesis, regeneration

and cancer, *Nat Rev Mol Cell Biol* **10**, 445–457.

20. Rorth, P. (2009) Collective cell migration, *Annu Rev Cell Dev Biol* **25**, 407–429.
21. Theveneau, E., Marchant, L., Kuriyama, S., Gull, M., Moepps, B., Parsons, M., and Mayor, R. (2010) Collective chemotaxis requires contact-dependent cell polarity, *Dev Cell* **19**, 39–53.
22. Zigmond, S. H., and Hirsch, J. G. (1973) Leukocyte locomotion and chemotaxis. New methods for evaluation, and demonstration of a cell-derived chemotactic factor, *J Exp Med* **137**, 387–410.
23. Orr, W., and Ward, P. A. (1978) Quantitation of leukotaxis in agarose by three different methods, *J Immunol Methods* **20**, 95–107.
24. Pujic, Z., Giacomantonio, C. E., Unni, D., Rosoff, W. J., and Goodhill, G. J. (2008) Analysis of the growth cone turning assay for studying axon guidance, *J Neurosci Methods* **170**, 220–228.
25. Xu, X., Brzostowski, J. A., and Jin, T. (2009) Monitoring dynamic GPCR signaling events using fluorescence microscopy, FRET imaging, and single-molecule imaging, *Methods Mol Biol* **571**, 371–383.
26. Hynes, R. O. (2009) The extracellular matrix: not just pretty fibrils, *Science* **326**, 1216–1219.
27. He, L., and Niemeyer, B. (2003) A novel correlation for protein diffusion coefficients based on molecular weight and radius of gyration, *Biotechnol Prog* **19**, 544–548.
28. Young, M. E., Carroad, P. A., and Bell, R. L. (1980) Estimation of Diffusion-Coefficients of Proteins, *Biotechnology and Bioengineering* **22**, 947–955.

INDEX

Claire M. Wells and Maddy Parsons (eds.), *Cell Migration: Developmental Methods and Protocols*, Methods in Molecular Biology, vol. 769, DOI 10.1007/978-1-61779-207-6,

W

X

Y

Z

MIX
Papier aus verantwortungsvollen Quellen
Paper from responsible sources
FSC® C105338

If you have any concerns about our products,
you can contact us on
ProductSafety@springernature.com

In case Publisher is established outside the EU,
the EU authorized representative is:
Springer Nature Customer Service Center GmbH
Europaplatz 3, 69115 Heidelberg, Germany

Printed by Libri Plureos GmbH
in Hamburg, Germany